SOCIETÀ ITALIANA DI FISICA

RENDICONTI
DELLA
SCUOLA INTERNAZIONALE DI FISICA
«ENRICO FERMI»

CVIII Corso

a cura di M. Campagna e R. Rosei
Direttori del Corso
VARENNA SUL LAGO DI COMO
VILLA MONASTERO
12-22 Luglio 1988

Fotoemissione e spettroscopia di assorbimento di solidi e interfacce con radiazione di sincrotrone

1990

SOCIETÀ ITALIANA DI FISICA
BOLOGNA-ITALY

ITALIAN PHYSICAL SOCIETY

PROCEEDINGS
OF THE
INTERNATIONAL SCHOOL OF PHYSICS
«ENRICO FERMI»

COURSE CVIII
edited by M. CAMPAGNA and R. ROSEI
Directors of the Course
VARENNA ON LAKE COMO
VILLA MONASTERO
12-22 July 1988

Photoemission and Absorption Spectroscopy of Solids and Interfaces with Synchrotron Radiation

1990

NORTH-HOLLAND
AMSTERDAM · OXFORD · NEW YORK · TOKYO

Copyright ©, 1990, by Società Italiana di Fisica

PUBLISHED BY:

North-Holland
Elsevier Science Publishers B.V.
P.O. Box 211
1000 AE Amsterdam
The Netherlands

SOLE DISTRIBUTORS FOR THE USA AND CANADA:

Elsevier Science Publishing Company, Inc.
655 Avenue of the Americas
New York, N. Y. 10010
U.S.A.

Technical Editor
P. PAPALI

Library of Congress Cataloging-in-Publication Data

International School of Physics "Enrico Fermi" (1988 : Varenna,
  Italy)
    Photoemission and absorption spectroscopy of solids and interfaces
  with synchroton radiation : Varenna on Lake Como, Villa Monastero,
  12-22 July 1988 / edited by M. Campagna and R. Rosei.
       p. cm. -- (Proceedings of the International School of Physics
  "Enrico Fermi" ; course 108)
    At head of title: Italian Physical Society.
    Title on added t.p.: Fotoemissione e spettroscopia di assorbimento
  di solidi e interfacce con radiazione di sincrotrone.
    ISBN 0-444-88470-X
    1. Energy-band theory of solids--Congresses. 2. Solids--Surfaces-
  -Congresses. 3. Synchrotron radiation--Congresses.
  4. Photoemission--Congresses. 5. Atomic absorption spectroscopy-
  -Congresses.   I. Campagna, M. II. Rosei, R. III. Societa italiana
  di fisica. IV. Title. V. Title: Fotoemissione e spettroscopia di
  assorbimento di solidi e interfacce con radiazione di sincrotrone.
  VI. Series: International School of Physics "Enrico Fermi."
  Proceedings of the International School of Physics "Enrico Fermi" ;
  course 108.
  QC176.8.E4I536 1988
  530.4'16--dc20   90-14268
                                                         CIP

Proprietà Letteraria Riservata
Printed in Italy

INDICE

Avger 128

B. LENGELER – X-ray absorption and reflection in the hard-X-ray range.

F. J. HIMPSEL, B. S. MEYERSON, F. R. McFEELY, J. F. MORAR, A. TALEB-IBRAHIMI and J. A. YARMOFF – Core level spectroscopy at silicon surfaces and interfaces.

G. K. WERTHEIM – Core level spectroscopy of metals, alloys and intermediate-valence compounds.

V. DOSE – Inverse photoemission spectroscopy.

J. HAASE – X-ray absorption fine structure of adsorbates: SEXAFS and NEXAFS.

G. Margaritondo – Photoemission spectroscopy of interfaces.

H. C. Siegmann and H. Burtscher – Photoemission experiments on small particles in gas suspension.

F. MEIER, M. AESCHLIMANN, M. STAMPANONI and A. VATERLAUS –
Laser photoemission.

U. HEINZMANN – Photoemission with circularly polarized photons.

Preface.

Synchrotron radiation has become a very important tool for investigating condensed matter. Many different techniques have been developed which can give information on the electronic and the geometric structure of solids and surfaces. At the heart of all of them is the ubiquitous phenomenon of interaction of electromagnetic radiation with condensed matter.

At a time when new extremely powerful third-generation synchrotron radiation sources like ELETTRA in Trieste, ALS in Berkeley and ESRF in Grenoble are on the drawing boards around the world, it has been felt that a review of the latest major developments and of the leading areas of current interest was in order.

This volume originates from the lectures delivered at the CVIII Course of the Enrico Fermi Summer School, which was held in Varenna from July 12 to 22, 1988.

The field encompassed by the theory of the electromagnetic-wave interaction with condensed matter and by the various experimental techniques which harness synchrotron radiation is very broad indeed, so that no attempt to a complete coverage was made. The lecturers, who are among the people who have mostly contributed to each individual topic and also to the field as a whole, have given, through the Course, very lively and interactive lectures, providing the necessary background for a truly intercommunicating environment. They have also kindly agreed to prepare texts of their lectures for final publication.

The result which is now appearing in this volume is an overview of the present status of synchrotron radiation research and includes the background fundamental theoretical aspects of electronic structure of condensed matter and of condensed-matter interaction with photons.

The foundations of the theory of the electronic structure of solids and surfaces are presented by A. LIEBSCH, while B. DELLEY treats in some depth screening and correlation effects. G. D. MAHAN gives a comprehensive review of the theory of photoemission and finally O. BISI completes the theoretical part by reviewing the electronic structure of interfaces.

The properties of synchrotron radiation are introduced by C. KUNZ, who covers also aspects of synchrotron radiation instrumentation, including some of the latest developments.

The absorption spectroscopy experiments and phenomenology are presented in the lively lectures of J. C. FUGGLE (soft-X-ray range) and by B. LENGELER (hard-X-ray range and EXAFS).

F. HIMPSEL gives a very clear overview of the by now classical experiments of UV photoemission spectroscopy, including band mapping, while G. K. WERTHEIM concentrates on the subject of core level photoemission spectroscopy especially in the XPS regime.

Inverse photoemission is not a synchrotron radiation technique but gives information which is largely complementary and it was felt important to include it in the program. V. DOSE, a pioneer in this field, was in charge of these lectures.

J. HAASE, G. MARGARITONDO, H. C. SIEGMANN and R. JONES give ample examples of specific applications of synchrotron radiation experimental techniques in the fields of structural determination of adsorbed species, interfaces, environmental pollution and of clusters, respectively.

E. KISKER reviews the state of surface magnetism and spin-polarized photoemission spectroscopy.

Finally, very lively seminars were given by F. HIMPSEL, U. HEINZMANN and F. MEIER.

R. ROSEI

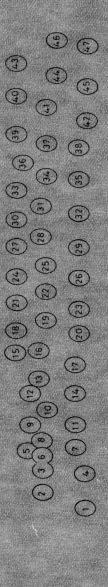

1) C. Ottaviani
2) F. Pittino
3) M. L. Rossi
4) A. Santaniello
5) N. Barret
6) M. E. Sabadini
7) E. Mazzi
8) J. Bonnet
9) F. J. Himpsel
10) J. Haase

11) R. Jones
12) M. Canepa
13) K. Hricovini
14) J. C. Fuggle
15) A. Campo
16) G. Paolucci
17) B. Lengeler
18) A. Morgante
19) N. Zema
20) L. P. De Padova

21) A. Di Cioco
22) S. D'Addato
23) A. Liebsch
24) M. Bertolo
25) P. Weijs
26) P. Perfetti
27) G. Comelli
28) J. Scarfe
29) R. Rosei
30) P. Rudolf

31) G. E. Van Dorssen
32) A. Santoni
33) R. Cimino
34) O. Bisi
35) L. Sorba
36) C. Astaldi
37) V. Dose
38) G. Reverberi
39) A. Terrasi
40) L. Soriano

41) G. K. Wertheim
42) R. Magri
43) M. Sanchez Del Rio
44) P. Heimann
45) M. Pedio
46) W. Dolinski
47) C. Kunz

SOCIETÀ ITALIANA DI FISICA

SCUOLA INTERNAZIONALE DI FISICA «E. FERMI»

CVIII CORSO - VARENNA SUL LAGO DI COMO - VILLA MONASTERO - 12-22 Luglio 1988

Electronic Structure of Solids and Surfaces and the Relation to Surface Spectroscopies.

A. LIEBSCH

Institut für Festkörperforschung, Kernforschungsanlage - 517 Jülich, B.R.D.

Introduction.

The understanding of the electronic properties of solids and surfaces has deepened appreciably over the past two decades for three reasons: First, electronic-structure calculations have become sufficiently detailed that quantitative predictions of ground-state properties such as binding energies, equilibrium geometries, vibration frequencies, work functions, surface energies, etc. are nearly routine. Second, a host of experimental techniques is now available that provide a detailed picture of the various structural, electronical and vibrational properties of well-controlled clean and adsorbate-covered surfaces. Third, the crucial link between electronic properties and spectroscopies, the characteristic changes in the electron system associated with the perturbing probes, are now, at least in some cases, much better understood than only a few years ago.

The aim of this lecture is to discuss some of the important aspects of electronic structures of solids and surfaces as well as their manifestations in surface spectroscopies. The first section deals with ground-state electronic properties. The theoretical description in terms of the density functional approach will be outlined. Particular emphasis is given to the phenomenon of surface reconstruction at metal surfaces. Several different types of atomic rearrangement have been studied in detail experimentally and theoretically and are now rather well understood. In the second section, the concept of quasi-particles will be introduced which is fundamental for the interpretation of electronic excitations seen, for example, in photoemission and optical spectra. Basic, long-standing issues such as the band gap problem in semiconductors and insulators and the band width problem in simple metals have recently been resolved by quantitative evaluation of quasi-particle energies. Phenomena such as relaxation shifts, lifetime broadening and satellites all go beyond the ground-state picture and follow naturally from the quasi-particle concept. The third section addresses a related topic, namely that of the dynamical response of electrons at surfaces to an external perturbation. We will demonstrate that experimental techniques using different probes such as photons, electrons, neutral atoms or ions share common features and require a detailed knowledge of the spatial and frequency dependence of the electronic screening charge distribution induced near the surface. Examples of both the linear and nonlinear dynamic response properties of simple metal surfaces will be discussed.

1. – Ground-state electronic properties.

1`1. *Density functional theory.* – The theoretical concept underlying nearly all calculations today of the ground-state electronic properties of solids and their surfaces is the density functional approach[1-3]. The central quantity in this theory is the electronic density $n(\mathbf{r})$ on which other properties such as the ground-state energy depend in a functional form. A key feature of this theory, that is important for practical calculations, is that the exact density of the interacting «many-body» electron gas is represented in terms of the single-particle solutions $\psi_i(\mathbf{r})$ of a (fictitious) noninteracting electron system:

$$(1.1) \qquad n(\mathbf{r}) = \sum_i \psi_i^*(\mathbf{r})\,\psi_i(\mathbf{r})\,.$$

The ground-state energy is then given by the expression (atomic units)

$$(1.2) \qquad E[n] = T[n] + \int \mathrm{d}^3 r\, V_{\mathrm{ion}}(\mathbf{r}) + \frac{1}{2}\int \mathrm{d}^3 r \int \mathrm{d}^3 r'\,\frac{n(\mathbf{r})\,n(\mathbf{r}')}{|\mathbf{r}-\mathbf{r}'|} + E_{\mathrm{xc}}[n]\,,$$

where $T[n]$ represents the kinetic energy

$$(1.3) \qquad T[n] = \sum_i \int \mathrm{d}^3 r\, \psi_i^*(\mathbf{r})\left(-\frac{1}{2}\nabla^2\right)\psi_i(\mathbf{r})$$

and $E_{\mathrm{xc}}[n]$ is the so-called exchange-correlation functional. Since $T[n]$ is the kinetic energy of the noninteracting electron gas, $E_{\mathrm{xc}}[n]$ must include also part of the many-body kinetic energy.

The single-particle wave functions $\psi_i(\mathbf{r})$ are to be calculated from the Hartree-like Schrödinger equation:

$$(1.4) \qquad \left\{-\frac{1}{2}\nabla^2 + V_{\mathrm{eff}}(\mathbf{r})\right\}\psi_i(\mathbf{r}) = E_i\,\psi_i(\mathbf{r})\,,$$

and the effective one-electron potential V_{eff} is given by

$$(1.5) \qquad V_{\mathrm{eff}}(\mathbf{r}) = V_{\mathrm{ion}}(\mathbf{r}) + \int \mathrm{d}^3 r'\,\frac{n(\mathbf{r}')}{|\mathbf{r}-\mathbf{r}'|} + \frac{\delta E_{\mathrm{xc}}[n]}{\delta n(\mathbf{r})}\,.$$

Up to this point the theory outlined above is exact. Since the true functional $E_{xc}[n]$ is in general not known, approximate forms must be employed. Formally, the following expansion holds:

$$(1.6) \qquad E_{xc}[n] = \int d^3 r \left\{ \varepsilon_{xc}(n(r)) n(r) + \varepsilon_{xc}^{(2)}(n(r)) |\nabla n(r)|^2 + \ldots \right\},$$

where $\varepsilon_{xc}(n)$ is the exchange-correlation energy per electron of a uniform electron gas of density n. If gradient corrections are neglected and only the first term in (1.6) is retained, one has the so-called local-density approximation (LDA) of the exchange-correlation energy functional. The last term in (1.5) then becomes the local exchange-correlation potential

$$(1.7) \qquad V_{xc}(n(r)) \equiv \frac{\delta E_{xc}[n]}{\delta n(r)} = \frac{d}{dn} (\varepsilon_{xc}(n) n).$$

The local-density approximation is exact in two limiting cases: for very slowly varying densities and at very high densities. Within the LDA, the total energy can be calculated from the expression

$$(1.8) \quad E[n] = \sum_i E_i - \frac{1}{2} \int d^3 r \int d^3 r' \frac{n(r) n(r')}{|r - r'|} - \int d^3 r V_{xc}(r) n(r) + E_{xc}[n].$$

The self-consistent solution of these one-electron-type equations can be obtained by using an iterative procedure.

The above formalism has been used to determine the ground-state properties of many metals, semiconductors and insulators. By varying the atomic geometries, the equilibrium configurations can be found. To determine surface-specific quantities, such as work function, surface energy and the atomic structure at surfaces, calculations for slabs of finite thickness are usually carried out. Binding energies and positions of isolated adsorbed atoms or molecules or of adsorbed monolayers can also be obtained from analogous calculations. Furthermore, by displacing specific atoms or atomic planes from their equilibrium positions, the interatomic forces and the associated vibration frequencies can be found. Surface phonons, for example, play an important role in phenomena such as phase transitions on clean or adsorbate-covered surfaces, desorption processes, energy dissipation at surfaces, etc.

The comparison with many experimental results has shown that the predictions of the local-density approximation for extended systems are in general rather good. Cohesive energies are typically overestimated by 10 to 20%. For example, the measured cohesive energy of diamond is 7.37 eV/atom, whereas the LDA predicts 8.63 eV/atom. Recent quantum Monte Carlo calculations [4] give instead 7.45 eV/atom in better agreement with experiment. The principal problem with the local-density approximation is that the different angular symmetries of orbitals are not distinguished in the exchange-correlation potential. This is particularly serious in systems where hybridization between s and p or s and d orbitals is important. A detailed discussion of this problem has been given by JONES and GUNNARSSON [5] for the case of atoms and molecules, where similar problems arise.

In contrast to the cohesive energies, the structural and vibrational properties of extended systems are generally rather accurately reproduced by the LDA. Equilibrium atomic distances agree often to within 1% or better with measured spacings. In the following we discuss three examples of progressively more complex structural rearrangements at metal surfaces which are now physically quite well understood and where excellent agreement between ground-state LDA total-energy calculations and experiments has been achieved.

1·2. Surface reconstruction.

1·2.1. Contraction. If we consider the fact that surface atoms are surrounded by a smaller number of nearest neighbours than bulk atoms, it seems almost surprising that many crystal surfaces have atomic structures so close to the ideal bulk geometry. For example, on the (111), (100) and (110) faces of a f.c.c. crystal, surface atoms have 9, 8 and 7 neighbours, respectively. Since these numbers almost coincide with the coordination numbers of bulk atoms in b.c.c. (8) and s.c. (6) crystals, the surface atoms appear to be far out of equilibrium indeed. A hint at the kind of rearrangement we might expect at surfaces can be obtained regarding the surface situation as intermediate between bulk and diatomic molecule. Typically, the interatomic spacings in diatomic molecules are smaller than bulk nearest-neighbour distances. This suggests that surface atoms should relax inward, i.e. that the interplanar spacings at the surface should be reduced.

This is indeed what is observed on many metal surfaces. The physical mechanism that drives this contraction was discussed by SMOLUCHOWSKI [6] and FINNIS and HEINE [7]. As illustrated in fig. 1, sp electrons can reduce their kinetic energy by flowing towards the holes between surface atoms, thereby making the electronic density smoother. The resulting increase in electrostatic

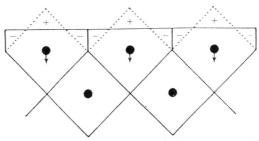

Fig. 1. – Wigner-Seitz cells at (110) surface of a simple cubic crystal. The surface layer contracts as indicated by the arrows. (Ref. [7].)

energy can be reduced if the outermost plane moves inwards as indicated by the arrows. Clearly, for a f.c.c. crystal, this effect is most pronounced for the open (110) face and weakest for the close-packed (111) face. Once the first atomic plane has been allowed to relax, there is no reason why the subsequent planes should not also adjust themselves. In fact, it has been shown that alternating contractions and expansions of diminishing magnitude are to be expected [8].

Precisely such a sequence of interplanar rearrangements was found by Ho and BOHNEN [9] in their self-consistent ground-state total-energy calculation of the Al(110) surface. First-principles pseudopotentials were employed and exchange-correlation was treated within the LDA. The calculations were carried out for a slab of fifteen layers in order to obtain converged results. Table I lists the calculated changes of spacings between the first four atomic planes, the work function, surface energy and several phonon frequencies together with the corresponding experimental values. The overall agreement for this set of ground-state structural, electronic and vibrational properties is excellent.

TABLE I. – *Calculated* [9] *and measured* [10] *ground-state properties of* Al(110) *surface.* (d_{ij} denotes the change of spacing between planes i and j.)

	Theory	Experiment
d_{12}	$-6.8 \pm 0.5\%$	$-8.5 \pm 1.0\%$
d_{23}	$+3.5 \pm 0.5\%$	$+5.5 \pm 1.5\%$
d_{34}	$-2.0 \pm 0.5\%$	≈ 0.0
work function	4.32 eV	4.28 eV
surface energy	0.77 eV	0.81 eV
surface phonon at $\bar{Y}$	7.9 meV	8.9 meV
surface phonon at $\bar{Y}$	14.1 meV	13.5 meV
surface phonon at $\bar{X}$	17.0 meV	14.6 meV
surface phonon at $\bar{X}$	17.4 meV	14.6 meV

1'2.2. Missing row: Au(110). A more complicated reconstruction is known to occur on the Au(110) surface where alternating rows in the first atomic plane are missing (see fig. 2). Similar lateral rearrangements are observed on the (110) faces of Ir and Pt. In contrast, the corresponding $4d$ metals Rh, Pd and Ag do not show this «missing-row» reconstruction; however, it can be induced on the (110) faces of these metals by small amounts of adsorbed alkali atoms.

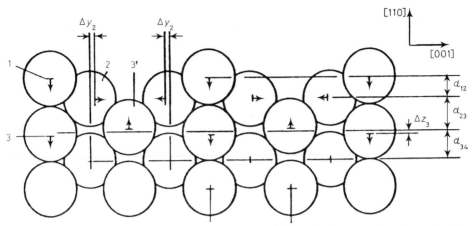

Fig. 2. – Schematic drawing of the geometry of the missing-row structure for the reconstructed Au(110) surface. (Ref. [9].)

In order to analyse this surface structure, HO and BOHNEN [11] have performed first-principles pseudopotential calculations for the Au(110) surface. Again the LDA was used to calculate the total energy, in this case for a finite slab of seven atomic layers. Apart from the missing rows, the remaining rows in the first plane are found to contract, the second-layer atoms are displaced laterally and the third plane is slightly buckled. These structural changes are also observed experimentally using electron diffraction [12] and ion scattering [13].

The different behaviour of the $4d$ and $5d$ metals indicates that the competition between sp and d electrons must play a crucial role. Although the $5d$ wave functions are much less localized than the $4d$ orbitals, the lattice spacings of Au and Ag are almost identical. Thus the sp charge in Au is significantly compressed. The missing rows on the Au(110) surface permit the sp charge to spread out and lower its kinetic energy, while the number of broken d bonds remains the same as for the unreconstructed surface.

In the case of the $4d$ metals, the sp charge is much less compressed, $i.e.$ little kinetic energy would be gained by a missing-row reconstruction. On the other hand, at low coverages alkali atoms on metal surfaces are known to be partially ionized. This charge donation increases the sp electron density in the surface region. Now the situation is similar to that of the $5d$ metals, and, indeed, the same kind of missing-row reconstruction has been observed on the (110) faces of Ag, Pd and Rh after submonolayers of alkali atoms were deposited [14]. This kind of alkali-induced surface reconstruction might be related to the fact that the alkalies serve as promotors in heterogeneous catalysis.

1˙2.3. Rippling: NiAl(110). As a third example of a surface structure determination based on a first-principles total-energy calculation we discuss the relatively complex NiAl(110) surface (see fig. 3). Since this alloy has a CsCl structure, the ideal (110) surface is composed of atomic planes that contain half Ni and half Al atoms. Both electron diffraction [15] and ion scattering measurements [16] indicate that the actual NiAl(110) surface shows a large rippled relaxation where Al atoms move outwards by 4.6% and Ni atoms move inwards by 6.0% of the ideal interlayer spacing.

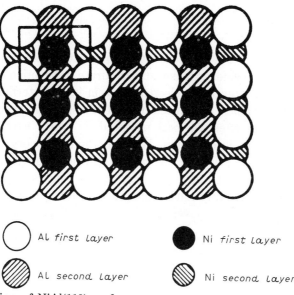

○ Al *first layer*   ● Ni *first layer*

◔ Al *second layer*   ◕ Ni *second layer*

Fig. 3. – Top view of NiAl(110) surface.

KANG and MELE [17] have performed ground-state total-energy calculations for a 5-layer slab of this alloy surface. As in the previous examples, *ab initio* pseudopotentials are used and the exchange-correlation is treated within the LDA. They find that the surface energy is minimized if the Al atoms expand out to the vacuum region by 6.6% of an interlayer spacing and the Ni atoms relax inward by 6.9%. These values are in good agreement with the experimental results. In addition, they obtain a surface phonon at $\bar{\Gamma}$ at 27 meV in excellent agreement with electron loss data taken by LIU *et al.* [18].

As KANG and MELE point out in the interpretation of their results, the average atomic volume in NiAl (12.2 Å³) is similar to that of bulk Ni (10.9 Å³) but is significantly smaller than that of bulk Al (16.6 Å³). Thus the *sp* charge at Al sites is very much denser in the alloy than in pure Al. This compression can be released by an outward expansion of the first-layer Al atoms. This effect seems to be stronger than the competing mechanism which makes the first layer relax inward as in the case of the pure Al (110) surface discussed above. The kinetic energy of the Al *sp* charge can be further reduced if it is allowed to flow into the region of the Al-Al bridges. This flow, however, leads to an enhanced electrostatic repulsion between Al sp electrons and Ni d_{xz} and d_{yz} orbitals whose lobes also point toward the Al-Al bridges. This repulsion can be diminished if the Ni atoms move inwards and if the Al atoms move outwards. Evidence for such an energy gain is the fact that Ni-derived surface states of d_{xz} and d_{yz} character lower their energy considerably upon inward relaxation of the Ni atoms.

2. – Quasi-particles.

2`1. *Formal theory.* – As pointed out in the previous section, the density functional formalism provides a reasonably accurate description of the ground-state properties of an interacting electron system in an external potential. In fact, the many-body problem is reduced to a set of self-consistent single-particle equations involving an effective local potential by which exchange and correlation effects are taken into account. Whereas the total energy of the ground state of a system can, in principle, be calculated exactly in this approach, the single-particle energies that appear do not have a concise physical meaning. Thus there is no formal justification for the interpretation of, say, photoemission spectra in terms of the energy levels obtained from a (local or nonlocal) density functional calculation.

Photoemission and optical spectra convey information on the electronic excitations of a system whose theoretical description must go beyond a mere ground-state theory. Obvious manifestations of such excitations are, for example, their screening and their decay, which lead to characteristic line shifts and line broadening, none of which can be obtained from a gound-state formalism. In addition, it is now known that the measured band gaps of semiconductors and insulators as well as the band widths of simple metals can be reproduced by going beyond the single-particle picture.

A created hole or an added electron in contact with the remaining electrons of the system is referred to as *quasi-particle* whose energy can be calculated rigorously from the so-called self-energy [19, 20]. Instead of the single-particle Schrödinger equation

$$(2.1) \qquad \left\{ -\frac{1}{2} \nabla^2 + V_{\text{ion}}(r) + V_{\text{Hartree}}(r) + V_{\text{xc}}(r) \right\} \psi_i(r) = E_i \psi_i(r) ,$$

which gives the one-electron energies and wave functions discussed in the previous section, the quasi-particle energies $E_{i,\text{qp}}$ and wave functions $\psi_{i,\text{qp}}$ follow from the analogous equation

$$(2.2) \quad \left\{ -\frac{1}{2} \nabla^2 + V_{\text{ion}}(r) + V_{\text{Hartree}}(r) \right\} \psi_{i,\text{qp}}(r) + \int d^3 r' \Sigma(r, r', E_{i,\text{qp}}) \psi_{i,\text{qp}}(r') = E_{i,\text{qp}} \psi_{i,\text{qp}}(r) .$$

Here, Σ is the nonlocal self-energy operator which contains the effect of exchange and correlation among the electrons. Since Σ is in general non-Hermitian, the quasi-particle energies $E_{i,\mathrm{qp}}$ are complex; the imaginary part specifies the quasi-particle lifetime.

As shown by HEDIN [19], an exact expression for the self-energy is given by

$$(2.3) \qquad \Sigma(1,2) = i \int \mathrm{d}(3) \int \mathrm{d}(4) \, G(1,3) \, W(1,4) \, \Gamma(4,2;3),$$

where we use the compact notation $1 = (\boldsymbol{r}, \boldsymbol{\sigma}, t)$, etc. G is the single-particle Green's function, W the dynamically screened Coulomb interaction, and Γ the vertex function. Since all three quantities depend on the quasi-particle spectrum, a self-consistent solution of these equations is exceedingly difficult. If vertex corrections are neglected (*i.e.* $\Gamma = 1$), the evaluation of Σ simplifies considerably. This is the so-called GW approximation:

$$(2.4) \qquad \Sigma(\boldsymbol{r}, \boldsymbol{r}', E) = \frac{i}{2\pi} \int \mathrm{d}\omega \exp[-i\delta\omega] \, G(\boldsymbol{r}, \boldsymbol{r}', E - \omega) \, W(\boldsymbol{r}, \boldsymbol{r}', \omega),$$

where δ is a positive infinitesimal.

Once the self-energy is known, the spectral weight function is given by the imaginary part of the Green's function:

$$(2.5) \qquad A(\boldsymbol{r}, \boldsymbol{r}', E) = \frac{1}{\pi} \mathrm{Im}\, G(\boldsymbol{r}, \boldsymbol{r}', E).$$

Thus, if nondiagonal elements of the self-energy can be neglected, the Fourier transform of the spectral function assumes the simple form

$$(2.6) \qquad A(k, E) = \frac{1}{\pi} \frac{\Gamma_k}{(E - E_k)^2 + \Gamma_k^2},$$

where E_k and Γ_k are the real and imaginary parts of the solution of eq. (2.2).

In the following we discuss three cases where the differences between independent-particle and quasi-particle energies are large and where detailed self-energy calculations have led to an excellent understanding of observed photoemission spectra: the band gaps of semiconductors and insulators, the band widths of simple metals and the appearance of satellites in the valence spectra of nickel.

2'2. *Semiconductors: the band gap problem.* – Whereas the local-density approximation provides an excellent description of the ground-state properties of many semiconductors and insulators, the calculated band gaps are typically too small by about a factor of two (see table II). Until quite recently, it was not clear whether this error is a consequence of the local-density approximation or

TABLE II. – *Calculated and measured band gaps of* Si, Ge *and diamond (in* eV) (*ref.* [21]).

	Si	Ge	Diamond
experiment	1.17	0.74	5.5
LDA ground state	0.52	∼0	3.9
quasi-particle	1.21	0.75	5.6

whether (local and nonlocal) density functional theory simply cannot adequately describe excited states such as the optical transitions between valence and conduction bands of semiconductors and insulators.

In order to resolve this issue for the case of silicon, GODBY *et al.* [22] have carried out a total-energy calculation using the full nonlocal exchange-correlation potential. They demonstrate that the nonlocal density functional band structure of Si is almost exactly the same as that obtained from a local-density calculation. Thus the «error» in the Si band gap is inherent in the ground-state density functional approach. To describe the observed band gap, it is, therefore, necessary to take into account the quasi-particle nature of the states in the valence and conduction bands.

Such quasi-particle calculations have recently been performed by HYBERTSEN and LOUIE [21] for Si, Ge and diamond within the framework of the *GW* approximation discussed above. The self-energy is evaluated using eq. (2.4). The full crystalline Green's function G, which depends, in principle, on the quasi-particle energies and wave functions, is constructed from the corresponding LDA quantities

$$(2.7) \qquad G(r, r', E) = \sum_{nk} \frac{\psi_{nk}(r)\,\psi_{nk}^*(r')}{E - E_{nk} - i\delta}.$$

The screened Coulomb interaction is given by

$$(2.8) \qquad W(r, r', \omega) = \int d^3 r'' \, \varepsilon^{-1}(r, r'', \omega) \, V_C(r'' - r'),$$

where V_C is the bare Coulomb interaction and ε^{-1} the full inverse dielectric function. This latter quantity is obtained in two steps: First, a RPA calculation of the static dielectric matrix $\varepsilon_{GG'}^{-1}(q, \omega = 0)$ is carried out. A generalized plasmon pole model is then used to extend the dielectric matrix to finite frequencies.

The band gaps of Si, Ge and diamond that result from these quasi-particle calculations are in very good agreement with the measured values (see table II). Similar agreement is achieved for all other measured optical transitions between the valence and conduction bands of these three elements.

The important conclusion from this work, therefore, is that the LDA is reasonably adequate as far as the description of genuine ground-state properties of semiconductors is concerned. Electronic excitation energies, however, such as the band gap, differ strongly from the single-particle energies and can be described only by accounting for the quasi-particle nature of the states involved in the optical transitions.

2˙3. *Simple metals: the band width problem.* – According to the quasi-particle calculations by HEDIN [19] for the homogeneous electron gas, the band width of the occupied part of the conduction bands of the alkali metals should be about 10% smaller than the free-electron band width. The self-energy in these calculations was evaluated within the GW approximation as discussed above. In addition, the Green's function appearing in eq. (2.4) was replaced by the free-electron Green's function and the dielectric function which accounts for the dynamical screening of the bare Coulomb interaction (see eq. (2.8)) was evaluated within the RPA. Band structure effects that play a crucial role in the case of semiconductors are negligible here, since the bands of the alkali metals are nearly-free-electron-like.

Since the average electron density in the alkali is rather low, however, exchange and correlation are more important than in the higher-density semiconductors. This was demonstrated in a recent calculation of the quasi-particle energies of Na by NORTHRUP *et al.* [23], who went beyond the RPA in the dynamic screening of the Coulomb potential. The free-electron band width of Na is 3.2 eV and the RPA leads to a narrowing of about 0.31 eV. If exchange and correlation are included in the dielectric function, the band narrowing increases by almost a factor of two to 0.57 eV. The expression used for the static dielectric function was

$$(2.9) \qquad \varepsilon(q) = 1 - \frac{v(q)\chi_0(q)}{1 - K_{xc}(q)\chi_0(q)},$$

where $v(q)$ is the Fourier transform of the bare Coulomb interaction and $\chi_0(q)$ the Lindhard independent-particle susceptibility. In the LDA, $K_{xc} = \delta V_{xc}/\delta\varrho$, whereas this term is dropped in the RPA. To describe the screening at finite frequencies, a generalized plasmon pole model was used.

The band narrowing becomes even stronger if the Green's function in expression (2.4) for the self-energy is calculated self-consistently from the quasi-particle spectrum rather than from the LDA band energies. Thus the total reduction of the band width of Na in the calculation by NORTHRUP *et al.* is 0.71 eV or 22%. This is in reasonable agreement with the most recent photoemission data which show a narrowing of about 18% [24]. Therefore, although the energy bands of Na are almost perfect free-electron bands, the creation of a hole state involves considerable correlation effects which lead to rather large self-energy corrections to the single-electron energies.

Figure 4 summarizes the calculated and observed self-energy corrections of the band width for several metals. The overall agreement suggests that a quantitative interpretation of the photoemission spectra of these systems can be achieved by taking into account the quasi-particle nature of the conduction bands.

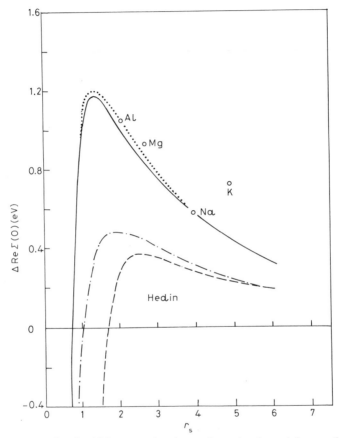

Fig. 4. – Self-energy band width correction for various simple metals as a function of r_s. The open circles are experimental results. (Ref. [24].)

It should be noted at this point that the interpretation of the Na photo-emission spectra is still controversial: Whereas NORTHRUP et al. [23] and LYO and PLUMMER [24] associate the observed band narrowing with the quasi-particle nature of the measured hole spectrum, SHUNG et al. [25] emphasize that an effective or apparent narrowing can also result from the interference of bulk and surface contributions to the transition matrix element. This interference is particularly strong in the case of Na, since bulk transitions are relatively weak, so that surface-barrier-induced terms become of comparable importance. To distinguish between these different physical mechanisms, data taken at various exit angles might be useful, since this should have a strong influence on the relative magnitude and sign of bulk and surface contributions.

2‘4. *Transition metals: satellites.* – The self-energy corrections discussed in the previous two subsections for semiconductors and simple metals lead to broadening and shifts of the band states, so that there remains a one-to-one correspondence between single-electron and quasi-particle energies. More dramatic self-energy modifications can, however, occur if the orbitals from which the band states are formed are more localized. Of particular interest in this regard is the case of nickel whose valence spectrum has been carefully studied experimentally and theoretically almost ten years ago. Photoemission data[26] clearly show an extra feature below the bottom of the 3*d* band which cannot be understood in terms of the calculated single-electron band structure. Also, the line widths of spectral features in angle-resolved spectra are much larger than, for example, in the case of Cu, the overall width of the Ni 3*d* band is much smaller than calculated and the exchange splitting between majority and minority spin states at the Fermi energy is only about half as large as predicted by ground-state band structure calculations.

Various self-energy calculations have been performed in order to interpret the measured spectra[27, 28]. Although they differ in detail, the basic common physical mechanism that causes the observed satellite structure is the following: A hole created in the photoemission from the nickel 3*d* band interacts with the holes that are present in this partially filled band. (The *sp* band can be neglected in a first approximation.) Because of the near-atomic character of the *d* states, the intra-atomic Coulomb repulsion between two *d* electrons (or holes) is large (of the order of several eV). Thus, if the created hole and the hole that is already present in the ground state are located at the same lattice site, the solid is in a state of higher energy than if these *d* holes were on different sites. Since the overall energy is conserved in the photoemission process, such an excitation reduces the kinetic energy of the emitted photoelectron. In the measured spectrum, therefore, this excitation manifests itself as a peak at a higher binding energy below the 3*d* band.

The self-energy can in this case also be obtained within the *GW* approximation, eq. (2.4), where *G* is now the *d* band Green's function and the screened Coulomb interaction is essentially the intra-atomic *d-d* repulsion *U*. (Interatomic *d-d* interactions tend to be much smaller.) Screening of the *d-d* repulsion via the *sp* bands is assumed to be included in *U*. Additional screening can, however, occur within the 3*d* band. It we take advantage of the fact that this band is nearly filled, the effective dynamically screened *d-d* Coulomb interaction is given by

$$(2.10) \qquad W(\omega, k) = \frac{U}{1 + UG^{(2)}(\omega, k)} \equiv t(\omega, k),$$

which is the so-called t-matrix originally introduced by KANAMORI [29]. $G^{(2)}$ is the two-particle Green's function which describes the propagation of two d holes. For simplicity, eq. (2.10) is written here for the single-band case; in the case of the 5-fold degenerate d band a matrix notation must be used [28].

Equation (2.10) shows that, if the on-site repulsion is sufficiently large, $W(\omega, k)$ exhibits a pole at an energy below the continuum of the band states. According to (2.4) and (2.6), such a singularity leads to an extra peak in the spectral function. In the case of nickel, U is of the order of the $3d$ band width, so that the self-energy calculations based on the mechanism described above do indeed give a satellite structure below the $3d$ band, as shown in fig. 5. In

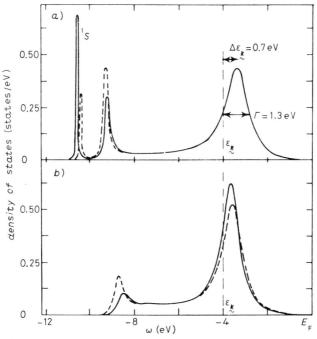

Fig. 5. – Spectral function of majority (a)) and minority (b)) spin states of Ni at 4 eV below E_{F}. The single-particle approximation gives a sharp feature as indicated by the dashed line. (Ref. [28].)

addition, the d-d interactions lead to a sizable narrowing of the d band and to a large broadening of the individual band states, both of which are also observed experimentally. Moreover, the exchange splitting is reduced to about one-half of the value obtained in band structure calculations and agrees well with the value observed in normal and spin-polarized angle-resolved photoemission.

In order to achieve a consistent description of the various experimental observations, in particular of the energy position of the satellite and the amount of band narrowing, it is actually necessary to go beyond the d-d interactions included in expression (2.10) [30]. If both hole-hole and electron-hole ladders are taken into account, the bare two-pole propagator $G^{(2)}$ in (2.10) is effectively renormalized because of these additional interaction channels. The resulting expression for the self-energy closely resembles the one that is obtained if multiple hole-magnon interactions [31] are considered as the main mechanism that determines the $3d$ quasi-particle spectrum.

3. – Electronic response at surfaces.

3˙1. *Relation to surface spectroscopies.* – In the previous section, we have seen that the energies and line widths of optical transitions are given by the quasi-particle spectra of valence and conduction bands. In order to interpret the observed intensities, it is necessary to consider the effective local field that arises from the superposition of the applied field and the induced field generated at the surface by the dynamical response of the system electrons. It is this effective or «local» field which determines the magnitude of transition matrix elements. The linear and nonlinear response characteristics of metal surfaces have been studied extensively over the recent years with the aim of arriving at a more detailed understanding of surface spectroscopies. Because of the rapid variation of the local field in the surface region, an accurate description of the electron distribution at the surface is essential. Moreover, since the response occurs typically on the scale of an electronic screening length, the nonlocal nature of the response must be taken into consideration.

The electrostatic potential of an external perturbing charge can be expressed in terms of the Fourier components

$$(3.1) \qquad \phi_{\text{ext}}(\boldsymbol{r}, \omega) = -\frac{2\pi}{q} \exp\left[i\boldsymbol{q}_{\parallel} \cdot \boldsymbol{r}_{\parallel} + qz + i\omega t\right],$$

with $q = |\boldsymbol{q}_{\parallel}|$. For simplicity, we consider a semi-infinite jellium metal whose uniform positive charges are located in the half-space $z \leqslant 0$. If we denote the Fourier components of the induced surface charge by $\delta n(z, q, \omega)$, the associated Coulomb potential is given by

$$(3.2) \qquad \delta\phi(z, q, \omega) = \frac{2\pi}{q} \int dz' \exp\left[-q|z - z'|\right] \delta n(z', q, \omega).$$

Thus, far from the surface, $\delta\phi$ has the form

$$(3.3) \qquad \delta\phi(z, q, \omega) = \frac{2\pi}{q} \exp\left[-qz\right] g(q, \omega), \qquad\qquad z \gg 0,$$

where the amplitude factor $g(q, \omega)$ is defined as

(3.4)
$$g(q, \omega) = \int dz \exp[qz] \, \delta n(z, q, \omega).$$

This amplitude is referred to as surface response function. It plays a central role in the description of various surface spectroscopies. For example, the rate of exciting electron-hole pairs at the surface is proportional to the imaginary part of $g(q, \omega)$. Thus, by measuring the intensity of inelastically scattered electrons, direct information on the surface response function can be obtained[32]. Also, the lifetime of adsorbate vibrations or the slowing-down of ions near a surface due to electron-hole pair excitation in the metal are related to the imaginary part of $g(q, \omega)$[33]. In the case of the interaction between a neutral atom and a metal, no energy is transferred; only virtual excitations can take place. Thus the attractive atom-solid Van der Waals interaction involves the surface response function at imaginary frequencies[34, 35]

(3.5)
$$E_{\mathrm{VdW}}(z) = -\frac{1}{2\pi} \int_0^\infty du \int_0^\infty dq \exp[-2qz] \, \alpha(q, iu) g(q, iu),$$

where z is the position of the atom and α its polarizability.

In the long-wavelength limit, $g(q, \omega)$ can be shown to have the expansion[36]

(3.6)
$$g(q, \omega) = \sigma(\omega) \left(1 + 2q \frac{\varepsilon(\omega)}{\varepsilon(\omega) + 1} d(\omega) + O(q^2)\right),$$

where

(3.7)
$$\sigma(\omega) = \int dz \, \delta n(z, 0, \omega) = \frac{\varepsilon(\omega) - 1}{\varepsilon(\omega) + 1},$$

(3.8)
$$d(\omega) = \int dz \, z \, \delta n(z, 0, \omega)/\sigma(\omega),$$

and $\varepsilon(\omega)$ is the bulk dielectric function. This long-wavelength limit is of interest since it determines the electronic response to a uniform electronic field perpendicular to the surface and varying in time like $\exp[i\omega t]$. This is precisely the response that is relevant for the variation of the normal component of the field of an electromagnetic wave in the surface region. Thus the nonlocal corrections to the Fresnel formula for the reflection coefficient of an incident p-wave are determined by the function $d(\omega)$[37]. Also, the photoabsorption cross-section at a metal surface for p-polarized light is given by[36]

(3.9)
$$P(\omega) \sim \omega(1 - \omega^2/\omega_{\mathrm{p}}^2) \operatorname{Im} d(\omega),$$

where ω_p is the bulk plasma frequency. Finally, the small-$q_\parallel$ dispersion of the surface plasmon mode is of the form

$$(3.10) \qquad \omega_s(q) = \frac{\omega_p}{\sqrt{2}} \left(1 - \frac{q}{2} d(\omega_s) + O(q^2) \right),$$

i.e. the linear coefficient is determined by the centroid of the dynamic screening charge, evaluated at the surface plasmon frequency.

These examples demonstrate the wide-ranging significance of the surface response function $g(q, \omega)$ [38, 39]. In order to calculate it, the spatial distribution of the dynamical induced screening charge at the surface must be found. This can be done by using an extension of the density functional approach to finite frequencies, which will be discussed in the following subsection.

3‘2. *Time-dependent density functional theory.* – In principle, the surface electron density induced by a weak external potential of the form (3.1) can be obtained from the expression

$$(3.11) \qquad \delta n_1(z, q, \omega) = \int dz' \, \chi(z, z', q, \omega) \, \phi_{\text{ext}}(z', q, \omega),$$

where χ is the exact linear density-density response function of the semi-infinite electron system. Since this response function which includes all many-body effects is not known, approximate procedures for the evaluation of δn_1 must be sought. Within the so-called time-dependent local-density approximation (TDLDA) [40, 41] (3.11) is replaced by

$$(3.12) \qquad \delta n_1(z, q, \omega) = \int dz' \, \chi_1(z, z', q, \omega) \, \phi_{\text{1SCF}}(z', q, \omega),$$

where χ_1 is the linear independent-particle susceptibility and ϕ_{1SCF} is given by

$$(3.13) \qquad \phi_{\text{1SCF}} = \phi_{\text{ext}} + L \, \delta n_1 .$$

Here, L is defined as $L = K + V'_{\text{xc}}$, where K denotes the Coulomb integral operator:

$$(3.14) \qquad (K \delta n_1)(z, q, \omega) = \frac{2\pi}{q} \int dz' \, \exp\left[-q |z - z'| \right] \delta n_1(z', q, \omega),$$

and $V'_{\text{xc}} = \delta V_{\text{xc}}[n]/\delta n$ involves the same local exchange-correlation functional that is used for the calculation of the ground state.

The TDLDA had originally been developed to interpret photoabsorption spectra of atoms. It has since been applied to a variety of other problems involving time-dependent perturbations, such as in small metal particles[42], slabs of finite thickness[43] and at surfaces of semi-infinite metals[44]. It is equivalent to the so-called modified Sternheimer approach which has been used by MAHAN[45] in his calculations of atomic polarizabilities.

The formal structure of this response approach is similar to time-dependent Hartree-Fock theory. The only difference between these schemes lies in the description of electron-electron interactions: instead of the local exchange-correlation potential, Hartree-Fock treats the nonlocal exchange exactly but neglects correlation. The random-phase approximation (RPA) is equivalent to the time-dependent Hartree theory, *i.e.* exchange and correlation are ignored altogether. It is important to use the same interaction picture in the ground state and in the presence of the external perturbation. This guarantees the satisfaction of various exact sum rules and ensures that the correct adiabatic limit is obtained at low frequencies[46]. In the case of extended systems, Hartree-Fock theory is generally not acceptable, since the neglect of correlation leads, for example, to a vanishing density of states at the Fermi energy. In contrast, the density functional approach provides a rather good description of the ground-state electronic properties as we have discussed in sect. 1. For consistency reasons, it is, therefore, also the preferred method to determine the response to a time-dependent perturbation.

3`3. *Linear response: surface photoelectric effect.* – In this subsection, we discuss the frequency dependence of the screening charge distribution at a metal surface in the long-wavelength limit, $q_{\parallel} = 0$. As pointed out above, this limit is of interest for the behaviour of an electromagnetic wave in the vicinity of a surface. The solution of the self-consistent set of response equations (3.12)-(3.14) is discussed in detail for this case in ref. [47]. Figure 6 shows the real part of the normalized induced density $\delta n_1(z, \omega)/\sigma(\omega)$ for various frequencies below ω_p. The bulk density corresponds to that of sodium ($r_s = 4$). The main peak of these distributions remains localized near the jellium edge ($z = 0$). This implies that the screening of the external uniform electric field takes place in the region where the ground-state electron density $n_0(z)$ varies most rapidly at the surface. Of course, at ω_p the metal becomes transparent, so that the induced density must spread out over the entire crystal.

In the interior, the density shows the characteristic Friedel oscillations related to the cut-off at the Fermi momentum k_F. In the static limit, only one oscillation with wavelength π/k_F is obtained. At finite ω, dn_1 consists of a complicated superposition of several such oscillations. The distribution at $\omega = 0.7\,\omega_p \approx \omega_p/\sqrt{2}$ illustrates the spatial form of the surface plasmon in the nonretarded long-wavelength limit. At $0.8\,\omega_p$, an additional «surface mode» is

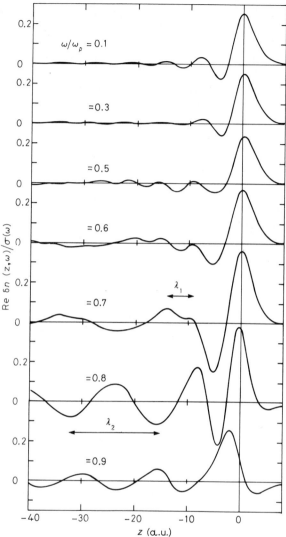

Fig. 6. – Real part of electron density induced at metal surface by a uniform electric field oriented normal to the surface, $r_s = 4$, $\omega/\omega_p = 0.1$ (Ref. [47].)

found which can be seen more clearly by plotting the centroid $d(\omega)$ of the density profiles shown in fig. 6.

The frequency dependence of $d(\omega)$ which determines the nonlocal corrections to the Fresnel reflection coefficient and the surface photoelectric effect (see eq. (3.9)) is given in fig. 7 for various bulk densities [47]. $\operatorname{Im} d(\omega)$ is seen to exhibit a shoulder near the threshold for electron emission (the work function Φ is marked by the arrows). Apparently, the rate of exciting electron-hole pairs increases if final states near the top of the surface barrier potential can be reached. A more pronounced, resonancelike spectral feature is located near $0.8\,\omega_p$. It is caused by

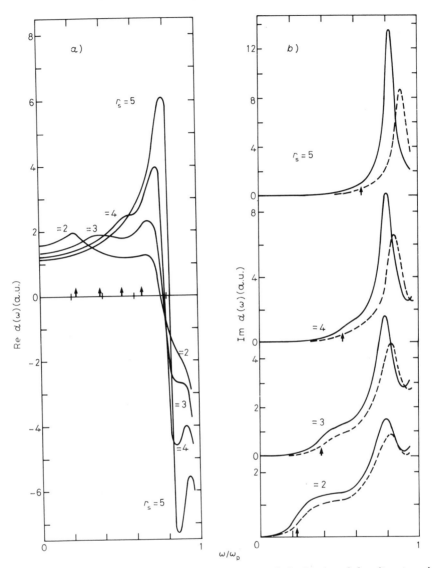

Fig. 7. – Real (*a*)) and imaginary (*b*)) parts of centroid $d(\omega)$ of induced density at various simple metal surfaces. The work functions are indicated by the arrows. Solid lines: TDLDA; dashed lines in *b*): LDA-based RPA. (Ref. [47].)

the screening of the bare induced density via the induced potential, *i.e.* by the second term in eq. (3.13). The main effect here comes from the Hartree contribution, whereas exchange-correlation effects lead to a small shift of this peak to lower frequencies. (Both results denoted by LDA and RPA are based on a LDA ground-state calculation.) Since $d(\omega)$ is a causal response function, the frequency dependence of the real part is related to that of $\operatorname{Im} d(\omega)$ via the Kramers-Kronig relations. Thus, close to the resonance at $0.8\,\omega_p$, $\operatorname{Re} d(\omega)$ shows the typical oscillation whose amplitude depends mainly on the sharpness of the peak in $\operatorname{Im} d(\omega)$.

As pointed out above, $\operatorname{Im} d(\omega)$ is directly proportional to the surface photoabsorption cross-section for p-polarized light (see eq. (3.9)). For $Al\,(r_s = 2)$, the calculations by FEIBELMAN [36], which are nearly identical with the curves labelled RPA in fig. 7 [48], are in excellent agreement with photoemission data by LEVINSON and PLUMMER [49]. Also, according to eq. (3.10), the quantity $d(\omega_s)$ should give the linear coefficient of the surface plasmon dispersion relation. However, the slopes obtained from electron loss measurements [50] on several simple metals are much smaller than predicted by either RPA or LDA calculations [36, 47]. This discrepancy is presumably related to the fact that bulklike loss processes, which are not included in the theory, play an important role. The large width of the observed loss peaks seems also to provide evidence for the relevance of such loss mechanisms.

3˙4. *Nonlinear response: second-harmonic generation.* – The examples discussed above typically involve sufficiently weak perturbations that the induced charge density and potential can be obtained from a linear-response treatment. If lasers are used, however, to investigate the electronic properties of surfaces, nonlinear effects can readily be detected. One particular technique that has been demonstrated to be extremely surface sensitive is second-harmonic generation (SHG): Instead of measuring the reflected intensity at the same frequency ω as that of the incident light, the signal at 2ω is detected which leaves the crystal in the same direction as the linearly reflected radiation.

Second-harmonic generation at simple metal surfaces is known to originate from three nonlinear sources [51-54]: a bulk source of magnetic-dipole character and two surface sources caused by the variation of the electric field near the metal-vacuum interface. Additional nonlinear bulk and surface source terms may arise as a result of interband transitions. In the case of a semi-infinite free-electron system, the weights of the bulk and parallel surface SH currents are determined by the macroscopic bulk dielectric properties of the metal. The normal surface SH current, on the other hand, depends sensitively on the dynamical nonlinear electronic screening properties in the surface region. For the jellium model (the positive background occupies the half space $z \leqslant 0$) and using the time-dependent local-density approximation [40, 41], the nonlinear induced density is given by [55-57]

$$(3.15) \quad \delta n_2(z, \omega) = \int \mathrm{d}z' \int \mathrm{d}z'' \chi_2(z, z', z'', \omega)\, \phi_{1\mathrm{SCF}}(z', \omega)\, \phi_{1\mathrm{SCF}}(z'', \omega) +$$

$$+ \int \mathrm{d}z' \chi_1(z, z', 2\omega) 0.5 V_{\mathrm{xc}}''(z')(\delta n_1(z', \omega))^2 + \int \mathrm{d}z' \chi_1(z, z', 2\omega) \int \mathrm{d}z'' L(z', z'') \delta n_2(z'', \omega),$$

where χ_1 and χ_2 represent the linear and second-order independent-particle susceptibilities. δn_1 and $\phi_{1\text{SCF}}$ denote the linear induced density and self-consistent complex potential, respectively, which were discussed in the previous subsection. L denotes again the (static) response kernel consisting of Hartree and exchange-correlation terms (see eq. (3.13)). δn_2 is, therefore, determined by a similar response equation as δn_1: Whereas the «driving» term in linear response depends on the bare external potential, the nonlinear driving terms involve the linear self-consistent potential and induced density (see the first two terms in (3.15)). The linear screening of these bare or unscreened quantities is described in both cases by the same Coulomb and exchange-correlation operator L.

Once the induced density $\delta n_2(z, \omega)$ is found for a given frequency ω, the complex parameter $a(\omega)$ that determines the weight of the normal SH surface current is related to the first moment of δn_2 as follows [52-54, 58, 59]:

$$(3.16) \qquad a(\omega) = -4n \int dz\, z\, \delta n_2(z, \omega)/\sigma(\omega)^2,$$

where $\sigma(\omega)$ is defined in eq. (3.7) and n is the bulk density.

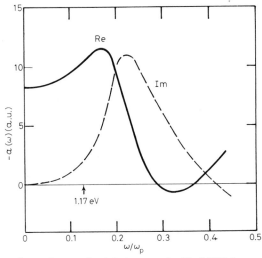

Fig. 8. – Frequency dependence of $a(\omega)$ for $r_s = 3$. (Ref. [60].)

Figure 8 shows the calculated frequency dependence of $a(\omega)$ for $r_s = 3$ corresponding to the free-electron density of Ag[60]. At the laser frequency of 1.17 eV, the real part of $a(\omega)$ is seen to be somewhat larger than the static value, while the imaginary component is still rather small. As ω increases, Im $a(\omega)$ becomes rapidly important and exhibits a resonance near $\omega \approx 0.23\,\omega_p$. Re $a(\omega)$ reaches a maximum below the peak of Im $a(\omega)$ and then shows the typical resonance-type behaviour expected from Kramers-Kronig relations. We have found qualitatively similar behaviour for various other other bulk densities. The overall magnitude of $a(\omega)$ for these systems scales roughly like in the static limit [58]. Remarkably, the resonance in Im $a(\omega)$ occurs in all cases whenever 2ω lies just above the threshold for electron emission. The resonance persists even if ϕ_{1SCF} in the driving term is replaced by its static form. It may be viewed as a damped nonlinear surface collective mode in analogy to the resonance that is seen in linear response near $0.8\,\omega_p$ and that gives rise to the local field enhancement observed in the surface photoemission from nearly free-electron metals (see preceding subsection).

Preliminary measurements of the polarization dependence of the SH signal from Ag(110) by SONG and PLUMMER [61] are consistent with these theoretical predictions for the normal surface second-harmonic polarization. Previous data taken on Ag and Al that seemed to suggest much smaller values of $a(\omega)$ [62] were most likely influenced by the state of the surface. Since they were not collected under ultra-high-vacuum conditions, some adsorbates were presumably present. These tend to bind the nearly free conduction electrons of the metal and thereby reduce their polarizability at the surface. As a result, the magnitude of $a(\omega)$ will be reduced from its clean-surface value.

REFERENCES

[1]   P. HOHENBERG and W. KOHN: *Phys. Rev.*, **136**, B864 (1964).
[2]   W. KOHN and L. J. SHAM: *Phys. Rev.*, **140**, A1133 (1965).
[3]   For an excellent review, see N. D. LANG: *Solid State Phys.*, **28**, 225 (1973).
[4]   S. FAHY, X. W. WANG and S. G. LOUIE: *Phys. Rev. Lett.*, **61**, 1631 (1988).
[5]   R. O. JONES and O. GUNNARSSON: *Phys. Rev. Lett.*, **55**, 107 (1985); O. GUNNARSSON and R. O. JONES: *Phys. Rev. B*, **31**, 7588 (1985).
[6]   R. SMOLUCHOWSKI: *Phys. Rev.*, **60**, 661 (1941).
[7]   M. W. FINNIS and V. HEINE: *J. Phys. F*, **4**, L37 (1974).
[8]   R. N. BARNETT, U. LANDMAN and C. L. CLEVELAND: *Phys. Rev. B*, **28**, 1685 (1983).
[9]   K. M. HO and K. P. BOHNEN: *Phys. Rev. Lett.*, **56**, 934 (1986); *Phys. Rev. B*, **32**, 3446 (1985).
[10] The references to the various experimental results can be found in ref. [9].
[11] K. M. HO and K. P. BOHNEN: *Phys. Rev. Lett.*, **59**, 1833 (1987).
[12] W. MORITZ and D. WOLFF: *Surf. Sci.*, **163**, L655 (1985); **88**, L29 (1979).
[13] M. COPEL and T. GUSTAFSSON: *Phys. Rev. Lett.*, **57**, 723 (1986).
[14] J. W. M. FRENKEN, R. L. KRANS, J. F. VAN DER VEEN, E. HOLUB-KRAPPE and K. HORN: *Phys. Rev. Lett.*, **59**, 2307 (1987).
[15] H. L. DAVIS and J. R. NOONAN: *Phys. Rev. Lett.*, **54**, 566 (1985).
[16] S. M. YALISOVE and W. R. GRAHAM: *Surf. Sci.*, **183**, 556 (1987).
[17] M. K. KANG and E. J. MELE: *Phys. Rev. B*, **37**, 7371 (1987).
[18] SIU-CHING LIU, M. H. KANG, E. J. MELE, E.W. PLUMMER and D. M. ZEHNER: *Phys. Rev. B*, **39**, 13149 (1989).
[19] L. HEDIN: *Phys. Rev.*, **139**, A796 (1965).
[20] L. HEDIN and S. LUNDQVIST: *Solid State Phys.*, **23**, 1 (1969).

[21] M. S. HYBERTSEN and S. G. LOUIE: *Phys. Rev. Lett.*, **55**, 1418 (1985).

[22] R. W. GODBY, M. SCHLÜTER and L. J. SHAM: *Phys. Rev. Lett.*, **56**, 2415 (1986).

[23] J. E. NORTHRUP, M. S. HYBERTSEN and S. G. LOUIE: *Phys. Rev. Lett.*, **59**, 819 (1987). See also ref. [24].

[24] I. W. LYO and E. W. PLUMMER: *Phys. Rev. Lett.*, **60**, 1558 (1988).

[25] K. W. K. SHUNG and G. D. MAHAN: *Phys. Rev. Lett.*, **57**, 1076 (1987); K. W. K. SHUNG, B. E. SERNELIUS and G. D. MAHAN: *Phys. Rev. B*, **36**, 4499 (1987).

[26] W. EBERHARDT and E. W. PLUMMER: *Phys. Rev. B*, **21**, 1558 (1988).

[27] D. PENN: *Phys. Rev. Lett.*, **42**, 921 (1979).

[28] A. LIEBSCH: *Phys. Rev. Lett.*, **43**, 1431 (1979).

[29] J. KANAMORI: *Prog. Theor. Phys.*, **30**, 235 (1963).

[30] A. LIEBSCH: *Phys. Rev. B*, **23**, 5203 (1981).

[31] J. A. HERTZ and D. M. EDWARDS: *J. Phys. F*, **3**, 2174 (1973).

[32] B. N. J. PERSSON and S. ANDERSSON: *Phys. Rev. B*, **29**, 4382 (1984).

[33] B. N. J. PERSSON and W. L. SCHAICH: *J. Phys. C*, **14**, 5583 (1981).

[34] B. N. J. PERSSON and E. ZAREMBA: *Phys. Rev. B*, **30**, 5669 (1984).

[35] A. LIEBSCH: *Phys. Rev. B*, **33**, 7249 (1986); **35**, 9030 (1987).

[36] P. J. FEIBELMAN: *Prog. Surf. Sci.*, **12**, 287 (1982), and references herein.

[37] W. L. SCHAICH and K. KEMPA: *Phys. Scr.*, **35**, 204 (1987).

[38] P. APELL: *Phys. Scr.*, **24**, 795 (1981).

[39] A. LIEBSCH: *Phys. Scr.*, **35**, 354 (1987).

[40] A. ZANGWILL and P. SOVEN: *Phys. Rev. A*, **21**, 1561 (1980).

[41] M. J. STOTT and E. ZAREMBA: *Phys. Rev. A*, **21**, 12 (1980).

[42] W. ECKHARDT: *Phys. Rev. Lett.*, **52**, 1925 (1984).

[43] A. EGUILUZ: *Phys. Scr.*, **36**, 115 (1987).

[44] A. LIEBSCH: *Phys. Rev. Lett.*, **54**, 67 (1985).

[45] G. D. MAHAN: *Phys. Rev. A*, **22**, 1780 (1980).

[46] A. LIEBSCH: *Phys. Rev. B*, **32**, 6255 (1985).

[47] A. LIEBSCH: *Phys. Rev. B*, **36**, 7378 (1987).

[48] K. KEMPA and W. L. SCHAICH: *Phys. Rev. B*, **37**, 6711 (1988); K. KEMPA, A. LIEBSCH and W. L. SCHAICH: *Phys. Rev. B*, **38**, 12645 (1988).

[49] H. LEVINSON, E. W. PLUMMER and P. J. FEIBELMAN: *Phys. Rev. Lett.*, **43**, 952 (1979).

[50] K. D. TSUEI, E. W. PLUMMER and P. J. FEIBELMAN: *Phys. Rev. Lett.*, **63**, 2256 (1989).

[51] N. BLOEMBERGEN, R. K. CHANG, S. S. JHA and C. H. LEE: *Phys. Rev.*, **174**, 813 (1968).

[52] J. RUDNICK and E. A. STERN: *Phys. Rev. B*, **4**, 4272 (1971).

[53] J. E. SIPE and G. I. STEGEMAN: in *Surface Polaritons*, edited by V. M. AGRANOVICH and D. L. MILLS (North Holland, New York, N.Y., 1982), p. 661.

[54] M. CORVI and W. L. SCHAICH: *Phys. Rev. B*, **33**, 3688 (1986).

[55] A. ZANGWILL: *J. Chem. Phys.*, **78**, 5926 (1983).

[56] G. SENATORE and K. R. SUBBASWAMY: *Phys. Rev. A*, **35**, 2440 (1987).

[57] A. LIEBSCH and W. SCHAICH: *Phys. Rev. B*, **40**, 5401 (1989).

[58] M. WEBER and A. LIEBSCH: *Phys. Rev. B*, **35**, 4711 (1987); **36**, 6411 (1987).

[59] W. L. SCHAICH and A. LIEBSCH: *Phys. Rev. B*, **37**, 6187 (1988).

[60] A. LIEBSCH: *Phys. Rev. Lett.*, **61**, 1233 (1988).

[61] K. J. SONG: Ph. D. Thesis, University of Pennsylvania (1988).

[62] J. C. QUAIL and H. J. SIMON: *Phys. Rev. B*, **31**, 4900 (1985).

Theory of Photoemission.

G. D. Mahan

Solid State Division, Oak Ridge National Laboratory
P.O. Box 2008 - Oak Ridge, TN 37831-6030
Department of Physics and Astronomy, University of Tennessee
Knoxville, TN 37996-1200

1. – Introduction.

Photoemission is an old science. The theoretical development started with Einstein[1], and the experimental work began even earlier. The experiments were difficult until 25 years ago[2]. Then modern surface science began with the development of high-vacuum technology. Experimental techniques have become increasingly sophisticated, which has stimulated more theoretical work.

Many of the earlier theoretical papers were on the surface effect[3-6]. Today most theories are concerned with the volume effect[7-13]. That both surface and volume processes contribute was recognized long ago by Tamm and Schubin[14].

In all optical absorption processes, the electron absorbs the photon which increases the energy of the electron. However, if the electrons are unbound, then they must also increase their momentum. The solid has numerous sources of momentum, and each causes photon absorption. Bragg scattering from the crystal potential we call «interband absorption». The scattering by phonons and impurities is called «Drude absorption». Since photoemission is a surface experiment, the electrons are near the surface. The surface itself provides a source of momentum, since electrons reflect which come toward it. The absorption of photons by the surface momentum is called the *surface effect* in photoemission. Once it was thought to be the dominant process in photoemission. Now we regard it as a relatively weak process. However, it certainly does occur, and is frequently observed and sometimes studied by itself[15].

The *volume effect* in photoemission describes the process in which the electron absorbs a photon while undergoing an interband transition. Some of the excited electrons find themselves headed towards the surface, and some of them escape. This view of photoemission is called the *three-step model*. The three steps are photon absorption, migration towards the surface and escape through

the surface into the external vacuum. The three-step model was first used by
SPICER [16]. Prior theoretical work utilized the two-step model, where the two
steps are optical absorption and emission from the surface. This model is correct
for the surface effect, where it was mostly applied [3-5]. However, it was also
used earlier for the volume effect [17, 18]. MAHAN [7] was the first to derive the
three-step model from scattering theory, although this derivation came well
after the model was in widespread usage.

2. – One-electron formalism.

The one-electron theory of photoemission assumes all electrons are in perfect
eigenstates. It ignores damping and scattering. This situation is far from reality.
Later we shall derive a many-electron theory which includes scattering and the
decay of states. However, the one-electron theory is a good place to begin. With
this model we can introduce a number of the key ideas of photoemission theory.

«One-electron theory» means that we can treat each electron separately.
Each one has its motion described by a Hamiltonian which contains a kinetic-
energy and a potential-energy term

$$(2.1) \qquad\qquad H = p^2/2m + V(r).$$

Choosing the potential $V(r)$ determines the complexity of the calculation. The
more realistically V is chosen, then the more realistic are the electron
eigenstates. After solving for these states, we use the golden rule to obtain the
rate of electron transitions

$$(2.2) \qquad\qquad W = 2\pi \sum_{if} |M|^2 \delta(E_i + \omega - E_f),$$

$$(2.3) \qquad\qquad M = \int d^3r\, \psi_f^+ \delta V \psi_i, \qquad \delta V = (e/2mc)[\boldsymbol{p} \cdot \boldsymbol{A} + \boldsymbol{A} \cdot \boldsymbol{p}],$$

where A is the vector potential from the photons. We use units where Planck's
constant $h = 2\pi$. The quantity W is the number of electrons each second which
absorb a photon and are ejected from the surface of the solid. Later we shall see
that it is proportional to the rate at which photons are incident upon the solid.
However, now we examine the eigenstates $\psi_{i,f}$ of the Hamiltonian H. The initial
state ψ_i is a bound state of the electron in the solid, and as such is neither ingoing
nor outgoing. However, how do we construct an eigenstate ψ_f for an electron
which is leaving the surface of the solid?

In order to treat the case of an electron leaving, then ψ_f^* must be an *incoming*
wave function. This surprising result was first derived in photoemission by
MACKINSON [5], who treated the surface effect. The first correct treatment of
the volume effect was by MAHAN [7].

The use of an incoming wave function to describe an outgoing particle is also well known in nuclear physics. There it is called the distorted-wave Born approximation (DWBA). They divide the potential term into $V = V_1 + V_2$. First solve the eigenstates of $H_1 = KE + V_1$. With these eigenstates, the transition matrix element is given in the DWBA to be

$$M = M_1 + \langle f|V_2|i \rangle,$$

where M_1 is the matrix element deduced from H_1. We are doing the same, where $V_2 = \delta V$ and V_1 is the V in (2.1). However, in our case $M_1 = 0$ since there is no transition without the photon field. In DWBA $|i\rangle$ is a bound initial state, while $\langle f|$ is an incoming wave.

2'1. Surface effect. – The surface effect is the simplest case to study. We further simplify the example by treating the solid as a homogeneous electron gas with no ions. Then the potential $V(r)$ for the electron depends only on the distance from the surface $V(z)$. The surface potential is taken to be a step at $z = 0$ between the vacuum on the left and the solid on the right. Take $V(z) = 0$ for $z < 0$ and $V(z) = -V_0$ for $z > 0$. The sample is taken to have an area S and a thickness L. The initial wave function has wave vectors $(k_z, \boldsymbol{k})$ and an initial energy $E_i = (k_z^2 + k^2)/2m - V_0$. The wave function is

(2.4) $\psi_i = (2/\sqrt{SL})\,\phi_i(z)\exp[i\boldsymbol{k}\cdot\boldsymbol{\rho}], \qquad \phi_i = \begin{cases} \sin(k_z z + \delta), & z > 0, \\ \sin\delta\exp[\alpha z], & z < 0, \end{cases}$

where

$$\alpha = [2mV_0 - k_z^2]^{1/2}, \qquad \mathrm{tg}\,\delta = k_z/\alpha.$$

The expression for the surface phase shift δ is derived by requiring that ϕ_i be continuous, and have continuous slope, at $z = 0$.

Next is found the incoming wave function. In the photoemission experiment, the electron leaves the solid surface with a momentum p at an angle θ from the normal to the surface. We define $p_z = p\cos\theta$ and $p = p\sin\theta$ to be the components in vacuum. We will always use p for wave vectors outside the solid, and k for inside. The incoming wave function is that for a plane wave coming towards the surface at the same angle θ from the normal

(2.5) $\begin{cases} \psi_f^+ = (1/\sqrt{SL})\,\phi_f^{\gtrless}(z)\exp[-i\boldsymbol{p}\cdot\boldsymbol{\rho}], \\[2mm] \phi_f^{\gtrless} = \begin{cases} \exp[ip_z z] + R\exp[-ip_z z], & z < 0, \\ T\exp[ik'z], & z > 0, \end{cases} \end{cases}$

where

$$k' = [p_z^2 + 2mV_0]^{1/2}, \quad R = -(k' - p_z)/(k' + p_z), \quad T = 2p_z/(k' + p_z).$$

The incoming wave is incident from the left. There is a reflected wave with amplitude R going to the left. Inside the solid, there is a wave going right with amplitude T. Note that the incoming wave does not have a complex conjugate or Hermitian conjugate symbol. The function as written in (2.5) is already conjugated. The superscript $>$ is used to remind us that this function is an incoming wave. A useful check on our understanding of these two wave functions is to show that they are orthogonal. For this proof one must evaluate the integral

(2.6)
$$0 = \int_{-\infty}^{\infty} dz \, \phi_f^> \phi_i.$$

One must separately do the integrals ($-\infty$ to 0) and (0 to ∞) and show these two terms sum to zero. For this proof one needs the definition of R, T and δ.

For the photoemission process we need the matrix element $\langle f | \delta V | i \rangle$. In (2.3) for δV the vector potential A has components perpendicular A_z and parallel A to the surface. The symbol p is a gradient of position. The term in A has a zero matrix element since the z part of the integrand is just the integral in (2.6) which is zero. We also assume the vector potential of the photon is a constant. Of course, it varies over the wavelength of the light, but these variations are small and are neglected. With all of these assumptions, the photoemission matrix element becomes

$$\langle f | \delta V | i \rangle = - i(eA_z/m) \, \delta_{kp} \int_{-\infty}^{\infty} dz \, \phi_f^> \frac{\partial}{\partial z} \phi_i.$$

The integral over the plane parallel to the surface is zero unless $p = k$. The integral over z can be done. One way to express the answer is

$$\langle f | \delta V | i \rangle = - i(eA_z/m\omega) \, V_0 \, \delta_{kp} \, \phi_f^>(0) \, \phi_i(0).$$

The integral is given by the amplitude of the two wave functions at the point $z = 0$. Of course, this is the same as $T \sin \delta$. There is a much easier way to evaluate the above integral. One uses a trick. The trick is to use the expression for the time derivative of the momentum

(2.7)
$$\partial p / \partial t = i[H, p],$$

where H is given in (2.1). The kinetic-energy term commutes with p. Only the potential energy does not commute with momentum. On the left-hand side of the equal sign we only want responses at frequency ω, so the time derivative must give $i\omega$. Thus we can specialize the above result to

$$i\omega p_z = -\partial V(z)/\partial z, \quad \langle f|p_z|i\rangle = \frac{i}{\omega} \int_{-\infty}^{\infty} dz\, \phi_f^{\ge} \frac{\partial V(z)}{\partial z} \phi_i.$$

In the present example, $V(z)$ is a step function, and its derivative is a delta-function at $z = 0$, or $\partial V/\partial z = V_0 \delta(z)$. Then the integral does give the value of the wave functions at $z = 0$.

The delta-function at $z = 0$ occurs because we assumed that the surface potential was a step. A more realistic model would have $V(z)$ a smooth function which goes from a value of zero to $-V_0$ over a distance of a few ångströms. In this case $\partial V/\partial z$ is a curve of a finite width, whose area equals $-V_0$.

Now we have accumulated enough results to evaluate the expression (2.2) for the rate that electrons are ejected from the surface. The photon flux F is given in terms of the vector potential and the fine-structure constant $\alpha = e^2/c$:

$$(eA/c)^2 = 2\pi\alpha F/n\omega S,$$

where n is the refractive index. Since the surface effect depends only upon the component of the vector potential A_z which is perpendicular to the surface, we insert a factor of $(\varepsilon \cdot z)^2$, where ε is the polarization vector of the incident photon. Also, since we assume that the vector potential is unchanged at the surface, then the photon flux F is the same inside and out—this is the same as assuming the reflectance is zero. Later, when we talk about the volume effect, we shall take F to be the flux inside of the solid.

In (2.2) there is a summation over initial and final states. In this case it is a summation over the wave vectors k and p. The three-dimensional integral over p we write as $d^3p = p^2\, dp\, d\Omega$, where $d\Omega$ is the unit of solid angle in the final state. This factor is taken to the left of the equal sign, and $W/d\Omega = dI/d\Omega$ is the number of electrons per second emitted into that unit of solid angle.

Furthermore, the delta-function for energy conservation eliminates the integral over dp. This brings us to the expression

(2.8)
$$\frac{dI}{d\Omega} = \frac{e\alpha F(\varepsilon \cdot z)^2}{2\pi^2 \omega^3 mn} \int d^3k\, \delta(\boldsymbol{k} - \boldsymbol{p})\, p\, \langle f| \frac{\partial V(z)}{\partial z} |i\rangle^2.$$

The evaluation of the matrix element was discussed above. The above integral can be evaluated analytically when $V(z)$ is a step function. The above expression is the basic formula for the intensity of photoemitted electrons from the surface effect. Often this formula must be altered to account for special cases: e.g., when there is a surface electronic state [19] which provides a significant contribution to the absorption. The above formula only applies to the absorption caused by three-dimensional-band electrons.

2·2. *Approximations.* – The above derivation made a number of approximations. The intent was to obtain a simple answer. Here we examine several of these approximations in order to see which can be eliminated, or made less severe.

We have already noted that the surface potential $V(z)$ need not be a step. One can use a realistic shape such as those calculated for real surfaces by LANG and KOHN [20]. This choice changes the matrix element $\langle f|\partial V/\partial z|i\rangle$. It affects $V(z)$, and the wave functions $\langle f|$ and $|i\rangle$. Generally they and the matrix element must be evaluated on the computer.

The vector potential $A_z(z)$ is not a constant in the surface region. Classical electromagnetic theory shows that the z-component of the vector potential changes by a factor of the dielectric function $\varepsilon(\omega)$ between the outside and inside of the solid. Quantum mechanics shows that this difference is achieved asymptotically far from the surface. Near the surface the vector potential has Friedel oscillations with a wave vector of $2k_F$ [21]. As a consequence, the z-dependence of the vector potential must be included when evaluating the z-integral

$$(2.9) \qquad \langle f|\partial V|i\rangle = \frac{e}{2imc}\,\delta_{k=p}\int dz\, \hat{\phi}_f^* \left\{\frac{\partial}{\partial z}A_z(z) + A_z(z)\frac{\partial}{\partial z}\right\}\phi_i.$$

When the vector potential has significant z-dependence, then the approximation of replacing the matrix element of p_z by $\partial V/\partial z$ is also not an accurate approximation. Instead, one must evaluate the above integral.

At high photon frequency the dielectric function approaches unity according to $\varepsilon(\omega) \approx 1 - \omega_p^2/\omega^2$, where ω_p is the plasma frequency. For photon frequencies which are greater than about three times the plasma frequency, the dielectric function $\varepsilon(\omega)$ is nearly unity. In that case the vector potential has no significant z-variation at the surface. For photoemission at high frequency one can treat $A_z(z)$ as a constant and use (2.8) for the evaluation of the surface effect. This formula is valid at high frequencies. The range of valid photon energies is above 50 eV for aluminum, and above 15 eV for an alkali metal.

Another approximation was using $\nabla V/\omega$ for p in the matrix element. ALMBLADH [13] has investigated this approximation and concluded that it is generally valid. The actual potential energy of the electron has terms from the surface, the host ions in the solid, the impurity ions and the ion vibrations (phonons). A rigorous theory would include all of these various potential terms in V, and in ∇V. We will only use this approximation at high frequencies. There the photon absorption is caused by the potentials from the surface and the host ions, while impurity and phonon contributions are negligible. Photoemission from the surface potential is the surface effect, and from the host ions is called the volume effect.

2'3. *Volume effect.* – The volume effect in photoemission comprises the following processes. An electron in the solid absorbs a photon and increases its energy by ω. Simultaneously it Bragg scatters from the crystalline potential and changes its wave vectors by one of the reciprocal-lattice vectors G. If the initial wave vector is $k - G$, and the final wave vector is k, energy conservation is

$$\frac{k^2}{2m} - \omega = \frac{(k - G)^2}{2m} = \frac{k^2}{2m} + E_G - \frac{kG \cos \theta}{m},$$

where $E_G = G^2/2m$. Thus we derive an expression for the final energy of the electron after it has absorbed the photon and had its Bragg scattering

(2.10) $$\varepsilon_k = k^2/2m = (\omega + E_G)^2/4E_G \cos^2 \theta.$$

The final energy depends upon the photon frequency ω, the Bragg wave vector G and the angle θ that k makes with G. All electrons at the same angle θ have the same energy. In real space the electrons of constant energy form a conical distribution where the direction G is the center of the cone. The initial electrons have a kinetic energy less than the Fermi energy E_F, so the excited electrons have energy less than $\varepsilon_k < E_F + \omega$. Using (2.10) one can show that E_F defines a maximum cone angle. This prediction of mine has been verified by numerous experiments[22-24]. These conical distributions are inside of the solid. The electron must escape from the surface in order to be measured in a photo-emission experiment. The surface potential causes the electron wave to refract at the surface. This refraction causes significant distortion of these conical shapes. These cones of electrons are called *primary cones*[7].

There are also *secondary cones* caused by Bragg scattering at the surface[7]. Low-energy electron diffraction (LEED)[25] is an experiment where electrons are shot at the surface, usually along a trajectory normal to the surface. Some of the electrons are reflected along a specular trajectory. Others are reflected in other directions because they Bragg scatter from the surface of the solid. The two-dimensional periodic structure of the surface defines a set of lattice vectors G. These surface lattice vectors are the projections onto the surface of the three-dimensional lattice vectors G. In LEED the electron whose specular wave vector would be p has additional scattering channels with the wave vectors $p + G$. In photoemission, the electron approaches the surface from inside, and has a probability of passing through the surface into the vacuum outside. During this passage it may Bragg scatter from the surface potential, which scatters it by G. This reciprocal-lattice vector is totally independent from the one which caused the original optical absorption. The electron can undergo two Bragg scatterings. One inside the volume causes the absorption of the photon. The second is a Bragg scattering from the surface as it exits the surface region. The primary cones cause distributions of electrons outside of the surfaces. The secondary cones are due to the Bragg scattering of this distribution by the surface potential.

The connection between photoemission and LEED is very strong. Remember that photoemission uses the incoming wave function. That is just the wave function calculated in LEED, where the electron is shot at the surface. In LEED one needs the amplitude of the electron wave reflected from the surface, while in photoemission one needs the amplitude of the wave transmitted into the surface. Both can be obtained from the same calculation.

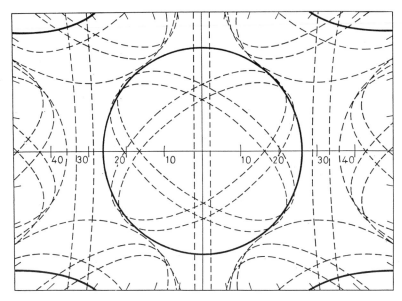

Fig. 1. – Primary cones (solid lines) and secondary cones (dashed lines) of photoemitted electrons from the sodium (110) surface. The photon energy is 18 eV.

An illustration of secondary and primary cones is shown in fig. 1 for the sodium (110) surface when the photon energy is 18 eV [26, 27]. This example is selected because Plummer's group [28-30] has done many measurements of angular photoemission at this frequency and geometry. The solid lines are the boundaries of primary cones obtained from the maximum cone angles. One at a polar angle of 27° is from the reciprocal-lattice vector which is normal to the surface. The other solid lines are from other directions of G, and these distributions are distorted by surface refraction. The dashed lines are secondary cones. There are many of them, which have odd shapes. They are found by just adding all possible values of G to the excited electrons, and computing at which angles and energies they will leave the solid.

Plummer's group has measured the azimuthal angular dependence of the photoemission at various polar angles. Primary cones predict no azimuthal dependence for polar angles less 50°. However, the secondary cones predict much azimuthal dependence. PLUMMER does see some azimuthal dependence.

So far it has been impossible to calculate the intensity of the secondary cones in alkali metals. One needs to know accurately the surface potential $V(r)$ of the electron. This has never been determined for sodium because LEED experiments are too difficult. The surface is very reactive chemically, which makes it impossible to have the high-quality surface needed for LEED. So there are no experimental data on the surface potential. One cannot even obtain theoretical estimates. Existing LEED computer codes assume the incoming electrons have very high energy, and ignore surface refraction. These codes are inaccurate for electrons whose kinetic energy is $(10 \div 20)\,\text{eV}$.

Our discussion of the volume effect has been concerned with the directions the electrons leave the solid. Also of great importance is the intensity of the photoemitted electron beam as a function of photon energy ω and the emitted direction. This intensity cannot be accurately calculated in the one-electron model. Consider the following cases:

a) The electron and photon both have infinite mean free paths (m.p.f.). Then the photons penetrate evenly throughout the volume of the solid. All electrons excited in the solid get to a surface and may be photoemitted. The effective volume for photoemission is SL, where S and L are the sample area and thickness. This model is unrealistic, but is the answer obtained in a simple one-electron theory.

b) The photons have a finite m.f.p. λ, but the electrons have infinite m.f.p. Then the photons will be absorbed within a distance λ of the surface. Only electrons in this volume participate in the photoemission process. The effective volume for photoemission is λS.

c) The photons have a m.f.p. of λ, while the electrons have a m.f.p. of l. Then the effective volume for photoemission is $\lambda l S/(\lambda + l)$. In nearly every experimental case one has the inequality $\lambda \gg l$, which means the effective volume is lS. Thus a realistic theory must include the finite m.f.p. of the electron. This requires a many-electron description.

3. – Many-electron theory.

A many-electron theory is needed for volume photoemission. The electron mean free path (m.f.p.) is relatively short for electrons with an energy well above the chemical potential. Most of the photoemitted electrons are excited by the optical absorption within the m.f.p. of the surface. The value of the m.f.p. in metals is entirely determined by electron-electron scattering. It is calculated from the imaginary part of the self-energy of the electron. This self-energy also has a real part, which shifts the total energy of the electron. This real part is also important for the calculation of the photoemission, as is shown below.

The self-energy of an electron in a metal has numerous contributions. There are terms from scattering the electron by the host ions, by impurities, by phonons and by other electrons. The first of these contributions defines energy bands. The next two make only small contributions to the electron's energy. They are important for electrons near the Fermi energy, where small contributions are relatively large compared with thermal energies $(k_B T)$. They are important for electrical resistivity. However, for electrons of large kinetic energy, only electron-electron interactions provide a significant contribution. The most important term is called the screened exchange interaction. Its evaluation is discussed elsewhere [31]. Figure 2 shows a recent calculation of the

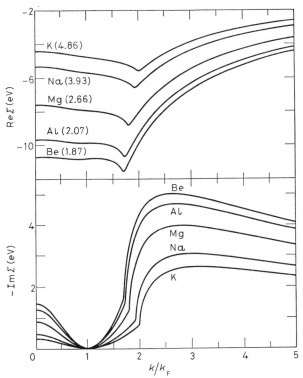

Fig. 2. – Real and imaginary part of the self-energy of an electron from the screened exchange interaction.

screened exchange energy $\Sigma(k, \xi_k)$ for some simple metals [32]. The imaginary self-energy is zero at the Fermi energy. At large wave vector it achieves a large value. In this region the m.f.p. of electrons in sodium is about $(5 \div 6)$ Å. Many of Plummer's experiments [28-30] excite electrons to the states with these short m.f.p. Thus his photoemitted electrons are coming from the first one or two atomic layers at the surface.

The real part of the self-energy is also important. For the occupied electron states $(k < k_F)$ the real part is large and negative. For the excited electron states

$(k \sim 2k_F)$ the self-energy is declining in value. If the real part of the self-energy is defined as $\sigma(k)$, then energy conservation is

(3.1) $$\varepsilon(k_i) + \sigma(k_i) + \omega = \varepsilon(k_f) + \sigma(k_f),$$

where $k_f = k_i + G$ for interband transitions. Since $\sigma(k)$ changes between initial and final state, then the final kinetic energy is changed by self-energy effects. This is not a small effect. Figure 2 shows that the changes can be as large as $(2 \div 3)$ eV. This change in the final kinetic energy shows up dramatically in shifting the apparent position of band gaps and other critical points in the final density of states[32, 33].

Another important question is our use of free-electron kinetic energy $\varepsilon(k) = k^2/2m$ for the final state. Band structure calculations show that the bands are distorted near critical points. They appear far from free-electron-like in these calculations. However, most band structure calculations make a serious approximation. They assume that the potential energy acting upon the electron is real. They neglect damping and effects associated with the finite m.f.p. of the electron. When these effects are included, the bands become very parabolic, and band gaps disappear. It takes a surprisingly small amount of damping to render the upper bands parabolic. This point was first shown by PENDRY[25]. Today most experimentalists assume the final bands are parabolic with the mass of the noninteracting electron.

We begin the discussion of many-electron effects by examining the following expression for the wave function of the electron of energy E outside of the metal:

$$\psi_E(\boldsymbol{R}) = \int d^3r \, G_E(\boldsymbol{R}, \boldsymbol{r}) \, \delta V \psi_i(\boldsymbol{r}).$$

The initial wave function $\psi_i(\boldsymbol{r})$ has the meaning used earlier—the one-electron wave function. The symbol G stands for the Green's function of the electron in the final state. We can use a fully interacting Green's function here, which includes many-electron effects in the final state. Later we will discuss the technique of including many-electron effects in the initial state also.

There are numerous ways of including many-electron effects in the final state. Here we adopt an intermediate approach. We will derive a formula which is easy for computation, yet includes the most important processes. First we take the limit that R becomes large as the electron leaves the surface towards the vacuum. Since there are no many-electron effects in the vacuum, the Green's function must have the form

$$\lim_{R \to \infty} G_E(R) = \frac{m \exp[ipR]}{2\pi R} \, \psi_f^>(\boldsymbol{r}),$$

where $\psi^>$ is again an incoming plane wave. Outside of the surface it has exactly

the same form (2.5) as in the noninteracting case. As we said, outside there are
no interactions. Inside of the surface, it has the form

$$(3.2) \qquad \psi_f^>(r) = T \exp[i\boldsymbol{p} \cdot \boldsymbol{\rho}] \exp[z(ik' - 1/2l)].$$

The mean free path l has been added to the z-part of the wave function. It is
important to note that we do not add it to the wave motion parallel to the
surface. The incoming wave has a parallel wave vector $\boldsymbol{p}$ which must be real since
it is in the vacuum. The wave in the solid must also have the same parallel wave
vector in order to match it to the incoming wave. Since the outside parallel wave
vector is real, so it must be inside. Inside the surface, the incoming electron must
have a parallel wave vector which is real. Only the z-component can have a
complex wave vector due to electron-electron interactions. Incidentally, the
same thing happens when photons are reflected from an absorbing surface.
Inside the surface the electromagnetic wave has a parallel wave vector which is
real, and a z-wave vector which is complex.

The above wave function can be used to calculate the external current in
photoemission

$$(3.3) \qquad \frac{dI}{d\Omega} = R^2(e/m) \operatorname{Im} \{\psi_E^* \nabla \psi_E\} = \frac{emp}{(2\pi)^2} \left| \int d^3r \, \psi_f^>(r) \, \delta V \, \psi_i(r) \right|^2.$$

Equation (3.3) contains the basic ingredients of the three-step model. The
incoming wave function in (3.2) contains the factor T which is the transmission
through the surface. The factor of the mean free path provides the probability of
transporting the electron to the surface in an unscattered state. The other
factors provide the rate of optical absorption. The expression contains the three
steps: absorption, transport to the surface and transmission through the surface.
The factor T is actually the probability of an outside electron getting into the
solid. However, the probability T' of getting out is proportional to T, as can be
shown by time-reversal invariance.

In doing the above integral $d^3r = d^2\rho \, dz$, the area integral $d^2\rho$ just forces the
initial and final parallel wave vectors to be equal. Thus the only interesting
integral is over dz:

$$(3.4) \qquad \frac{dI}{d\Omega} = \frac{e\alpha F}{(2\pi)^2 m\omega^3} \int dk_z \, p \left| \int dz \, \phi_f^>(z) \frac{\partial V(z)}{\partial z} \phi_i(z) \right|^2,$$

where

$$V(z) = V_s(z) + \sum_G \exp[izG] V_G.$$

Equation (3.4) was evaluated by SHUNG and MAHAN in their original calculation
of the photoemission of sodium[26]. The two potential terms are from the surface

effect and the volume effect. This particular form is valid for the case in which the reciprocal-lattice vector G is normal to the surface, so there is interference effect between the surface and volume terms. The incoming wave function in (3.2) was used, with the energy-dependent m.f.p. This simple calculation explained most of the results reported in ref. [28]. It explained line shapes perfectly, particularly those peaks which appeared in the forbidden gap of final states. They were found to arise from effects associated with the finite m.f.p. PLUMMER also provided us with many unpublished measurements of angular and frequency-dependent EDC's. These we also fitted very well by our theory.

There were several features of Plummer's measurements that our theory does not explain. Before we dwell on the disagreements between theory and experiment, it should be emphasized that we do explain about 95% of his results. Nonetheless, physicists love to focus on disagreements rather than agreements.

One disagreement was the position of the energy gaps in the excited states. The original calculation of Shung and Mahan did not include the real part of the self-energy. Because of this omission, the predicted gaps occurred at the wrong frequency with an error of 2 eV. SHUNG et al. later pointed out that including the real part of the electron-electron self-energy moved the theoretical gaps to the right place—to the experimental ones [32]. The correct physical model for the final states is to take them as parabolic bands which are corrected by the self-energy from electron-electron interactions. This latter correction shifts final-state energies by several electronvolt. The photoemission data provide a direct confirmation that the screened exchange energy declines with energy, and at the rate calculated by simple theories [32, 33].

The second disagreement is that our theory is unable to explain a second peak in the EDC's which are observed at photon energies above 40 eV. Our one-electron calculations have no mechanism for producing this peak. There have been two other proposed explanations for these peaks: 1) OVERHAUSER suggested they are due to spin or charge density waves [34], and 2) INGLES-FIELD et al. suggested they are due to empty surface states [35]. Neither group has produced a calculation which shows such a peak at the correct photon frequency and electron energy. Until they do, we regard this peak as unexplained.

Another important feature of Plummer's results are the band widths. Plummer's group reported that the energy width of the occupied bands in sodium is about 2.5 eV. This value is at least 0.5 eV smaller than expected from previous many-electron calculations. Previous X-ray emission data also show a band width of 3.1 eV, in great disagreement to Plummer's measurements. However, LYO and PLUMMER [29] showed that a calculation of the screened exchange energy with the best dielectric function predicts a band width in agreement with their measurements. The current thinking is that they are right and the band width of sodium and other metals is much narrower than thought previously.

Our band width calculations include many-electron effects in the initial state as well as in the final state. They are included by using techniques well-known from prior theories [8, 36]. Now we describe how to include many-electron effects in the initial state.

If we add a delta-function for energy conservation, then the summation over initial states is just a spectral function

$$A(r, r', E) = \sum \psi_i(r) \, \psi_i^*(r') \, 2\pi \, \delta(E_i - E),$$

$$\frac{\mathrm{d}^2 I}{\mathrm{d}\Omega \, \mathrm{d}E} = \frac{emp}{(2\pi)^2} \int \mathrm{d}^3 r \int \mathrm{d}^3 r' \, \psi_f^>(r) \, \delta V(r) \, A(r, r', E - \omega) \, \delta V(r') \, \psi_f^>(r')^*,$$

where we have written out the square of the matrix element in (3.3) as a double integral. One way to include many-electron effects in the initial states is to use a many-electron form for the spectral function A. This spectral function is just the imaginary part of the Green's function. One calculates the self-energy of the electron, as shown in fig. 2, and from it one obtains the spectral function. Then the above integral is done numerically [36].

The integrals over $\mathrm{d}^2\rho$ and $\mathrm{d}^2\rho'$ just require the conservation of parallel wave vector. Numerical integrals only need to be done over the two z-integrals. That is, we obtain an expression of the form

$$(3.5) \quad \frac{\mathrm{d}^2 I}{\mathrm{d}\Omega \, \mathrm{d}E} = \frac{e\alpha pF}{(2\pi)^3 \, m\omega^3} \int \mathrm{d}z \int \mathrm{d}z' \, \phi_f^>(z) \frac{\partial V(z)}{\mathrm{d}z} A(z, z', E - \omega) \frac{\partial V(z')}{\mathrm{d}z'} \, \phi_f^>(z')^*,$$

where

$$A(z, z', E) = \int \mathrm{d}k \, \phi_k(z) \, \phi_k^*(z') \frac{-2 \, \mathrm{Im}\, \Sigma(k, E)}{(E - \varepsilon(k) - \mathrm{Re}\, \Sigma(k, E))^2 + (\mathrm{Im}\, \Sigma(k, E))^2}.$$

The initial-state wave functions ϕ_k are shown in (2.4). This integral is also done numerically.

4. – Concluding remarks.

This lecture has presented a simple theory of photoemission in a free-electron metal such as sodium. The photoemission spectra of the alkali metals have been measured many times. There have been numerous theoretical discussions [37-40]. Yet the subject still remains interesting, and new aspects are continually being discovered. The alkali metals seem not to be so simple.

Photoemission is measured in many different kinds of materials. There have been many theories which can be applied to these more complex solids [37-40]. Some of these topics will be covered by the other lecturers.

* * *

I wish to thank K. SHUNG for his collaboration during the calculations used to illustrate this lecture. Research support is gratefully acknowledged from the National Science Foundation NSF-DMR 87-04210, from the University of Tennessee and from the U.S. Department of Energy under contract DE-AC05-8421400 with Martin Marietta Energy Systems.

REFERENCES

[1] A. EINSTEIN: *Ann. der Phys.*, **17**, 132 (1905).
[2] For a history of early work see G. MARGARITONDO: *Phys. Today*, p. 66 (April, 1988).
[3] K. MITCHELL: *Proc. R. Soc. London, Ser. A*, **146**, 442 (1934).
[4] A. G. HILL: *Phys. Rev.*, **53**, 184 (1938).
[5] R. E. B. MACKINSON: *Phys. Rev.*, **75**, 1908 (1949).
[6] I. ADAWI: *Phys. Rev.*, **134**, A788 (1964).
[7] G. D. MAHAN: *Phys. Rev. B*, **2**, 4334 (1970).
[8] W. L. SCHAICH and N. W. ASHCROFT: *Solid State Commun.*, **8**, 1959 (1970); *Phys. Rev. B*, **3**, 2452 (1971).
[9] G. PAASCH: *Phys. Status Solidi B*, **87**, 191 (1978).
[10] N. J. SHEVCHIK and D. LIEBOWITZ: *Phys. Rev. B*, **18**, 1618 (1978).
[11] B. C. MEYERS and T. E. FEUCHTWANG: *Phys. Rev. B*, **27**, 2030 (1983).
[12] W. BARDYSEWSKI and L. HEDIN: *Phys. Scr.*, **32**, 439 (1985).
[13] C.-O. ALMBLADH: *Phys. Rev. B*, **34**, 3798 (1986).
[14] I. TAMM and S. SCHUBIN: *Z. Phys.*, **68**, 97 (1931).
[15] H. J. LEVINSON and E. W. PLUMMER: *Phys. Rev. B*, **24**, 628 (1981).
[16] W. E. SPICER: *Phys. Rev.*, **112**, 114 (1958); C. N. BERGLAND and W. E. SPICER: *Phys. Rev.*, **136**, A1030 (1964); E. O. KANE: *Phys. Rev.*, **127**, 131 (1962); G. W. GOBELI and F. G. ALLEN: *Phys. Rev.*, **127**, 141 (1962).
[17] L. APKER, E. TAFT and J. DICKEY: *Phys. Rev.*, **74**, 1462 (1948).
[18] L. APKER and E. TAFT: *Phys. Rev.*, **79**, 964 (1950).
[19] P. O. GARTLAND and B. J. SLAGSVOLD: *Solid State Commun.*, **25**, 489 (1978).
[20] N. D. LANG and W. KOHN: *Phys. Rev. B*, **1**, 4555 (1970).
[21] P. J. FEIBELMAN: *Surf. Sci.*, **46**, 558 (1974).
[22] T. GUSTAFSSON, P. O. NILSSON and L. WALLDEN: *Phys. Lett.*, **37**, 121 (1971).
[23] L. WALLDEN and T. GUSTAFSSON: *Phys. Scr.*, **6**, 73 (1972).
[24] W. L. SCHAICH: *Phys. Status Solidi B*, **66**, 527 (1974).
[25] J. B. PENDRY: *Low Energy Electron Diffraction* (Academic Press, London, 1974).
[26] K. W. K. SHUNG and G. D. MAHAN: *Phys. Rev. Lett.*, **58**, 960 (1987).
[27] K. W. K. SHUNG and G. D. MAHAN: *Phys. Rev. B*, **38**, 3856 (1988).
[28] E. JENSEN and E. W. PLUMMER: *Phys. Rev. Lett.*, **57**, 1076 (1986).
[29] IN-WHAN LYO and E. W. PLUMMER: *Phys. Rev. Lett.*, **60**, 1558 (1988).
[30] E. W. PLUMMER: private communication.
[31] G. D. MAHAN: *Many-Particle Physics* (Plenum Press, New York, N.Y., 1981), Chapt. 5.
[32] K. W. K. SHUNG, B. E. SERNELIUS and G. D. MAHAN: *Phys. Rev.*, **36**, 4499 (1987).
[33] P. O. NILSSON and C. G. LARSSON: *Phys. Rev. B*, **27**, 6143 (1983).

[34] A. W. OVERHAUSER: *Phys. Rev. Lett.*, **55**, 1916 (1985).

[35] J. H. KAISER, J. E. INGLESFIELD and G. C. AYERS: *Solid State Commun.*, **63**, 689 (1987).

[36] C. CAROLI, D. LEDERER-ROZENBLATT, B. ROULET and D. SAINT-JAMES: *Phys. Rev. B*, 8, 4552 (1973).

[37] D. R. PENN: *Phys. Rev. Lett.*, **42**, 921 (1979).

[38] L. C. DAVIS and L. A. FELDCAMP: *Phys. Rev. B*, 23, 6239 (1981).

[39] G. WENDIN: *Structure and Bonding* (Springer-Verlag, Berlin, 1981).

[40] C.-O. ALMBLADH and L. HEDIN: in *Handbook on Synchrotron Radiation*, edited by E. E. KOCH (North-Holland, Amsterdam, 1983), Chapt. 8.

Dynamical Effects, Screening and Correlation in Mixed-Valence Compounds.

B. DELLEY

Paul Scherrer Institut, c/o Laboratories RCA - Badenerstrasse 569, CH-8048 Zürich

1. – Introduction.

Dilute magnetic-impurity compounds have been studied for a long time, motivated by hopes that they might prove to be simple model systems for understanding concentrated ferromagnets. A model Hamiltonian created by ANDERSON[1] helps to understand the appearance of local moments when treated in an unrestricted Hartree-Fock approximation. Mixed-valence rare-earth compounds are simpler in having a small hybridization coupling as compared to transition metals. The existence of a very small hopping term has been proposed to explain thermodynamic and transport properties which depend on a large density of states at the Fermi energy ε_F. Lattice parameter studies suggest that the f occupation n_f depends on chemical environment. The now outdated promotional model assumes that a virtual f-level is placed at ε_f within 0.1 eV of ε_F with a very weak coupling expressed by a resonance width $\Delta \approx 0.01$ eV. The ingredients of this simple model produce a large density of states near ε_f and explain the sensitivity to the environment acting on ε_f.

A different picture, however, has emerged from photoelectron spectroscopic studies of mixed-valence compounds. Core XPS spectroscopy showing rather intense satellite structures is most naturally explained by a sizable coupling strength. Valence photoemission suggests, in the case of Ce, that $\varepsilon_f \approx 2$ eV below ε_F and there are indications that there is more than one f-related feature. This latter fact points to gross effects due to electron correlation. Bremsstrahlung isochromatic spectroscopy showing f-related structures at ε_F and another f-related peak at higher energy gives another important hint to correlated electron states.

These facts call for a coherent theory explaining both thermodynamic and spectroscopic properties. In fact the Anderson model provides a suitable framework for such a unified treatment. Solution of the model was, however, a difficult problem until it was realized by RAMAKRISHNAN[2] and ANDERSON that a large f-degeneracy simplifies the problem by making an expansion in the small parameter $1/N_f$ possible. It was realized by GUNNARSSON and SCHÖNHAMMER [3] that such a $1/N_f$ expansion can also be used to calculate spectral functions.

2. – Model.

In this section I would like to discuss the choice of a model that contains enough physics to produce the most prominent effects seen in these interesting compounds. In the mixed-valence Ce compounds, configurations with zero f^0 and with one f-electron f^1 play a most important role. Figure 1 shows some single-

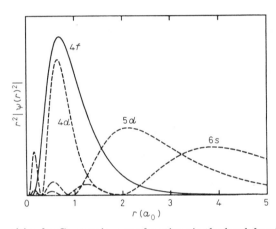

Fig. 1. – Radial densities for Ce atomic wave functions in the local density approximation.

particle functions for Ce according to a calculation in the local density functional approximation. Clearly the partially occupied $5d$ and $6s$ functions have fairly long tails that overlap significantly with neighbouring atoms. The $4f$ function has a radial extent comparable to a typical core orbital, it derives its interesting properties from the fact that, in the rare-earth elements, the nuclear charge is not large enough to overcome the electrostatic repulsion between f-electrons and permit full occupation. In contrast to core electrons, the Coulomb interaction U is comparable with the binding energy and it is thus important to include this interaction in the model. With some regret we will, however, neglect the full details of the electrostatic interaction between the f-electrons, which gives rise to the multiplet splitting of configurations with more than one f-electron or with a core hole. Since the f-electron has rather little overlap with orbitals centred on other atoms because of its corelike shape, the hybridization matrix element remains relatively small. One the one hand, it is still essential to include this hopping matrix element in the model in order to be able to get nonintegral occupation of the f-level. On the other hand, the relative smallness of this coupling in the rare-earth compounds as compared to transition metal compounds has helped a lot to ease understanding of magnetic-impurity systems. One is almost naturally led to start from the dilute limit and to study the Anderson impurity Hamiltonian[1]:

$$(1) \quad H = \sum_{kn\sigma} \varepsilon_{kn\sigma} n_{kn\sigma} + \varepsilon_f \sum_{m\sigma} n_{m\sigma} + \sum_{knm\sigma} (V_{knm} \psi_{m\sigma}^\dagger \psi_{kn\sigma} + \text{h.c.}) + \frac{1}{2} U \sum_{m \neq m'} n_m n_{m'},$$

where $\varepsilon_{kn\sigma}$ refers to a conduction band state with wave vector $\mathbf{k}$, band index n and spin σ. ε_f is the energy of the degenerate localized levels with orbital index m. V_{knm} is the hybridization matrix element between the f-level and a conduction band state. The last term describes the repulsive Coulomb interaction between f-electrons.

This has been a very popular model in the field of magnetic impurities and mixed-valence compounds and variants of this model have been used to calculate thermodynamic, spectroscopic and transport properties. Despite appearances, the least difficult case to solve is not obtained when the localized level has only a spin index and no orbital index m. It has rather been found[2] that a large degeneracy N_f of the f-level helps to make perturbation expansions converge. On the basis of $1/N_f$ expansions it has also been shown[4] that impurity-impurity interactions are of the order $1/N_f$ as compared to single-site terms which have leading order 1. This helps to explain the success of the model for many properties of mixed-valence compounds which are by no means dilute!

It would be nice to deduce the model formally from first principles which would then also offer a possibility to calculate all the parameters entering the model. However, the model is rather a caricature of the full physics of the problem and a justification form first principles is probably not feasible.

To a large extent, it is possible to calculate the parameters entering the model from first-principles single-electron calculations. The energies for configurations with zero, one and two localized electrons, which are parametrized by the parameters ε_f and U, should be calculated including the renormalization effects which arise from the interactions that are neglected in the model. In particular, ε_f and U should contain in an average sense the Coulomb interactions with the conduction electrons. Such calculations have actually been realized. The energy averages of the configurations with an integral number of localized electrons for an atom immersed in a solid have been obtained from renormalized atom[5] and cluster calculations[6] and constrained band structure impurity calculations[7]. In these calculations, the hopping matrix elements between the localized electron and the conduction electrons are set to zero; the total number of electrons is held constant since the solid acts as a reservoir. This procedure includes the screening of the f-orbital by the conduction electrons fully up to a truncation of the charge density oscillations at the atomic sphere, respectively the cluster boundary. In this way one obtains, for example in the case of cerium,

$\varepsilon_f \approx -2\,\mathrm{eV}$ and $U \approx 7\,\mathrm{eV}$. This is quite different from what one gets from normal band structure calculations where the f-resonance is pinned at the Fermi energy and in the case of Ce the centre of gravity is even above the Fermi energy. These calculations show also that U is renormalized substantially from the bare Coulomb interaction for f-electrons which is given by the F_0 integral which is of the order 25 eV. This renormalization procedure breaks down if the conduction electrons cannot adjust to the number of f-electrons. This is to be expected if the Coulomb repulsion between the conduction electrons U_{dd} or the one between f- and conduction electrons becomes comparable to or larger than the conduction bandwidth.

The sum over the $\varepsilon_{kn\sigma}$ extends over the single-particle bandwidth of both occupied and unoccupied conduction electron states. Quite good estimates of the conduction bandwidths are usually obtained from the density functional band structure calculations. Hopping matrix elements V_{knm} have been extracted from band structure calculations by SAKAI et al. [8] for Ce compounds and by MONNIER et al. [9] for Yb compounds. In these calculations, a local density functional (LSD) all-electron band structure calculation for the compound under consideration was done as a first step. The dispersion of the bands was then fitted by a Slater-Koster tight-binding fitting procedure. Finally, matrix elements between the localized orbital and the tight-binding single-particle functions were calculated. GUNNARSSON et al. [10] have pointed out that the coupling strength $\Delta(\varepsilon) = \pi \sum_{kn} V^*_{knm} V_{knm'} \delta(\varepsilon - \varepsilon_{kn})$ can be expressed entirely in terms of the density of states projected onto the localized impurity orbital. Again, this approach assumes that there are no renormalization effects beyond the ones already contained in a local density functional calculation.

It is interesting to ponder the fact that all the parameters entering the Anderson impurity model can be calculated with good accuracy from LSD calculations. On the other hand, dynamical properties of the model are hard to get at by density functional methods. In fact, the Hohenberg-Kohn theorem guarantees that, once the ground-state density is known, all properties are functionals of the density. However, these density functionals for dynamical properties are not, or only badly, known. Using the model Hamiltonian it has recently been shown that dynamical properties can be calculated.

To describe core level spectroscopies the Hamiltonian (1) is complemented by a term studied first by KOTANI and TOYOZAWA [11]:

$$(2) \qquad\qquad H_{\mathrm{core}} = \varepsilon_c n_c - U_{fc}(1 - n_c) \sum_{m\sigma} n_{m\sigma} .$$

ε_c is minus the binding energy of the core level and U_{fc} is the core-hole localized-electron attraction. The core Hamiltonian describes that the localized level is pulled down to higher binding energy in the presence of a core hole by the term with the parameter U_{fc}. The effect of the renormalized U_{fc} is similar to going to the next element in the rare-earth series; it changes the ordering of the configurations. For instance, in Ce metal the f^2-configuration is the lowest in the presence of a core hole. The attraction of the core hole to the extended states is much smaller and entirely neglected in the model. One has to expect that U_{fc} is comparable to U and somewhat larger than U owing to a stronger localization of the core orbital.

The coupling of the localized level to the conduction band is relatively small; we have seen, however, that it is an essential part of the Hamiltonian which cannot be neglected. Nevertheless, it seems exaggerated to include all the detailed k-dependence of V_{knm} in the model. To justify this on a formal basis one has to make a unitary transformation of the band states to the point group representation of the impurity [3, 12]. In the case of an atom in a jellium one would have that

(3)
$$\sum_{kn} V^*_{knm} V_{knm'} \delta(\varepsilon - \varepsilon_{kn}) = |V_m(\varepsilon)|^2 \delta_{mm'} .$$

This result can also be obtained rigorously by a unitary transformation if the the crystal point group symmetry is high enough. In the other cases (3) will be assumed as an approximation. It is then possible to construct orthogonal continuum states [3]:

(4)
$$|\varepsilon m \sigma\rangle = \frac{1}{V_m(\varepsilon)} \sum_{knm} \delta(\varepsilon - \varepsilon_{knm}) |kn\sigma\rangle .$$

The continuum states, which do not fall into the point group representation of the localized level, cannot couple to it by symmetry. These electrons contribute a constant energy in the model. Note also that $V(\varepsilon)$ already contains the density of states. The model Hamiltonian can now be rewritten with an energy-dependent band and coupling term:

(5)
$$H = \sum_{\nu} \left(\int \varepsilon n_{\varepsilon\nu} \, d\varepsilon + \varepsilon_f n_{\nu} + \int (V_{\nu}(\varepsilon) \psi^{\dagger}_{\nu} \psi_{\nu\varepsilon} + \text{h.c.}) \, d\varepsilon + \frac{1}{2} U \sum_{\nu' \neq \nu} n_{\nu} n_{\nu'} \right),$$

where the orbital index m and the spin index σ have been combined into a single index ν. The properties that we will study do not depend terribly on the details of $V_{\nu}(\varepsilon)$, so we follow the general custom to substitute a simple functional form for V, neglecting also any ν dependence. We call $\pi|V(\varepsilon)|^2 = \Delta(\varepsilon)$ the coupling strength. Δ is the width of a virtual level coupled to a band. A constant with cut-off, semi-elliptic, Lorentzian and Gaussian etc. were considered as simple functional forms for $\Delta(\varepsilon)$.

3. – Ground state.

The ground state can be studied by a variational approach by introducing suitable basis states. The subsequent calculation of a variational ground state

and dynamical properties actually closely follows the derivation given by
GUNNARSSON and SCHÖNHAMMER [3]. Clearly in such an approach one should
consider a basis state with all band states below the Fermi energy occupied. If a
model containing core states (2) is studied, this basis state has the core states
occupied too. This state is denoted by $|0\rangle$ in the following. It is a singlet state. By
letting H act on $|0\rangle$ one led to introduce basis states

(6)
$$|\varepsilon\rangle = \frac{1}{\sqrt{N_f}} \sum_{\nu} \psi_{\nu}^{\dagger} \psi_{\nu\varepsilon} |0\rangle .$$

In this state a conduction electron of energy ε has hopped onto the f-level. If we
set the origin of energy at the Fermi energy, ε takes values smaller than 0 and
ranges from B^- to 0.

The hybridization term couples these states to states with two f-electrons

(7)
$$|\varepsilon\varepsilon'\rangle = \frac{1}{\sqrt{N_f(N_f-1)}} \sum_{\nu\neq\nu'} \psi_{\nu}^{\dagger} \psi_{\varepsilon\nu} \psi_{\nu'}^{\dagger} \psi_{\varepsilon\nu'} |0\rangle , \qquad \varepsilon > \varepsilon',$$

and to states with a conduction electron conduction hole pair

(8)
$$|E\varepsilon\rangle = \frac{1}{\sqrt{N}} \sum_{\nu} \psi_{E\nu}^{\dagger} \psi_{\varepsilon\nu} |0\rangle .$$

E refers to a state above the Fermi energy. The states $|E\varepsilon\rangle$ couple to states with
one f-electron, one conduction electron and two conduction holes:

(9)
$$|E\varepsilon\varepsilon', 1\rangle = \frac{1}{\sqrt{N_f(N_f-1)}} \sum_{\nu\neq\nu'} \psi_{E\nu}^{\dagger} \psi_{\varepsilon\nu} \psi_{\nu'}^{\dagger} \psi_{\varepsilon\nu'} |0\rangle$$

and

(10)
$$|E\varepsilon\varepsilon', 2\rangle = \frac{1}{\sqrt{N_f}} \sum_{\nu} \psi_{E\nu}^{\dagger} \psi_{\varepsilon\nu} \psi_{\nu}^{\dagger} \psi_{\varepsilon\nu} |0\rangle , \qquad \varepsilon > \varepsilon'.$$

This can be continued *ad infinitum*. A set of such basis states is shown
schematically in fig. 2. To perform a calculation the matrix elements of H are
needed. One finds

(11)
$$\langle \varepsilon | H | 0 \rangle = \sqrt{N_f} V(\varepsilon),$$

(12)
$$\langle \varepsilon\varepsilon' | H | \varepsilon'' \rangle = \sqrt{N_f-1} \left[V(\varepsilon') \, \delta(\varepsilon - \varepsilon'') + V(\varepsilon) \, \delta(\varepsilon' - \varepsilon'') \right],$$

(13)
$$\langle E\varepsilon | H | \varepsilon' \rangle = V(E) \, \delta(\varepsilon - \varepsilon'),$$

(14) $$\langle E\varepsilon\varepsilon', 1|H|E'\varepsilon''\rangle = \sqrt{N_f - 1}\, V(\varepsilon')\,\delta(E - E')\,\delta(\varepsilon - \varepsilon'),$$

(15) $$\langle E\varepsilon\varepsilon', 2|H|E'\varepsilon''\rangle = [V(\varepsilon')\,\delta(\varepsilon - \varepsilon'') - V(\varepsilon)\,\delta(\varepsilon' - \varepsilon'')]\,\delta(E - E').$$

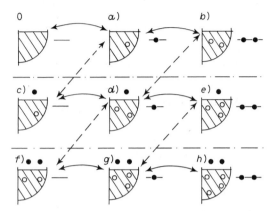

Fig. 2. – Schematic representation of the basis states. Solid circles show electrons and open circles show holes. The hatched part indicates the filled conduction band and the horizontal line the f-level. The arrows show which states couple to each other. A solid line indicates the strength $\sqrt{N_f}V$ and a dashed line the strength V (cited from [13]).

These matrix elements illustrate that there are states which couple with the strength $\sqrt{N_f}V(\varepsilon)$ and others with $V(\varepsilon)$. Normalizing to the largest coupling strength shows that the coupling (13) and (15) etc. vanishes in a large-N_f limit. Figure 2 shows that there is a hierarchy of states which are coupled to $|0\rangle$ by the orders 1, $1/N_f$, $(1/N_f)^2$... in the large-N_f limit.

Similarly one might have started with basis states with a filled Fermi sea and containing one f-electron thus having a symmetry characterized by ν:

(16) $$|\nu\rangle = \psi_\nu^\dagger|0\rangle.$$

This state couples to

(17) $$|E\nu\rangle = \psi_{E\nu}^\dagger|0\rangle,$$

with the strength

(18) $$\langle E\nu|H|\nu\rangle = V(E).$$

Also this sequence of basis states can be carried on. Evidently the first and largest coupling that enters here is one order in $1/N_f$ weaker than the one in eq. (11).

Let us consider the case where $U = \infty$ and include basis states up to order $1/N_f$. The singlet wave function then has the expansion

(19) $|E_0\rangle = a_0|0\rangle + \int d\varepsilon\, a_\varepsilon|\varepsilon\rangle +$

$$+ \int d\varepsilon \int dE\, a_{E\varepsilon}|E\varepsilon\rangle + \int d\varepsilon \int d\varepsilon' \int dE\, a_{E\varepsilon\varepsilon'}|E\varepsilon\varepsilon', 1\rangle \ldots .$$

The singlet energy relative to the energy of filled band to order $1/N_f$ is

(20) $$E_0^{(1)} = N_f \int\limits_{B^-}^0 d\varepsilon\, \frac{V(\varepsilon)^2}{E_0^{(1)} - \varepsilon_f + \varepsilon - \Sigma(E_0^{(1)} + \varepsilon)}.$$

Where we have made use of the functions

(21) $$\Sigma(z) = \int\limits_0^{B^+} dE\, \frac{V(E)^2}{z - E - \Pi(z - E)}$$

and

(22) $$\Pi(z) = \int\limits_{B^-}^0 d\varepsilon\, \frac{N_f V(\varepsilon)^2}{z - \varepsilon_f + \varepsilon}.$$

In writing eq. (22), a factor $\sqrt{N_f}$ was used in (14) and the complications from the coupling (15) which require matrix inversion could be neglected altogether leaving the result correct to order $(1/N_f)^1$. For a calculation of $E_0 \approx E_0^{(0)}$ to order $(1/N_f)^0$ one should rewrite eq. (20) without the Σ term. Extensive discussions of the accuracy of finite-order calculations have been given by GUNNARSSON and SCHÖNHAMMER [3].

If the same type of calculation is applied to the magnetic states with symmetry ν, one obtains

(23) $$E_{0,\nu}^{(1)} = \varepsilon_f + \Sigma(E_{0,\nu}^{(1)}).$$

The actual ground state is the singlet state approximated by (19) and (20) except for cases with a very special choice of $V(\varepsilon)$.

If $V(\varepsilon)$ is a constant for $|\varepsilon| < B$ and zero otherwise one finds $E_0^{(0)} \approx$ $\approx N_f V^2 \log|\delta/B|$. It was assumed that $\delta = \varepsilon_f - E_0$ is small compared to B. $\delta \approx B \exp[\varepsilon_f/N_f V^2]$. In this limit the average population of the f-level n_f is very near to unity: $n_f = N_f V^2/(\delta + N_f V^2)$. δ sets a small energy scale which is important for dynamical properties. The energy δ depends exponentially on a coupling constant $V^2/|\varepsilon_f|$. This suggests that one can interpret δ as the Kondo temperature.

4. – Spectroscopies.

4'1. *General considerations*. – Various spectroscopies will be studied in the following subsections. Since most of the spectra are not very much dependent on sample temperature, a zero-temperature theory presents the most direct way for a discussion. The intermediate-state method put forward by GUNNARSSON and SCHÖNHAMMER [3] has been worked out for many spectroscopies and lends also to visualization of final states. However, in the case of valence spectra finite-temperature effects have been seen and understood. There the so-called noncrossing approximation to the slave-boson approach will form the basis for discussions.

Various spectroscopies can be related to one-electron Green's functions by using the sudden approximation. As an example, consider valence photo-emission. An operator T is introduced which describes transitions where one photon is absorbed and an electron from orbital i is promoted to a continuum state x:

$$(24) \qquad T = \sum_{x,i} \tau_{x,i} \psi_x^\dagger \psi_i .$$

Here τ is a matrix element of the dipole operator between orbitals i and x. Part of the sudden approximation is the assumption that the many-electron final state can be written as $\psi_x^\dagger |\Phi_{N-1,n}\rangle$, where $|\Phi_{N-1,n}\rangle$ is an eigenstate of the Hamiltonian. In other words, interactions of the outgoing electron with the system left behind are systematically neglected. This is a good approximation in surprisingly many cases. However, near threshold corrections can be substantial enough to be measurable [14]. In resonance photoemission interaction of the outgoing electron with the system left behind is essential. The coupling is due to the Auger matrix element coupling the outgoing electron state with core hole states. A treatment of these effects has been given by GUNNARSSON and LI [15]. According to the Fermi golden rule the energy distribution function of the emitted electrons is proportional to

$$(25) \qquad \rho(\varepsilon_x) = \sum_n \left| \langle \Phi_{N-1,n} | \sum_i \tau_{x,i} \psi_i | E_0 \rangle \right|^2 \delta(\varepsilon_x + E_{N-1,n} - \hbar\omega - E_0) ,$$

where the continuum state x has been eliminated from the problem, since it was assumed that the outgoing electron state is unoccupied in the ground state: $\psi_x |E_0\rangle = 0$. The sum over i runs over all single-electron functions of the system. Interference terms are potentially of importance, especially in the valence spectroscopy. In order to obtain simple formulae, this will be neglected in the following. The most interesting contribution to the valence spectrum would then come from the f-electrons in our case. If the energy dependence of $\tau_{x,i}$ is neglected, one can write[3, 14]

$$(26) \qquad \rho_\nu(\varepsilon) = \frac{1}{\pi} \mathrm{Im}\, g^<(\varepsilon - i0) = \frac{1}{\pi} \mathrm{Im}\, \langle E_0 | \psi_\nu^\dagger \frac{1}{\varepsilon - i0 + H(N-1) - E_0} \psi_\nu | E_0 \rangle .$$

It remains to calculate the ground state $|E_0\rangle$ and either to evaluate the Green's function in eq. (26) or to also explicitly calculate the final states which diagonalize $H(N-1)$.

  4‘2. *f-photoemission.* – For the case $U = \infty$ and working to the lowest order in $1/N_f$, the ground state is expanded in the states $|0\rangle$ and $|\varepsilon\rangle$. Given that $\psi_\nu|0\rangle = 0$ and $\psi_\nu|\varepsilon\rangle = 1/\sqrt{N_f}\psi_{\varepsilon\nu}|0\rangle$, suitable basis states are

$$(27) \qquad\qquad\qquad |\varepsilon\nu\rangle = \psi_{\varepsilon\nu}|0\rangle .$$

These are the only states that overlap with ψ_ν applied to the lowest-order ground-state wave function. For a calculation to lowest order it is necessary not to forget the coupling of (27) with strength $\approx \sqrt{N_f}V(\varepsilon')$ to states

$$(28) \qquad\qquad\qquad |\varepsilon\varepsilon'\nu; 1\rangle = \frac{1}{\sqrt{N_f - 1}} \sum_{\nu' \neq \nu} \psi_{\nu'}^\dagger \psi_{\varepsilon'\nu'} \psi_{\varepsilon\nu} .$$

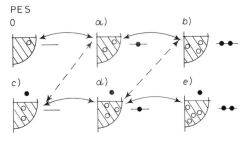

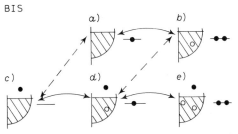

Fig. 3. – Schematic representation of basis states used for valence photoemission and BIS calculations. Symbols as in fig. 2 (cited from[13]).

For a higher-order calculation one would also need to include further states, as shown schematically in fig. 3. It follows from eq. (26) that we need to calculate $g(z) = \langle \varepsilon v | (z - E_0^{(0)} + H)^{-1} | \varepsilon' v' \rangle$. It is useful to define the function

$$(29) \qquad\qquad g(z) = \frac{1}{z + \Pi(-z)},$$

where Π has been defined in (22). Together with (19), the valence spectral density becomes [3]

$$(30) \qquad\qquad \rho_f(\varepsilon) = \sum_v \rho_v(\varepsilon) = \frac{1}{\pi} \operatorname{Im} \int_{B^-}^{0} d\varepsilon' \, a_{\varepsilon'}^2 \, g(\varepsilon - i0 - E_0^{(0)} - \varepsilon').$$

It is interesting to look at the structure of the lowest-order result. The function $\operatorname{Im} g(\varepsilon - i0)$ has a continuum part for $B^- - \varepsilon_f \leqslant \varepsilon \leqslant -\varepsilon_f$ because of the imaginary part of $\Pi(\varepsilon - i0)$. For small coupling strength there is a peak with width $\pi N_f V(\varepsilon)^2$ at $\varepsilon \approx \varepsilon_f$ (fig. 4). We may consider this peak as an ionization peak corresponding to the removal of an electron out of a virtual f-level.

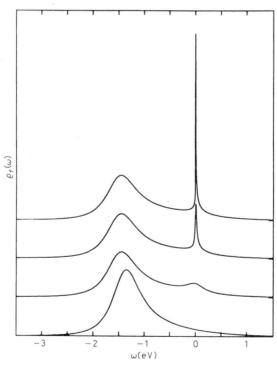

Fig. 4. – f-spectrum at various temperatures for $\varepsilon_f = -1.2$ eV, $N_f = 14$ and $\varDelta = 0.04$ eV at the maximum of a Gaussian band. The band is centred at $\varepsilon = 0$ and has a width of 3.33 eV (FWHM). From top to bottom $kT = 0.001$, 0.01, 0.1, 1 eV. The top curve is characteristic of the low-temperature regime $kT < \partial$.

Single-electron theory would place such a peak off the Fermi energy only if the level is almost completely filled. For example, this may occur for a spin-orbit split f-level where, $e.g.$, the $f_{5/2}$ level might be filled. Another case would be a spin split f-level where a majority spin level moves off the Fermi energy in spin-unrestricted theory when it is filled with ≈ 7 electrons. The noninteger part in the occupation is entirely due to the unoccupied part of the Lorentzian tail in single-electron theory.

$g(\varepsilon - i0)$ has a pole at $\varepsilon = -E_0^{(0)}$ from the definition of $E_0^{(0)}$. The pole strength is $1 - n_f$. The corresponding part of the spectrum can be written [3]

$$(31) \qquad \rho_{f,\text{pole}}(\varepsilon) = N_f (1 - n_f)^2 V(\varepsilon)^2 / (\partial - \varepsilon)^2 .$$

Equation (31) describes a sharp rise in the spectrum on the energy scale ∂ just below $\varepsilon_F \equiv 0$ (fig. 4). This sharp rise and a corresponding peak in the BIS spectrum are a result of the dynamics produced by a large U. This peak is the «Kondo» peak, anticipated long before the spectrum had been actually calculated. In order to understand the origin of this feature, it is useful to investigate the character of the final states contributing to the resonance. Actually, to lowest order in $1/N_f$, the final states up to an energy ∂ above threshold are closely related to the ground state:

$$(32) \qquad |\Phi_\nu(\varepsilon)\rangle = \psi_{\varepsilon\nu} |E_0^{(0)}\rangle .$$

So the final states in that energy range are replicas of the ground state with an additional hole at energy ε. The spectrum measures (31) the weight of states with a hole at ε. The distribution of such hole states is the reason for the steep rise of the spectrum near ε_F. This is a many-body effect, it disappears if the spectral function is calculated for $U = 0$ or equivalently $N_f = 1$. The properties and convergence of such spectra have been discussed extensively by GUNNARSSON and SCHÖNHAMMER.

4`3. *Bremsstrahlung isochromatic spectroscopy*. – In inverse photoemission, high-energy electrons are absorbed in the sample and the photons emitted in a radiative transition are detected. In the BIS the electron beam energy is varied and photons of fixed energy are detected. The Green's function [3]

$$(33) \qquad g_f^>(z) = \sum_\nu \langle E_0| \psi_\nu \frac{1}{z + E_0 - H} \psi_\nu^\dagger |E_0\rangle$$

is closely related to the BIS spectral density. The integrated weight of the BIS spectrum is of order N_f and $N_f(1 - n_f)$ in the limit $U = \infty$. In order to obtain results accurate to order 1 it is necessary to include basis states of order $1/N_f$. In the limit $U = \infty$ basis states to order $1/N_f$ are [16]

(34)
$$|v\rangle = \psi_v^\dagger |0\rangle ,$$

(35)
$$|Ev\rangle = \psi_{Ev}^\dagger |0\rangle ,$$

(36)
$$|E\varepsilon v\rangle = \frac{1}{\sqrt{N_f}} \sum_{v'} \psi_{Ev}^\dagger \psi_{v'} \psi_{\varepsilon v'} |0\rangle .$$

This leads to

(37)
$$g_f^>(z) = \frac{(1 - n_f) N_f}{z + E_0^{(0)} - \varepsilon_f - \Sigma(z + E_0^{(0)})} ,$$

yielding a steep rise of the spectral density above ε_F culminating in a narrow peak at $\varepsilon \approx \delta$. For a comparison with experiments for Ce at higher energies it is crucial to treat U as finite and to include basis states with double f-occupation, but this necessitates numerical inversion of a matrix with continuum indices. The f^2 peak has a broadening of order 2Δ (the f^2 peak is not shown in fig. 4 since $U = \infty$) because either of the two f-electrons can decay into the continuum. In experimental spectra, broadening by multiplet effects and lifetime dominates the picture.

4`4. *f-spectra at finite temperature.* – The extension of the large-N_f approach to finite temperatures has been carried out by a number of authors [14, 17-21]. All these approaches lead to the same set of integral equations in the so-called noncrossing approximation. In this approximation only the case $U = \infty$ is treated and complications arising from terms like the one in (15) of order $(1/N_f)^2$ are neglected. According to COLEMAN, the Hamiltonian (5) is rewritten with a boson bookkeeping device for the number of f-electrons:

(38)
$$H' = \sum_v \left(\int \varepsilon n_{\varepsilon v} \, d\varepsilon + \varepsilon_f n_v + \int (V_v(\varepsilon) \, \psi_v^\dagger \psi_{v\varepsilon} b + \text{h.c.}) \, d\varepsilon + \frac{1}{2} U \sum_{v' \neq v} n_v n_{v'} \right) + \varepsilon_b b^+ b .$$

It is equivalent to the Hamiltonian (5) if ε_b is set to 0. The advantage of this technique is that standard finite Green's function approaches can be used since the operators all observe standard commutation or anticommutation rules. The modified mixing term converts the boson into an f-electron and *vice versa*, which conserves a quantity Q:

(39)
$$Q = b^+ b + \sum_v n_v .$$

The subspaces F_Q with different Q are thus disjoint. The relevant subspace to be considered is the one with $Q = 1$. In order to project onto the physical subspace

$Q = 1$, a chemical potential $-\lambda$ is associated with Q and the limit $\lambda \to \infty$ is taken at the end of the diagrammatic calculation. This leads to the coupled integral equations

$$(40) \qquad D(z) = \left(z - N_f \int V(\varepsilon)^2 G_v^{\mathrm{ps}}(z + \varepsilon) f(\varepsilon)\, d\varepsilon \right)^{-1},$$

$$(41) \qquad G_v^{\mathrm{ps}}(z) = \left(z - \varepsilon_f - \int V(\varepsilon)^2 D(z - \varepsilon)(1 - f(\varepsilon))\, d\varepsilon \right)^{-1}.$$

The real f-electron spectral function $\rho_v(\varepsilon)$ is given as a convolution of the boson spectral function $B(\varepsilon) = 1/\pi \operatorname{Im} D(\varepsilon - i0)$ and the pseudofermion spectral function $A_v(\varepsilon) = 1/\pi \operatorname{Im} G_v^{\mathrm{ps}}(\varepsilon - i0)$:

$$(42) \qquad \rho_v(\varepsilon) = \frac{1}{Z_{\mathrm{MV}}} \int A_v(\varepsilon - \varepsilon') B(\varepsilon')(\exp[-\beta(\varepsilon - \varepsilon')] + \exp[-\beta \varepsilon'])\, d\varepsilon',$$

where the mixed-valence partition function is defined as

$$(43) \qquad Z_{\mathrm{MV}} = \int (N_f A_v(\varepsilon) + B(\varepsilon)) \exp[-\beta \varepsilon]\, d\varepsilon.$$

The integral equations (39)-(43) have to be solved numerically. In addition to the f-spectral function, the temperature dependence of the specific heat and the magnetic susceptibility have been calculated[22, 23]. At temperatures much smaller than the Kondo temperature, spurious nonanalyticities mar the small-frequency range. The zero-temperature limit has been discussed carefully by MÜLLER-HARTMANN[24]. In the spin fluctuation limit the finite-temperature f-spectral functions show a narrow resonance of weight $N_f(1 - n_f)$ slightly above the Fermi level. This resonance indicates the presence of a cluster of $(n + 1)$-particle states at an energy δ above the ground state. These are f^1 magnetic states. Figure 4 shows f-spectral functions at various temperatures for parameters in the Kondo regime. Negative energies correspond to electron removal and positive energies to electron addition. The lower peak is very similar to the Lorentzian peak obtained from mean-field theory with level broadening of $N_f \Delta$. It arises from final states with an empty f-level (f^0). The ground state has a f-population of $n_f \approx 1$. The peak at the positive energy $\delta \approx \varepsilon_f - E_0$ is a many-body effect and cannot be interpreted with the independent-electron picture. It is commonly called the Kondo resonance. In contrast to single-particle features of the spectrum, the Kondo resonance is strongly temperature dependent as fig. 4 shows. The photoemission spectrum is obtained by multiplying $\rho_f(\varepsilon)$ by the Fermi function.

If the f-level is split by spin-orbit interaction and crystal field splitting, the Kondo resonance becomes structured. Figure 5 shows a case with spin-orbit

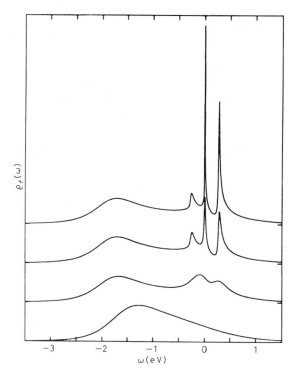

Fig. 5. – f-spectrum for a spin-orbit split level at various temperatures. $N_{f5/2} = 6$, spin-orbit splitting $\Delta_{so} = 0.28\,eV$, $N_{f7/2} = 8$ and $\Delta = 0.07\,eV$. Otherwise same as in fig. 4.

splitting where the coupling is relatively weak and the resonance is actually split. The ground state has predominantly $f_{5/2}$ character. Excited $(n - 1)$-electron states of $f_{7/2}$ character cause the resonance at energy $\approx -\Delta_{so}$. The highest resonance at $\approx +\Delta_{so}$ derives from $(n + 1)$-electron states again with mostly $f_{7/2}$ character. The central peak, the Kondo resonance, derives from states similar to the ground state except for an additional band electron or hole. At temperatures of order δ the peaks at positive energy become depleted since their intensity depends on the f^0 admixture in the ground state. At temperatures of order δ magnetic states become populated and the f^0 part in the initial states correspondingly reduced. The peak intensity at $-\Delta_{so}$ depends on the $f_{5/2}$ population in the initial state, which remains conserved up to a temperature of order Δ_{so}.

4˙5. *Core level* XPS. – In core XPS, a core electron is emitted. In the sudden approximation, the Green's function

(44)
$$g_c(z) = \langle E_0| \psi_c^\dagger \frac{1}{z - E_0 + H} \psi_c |E_0\rangle$$

is directly related to the core spectral function [3]. In the final states containing a core hole the core hole term (2) pulls down higher f^{n+1} configurations. For realistic values of N_f the $1/N_f$ method converges rapidly to the exact spectrum if a realistic lifetime broadening of the core hole is introduced. The exact spectrum (without broadening), however, has an infrared singularity at threshold due to excitation of an infinite number of electron-hole pairs which is not reproduced from finite-order $1/N_f$ expansions. The core spectroscopies can be used as a tool to determine the f-population n_f and the coupling strength $\Delta = \pi V^2$.

A qualitative discussion can be given in terms of the ligand field limit of the Anderson model where the conduction bandwidth is set to zero [3, 25]. The system then has only three basis states $|f^0\rangle$, $|f^1\rangle$ and $|f^2\rangle$, which represent the appropriate f-configurations. The ground state is given as

(45) $$|E_0\rangle = a_0|f^0\rangle + a_1|f^1\rangle + a_2|f^2\rangle \, .$$

If $\sqrt{N_f}V$ is chosen negative, all coefficients a are positive. According to eq. (25), final states in the presence of the core hole are needed. Since there are three basis states, the simple model yields three final states. As long as the mixing

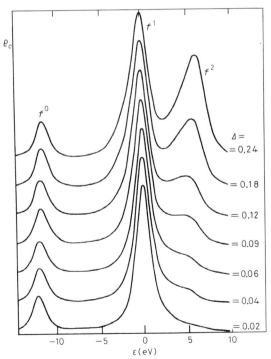

Fig. 6. – XPS core spectrum as a function of the hybridization strength Δ; $N_f = 14$, $U = 6.4$. The spectra are normalized to the height of the f^1 peak and a Lorentzian broadening of 1.8 eV (FWHM) has been introduced (cited from [3]).

term $N_f V^2$ is small compared to the energy separation of the uncoupled basis states, the final states can be labelled by the configuration contributing most. If the final state has coefficients b_m^n, the spectral weight calculated by (25) becomes

$$(46) \qquad p(n) = \left| \sum_{m=0}^{2} a_m b_m^n \right|^2 .$$

As an example, a Ce compound with $a_1 \approx 1$ is considered. The lowest final state will be the «f^2» state with $b_2^2 \approx 1$. Its weight depends mainly on the admixture of $|f^2\rangle$ in the ground state expressed by a_2 and the admixture in the final state expressed by b_1^1. As the coupling increases, both a_2 and b_1^2 increase. This also increases the weight $p(2)$. These highly simplified arguments remain qualitatively correct for the Anderson model with finite bandwidth. Figure 6 illustrates the increase of the f^2 weight in the core XPS spectrum for some values of the coupling strength Δ. In a similar way one may calculate the f^0 weight $p(0)$. This is an antibonding final state and the coefficients b_0^0 and b_1^0 have different signs (for $V < 0$, see above). This is illustrated in fig. 7 with weights obtained from the Anderson model with finite bandwidth.

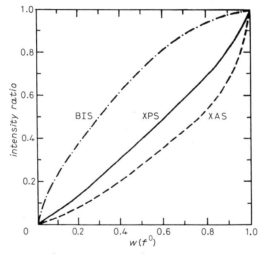

Fig. 7 – Intensity ratio $p(0)/(p(0) + p(1) + p(2))$ for the XPS core spectrum and the ratio $p(1)/(p(1) + p(2))$ for the $3d \rightarrow 4f$ XAS and BIS spectrum as a function of the f^0 weight $w(f^0)$ in the ground state (cited from [3]).

4˙6. X-ray absorption spectroscopy and inelastic electron scattering.

– In $L_{\mathrm{II,III}}$ edge absorption spectroscopy, a $2p$ core electron is excited into an unoccupied continuum state of s and d symmetry just above the Fermi level. These states do not couple to the f-electron system. The transition operator $\int \mathrm{d}\varepsilon \, w(\varepsilon) \, \psi_d^\dagger \psi_c$ contains single-particle density of states and dipole matrix elements through $w(\varepsilon)$. The Green's function to be studied is then

$$(47) \qquad g_L(z) = \langle E_0 | \int \mathrm{d}E'\, w(E')\, \psi_c^\dagger \psi_{d,E'} \frac{1}{z + E_0 - H - H_d} \int \mathrm{d}E''\, w(E'')\, \psi_{d,E''}^\dagger \psi_c | E_0 \rangle \,.$$

It is necessary to include a term describing the d-electrons in the Hamiltonian. The most basic possibility is to describe the d-electrons as noninteracting:

$$(48) \qquad H_d = \int \varepsilon n_d(\varepsilon)\, \mathrm{d}\varepsilon \,.$$

The single-particle properties that actually have to be known enter via the $w(\varepsilon)$ function in the transition operator. In this approximation, there is no interaction of the d-electrons with the other electrons beyond a possible renormalization of the model parameters. The transition $p \to s$ is neglected together with Coulomb and exchange terms. In this approximation [26] it is possible to write the spectral function as

$$(49) \qquad \rho_L(\hbar\omega) = \int\limits_0^\infty w^2(E)\, \rho_c(E - \hbar\omega)\, \mathrm{d}E \,.$$

This shows that the L-edge spectrum is closely related to the core XPS spectrum ρ_c for the $2p$ core level, which is expected to be almost identical to the $3d$ core spectrum if multiplet effects and core hole lifetime are neglected. The present convolution approximation relates the structures at the edge to the structures in the XPS core spectrum. At higher photon energy the core hole spectrum introduces an additional broadening to the rather large lifetime broadening of the continuum electron. In that spectral range, the extended absorption fine structure is due to the single-particle effects contained in $w^2(E)$.

In addition to L edges also the $M_{4,5}$ edges involving $3d \to 4f$ transitions have been studied [27]. The same transitions, however weighted by a different matrix element, have been studied with the electron energy loss spectroscopy (EELS). These spectroscopies have formal similarities to BIS, since an electron is added to the f-level. The spectra of this subsection, however, have to be calculated with a core hole attracting the f-level. In the formalism this enters as a shift $\varepsilon_f \to \varepsilon_f - U_{fc}$. The same types of basis states as in BIS calculations can be used.

5. – Case studies.

5`1. Ce: γ *and* α *phases.* – The isostructural $\gamma \to \alpha$ phase transition in Ce has traditionally been associated with some yet unknown major change in the electronic structure. Many experimental data have been accumulated in the past few years. The experimental data suggested that there is only a minor reduction of the f-occupancy n_f and that the hybridization would increase in going from the γ to the α phase.

PATTHEY *et al.* [28] studied both phases with the exceptionally high resolution of 0.02 eV. The spectra in fig. 8 show a double peak or shoulder structure with a separation of approximately 0.28 eV. This can be understood from a calculation

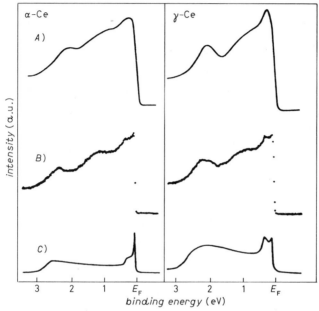

Fig. 8. – Ultraviolet photoemission spectra of α and γ Ce. A) spectrum from best-resolved electron spectroscopy with monochromatized synchrotron radiation [29], B) with He(II)-UPS and C) calculation from ref. [28].

where the f-level is split into a spin-orbit doublet. The calculation shows that in the case of γ-Ce the ground state has a f-population $n_f \approx 0.98$ derived almost entirely from the $4f_{5/2}$ level. This population is consistent with the intensity of the satellite in the XPS core spectrum. The $4f_{5/2}$ character is probed and confirmed by neutron scattering experiments, since the $4f_{5/2}$ form factor is clearly different from the $4f_{7/2}$ form factor. The calculated final states after emission of the f-electron corresponding to the peak right at ε_F resemble the ground state, but contain an additional hole in the conduction band. This means that an electron has hopped back onto the $4f_{5/2}$ level. The low-lying final state should then be split of the continuum, like the ground state is a singlet split of a continuum. The overlap part of the transition matrix element samples the distribution function for basis states with one f-electron in the ground state. The corresponding part of the spectrum appears near the threshold and includes the first peak which is well resolved in the spectra B) (and C)) in fig. 8. In the case of γ-Ce this peak can then be called the $4f_{5/2}$ peak. In a more general context one would identify it with the Kondo peak. As one may have presumed, the peak at 0.28 eV derives from resonances states with the $4f_{7/2}$ level. In the case of α-Ce, the coupling strength has increased, consistent with the smaller lattice constant. The ground state then has a smaller f-population $n_f \approx 0.88$ with some admixture of $4f_{7/2}$ character due to a 30% increase in the coupling strength. The effect on the spectrum is that much more weight is now found in the Kondo peak. It is also interesting to comment on the lattice parameter interpretations. The n_f values deduced here are substantially larger than the ones deduced from lattice parameter data. These older estimates assume that the f-contribution to the cohesion can be neglected. This leads to an underestimate of n_f. For the present values of the hopping matrix element, however, it is obvious that there is a significant f-contribution to the electronic energy, namely E_0 (eq. (19)) for a dilute system. The singlet energy lowering enters the Kondo volume collapse model [30] and actually drives the phase transition.

5˙2. CeCu₆ *a heavy fermion compound.* – The compound $CeCu_6$ shows one of the largest coefficients of specific heat ($\gamma \approx 1600$ mJ/mol K²) at low temperature. It also comes close to an impurity compound since each Ce atom is surrounded by a cage of 19 Cu atoms, resulting in a spacing 32% larger than in γ-Ce. In line with expectations of quite a weak coupling strength, the core XPS spectrum shows almost no satellite intensity. The BIS spectrum shows no discernible peak at ε_F, which points to an f-population of $n_f \approx 1$. The valence f-spectrum has been investigated by PATTHEY et al. [31] by high-resolution ultraviolet photoemission spectroscopy (UPS) (fig. 9). The measured spectrum shows no indication of a Kondo peak. However, there is a $4f_{7/2}$ resonance observed at 0.28 eV. It is not so surprising that f-electron-related structures near ε_F disappear for small hybridization, since obviously, for vanishing hopping matrix elements, only the atomiclike ionization peak at ε_f should remain. The relative weight of the spectral features near ε_F is a more complicated issue. Investigation of the Gunnarsson-Schönhammer theory, including a spin-orbit splitting, shows that there is an exponential crossover from a strong hybridization regime where $n_f(1 - n_f)$ weight is on the Kondo peak to a weak hybridization regime where the weight is on the (well-defined in that limit) $4f_{7/2}$ peak if the valence spectrum is normalized to n_f.

5˙3. CeSi₂ *and the temperature dependence of the Kondo resonance.* – The compound $CeSi_2$ is an interesting candidate for an investigation of the temperature dependence of many-body effects seen in photoelectron spectro-scopies. There appears to be no magnetic ordering at low temperatures. The linear coefficient of specific heat γ is enhanced 100 times as compared to the free-electron model but also an order of magnitude smaller than in extreme heavy fermion compounds. The crystal field splitting of the f-level has important

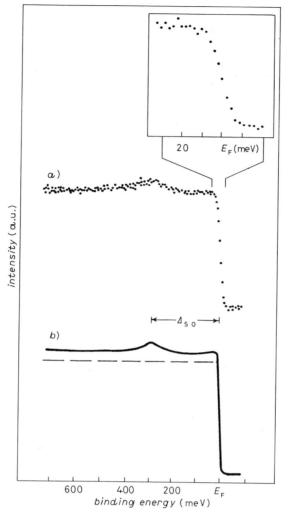

Fig. 9. – Ultraviolet photoemission spectrum of CeCu₆, $T = 15\,\mathrm{K}$, $h\nu = 40.8\,\mathrm{eV}$; *a*) measured spectrum with resolution 0.02 eV; insert: edge measured with 0.012 eV resolution; *b*) spectrum calculated by the Gunnarsson-Schönhammer approach (figure from ref. [31]).

consequences, undermining the large-N_f limit argument in favour of the singlet condensation. In the present case of a tetragonal symmetry, three doublets are formed from the $j = 5/2$ level. For $N_f = 2$ representing just spin degeneracy, there is still an excess of possibilities for an electron to hop into the degenerate f-level as compared to the single way a f-electron can hop into the conduction band. So one still has to expect that Kondo condensation should happen. The crystal field splitting between the ground state and the first excited doublet of $\Delta_{\mathrm{CF}} = 35\,\mathrm{meV}$ is important on the scale of the hybridization strength $\Delta = 95\,\mathrm{meV}$

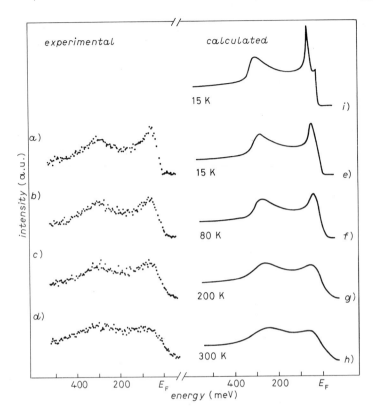

Fig. 10. – Comparison of experimentally derived and calculated spectra of $CeSi_2$ according to ref. [32]. The data a)-d) represent differences of spectra taken 40.8 eV and 21.2 eV photon energy. Curves e)-i) are calculated with the slave-boson approach.

as estimated from the UPS spectrum [32]. The calculated lowering of the singlet sets an energy scale of $T = 35$ K. At low temperature, spectral intensity is shifted away from ε_F onto the peak at Δ_{CF} and the $4f_{7/2}$ resonance. The true Kondo peak is very weak and is seen in the measured spectrum (fig. 10) just as a shoulder at the lowest temperature. At temperatures up to 80 K, the spectral weights near ε_F remain quite intact. One should recall that the Kondo peak is related to the probability of finding the final state in a ground-state-like singlet with the corresponding energy lowering. At finite temperature, magnetic states ≈ 3 meV above the ground state become populated. Kondo condensation can occur in the corresponding final states, thus contributing to the spectral intensity near ε_F. At even higher temperatures, on the order of room temperature in $CeSi_2$, spectral weight begins to fade out of the region within a few hundred meV of ε_F before actually thermal broadening is that large. This rests on the necessity of a sufficiently sharp Fermi edge for the occurrence of the singlet lowering.

5˙4. *Core level line shapes.* – The appearance of satellite lines in core level spectra can be discussed in terms of the impurity model complemented by the core hole terms (2). In the rare-earth compounds, the appearance of two sets of structures has been interpreted from early on in terms of two localized $4f$ configurations. It became clear that the most effective screening mechanism for the core hole would be filling of an f-level. The term «energy gain» and «shakedown» satellites were coined from the idea that the presence of a core hole can substantially lower the energy of an empty localized orbital. Later on it was

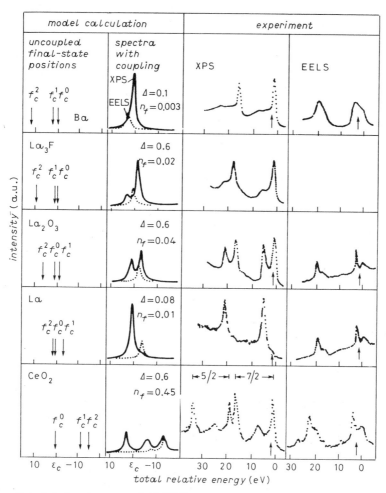

Fig. 11. – Calculated and measured 3d XPS and 3d EELS spectra [33]. The left-hand part shows the energies of the different configurations in the absence of a coupling. The calculated spectra have a Lorentzian broadening of 1.2 eV (FWHM) to include lifetime broadening. The two panels at the right show measured spectra. The full curves show the XPS and the dotted curves the EELS spectra. The arrows show the centre of gravity of the experimental EELS spectra.

recognized, on the basis of Hamiltonians containing conduction electrons and an empty localized level (2) attracted by the core hole [11, 34], that the role of different configurations of the localized level are more symmetrical and the terms «well screened» and «poorly screened» [35] were used.

In the series Ba, La, Ce compounds the f-level moves down to below ε_f; this can observed very nicely by BIS spectroscopy [33]. In the ground state of the beginning of this series, the f-level has a very small population and no f-contribution to the valence spectrum is discernible. Figure 11 shows calculated and measured 3d XPS and 3d EELS core spectra. The energies of the XPS spectrum are taken with reference to the observed Fermi edge. The energies of the EELS spectrum represent the energy loss suffered by the electron in the inelastic collision. In fig. 11 the zero mark is shifted by a constant «ε_c». The centre of gravity of the XPS spectra of the compounds with $n_f \approx 0$ should then fall near «0». The centre of gravity for the EELS spectrum should then be near the multiplet average energy of the f^1-configuration in the presence of a core hole. In the case of Ba, the coupling is too weak to lead to appreciable final-state f-configuration mixing. The model, therefore, predicts a Lorentzian peak of final-state f^0 character. The satellites seen in the experimental spectrum are due to plasmon excitations; the peak asymmetry is due to conduction electron-hole pair excitations [36]. In the case of pure La, it is justified to speak of the main line as the poorly screened peak and of the small satellite shoulder as the well-screened feature. In the cases of the La compounds, the XPS final states are calculated to be strongly mixed—a necessity to explain theoretically the strong satellite features. The large coupling strength can be rationalized from the short La-X distance ($X = $ O, F...). The separation of the peaks is a rather direct measure of the coupling strength in this case. Relatively little mixing of f-configurations occurs in the EELS final states. Two reasons help to explain this. i) In the La compounds in fig. 11, the f^1 and f^2 configurations have a sizable energy separation on the scale of the coupling strength. ii) In the case of EELS (and also XAS) a destructive interference tends to weaken the intensity of the «anti-bonding» structure if configuration mixing is appreciable.

The case of CeO_2 is illustrative for several reasons. It is not a mixed-valence metallic compound but rather an insulator with a sizable optical gap. No physics related to infrared divergences can occur for a Ce impurity in an insulator. In the impurity model, a singlet ground state similar to the large-N_f limit exists irrespective of the value of N_f. Rather strong hybridization results in an f-population $n_f \approx 0.5$. In the presence of a core hole, the f^0-configuration has now the highest energy and is well separated from the f^1 and f^2 configurations. In the XPS final states, the f^0-configuration remains almost pure. A large intensity testifies to the important f^0-component in the initial state. From the f^1 and f^2 configurations a continuum of strongly mixed states derives in the presence of the core hole which show up in the spectrum as an asymmetric peak and an additional hump. The EELS 3d core spectrum is dominated by $f^2 d^9$ multiplet structures, which are not fully resolved in the experimental spectrum. In the model calculation, multiplet splitting has not been included together with the configuration interaction. Interestingly, the calculation yields a small hump driven off the f^1 energy by the hybridization, which is in agreement with the experimental spectrum. The intensity for this structure remains small because of the destructive-interference effect already mentioned.

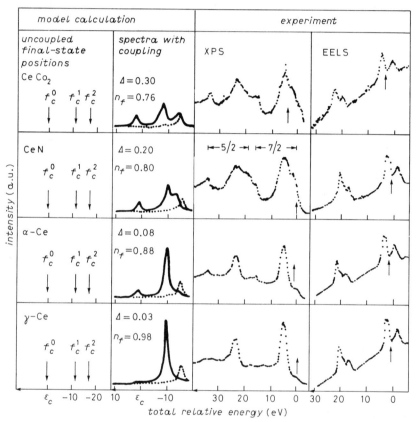

Fig. 12. – Calculated and measured 3d XPS and 3d EELS spectra for Ce compounds [33]. Panels as in fig. 11.

In the Ce compounds shown in fig. 12, n_f increases from 0.75 in $CeCo_2$ characteristic of a mixed-valence compound to 0.98 in γ-Ce, which serves as an example of the spin fluctuation case. This increase in f-population is mainly due to a decrease of the coupling strength Δ across this series. As we have seen in fig. 6, the intensity of the «f^2» satellite should be a measure of the coupling strength. The experimental spectra are slightly complicated, however, by the presence of multiplet splitting not included in the calculation. The multiplet splitting broadens both the f^2 and f^1 final-state configuration, the latter because of the d-

hole f-electron interaction. Another interesting test of the theory comes from the relative position of the XPS f^2 peak and the EELS f^2 peak. Especially in the case of CeN the two are clearly not coincident. Contrary to the case of La_2O_3, where one can say that the hybridization drives the XPS peaks apart, it is rather interference effects which influence the exact location of the f^2 XPS peak in the calculated spectrum.

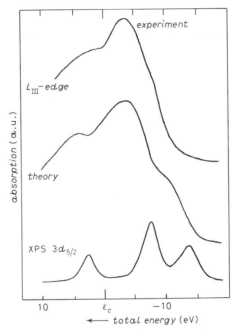

Fig. 13. – Calculated and measured L_{III} edge spectrum of $CeCo_2$ compared with calculated XPS spectrum [33, 37].

Figure 13 shows the calculated and measured L-edge absorption spectrum for $CeCo_2$ together with the calculated XPS core spectrum. Density of empty states and matrix element energy dependence are roughly approximated by $w^2(E) = \sqrt{E} \exp[-E/7\,\text{eV}]$. The three well-separated peaks show up as attenuated and broadened structures in the L-edge spectrum. A comparison with the measured XPS spectrum fig. 13 suggests that the separation between the f^1 and f^2 structures has been overestimated by the theory. Bringing them closer together would further improve the success of the convolution approximation for $CeCo_2$. The illustration in fig. 7 shows that the weight of the f^0 peak in the XPS spectrum is not a linear function of the admixture of the f^0-configuration in the ground state. This should warn the reader from a simple-minded deconvolution of the L-edge spectrum into Lorentzian lines and take the intensity ratios as a direct measure of configuration mixing in the ground state. Also there is little hope that a more complete theory of edge absorption including interaction of the continuum electron and possibly the breakdown of the sudden approximation would simplify interpretation of L-edge spectra.

* * *

I like to thank Y. BAER and W.-D. SCHNEIDER and co-workers for a stimulating and lasting collaboration from which I have learnt so much. I also like to thank H. BECK, O. GUNNARSSON and J. W. WILKINS for conversations from which I have greatly benefitted.

REFERENCES

[1] P. W. ANDERSON: *Phys. Rev.*, **124**, 41 (1961).
[2] T. V. RAMAKRISHNAN: in *Valence Fluctuations in Solids*, edited by L. M. FALICOV, W. HANKE and M. B. MAPLE (North Holland, Amsterdam, 1981), p. 13; P. W. ANDERSON: in *Valence Fluctuations in Solids*, edited by L. M. FALICOV, W. HANKE and M. B. MAPLE (North Holland, Amsterdam, 1981), p. 451.
[3] O. GUNNARSSON and K. SCHÖNHAMMER: *Phys. Rev. B*, **28**, 4315 (1983).
[4] N. READ and D. M. NEWNS: *Solid State Commun.*, **52**, 993 (1984).
[5] J. F. HERBST, R. E. WATSON and J. W. WILKINS: *Phys. Rev. B*, **17**, 3089 (1978).
[6] B. DELLEY and H. BECK: *J. Phys. C*, **17**, 4971 (1984).
[7] P. H. DEDERICHS, S. BLÜGEL, R. ZELLER and H. AKAI: *Phys. Rev. Lett.*, **53**, 2512 (1984).
[8] O. SAKAI, H. TAKAHASI, H. TAKESHIGE and T. KASUYA: *Solid State Commun.*, **52**, 997 (1984).
[9] R. MONNIER, L. DEGIORGI and D. D. KOELLING: *Phys. Rev. Lett.*, **56**, 2744 (1986).
[10] O. GUNNARSSON, O. K. ANDERSON, O. JEPSEN and J. ZAANEN: in *Proceedings of the 10th Taniguchi Symposium on Core Level Spectroscopy* (Springer, Berlin, 1988), p. 82.
[11] A. KOTANI and Y. TOYOZAWA: *J. Phys. Soc. Jpn.*, **35**, 1073 (1973).
[12] O. GUNNARSSON and K. SCHÖNHAMMER: *Phys. Rev. Lett.*, **50**, 604 (1983).
[13] O. GUNNARSSON and K. SCHÖNHAMMER: *Phys. Rev. B*, **31**, 4815 (1985).
[14] J. C. FUGGLE, R. LÄSSER, O. GUNNARSSON and K. SCHÖNHAMMER: *Phys. Rev. Lett.*, **44**, 1090 (1980).
[15] O. GUNNARSSON and T. C. LI: *Phys. Rev. B*, **36**, 9488 (1987).
[16] O. GUNNARSSON and K. SCHÖNHAMMER: in *Handbook on the Physics and Chemistry of Rare Earths*, Vol. **10**, edited by K. A. GSCHNEIDNER jr., L. EYRING and S. HÜFNER (Elsevier, Amsterdam, 1987), p. 103 ff.
[17] Y. KURAMOTO: *Z. Phys. B*, **53**, 37 (1983).
[18] H. KEITER and G. CZYCHOLL: *J. Magn. Magn. Mater.*, **31**, 477 (1983).
[19] N. GREWE: *Z. Phys. B*, **53**, 271 (1983).
[20] F. C. ZHANG and T. K. LEE: *Phys. Rev. B*, **28**, 33 (1983).
[21] P. COLEMANN: *Phys. Rev. B*, **29**, 3035 (1984).
[22] N. E. BICKERS, D. L. COX and J. W. WILKINS: *Phys. Rev. Lett.*, **54**, 230 (1985).
[23] N. E. BICKERS, D. L. COX and J. W. WILKINS: *Phys. Rev. B*, **36**, 2036 (1987).
[24] E. MÜLLER-HARTMANN: *Z. Phys. B*, **57**, 281 (1984).

[25] A. FUJIMORI: *Phys. Rev. B*, **27**, 3992 (1983).

[26] B. DELLEY and H. BECK: *J. Magn. Magn. Mater.*, **47/48**, 269 (1985).

[27] See lecture by J. C. FUGGLE: this volume, p. 127.

[28] F. PATTHEY, B. DELLEY, W.-D. SCHNEIDER and Y. BAER: *Phys. Rev. Lett.*, **55**, 1518 (1985).

[29] M. WIELICZKA, C. G. OLSON and D. W. LYNCH: *Phys. Rev. B*, **29**, 3028 (1984).

[30] J. W. ALLEN and R. M. MARTIN: *Phys. Rev. Lett.*, **49**, 1106 (1982).

[31] F. PATTHEY, W.-D. SCHNEIDER, Y. BAER and B. DELLEY: *Phys. Rev. B*, **34**, 2967 (1986).

[32] F. PATTHEY, W.-D. SCHNEIDER, Y. BAER and B. DELLEY: *Phys. Rev. Lett.*, **58**, 2810 (1987).

[33] W.-D. SCHNEIDER, B. DELLEY, E. WUILLOUD, J.-M. IMER and Y. BAER: *Phys. Rev. B*, **32**, 6819 (1985).

[34] O. GUNNARSSON and K. SCHÖNHAMMER: *Solid State Commun.*, **23**, 691 (1977).

[35] J. C. FUGGLE, M. CAMPAGNA, Z. ZOLNIEREK, R. LÄSSER and A. PLATAU: *Phys. Rev. Lett.*, **45**, 1597 (1980).

[36] See the lecture by G. D. MAHAN: this volume, p. 25.

[37] C. N. R. RAO, D. D. SARMA, P. R. SARODE, R. VIJAYARAGHAVAN, S. K. DHAR and S. K. MALIK: *J. Phys. C*, **14**, 451 (1981).

Electronic and Magnetic Properties of Interfaces.

O. Bisi and S. Ossicini

Dipartimento di Fisica dell'Università - Via Campi 213/A, I-41100 Modena, Italia

During the last years much attention has been devoted to the investigation of the electronic properties of surfaces and interfaces from first-principle calculations. The microscopic physics of a great number of systems has been investigated in details and many mechanisms and processes have been focused and understood. First-principle calculations are mainly based on the density functional theory (DFT) in its local density approximation (LDA)[1, 2]. Recently a substantial advancement in the calculation of the electronic states has been achieved by the computation of quasi-particle energies including self-energy correction[3].

The computational techniques used can be divided between all-electron and pseudopotential approaches. All-electron calculations are usually performed within the linear-method scheme that strongly simplifies the computational problem[4]. After a brief introduction of the theoretical framework and methods, we will present the application of linear methods to the investigation of *a*) the clean surface of a semiconductor, *b*) the metal-semiconductor interface and *c*) the magnetic properties of surfaces and interfaces. The pseudopotential approach to the investigation of surfaces and interfaces will not be discussed here. The reader interested is referred to the recent review of Calandra and Manghi[5].

1. – First-principle calculation of the electronic properties of solids.

The general aim of a theory of the electronic properties of solids, surfaces and interfaces is to solve the Schrödinger equation

$$(1) \qquad \left\{ -\frac{\hbar^2}{2m} \sum_{i=1}^{N} \nabla_i^2 + \sum_{i=1}^{N} V(\mathbf{r}_i) + \frac{1}{2} \sum_{i,j=1}^{N} \frac{e^2}{|\mathbf{r}_i - \mathbf{r}_j|} - E \right\} \Psi(\mathbf{r}_1, \ldots, \mathbf{r}_N) = 0$$

for an interacting many-electron system in an external potential $V(\mathbf{r})$. As a consequence of the many-body character of the Hamiltonian for a system of interacting electrons we cannot in principle write down a single-particle potential and single-particle states in a solid.

A very useful approach to the treatment of the many-body effects has been achieved by the density functional theory [1]. This formalism is based on a theorem by HOHENBERG and KOHN, which states that all ground-state properties can be expressed as a unique functional of the electron density $n(r)$. The total energy is such a functional and is written as the sum of four different terms

$$(2) \qquad E_v[n] = \int v(r)\, n(r)\, dr + \frac{e^2}{2} \int \frac{n(r)\, n(r')}{|r - r'|}\, dr\, dr' + T_s[n] + E_{xc}[n].$$

The first contains the external potential, the second is the electrostatic term, T_s is the kinetic energy of a system of noninteracting electrons with the same density $n(r)$ and $E_{xc}[n]$ is the exchange and correlation energy defined as the difference between the correct total energy and the first three terms. This exchange and correlation energy contains all many-body effects beyond the Hartree potential.

Furthermore, the Hohenberg and Kohn theorem states that the total energy attains its minimum value for the correct ground-state density. By using the formal similarity between the functional (2) and the Hartree energy, KOHN and SHAM [2] have shown that is possible to reduce the N-electron problem to the solutions of N simultaneous single-particle equations

$$(3) \qquad \{T + V_{ext} + V_H\}\, \psi_{n,k}(r) + v_{xc}(r)\, \psi_{n,k}(r) = E_{n,k}^{DF}\, \psi_{n,k}(r),$$

where

$$(4) \qquad n(r) = \sum_{n,k}^{occup} |\psi_{n,k}(r)|^2,$$

T is the kinetic-energy operator, V_{ext} is the external potential due to the ions, V_H is the Hartree potential due to the electrons and the exchange and correlation potential $v_{xc}(r)$ is the functional derivative

$$(5) \qquad v_{xc}(r) = \frac{\delta E_{xc}[n]}{\delta n(r)}.$$

Regarding the physical meaning of the one-particle eigenvalues in eq. (3) we notice that they are Lagrange parameters and consequently they have no formal justification as one-particle energies. Rather they must be considered as an ingredient to calculate the total energy of the system. The great advantage of this result is that, if the exact $E_{xc}[n]$ functional were known, eqs. (2)-(5) would give the exact density and total energy of the system under consideration. Clearly the limitation is related to the fact that we cannot give an exact functional for the exchange and correlation density.

The most widely used scheme for E_{xc} is the local density approximation (LDA)[1]:

$$(6) \qquad E_{xc}[n] = \int n(r) \, \varepsilon_{xc}(n(r)) \, dr \,,$$

where $\varepsilon_{xc}(n)$ is the exchange and correlation energy per electron of an homogeneous system.

Although this approximation is exact only in the limits of slowly varying electron density and fails to account for different features (for example, when applied to surface problems does not reproduce the long-range behaviour of the image potential, due to the neglect of the image charge formation), it has provided a very successful description of a great number of atomic, molecular and solid-state systems[6].

Attempts to go beyond the LDA using a form for the exchange and correlation potential which includes the long-range image behaviour have been based on the nonlocal exchange and correlation energy functional proposed by GUNNARSSON and JONES[7]. In this method the exchange and correlation energy is explicitly written in terms of a two-particle correlation factor and the shape of the exchange and correlation hole is modelled to satisfy at any point in space a normalization sum rule and to reproduce the exchange and correlation energy density in the homogeneous limit. This nonlocal approximation has been applied in band calculations for bulk semiconductors[8] and has been used by OSSICINI and co-workers[9-11] to study the behaviour of the valence charge density at metal surfaces using a jellium model and in atomic-structure calculations.

In the case of surfaces[10] the comparison between local and nonlocal approach confirms the reliability of the LDA in the determination of the charge density distribution for the surfaces. The main advantage of the nonlocal scheme is the possibility to evaluate the binding energy of the image states[12, 13].

Once one has described the exchange correlation part of the potential, one can solve the one-electron Schrödinger-like equation by expanding the wave functions either in a set of fixed basis functions or in a set of energy- and potential-dependent plane waves.

A different approach to the many-body problem can be expressed in terms of the one-particle Green's function G. For the general case of an inhomogeneous system, the quasi-particle energies and wave functions are obtained by solving[14]

$$(7) \qquad \{T + V_{ext} + V_H\} \, \psi_{n,k}(r) + \int dr' \, \Sigma(r, \, r', \, E^{qp}_{n,k}) \, \psi_{n,k}(r') = E^{qp}_{n,k} \, \psi_{n,k}(r) \,,$$

where Σ is the electron self-energy operator. It is in general a nonlocal, energy-dependent, non-Hermitian operator. Due to these last properties the eigenvalues of eq. (7) are in general complex; the imaginary part gives the lifetime of the quasi-particle. It is important to recognize the similarities and differences between eq. (3) and eq. (7). The self-energy operator Σ is approximated in the Kohn-Sham equation (3) by the local, energy-independent and Hermitian operator $v_{xc}(r)$. Therefore, the Kohn-Sham eigenvalues and eigenvectors differ from the quasi-particle energies and wave functions. Recently the quasi-particle problem (eq. (7)) has been solved by evaluating Σ at the first term in an expansion in the fully screened Coulomb interaction W (GW approximation for Σ)[3]. From this kind of calculation it is possible to check the approximation involved by considering the LDA (eq. (3)). Figure 1 shows the

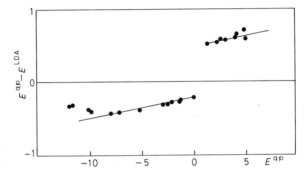

Fig. 1. – The difference between the quasi-particle energy, E^{qp}, and the eigenvalue in the LDA, E^{LDA}, for silicon is plotted against the quasi-particle energy for states at several high-symmetry points in the Brillouin zone. The straight lines are drawn as a guide to the eye. After[3].

difference between the GW and LDA eigenvalues as a function of the GW energies. The LDA energies of the filled states are found higher than the GW quasi-particle eigenvalues. The opposite happens in the conduction band, where the LDA energies are lower. The net result of the quasi-particle corrections to the LDA eigenvalues is to increase the minimum gap in semiconductor and insulator systems. In the case of Si the GW quasi-particle gap is 1.29 eV in good agreement with the experimental data (1.17 eV). The LDA gap for Si is 0.52 eV. The GW quasi-particle wave functions are essentially unchanged from the LDA one-particle wave functions.

Since GW calculations for the complex systems we will examine are not achievable at present, we will consider the solution of the one-particle equation (3) within the LDA scheme. This approach provides a very good description of the wave functions and an approximate treatment of the eigenvalues that in silicon differ of $\sim (0.2 \div 0.4)\,\mathrm{eV}$ from the quasi-particle energies.

2. – Linear methods for the investigation of surfaces and interfaces.

The calculation of the electronic properties of surfaces and interfaces may be considered one order of magnitude more difficult than the corresponding bulk problem. For this reason first-principle investigations for surfaces and interfaces are often based on the highly efficient *linear methods*, devised by ANDERSEN in 1975 [4]. The reader interested is referred to Andersen's lecture at the 1985 Varenna Course on *Highlights of Condensed-Matter Theory* [15].

In the first part of this section we will introduce our proposal to investigate the solid-vacuum and solid-solid interfaces through the linear muffin-tin orbital (LMTO) method in the atomic-sphere approximation (ASA), in the same version normally used for bulk problems.

Then we will examine the linear approaches used to study a two-dimensional (2D) thin slab, *i.e.* the linear augmented-plane-wave (LAPW) method [16] and full potential LAPW (no shape approximation for the potential: FLAPW) [17].

The possibility of treating the 2D surface problem as a bulk one relies on the supercell technique. The solid-vacuum (-solid) interface is simulated by an infinite sequence of solid and vacuum (solid) films, which recover the third dimension. The film thickness should be enough to avoid surface-surface interaction. In this way the solid-solid interface can be directly studied by using the standard procedures of bulk compound. In the section devoted to the semiconductor-metal interfaces we will show an application of this method. In case of a clean surface the presence of the vacuum region precludes a direct application of the LMTO-ASA approach. Our approach to avoid this drawback

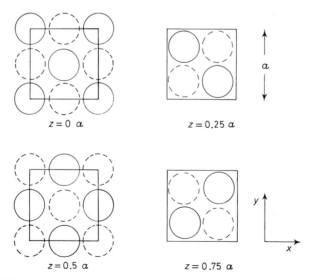

Fig. 2. – The different (001) planes of the Si cubic cell according to the empty-sphere technique. The silicon spheres (full line) and the empty spheres (dashed line) form a close-packed crystal with b.c.c. structure.

consists in filling up the vacuum region with empty spheres. Empty spheres have been firstly introduced in order to apply the muffin-tin approximation to loosely packed solids, as fictitious atoms with zero atomic number to be placed on void sites in order to obtain a close packing [18]. The muffin-tin potential of real atoms and empty spheres is, therefore, spherically symmetric within the spheres and constant in the interstitial region. The last with a suitable number of empty spheres may be reduced to values characteristic of close-packed solids (interstitial region $\leqslant 0.33$ of the total volume). In the case of Si (or Ge or C) the sublattice of empty spheres is equivalent to a diamond lattice shifted along the (100) direction by half the cube axis (fig. 2). This procedure yields a close-packed solid with a b.c.c. lattice, sixteen spheres of equal radius per unit cell, that is, eight Si atoms and eight empty spheres (void sites).

Close packing is a necessary requirement for the application of the atomic-sphere approximation (ASA), where the muffin-tin interstitial region is «annihilated» through the expansion of the muffin-tin spheres and the neglect of the slight overlap (fig. 3). This approximation greatly reduces the complexity of

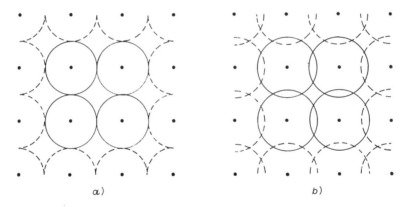

$a)$ $b)$

Fig. 3. – Panel $a)$ the muffin-tin approximation to the crystal potential. The potential of the solid is assumed spherically symmetric inside the spheres and constant in the interstitial region. Panel $b)$ the ASA potential. The muffin-tin sphere becomes the Wigner-Seitz (WS) sphere filling up the space. The crystal is regarded as an array of WS spheres and overlap and interstitial regions are neglected.

the band calculation. The LMTO-ASA approach and the empty-sphere technique have been successfully applied to the calculation of the energy bands, total energies and structural stability of group IV semiconductors [19, 20]. The result of our LMTO-ASA calculation of the electronic states of bulk Si is shown in fig. 4 [21]. It is evident that, in spite of the approximations of the approach, the description of the electronic states of Si is excellent. The energy gap between filled and empty states is $\sim 0.5\,\text{eV}$, nearly half the experimental value. This discrepancy it is not due to the above-mentioned approximations but to the local

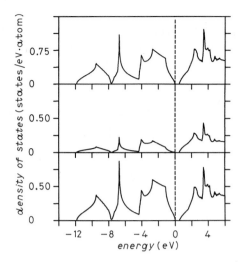

Fig. 4. – Density of states and site-projected density of states of Si bulk, according to this LMTO-ASA calculation. The density shown refer to Si (bottom), empty sphere (central) and total DOS (top). Energies (in eV) are referred to the valence band maximum.

density approximation to the exchange and correlation potentials. As explained in the previous section, the inclusion of self-energy corrections is necessary in order to obtain the correct energy gap value. The site-projected density of states (PDOS) of fig. 4 shows that within the empty spheres we have nonzero PDOS contribution. This effect, which is due to the tails of the wave functions centred on the Si atoms, is stronger in the conduction band, where the electron states are less localized. Since the empty spheres have $Z = 0$, their net charge is always negative and the potential at the void sites repulsive. In this way the nonsphericity of the potential in the diamond lattice is taken into account. In our model of filling up the vacuum region with empty spheres we expect a decreasing negative charge on the empty spheres on moving far from the surface, with the formation of a potential barrier at the solid-vacuum interface. We will see in the next section that the surface potential barrier created by the vacuum side empty spheres is realistic.

The great advantage of this new surface LMTO approach (SLMTO-ASA)[21] with respect to most accurate LAPW and FLAPW relies on the possibility of treating much more complex systems. While the last are limited to $5 \div 7$ layer slabs (see below), in our approach a 12-layer slab can be studied without particular problems. It is, therefore, possible even to investigate reconstructed surfaces. It looks powerful as the LMTO version for the slab geometry of fig. 5, recently developed and tested for a Fe(001) 5-layer slab[22].

The geometry considered in LAPW and FLAPW calculations in shown in fig. 5. The basis wave functions are constructed from the solutions of the Schrödinger equation for the appropriate symmetrized potentials: spherical

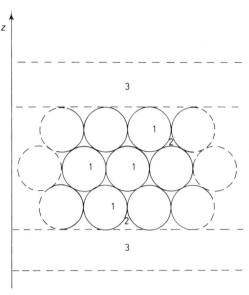

Fig. 5. – Thin-film geometry as used in LAPW and FLAPW methods. The spheres, the interstitial and the vacuum regions are indicated by 1, 2 and 3, respectively.

inside the sphere (region 1), plane wave in the interstitials (region 2) and 2-D plane wave, decaying normally to the surface in the vacuum region 3. The wave functions are continued smoothly in value and derivative across the boundaries. To ensure accurate results, the slab should be thick enough so that the electronic structure in the interior of the slab resembles closely the expected bulk spectrum. For metal films with well-localized surface states this requirement is satisfied for slabs as thin as 5 layers. If it is necessary to identify more extended surface states, thicker slabs must be used. The desirability of thicker slabs must be balanced, however, by the practical consideration that the magnitude of the calculation (computer memory and time) increases rapidly with the number N of layers in the slab (roughly between N^2 and N^3).

3. – The semiconductor-vacuum interface.

The purpose of this section is to show that reliable self-consistent density of states and energy bands at the semiconductor-vacuum interface can be obtained by the SLMTO-ASA method.

As model system we chose the Si(111)(1 × 1) surface; this surface is the most studied in surface science and its interest is still high. Moreover, this is a simple case and it is important to test the method in this system, before extending it to more complex systems, like reconstructed surfaces or ultrathin overlayer chemisorbed structures.

We will show that the semiconductor-vacuum interface can be well described if the semi-infinite vacuum region is simulated by a semi-infinite empty-sphere solid. In other words, we investigate if a lattice of empty spheres is able to simulate the vacuum region, near the surface.

In our calculation we simulate the Si-vacuum interface by a 12-layer Si(111) slab and a vacuum region, which repeats periodically in the (111) direction [21]. The empty spheres of the Si crystal constitute a second lattice of 12 (111) planes. With the abstraction of atom type the lattice is then b.c.c. with equally spaced (111) planes forming an $ABCABC...$ sequence. This b.c.c. lattice is continued into the vacuum region, where now all sites are occupied by empty spheres (of the same dimension). We found that a 6-layer slab with empty spheres is

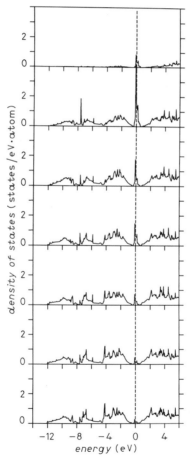

Fig. 6. – Site-projected density of states of the Si(111)(1 × 1) surface simulated by a 12 Si (111) + 6 layer slab with empty spheres according to the present SLMTO-ASA calculation. On going from the bottom to the top we move from the central Si layer to the first empty-sphere layer on the vacuum site. Each Si PDOS contains the contribution of the Si atom and of its neighbour empty sphere. Energies (in eV) are referred to the Fermi level.

sufficient to avoid surface-surface interaction. With this geometry we have a three-dimensional (3D) unit cell with one atom per (111) layer. The total number of atoms per unit cell is 30: 12 Si + 12 empty spheres for the silicon side + 6 empty spheres to simulate the vacuum.

The first result we will show is the calculated self-consistent densities of states, projected on the silicon and vacuum side (fig. 6). On moving from the bottom to the top of the figure, we go from the central Si atom, to that at the surface and to the empty sphere at the solid-vacuum interface.

We note that near and at the surface a large peak appears in the semiconductor gap. This peak is due to the dangling-bond surface state, which is half filled and is responsible of the metallic character of the (111) ideal surface of silicon. Other surface states and resonances are present and can be recognized by comparing these results to the corresponding bulk densities of states (fig. 4). A prominent surface feature is, for example, the structure located at $\sim -8\,\text{eV}$ in the hybridization gap of Si.

More detailed information on surface states and resonances can be obtained by looking at the projected bulk band structure (fig. 7) and at the two-dimensional band structure of the 12-layer Si slab (fig. 8). By comparing the projection of the bulk bands of Si onto the surface Brillouin zone (BZ) with the bands of fig. 8 all surface structures can be identified. Firstly we note two nearly degenerate dangling-bond surface bands in the semiconductor energy gap, one for each surface of our slab. As expected for this ideal surface, they are flat bands, *i.e.* with no dispersion parallel to the surface. A measure of the interaction between our surfaces is given by the fact that these two bands are

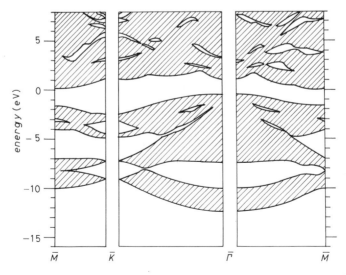

Fig. 7. – Projection of the Si bulk bands on the (111) surface. Hatched areas show regions where bulk states are found. For a better comparison with fig. 8 energies (in eV) are referred to the Fermi level of intrinsic Si (half gap).

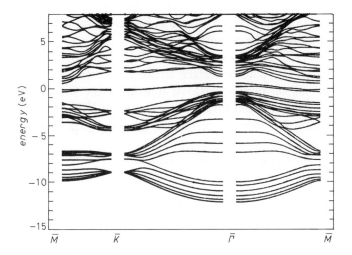

Fig. 8. – Two-dimensional bands of a 12-layer Si(111) slab. Energies (in eV) are referred to the Fermi level.

not exactly degenerate, as they should be when surface-surface interaction is zero. We may check that this effect is negligible.

Other various surface states are present but no more over the entire two-dimensional Brillouin zone. For example, the surface states located in the gap found near $\overline{K}$ and $\overline{M}$ around $-8\,\mathrm{eV}$. Moving away from this part of the BZ, they merge in the continuum of the projected bulk bands and become strong surface resonances.

Since the Si(111)(1 × 1) surface is well known and our main interest is focused on the reliability of the calculation, we do not go into a detailed analysis of our results. We simply note that they are in excellent agreement with the literature [23, 24]. In conclusion we have shown that the LMTO-ASA approach may be applied to investigate not only the internal interfaces of solids [25], but even the clean surfaces of semiconductors in the new SLMTO-ASA version. In this all-electron calculation, core level energies may be obtained [26] and significant information, *e.g.* surface core level energy shifts, is available. Furthermore, it is possible to apply a recent development of the LMTO method [27], *i.e.* transformation in a tight-binding basis to obtain accurate total energies and structural information.

4. – The semiconductor-metal interface.

Transition metals deposited onto silicon surfaces react, leading to the formation of silicide compounds [28]. Most of these compounds are metals, so that what the reaction does is to replace the transition metal-silicon interface with a rather stable silicide-silicon interface. Besides, it removes impurities from the interface, leaving a clean contact.

The silicon-silicide interface may be epitaxial if the relative difference between the silicon and silicide surface lattice constants is only a few percent. Well-established silicon-silicide epitaxial interfaces are those between Si(111) and $NiSi_2$(111), $CoSi_2$(111) and Pd_2Si(0001). On top of $NiSi_2$ or $CoSi_2$, epitaxially grown on Si(111), it was recently possible to grow epitaxially Si(111), leading to epitaxial Si-silicide-Si structures [29].

To achieve a description of the silicide-silicon interface on a microscopic scale is a very difficult task. We present here an investigation of the interface between Si(111) and $NiSi_2$(111). This system is interesting because two distinct interface orientations, called A type (perfect epitaxial orientation) and B type (180° rotated structure), have been experimentally found [30]. Furthermore, these two interfaces show different Schottky barrier height [31, 32]. Therefore, this system is a sort of model system.

We simulate the Si-$NiSi_2$ interface by a 21-atomic-plane slab, which repeats periodically in the direction normal to the interface, as shown in fig. 9 for type-A

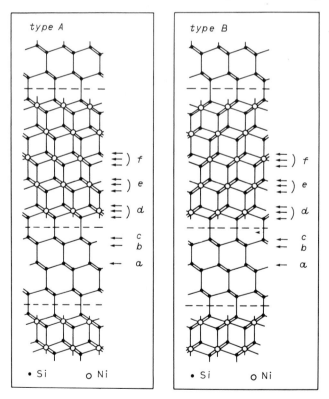

Fig. 9. – Atomic arrangements of the computed models for type-A (left panel) and type-B (right panel) orientation of the $NiSi_2$(111)-Si(111) interface. The unit cell contains 33 atoms: 16 Si + 5 Ni + 12 empty spheres (not shown).

and type-B orientations. In this approach the 21 atomic layers correspond to 6 Si(111) planes (of thickness 10.1 Å) plus 5 NiSi$_2$(111) planes (of thickness 15.5 Å). The number of planes is chosen so as to have a negligible interface-interface interaction. For this reason a small number of NiSi$_2$ layers, two in order to avoid stacking faults, cannot be used. The 6-layer Si slab has often been used to simulate the Si(111) surface, its thickness being sufficient to avoid a strong surface-surface interaction. With this geometry we have a three-dimensional (3D) unit cell with one atom per (111) layer. In order to guarantee good space filling as required by the LMTO-ASA approach, 12 empty spheres (*i.e.* sites with nuclear charge $Z = 0$) have been added to the unit cell. The total number of atoms per unit cell is 33: 16 Si + 12 empty spheres + 5 Ni atoms.

In the geometries investigated the silicide is terminated by planes of silicon and the coordination of the Ni atom in the last Ni plane is seven. This value has to be compared with the 8-fold coordination of Ni in bulk NiSi$_2$. If the silicide were terminated by planes of Ni, the coordination of these atoms would be five. The Ni atom 7-fold coordination at the interface has been recently found for both type-A and type-B orientations by X-ray standing-wave measurements [33] and shows the lowest computed total energy [34, 35]. By the same experimental

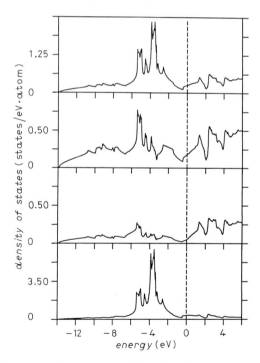

Fig. 10. – Density of states and site-projected density of states of NiSi$_2$, according to this LMTO-ASA calculation. From the bottom to the top the Ni, empty sphere and Si projected DOS and total DOS are displayed. Energies (in eV) are referred to the Fermi level.

technique the interplanar Si-NiSi$_2$ distance has been determined. For the perfect epitaxial type-A interface this distance is (2.31 ± 0.05) Å, for the rotated type B is slightly contracted, (2.24 ± 0.03) Å. These values must be compared with the unrelaxed 2.35 Å calculated values using bulk bond lengths.

It is important to note that our approach to the interface problem, with no solid-vacuum region, simulates a thick silicon-silicide interface. Therefore, the results will be appropriate to investigate the internal interface, when epitaxy is already established, not the early stages of metallic deposition, when the interface reaction takes place[36].

Before showing the results for the interface system it is interesting to concentrate our attention to the corresponding bulk results for Si (fig. 4) and NiSi$_2$ (fig. 10).

In the case of NiSi$_2$ we have one empty sphere per Ni atom and the crystal can be viewed as composed by Si atoms forming a simple cubic lattice with Ni and empty spheres occupying alternatively the centre of every cube. With abstraction of the atom types, the lattice of the NiSi$_2$ structure is then b.c.c.-like. Figure 10 shows our self-consistent result for the density of states of bulk NiSi$_2$. This result provides evidence of a relevant interaction between Si p and Ni d orbitals, in agreement with the proposed picture of the chemical bond in Ni silicides[28, 37].

Returning to the interface system, the relative self-consistent densities of states, projected on the silicon and silicide sites, are shown in fig. 11 and 12 for the type-A and type-B orientation, respectively. The densities of states of Si include the empty-sphere contribution and those of NiSi$_2$ are built up with 2 Si, 1 Ni and 1 empty sphere. The densities shown refer to the central Si atom (a), the interface Si atoms (b and c) and to the three NiSi$_2$ layers (d, e, f) of fig. 9.

The perturbation caused by the presence of the interface is evident on both silicon and silicide sides. Whereas the densities of states of the central silicon atom and of the second and third NiSi$_2$ layers are already similar to that of bulk silicon and bulk silicides, those related to the interface layers show well-defined features.

An important structure arises from the unpaired electrons of the 7-fold Ni at the interface. From the densities of states relative to the silicide d layer we see that the characteristic double peak of bulk silicide is lost and that the peak is shifted to lower binding energy. This is a consequence of the lower coordination of the Ni atoms at the interface.

Regarding the silicon interface atoms, we see that due to the presence of the interface they have a metallic character and that interface features are present in both A-type and B-type densities of states at an energy location of about $(0.7 \div 0.8)$ eV below the Fermi energy; this peak is clearly present also in the density of states relative to the silicide interface layer.

More information on the interface states can be obtained by looking at the two-dimensional band structure of the interface system both for A and B type

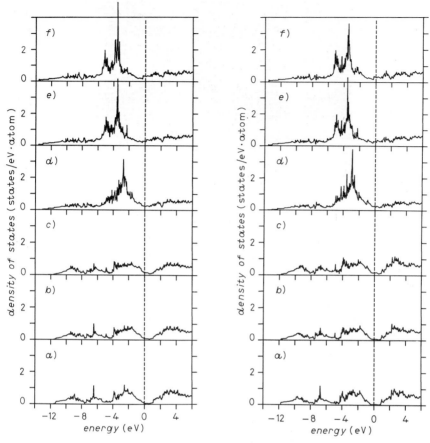

Fig. 11.   Fig. 12.

Fig. 11. – Site-projected density of states for the NiSi$_2$(111)-Si(111) interface with type-A orientation. The densities shown refer to the central Si atom (a)), the interface region Si atoms (b), c)) and to the three NiSi$_2$(111) layers (d), e), f)), as shown in fig. 9.

Fig. 12. – Site-projected density of states for the NiSi$_2$(111)-Si(111) interface with type-B orientation. The densities shown refer to the central Si atom (a)), the interface region Si atoms (b), c)) and to the three NiSi$_2$(111) layers (d), e), f)), as shown in fig. 9.

(fig. 13 and 14) and at the projected bulk band structures (see, for example, fig. 7 for the Si case). From the figure relative to the B-type bands we clearly see at the $\overline{M}$-$\overline{K}$ line two nearly degenerate interface bands located at 0.8 eV below the Fermi energy, which show practically no dispersion parallel to the surface. More pronounced dispersion is shown by the corresponding bands in the A-type case (fig. 13). For a more general discussion on these metal-induced gap states the reader is referred to ref. [38].

The metal-induced gap states can play an important role in the determination of the Schottky barrier height for these metal-semiconductor systems. The

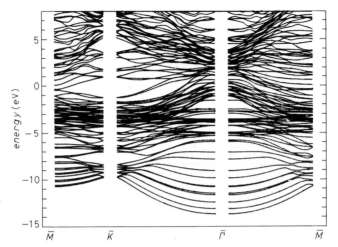

Fig. 13. – Two-dimensional bands of the NiSi$_2$(111)-Si(111) interface with type-A orientation. Energies (in eV) are referred to the Fermi level.

Schottky barrier height ϕ_B is given by the energy difference between the bottom of the semiconductor conduction band and the metal Fermi level. Due to the fact that the LDA underestimates the minimum gap, the Schottky barrier height ϕ_B is expected to be underestimated. This difficulty may be partially overcome by considering that in DFT the eigenenergy of the highest occupied orbital of a metallic system equals the Fermi energy. It follows that we correctly estimate the energy difference Δ between the silicide Fermi level E_F and the Si valence band maximum E_v and by adding the experimental Si gap value E_g (1.17 eV) we may compute the Schottky barrier height $\phi_B = E_g - \Delta$.

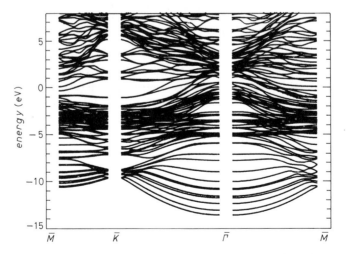

Fig. 14. – Two-dimensional bands of the NiSi$_2$(111)-Si(111) interface with type-B orientation. Energies (in eV) are referred to the Fermi level.

To do this correctly we use the potential alignment method developed by CHRISTENSEN for the calculation of the band offsets at semiconductor interfaces [39]. We calculate first the self-consistent potentials for the supercell geometry, then we extract the central-layer potentials and use these to perform a nonself-consistent band calculation in the bulk geometry. From these band structures the Si valence band maximum E_v and the silicide Fermi level E_F are computed in the same energy scale and Δ is determined. In this way we obtain for a calculation at the experimental interplanar distance [33] $\phi_B = 1.07$ eV (type A) and $\phi_B = 0.87$ (type B). The Schottky barrier of the ideal interface (type A) is practically as wide as the Si gap, while an extended defect like a 180° rotation (type B) lowers the barrier to 0.87 eV. Experimentally the value of ϕ_B is controversial. According to TUNG [31] the Schottky barrier is 0.65 and 0.79 for type A and type B, respectively. A different result has beeh obtained by Ho et al. [32], with a value of 0.78 eV for both type-A and type-B interfaces when they are almost perfect. Less perfect interfaces yielded the low barrier value of 0.66 eV.

The difference between our evaluation of ϕ_B for type-A orientation and the experimental data provides evidence of the complex nature of the Schottky barrier formation in silicon-silicide interfaces. We are led to suppose that even in the epitaxial type-A orientation some effect similar to the type-B extended defect is present and lowers the value of ϕ_B.

A possible mechanism which may modify the Schottky barrier height is the relaxation of the interplanar Si-NiSi$_2$ distance [33]. A recent work [40] finds this distance to be 2.20 Å in the case of A type, even more contracted with respect to the B-type value (2.24 Å). We found indeed that the interface relaxation modifies the Schottky barrier value. The value $\phi_B = 0.87$ eV found for type B at the measured interplanar distance increases to $\phi_B = 1.01$ eV for the calculation of the ideal unrelaxed interface. A complete discussion of this result will be published in ref. [38].

The possibility that even for the epitaxial Si(111)-NiSi$_2$(111) interface deviation from perfect epitaxy must be considered in order to obtain the experimental ϕ_B value can found some evidence in the very recent work of Bennett et al. [41], which shows that all the structures preceding the formation of NiSi$_2$ B type are epitaxial with a well-defined habit plane, while those preceding A type are not.

5. – Magnetic properties of surfaces and interfaces.

In the following we will concentrate on three specific issues of surface magnetism: i) surface magnetism of a nonmagnetic system, ii) surface magnetism of a magnetic system, iii) surface magnetism of an adsorbed overlayer.

i) *Surface magnetism of a nonmagnetic system.* Among the paramagnetic systems which may present a magnetic surface we will consider the case of vanadium. Metallic V has a b.c.c. crystal structure and is known experimentally to be nonmagnetic. However, the isolated vanadium atom in its ground state has a permanent magnetic moment of $3.0\,\mu_B$. Thus the solid would be expected to transform to a magnetic state at some (hypothetical) lattice constant larger than that found experimentally. Spin-polarized augmented-plane-wave (APW) calculations have been performed by HATTOX *et al.* [42] to determine the energy band structure of metallic vanadium in an assumed ferromagnetic b.c.c. structure as a function of lattice parameter. The calculations correctly predict the absence of magnetic moment of V at its equilibrium lattice constant. However, a nonmagnetic-to-magnetic transition is found to occur abruptly at the lattice constant of 3.70 Å, corresponding to a V-V distance of 3.20 Å. The corresponding V-V distance in the equilibrium lattice is 2.62 Å. The V energy bands in two symmetry directions are compared in fig. 15 for the equilibrium lattice constant

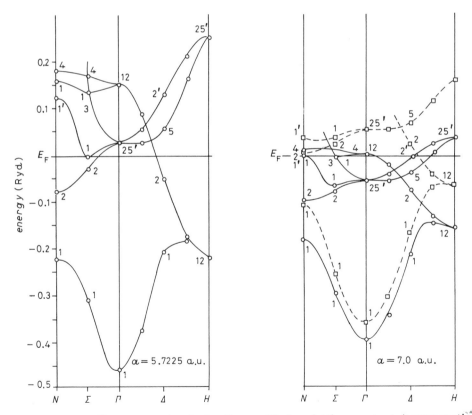

Fig. 15. – Vanadium energy bands for the equilibrium lattice constant (nonmagnetic behaviour, left), the expanded lattice constant ($a = 3.70$ Å, $\mu_B = 2.2$, right); —— α-spin, ––– β-spin. After [42].

and the first lattice constant ($a = 3.70$ Å) for which the bands exhibit spin polarization, leading to a magnetic moment of $2.22 \mu_B$. Figure 15 shows that, for the spin-polarized case, the $3d$ bands have narrowed in the vicinity of the Fermi energy to produce the high density of states required for a stable magnetic state. With increasing lattice parameter the bands become narrower and the splitting between the majority (α-spin) and minority (β-spin) bands becomes greater. Thus an increasing portion of the majority band and a decreasing fraction of the minority states lie below the common Fermi energy as the lattice parameter increases, resulting in an increase of the magnetic moment. These results are in good qualitative agreement with the appearance of a local magnetic moment in certain vanadium alloys where the V-V distance is greater than ~ 3.10 Å.

Since surface atoms may be considered as intermediate between bulk and free atoms, intense theoretical interest exists in the vanadium surfaces, mainly (001), because of the possibility of inducing ferromagnetism in its surface layer while retaining paramagnetic behaviour in the bulk [43-46]. Support of this hypothesis is derived from the experimental finding of magnetic behaviour of hyperfine particles (($100 \div 1000$) Å) of vanadium [47].

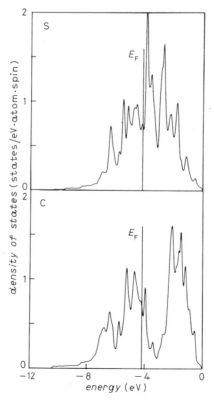

Fig. 16. – Site-projected partial DOS in the surface (S) and center (C) layers of V(001), according to the FLAPW calculation of Ohnishi *et al.* [46].

It is now established that a large magnetic moment $(3.09\,\mu_B)$ is present only in an isolated V(001) monolayer, while the (001) surface of vanadium does not present surface magnetism [46]. The stability of the paramagnetic V(001) surface can be understood by looking at the site-projected DOS of fig. 16. The DOS of the centre layer for a 7-layer slab is essentially bulklike with well-separated bonding and antibonding regions. For the surface layer, the sharp surface state DOS peak which is responsible of the occurrence of surface magnetism in Ni and Fe (see below) is located $0.3\,\mathrm{eV}$ above E_F. As a result, the paramagnetic state is stable.

The experimental result of ref. [47] may be interpreted by considering the presence of oxygen in the surface region, which is found to be related to the V(001)(5 × 1) reconstruction [48, 49]. The model proposed for this reconstructed surface consists of an outermost layer of V atoms of (001) symmetry and a

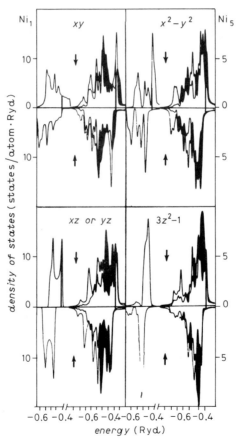

Fig. 17. – Site- and angular-momentum-projected DOS of Ni(001). To the left in each box is shown the result of the monolayer film and to the right are shown the results of the central and surface layers. The shaded areas indicate excess states on the surface. The Fermi level is indicated by a vertical line. After [50].

subsurface layer of oxygen atoms with (5×1) symmetry. If this model is correct, both the increase of the first interlayer spacing and the reduction of the coordination number of the V surface atoms may lead to a configuration similar to that of an isolated (001) V monolayer and to the formation of surface magnetism for V(001).

ii) *Surface magnetism of a magnetic system.* The magnetic metals we are considering are Ni and Fe, which have a bulk magnetic moment of $0.56\,\mu_B$ and $2.12\,\mu_B$, respectively. The study of surface magnetism of Ni and Fe began with the observation of *dead layers*, *i.e.* paramagnetic behaviour of the Ni and Fe surfaces[51]. These early results are now believed to be due to impurities present at the surface. It is now known that the presence of the surface increases the magnetic moment of Ni and Fe.

The density of states for a monolayer and a 5-layer slab Ni(001) surface, as obtained by a LAPW calculation[50], is shown in fig. 17. The main features of surface magnetization are already recognizable in the monolayer. The hybridization of the monolayer $(x^2 - y^2)$ band causes a three-peak structure with the main peak at the top of the band which is empty for the minority spins and full for the

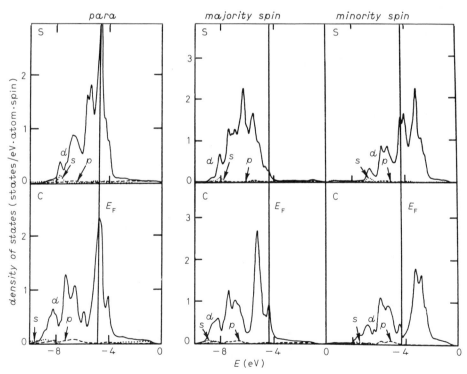

Fig. 18. – Site- and angular-momentum-projected DOS in the paramagnetic and ferromagnetic states of Fe(001). The surface and central layers are indicated with S and C, respectively. After[52].

majority spins. This structure contributes to the large magnetic moment
($0.95\,\mu_B$) of the monolayer. Even in the 5-layer slab, the increase of the magnetic
moment at the surface is mainly of $(x^2 - y^2)$ character. Figure 17 shows that the
$(x^2 - y^2)$ surface DOS is half full for the minority spins and full for the majority
spins, while the peak for the bulk is occupied for both spins. It is interesting to
observe that the (xz, yz) high-energy structure in the central layer lies above the
corresponding surface feature leading to a magnetic-moment contribution larger
in the bulk than at the surface. The resulting magnetic moment at the surface
layer is $0.65\,\mu_B$. A similar result has been obtained by a FLAPW calculation on a
7-layer Ni(001)[53].

The enhancement of the magnetization at the (001) surface of iron is clearly
shown in fig. 18[52]. It is evident that the DOS of the surface layer is
dramatically changed compared to the bulklike DOS of the central layer. Due to
the surface d-band narrowing the majority d band is now almost completely
filled. The minority DOS is characterized by a surface structure around the
Fermi level. The net result of the surface-induced d-band narrowing and the
occurrence of surface states is a larger spin unbalance corresponding to an
enhancement of the magnetic moment from $2.27\,\mu_B$ (central layer) to $2.96\,\mu_B$
(surface).

iii) *Surface magnetism of an adsorbed overlayer*. Many bimetallic interfaces and overlayers have been investigated up to now. This field has attracted
increasing interest and the results obtained are quite promising. Since it is out of
the purpose of the present lecture, we will simply examine one system, Fe on
Ag(001). The reader interested is referred to the review paper of Freeman *et
al.* [54]. An enhancement of 36% of the Fe magnetic moment has been found for
Fe monolayer epitaxially adsorbed on Ag(001) substrate[55]. From the DOS of
fig. 19 the large splitting between minority and majority spin d band of Fe is
evident. This splitting is slightly larger than the Fe d-band width, leading to a

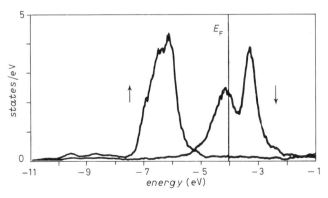

Fig. 19. – Fe monolayer epitaxially adsorbed on Ag(001) substrate: densities of states for
majority and minority bands as projected on the Fe monolayer. After[55].

separation of majority and minority spin bands. The resulting magnetic moment is $3.0\,\mu_B$.

$$* * *$$

Financial support by Consiglio Nazionale delle Ricerche and Ministero della Pubblica Istruzione is acknowledged. The calculations have been performed at the CICAIA, University of Modena.

REFERENCES

[1] P. HOHENBERG and W. KOHN: *Phys. Rev.*, **136** B, 864 (1964).

[2] W. KOHN and L. J. SHAM: *Phys. Rev.*, **140** A, 1133 (1965).

[3] M. S. HYBERTSEN and S. LOUIE: *Phys. Rev. B*, **34**, 5390 (1986), and references therein.

[4] O. K. ANDERSEN: *Phys. Rev. B*, **12**, 3060 (1975).

[5] C. CALANDRA and F. MANGHI: in *Semiconductor Interfaces: Formation and Properties*, edited by G. LE LAY, J. DERRIEN and N. BOCCARA (Springer-Verlag, Berlin, 1987), p. 162.

[6] U. VON BARTH and A. R. WILLIAMS: in *Theory of Inhomogeneous Electron Gas*, edited by S. LUNDQVIST and N. H. MARCH (Plenum Press, New York, N. Y., 1983), p. 186.

[7] O. GUNNARSSON and R. O. JONES: *Phys. Scr.*, **21**, 394 (1980).

[8] F. MANGHI, G. RIEGLER, C. M. BERTONI and G. B. BACHELET: *Phys. Rev. B*, **31**, 3680 (1985).

[9] S. OSSICINI and C. M. BERTONI: *Phys. Rev. A*, **31**, 3550 (1985).

[10] S. OSSICINI, C. M. BERTONI and P. GIES: *Surf. Sci.*, **178**, 244 (1986).

[11] S. OSSICINI, F. FINOCCHI and C. M. BERTONI: *Surf. Sci.*, **189/190**, 776 (1987).

[12] P. GIES: *J. Phys. C*, **19**, L209 (1986).

[13] F. MANGHI: *Phys. Rev. B*, **33**, 2554 (1986).

[14] L. HEDIN and S. LUNDQVIST: *Solid State Phys.*, **23**, 1 (1969).

[15] O. K. ANDERSEN, O. JEPSEN and D. GLÖTZEL: in *Highlights of Condensed-Matter Theory*, *Proc. S.I.F.*, Course LXXXIX, edited by F. BASSANI, F. FUMI and M. P. TOSI (North Holland, Amsterdam, 1985), p. 59.

[16] O. JEPSEN, J. MADSEN and O. K. ANDERSEN: *Phys. Rev. B*, **18**, 605 (1978); H. KRAKAUER, M. POSTERNAK and A. J. FREEMAN: *Phys. Rev. B*, **19**, 1706 (1979).

[17] A. WIMMER, H. KRAKAUER, M. WEINERT and A. J. FREEMAN: *Phys. Rev. B*, **24**, 864 (1981).

[18] J. KELLER: *J. Phys. C*, **4**, L85 (1971).

[19] D. GLÖTZEL, B. SEGALL and O. K. ANDERSEN: *Solid State Commun.*, **36**, 403 (1980).

[20] A. K. MCMAHAN: *Phys. Rev. B*, **10**, 5835 (1984).

[21] S. OSSICINI and O. BISI: *Surf. Sci.*, **211/212**, 572 (1989).

[22] G. W. FERNANDO, B. R. COOPER, N. V. RAMANA, H. KRAKAUER and C. Q. MA: *Phys. Rev. Lett.*, **56**, 2299 (1986).

[23] M. SCHLÜTER, J. R. CHELIKOWSKY, S. G. LOUIE and M. L. COHEN: *Phys. Rev. B*, **12**, 4200 (1975).

[24] F. CASULA, S. OSSICINI and A. SELLONI: *Solid State Commun.*, **30**, 309 (1979).

[25] O. BISI and S. OSSICINI: *Surf. Sci. B*, **189/190**, 285 (1987). Due to an error in the plotting code, the densities of the NiSi$_2$-Si interface drawn in this paper are slightly different from those presented here.

[26] W. R. L. LAMBRECHT: *Phys. Rev. B*, **34**, 7421 (1986).

[27] O. K. ANDERSEN and O. JEPSEN: *Phys. Rev. Lett. B*, **53**, 2571 (1984).

[28] C. CALANDRA, O. BISI and G. OTTAVIANI: *Surf. Sci. Rep.*, **4**, 271 (1985).

[29] J. DERRIEN and F. ARNAUD D'AVITAYA: *J. Vacuum Sci. Technol. A*, **5**, 2111 (1987).

[30] R. T. TUNG, J. M. GIBSON and J. M. POATE: *Phys. Rev. Lett.*, **50**, 429 (1983).

[31] R. T. TUNG: *Phys. Rev. Lett.*, **52**, 461 (1984).

[32] M. LIEHR, P. E. SCHMID, F. K. LEGOUES and P. S. HO: *Phys. Rev. Lett.*, **54**, 2139 (1984).

[33] E. VLIEG, A. E. M. J. FISCHER, J. F. VAN DER VEEN, B. N. DEV and G. MATERLIK: *Surf. Sci.*, **178**, 36 (1986).

[34] D. R. HAMANN: *Phys. Rev. Lett.*, **60**, 313 (1988).

[35] P. J. VAN HOEK, W. RAVENEK and E. J. BAERENDS: *Phys. Rev. Lett.*, **60**, 1743 (1988).

[36] D. R. HAMANN and L. F. MATTHEIS: *Phys. Rev. Lett.*, **54**, 2517 (1985).

[37] XU JIAN-HUA and XU YONG-NIAN: *Solid State Commun.*, **55**, 891 (1985); W. R. L. LAMBRECHT, N. E. CHRISTENSEN and P. BLÖCHL: *Phys. Rev. B*, **36**, 2493 (1987).

[38] S. OSSICINI, O. BISI and C. M. BERTONI: *Phys. Rev. B*, in press.

[39] N. E. CHRISTENSEN: *Phys. Rev. B*, **35**, 6182 (1987).

[40] K. G. HUANG, J. ZEGENHAGEN, W. M. GIBSON, B. D. HUNT and L. J. SCHOWALTER: *Bull. Am. Phys. Soc.*, **33**, 715 (1988).

[41] P. A. BENNETT, A. P. JOHNSON and B. N. HALAWITH: *Phys. Rev. B*, **37**, 4268 (1988).

[42] T. M. HATTOX, J. B. CONKLIN, J. C. SLATER and S. B. TRICKEY: *J. Phys. Chem. Solids*, **34**, 1627 (1973).

[43] G. ALLAN: *Phys. Rev. B*, **19**, 4774 (1979).

[44] D. R. GREMPEL and S. C. YING: *Phys. Rev. Lett.*, **45**, 1018 (1980).

[45] G. YOKOYAMA, H. HIRASHITA, T. OGUCHI, T. KAMBAYA and K. I. GONDAIRA: *J. Phys. F*, **11**, 1463 (1981).

[46] S. OHNISHI, C. L. FU and A. J. FREEMAN: *J. Mag. Magn. Mater.*, **50**, 161 (1985).

[47] H. AKOH and A. TASAKI: *J. Phys. Soc. Jpn.*, **42**, 791 (1977).

[48] J. S. FORD, A. P. C. REED and R. M. LAMBERT: *Surf. Sci.*, **129**, 79 (1983).

[49] V. JENSEN, J. N. ANDERSEN, H. B. NIELSEN and D. L. ADAMS: *Surf. Sci.*, **116**, 66 (1984).

[50] O. JEPSEN, J. MADSEN and O. K. ANDERSEN: *Phys. Rev. B*, **26**, 2790 (1982).

[51] L. N. LIEBERMANN, J. CLINTON, P. M. EDWARDS and J. MATHON: *Phys. Rev. Lett.*, **25**, 232 (1970).

[52] S. OHNISHI, A. J. FREEMAN and M. WEINERT: *Phys. Rev. B*, **28**, 6741 (1983).

[53] E. WIMMER, A. J. FREEMAN and H. KRAKAUER: *Phys. Rev. B*, **30**, 3113 (1984).

[54] A. J. FREEMAN, C. L. FU, S. OHNISHI and M. WEINERT: in *Polarized Electrons in Surface Physics*, edited by R. FEDER (World Scientific, Singapore, 1985), p. 3.

[55] R. RICHTER, J. G. GAY and J. R. SMITH: *Phys. Rev. Lett.*, **54**, 2704 (1985).

Properties of Synchrotron Radiation and Instrumentation for Photoemission and Absorption Spectroscopy.

C. KUNZ

II. Institut für Experimentalphysik der Universität Hamburg
Luruper Chaussee 149, D-2000 Hamburg 50, B.R.D.

1. – Introduction.

The application of synchrotron radiation from electron storage rings to photoemission and absorption spectroscopy of solids and interfaces is one of the most important experimental developments in recent years for characterizing electronic states. In several cases even contributions to the clarification of the geometrical arrangement of atoms were obtained by this technique and more of this is expected to occur in the future.

The spectrum of synchrotron radiation covers a very wide range of photon energies extending from the visible over the vacuum ultraviolet to the soft and hard X-ray regions. Due to the high cross-section for photoabsorption at lower photon energies most photoemission investigations up to date use photon energies below 1000 eV photon energy. Terms as «vacuum ultraviolet» (VUV), «soft X-ray» (SX), «extreme ultraviolet» (XUV) are used to characterize this spectral regime, but it should be noted that the borderline energies are not well defined due to the historical origin of these notations. In addition, in photoemission experiments the necessary resolution in the order of $(10 \div 100)$ meV forbids excitation of valence and outer core electrons to too high kinetic energies.

Monochromators for synchrotron radiation (SR) in the energy range up to 1000 eV are almost uniquely based on reflection gratings. Nevertheless, interesting developments occur where new components like, *e.g.*, crystals with wide lattice spacings and multilayers penetrate into this energy range. The other interesting frontier is the extension of photoemission spectroscopy to energies above 1000 eV into the X-ray region proper. Synchrotron radiation originally was used only from bending magnets in synchrotrons and storage rings. These machines were built and optimized for high-energy physics. In recent years a rapid development took place in designing storage rings dedicated to synchrotron radiation work. One optimization parameter in such a machine is the source size and source divergence, the product of which is closely related to the so-called «electron beam emittance». The other optimization goal is the provision of as many as possible straight sections around the ring which are long enough and especially tailored with respect to the electron optics to accommodate insertion devices. These are wigglers and undulators, sections with alternating magnetic fields where the electron beam is forced on a sinusoidal path. The overall intensity gain is given by the number of bends N in a wiggler, while through interference a factor of up to N^2 is gained in peaks of the spectral distribution (undulator effect).

These devices need monochromators with special components in certain places in order to stand, $e.g.$, the great power load and in order to take full advantage of the gain in intensity and more specifically in brilliance. Brilliance is the important parameter by which the quality of the new dedicated storage rings and the insertion devices is judged. Brilliance is the number of photons per given band width per unit source size and unit solid angle. With hypothetically perfect optical elements (focusing mirrors, etc.) brilliance is an invariant when the radiation is guided from the source through the monochromator onto the sample. Building improved optical elements (or inventing monochromator mountings which make optimum use of technically feasible elements) is another frontier in the field.

The enormous brilliance generated by undulators in dedicated storage rings with very small emittances can be used advantageously in many different ways. Quite obvious are: better spectral resolution, better energy resolution of the emitted photoelectrons, better accuracy through better photon statistics in order to determine small changes in spectra due to modulation of the sample properties (temperature, laser excitation, etc.) or due to oxidation, gas adsorption, interface reactions, surface phase transitions, etc. This requires other properties of the new dedicated storage rings which always are taken as granted on paper but practically are fulfilled only to a certain degree, these are: stability, reproducibility of their properties over long periods and reliability. These properties have to be fulfilled by the complete experimental station. A continuous battle needs to be fought there and unfortunately only the defeats are noted and criticized, while the victories go without notice. Certainly the users of synchrotron radiation have to learn that they deal with a light source of unprecedented complexity.

In addition to photoemission other secondary processes like photostimulated desorption and fluorescence will also greatly benefit from the increase in intensity on the sample, but they are not within the scope of this lecture. In this context it should be remembered that absorption spectroscopy is not only performed by transmission measurements on thin samples but can also be done in a photoemission measurement. The total photoelectric yield from a solid sample displays all the features of the absorption coefficient, since the yield is proportional to the power absorbed in a thin surface layer of $(20 \div 50)\,\text{Å}$ thickness. There is one direction which will profit from higher brightness more than anything else and this is scanning photoabsorption and photoemission with high spatial resolution. The goal is to focus the radiation down to a point on the sample with $0.1\,\mu\text{m}$ diameter or less with Fresnel zone plates or mirrors. Sufficiently high intensities to allow selection of specific electron energies with an electron energy analyser are required. Alternatively or additionally spin selection would allow the investigation of magnetic domains at the surface of ferromagnetic materials. Such a microprobe should reveal structures in a very easily interpretable way. Element specific contrast should be obtainable if photon and electron energies could be varied at choice.

The material included in this lecture is selected in such a way that emphasis is given to developments which point to the future. Examples are selected predominantly from projects with which the author himself was connected in one or another way. The past developments in the field of synchrotron radiation are covered in several reviews of which only three are cited here [1-3]. There is also available a very useful collection of equations and data [4]. The instrumentation of synchrotron radiation is regularly presented in a series of conference proceedings [5-17]. A full survey of *all* new developments is not within the scope of this lecture. It is anyhow difficult since several recent advances will be spared for presentation at the Conference on Synchrotron Radiation Instrumentation, SRI 88 to be held in Tsukuba, Japan, *August 29 - September 2, 1988*. The results will be published in a special issue of the *Review of Scientific Instruments* soon thereafter [14].

In the following the material is arranged in three sections. In sect. 2 the properties of SR are summarized including the main properties of wigglers and undulators. In sect. 3 the typical standard instrumentation for spectroscopy and an advanced instrument on a wiggler/undulator is presented with the FLIPPER monochromator at the W1 undulator at HASYLAB as a specific and operational example. Section 4 finally is devoted to the discussion of the present attempts and directions to obtain as much intensity as possible in a microfocus. This should allow to make photoemission experiments on microscopic samples and would allow a microscopy with element specific contrast.

2. – Properties of synchrotron radiation.

2`1. *Radiation from bending magnets.* – The properties of SR can be calculated by applying the methods of classical electrodynamics to relativistic electrons (or positrons) on circular orbits. The dipole characteristic is relativistically distorted in the laboratory frame into a narrow cone according to fig. 1. The width of the cone is in the order of $\gamma^{-1} = 1 - \beta^2$, where βc is the particle velocity, c the velocity of light and γ the ratio of the particle energy relative to the electron rest mass energy.

Already SCHOTT[18] treated this problem in connection with classical models of the atom at the beginning of the century, IVANENKO and POMERANCHUK[19], followed by SCHWINGER[20], were the first who predicted its importance for circular particle accelerators. Now the main equations are derived in standard textbooks, like those of SOMMERFELD[21] and JACKSON[22].

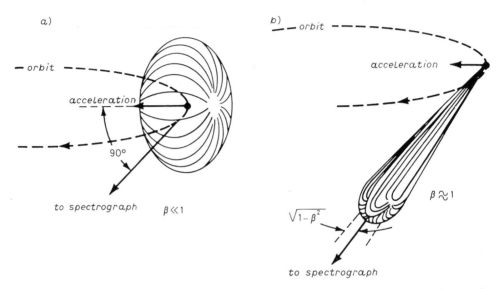

Fig. 1. – Angular distribution of synchrotron radiation a) as emitted in the rest frame of the electron, b) as emitted in the laboratory frame. (From ref.[23].)

The properties of SR are listed as follows:

1) continuous spectrum from the infrared to the region of hard X-rays;

2) strong collimation in the instantaneous direction of flight (typically 1 mrad);

3) linear polarization in the plane of the orbit;

4) circular polarization in the «wings» above and below the plane of the orbit;

5) pronounced time structure, which is a copy of the pulse structure of the electron beam (pulses as short as 100 ps);

6) absolute calculability of all the properties of the source, once parameters of the particle beam are given;

7) cleanliness and stability (in particular with storage rings) of the source, which, in contrast to gas discharge sources, exists in an extremely good vacuum.

The flux of radiation is shown for BESSY I and the planned storage ring BESSY II in fig. 2 [24]. The thus marked curves give the number of photons/s in the $\Delta\varepsilon/\varepsilon = 10^{-3}$ or $\Delta\lambda/\lambda = 10^{-3}$ bandwidth in a 1 mrad wide horizontal segment while vertically intensity is fully integrated. ε and λ are photon energy and wavelength, respectively. The assumed currents $j = 300$ mA (BESSY I) and $j = 100$ mA (BESSY II) explain most of the difference in the low-photon-energy region. The fall-off at high energies is determined by the characteristic energy ε_c (or wavelength λ_c). It depends on the radius of curvature R, the magnetic field B and the particle energy E.

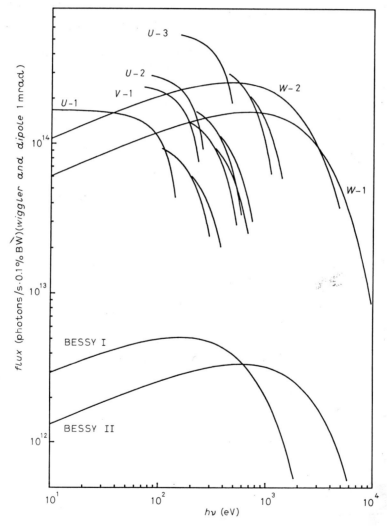

Fig. 2. – Flux of photons emitted from the bending magnets of BESSY I (300 mA) and II (100 mA) and from the planned wigglers W and U of BESSY II. (From ref. [24].)

The vertical angular spread $\Delta\psi$(FWHM) of intensity can be approximated as (see, *e.g.*, [25])

(2.1) $\Delta\psi \approx (2/\gamma) \cdot \{\varepsilon_c/\varepsilon\}^{1/3}$ for $\varepsilon \ll \varepsilon_c$,

(2.2) $\Delta\psi \approx (2/\gamma) \cdot \{\varepsilon_c/3\varepsilon\}^{1/2}$ for $\varepsilon \gg \varepsilon_c$.

Useful equations are:

(2.3) $B[\mathrm{T}] \cdot R[\mathrm{m}] = 3.34\,E[\mathrm{GeV}]$,

(2.4) $\gamma = 1957\,E[\mathrm{GeV}]$,
A2
(2.5) $\varepsilon_c[\mathrm{eV}] = 2218\,E^3[\mathrm{GeV}]/R[\mathrm{m}] = 665.1\,B[\mathrm{T}]\,E^2[\mathrm{GeV}]$,

(2.6) $\lambda_c[\text{Å}] = 5.59\,R[\mathrm{m}]/E^3[\mathrm{GeV}] = 18.64/B[\mathrm{T}]\,E^2[\mathrm{GeV}]$,

(2.7) $\varepsilon_c[\mathrm{eV}] = 12400/\lambda_c[\text{Å}]$,

(2.8) $I[\mathrm{phot}/(\mathrm{s} \cdot \mathrm{mrad} \cdot 0.1\%\,\mathrm{BW})] \approx 4.5 \cdot 10^9\,R^{1/3}[\mathrm{m}]\,\varepsilon^{1/3}[\mathrm{eV}]\,j[\mathrm{mA}]$ for $\varepsilon \ll \varepsilon_c$.

The brilliance $\eta(\psi, \varepsilon)$ is defined as the number of photons per second, per 0.1% bandwidth, per unit area A of the source (which is the cross-section of the electron beam) and per solid angle. In contrast to many classical sources $\eta(\psi, \varepsilon)$ is a very anisotropic quantity. The maximum brilliance in the plane of the orbit is given from eqs. (2.8) and (2.1) by

(2.9) $\eta(0, \varepsilon) \propto j\,R^{2/3}\,A^{-1}\,\varepsilon^{2/3}$ for $\varepsilon \ll \varepsilon_c$.

This quantity is plotted for BESSY I and BESSY II in fig. 3. It is easily recognized that the smaller source size A in BESSY II more than compensates for the lower current. The cross-section of the electron beam around the orbit (see, *e.g.*, [26]) is variable and characterized by two quantities: the position (Z)-dependent β-functions $\beta_x(Z)$ and $\beta_y(Z)$ and the invariable horizontal and vertical emittances of the storage ring ε_x and ε_y. With 5% coupling the vertical emittance would be $\varepsilon_y = 0.05\,\varepsilon_x$. For a given storage ring lattice the emittance varies with beam energy like E^2. Electrostatic and electromagnetic interactions of the particles within the beam will usually lead to an increase of $\varepsilon_{x,y}$ with j. The beam sizes (FWHM) are given by $2.35\,\sigma_{x,y}$ and are calculated according to

(2.10) $\sigma_x = \sqrt{\varepsilon_x\beta_x(Z)}$.

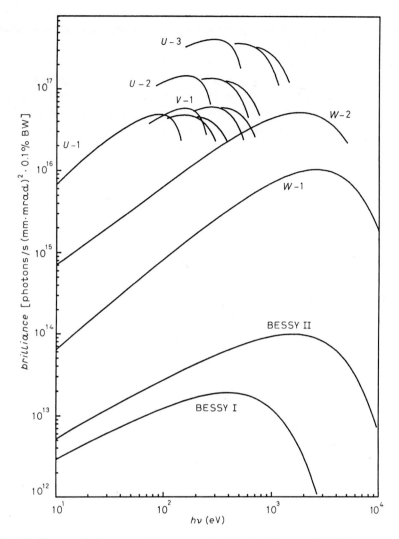

Fig. 3. – Brilliance of the same elements as in fig. 2. (From ref. [24].)

The beam divergence is given by

$$(2.11) \qquad \sigma_{x'} = \sqrt{\varepsilon_x/\beta_x(Z)} \, \sqrt{1 + \beta_x'(Z)^2/4},$$

$\beta_x'(Z)$ is the derivative of $\beta_x(Z)$ with respect to Z. Due to the second square root term $\sigma_x \cdot \sigma_{x'} = \varepsilon_x$ holds only at waists or bellies of the «envelope function», where $\beta'(Z) = 0$.

Due to this function and due to another one which is of lesser influence, the dispersion function $D(Z)$, the beam size and divergence can be varied at different positions around the orbit by the designers of storage rings according to the needs of the experiments.

The variation is possible within wide boundaries, but the general rule is that σ_x and $\sigma_{x'}$ are coupled through eqs. (2.10) and (2.11) and change in opposite directions. The size of the electron beam determines the source size and its divergence has to be convoluted with the natural divergence of synchrotron radiation. This does not only reduce the brilliance of the source but also mixes the different types of polarization and thereby reduces the actual degrees of polarization.

More information on the design and optimization of storage rings is now available in several reviews [7, 9, 11].

2'2. *Radiation from wigglers and undulators.* – Figure 4 shows different types of insertion devices [26] for which straight sections with lengths between 2 and 6 m will be provided in the next generation of dedicated storage rings (ALS (Berkeley), Sincrotrone Trieste (Triest), BESSY II (Berlin)). In these machines the electron beam optics is matched at the beginning and end of the straight sections in such a way that additional magnetic lenses (quadrupoles) can shape the β-function at the insertion devices in a large variety of ways without otherwise affecting the storage ring. Variations of the parameters of the inser-

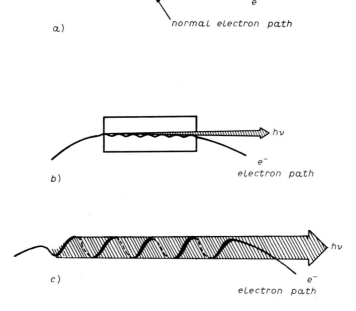

Fig. 4. – Three different types of insertion devices: *a*) wavelength shifter, *b*) multipole wiggler magnet, *c*) helical wiggler. (From ref. [25].)

tion devices like, *e.g.*, magnetic fields should also be decoupled from the rest of the storage ring. Up to now these are design goals for which solutions appear to exist in theory, not much practical experience has yet to be gained, however.

A wavelength shifter (fig. 4), typically a superconducting magnet at high magnetic field, is a device in which high photon energies are produced in low-energy storage rings according to eq. (2.5). Presently such devices are operated at the ADONE ring (Frascati), at the SRS (Daresbury), at the NSLS (Brook-haven), at the PHOTON FACTORY (Tsukuba), at Novosibirsk (USSR) and elsewhere.

A multipole wiggler deflects the beam to a sinusoidal path in a plane, while a helical wiggler generates a spiral path (see fig. 4). They need specially shaped magnetic configurations. In the first case linearly polarized radiation is produced, while the helical wiggler produces circular polarization. Circular polarization is of great interest in the investigation of magnetic phenomena and can successfully be combined with electron spin analysis.

A multipole wiggler produces N times the flux of a bending magnet if the peak magnetic field of the wiggler is the same as that of the bending magnet. More accurately a superposition of contributions from all points on the trajectory has to be calculated and results in a narrow angular distribution around the symmetry line of the wiggler. It turns out, however, to be not just a superposition of intensities but rather of amplitudes which interfere. Although typical period lengths λ_0 of wigglers are macroscopic and are in the order of $\lambda_0 = (2 \div 20)$ cm, the whole structure displays strong interference patterns for wavelengths down to $\lambda = 1$ Å. This undulator effect is explained in fig. 5.

There are three contributions to the phase shift between light and electrons while both travel a period length λ_0 of the wiggler/undulator. The sum of the three path differences l_i has to fulfil the condition of constructive interference

$$(2.12) \qquad\qquad l_1 + l_2 + l_3 = m\lambda.$$

Due to the small difference of the speed of the electron v to the speed of light c even if both travel on a straight line there is a shift l_1 given by

$$(2.13) \qquad \frac{l_1}{\lambda_0} = \frac{c - v}{c} \approx \frac{1}{2}\left(1 - \left(\frac{v}{c}\right)^2\right) = \frac{1}{2}\gamma^{-2}, \qquad l_1 = \frac{\lambda_0}{2\gamma^2}.$$

The second contribution l_2 arises from the detour which the electron makes on its curved path compared to a straight line. This path integral is readily evaluated for a sinusoidal path, it depends, of course, on the maximum magnetic field and on the electron energy parameter γ.

The result is

$$(2.14) \qquad\qquad l_2 = \frac{\lambda_0 K^2}{4\gamma^2},$$

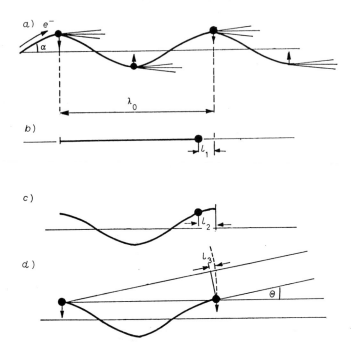

Fig. 5. – The undulator with period length λ_0 shown in a) fulfils the undulator condition whenever the sum of the three path length differences l_1, l_2 and l_3 explained in b)-d) is equal to a multiple of the photon wavelength λ. The angle α is relevant in eq. (2.15).

where K is called the undulator parameter

$$(2.15) \qquad\qquad K = 93.4\, B[\mathrm{T}]\,\lambda_0[\mathrm{m}] = \alpha\gamma$$

with α the angle of intersection of the sine wave and the symmetry line (see fig. 5).

If radiation from the undulator is observed from a direction which is at an angle θ with the symmetry line (this angle may have any azimuth), the typical path length change encountered in observing wave fronts originating from points λ_0 apart is

$$\lambda_0 - l_3 = \lambda_0 \cos\theta \approx \lambda_0 - \frac{\lambda_0}{2}\theta^2.$$

This yields

$$(2.16) \qquad\qquad l_3 = \frac{\lambda_0}{2}\theta^2.$$

From eqs. (2.12) to (2.16) we obtain the positions of the interference peaks in the spectrum as

(2.17)
$$\lambda_m = \frac{1}{m} \frac{\lambda_0}{2\gamma^2}\left(1 + \frac{K^2}{2} + \gamma^2\Theta^2\right).$$

Comparing the corresponding undulator peak energies with ε_c from eq. (2.5) we obtain

(2.18)
$$\frac{\varepsilon_m}{\varepsilon_c} = 1.33 \frac{m}{K\left(1 + \dfrac{K^2}{2}\right)}.$$

If $\varepsilon_m/\varepsilon_c$ is much larger than 1, there will not be much intensity in the undulator peak. An exact evaluation shows that the maximum intensities in the $m = 1, 3, 5, 7$ harmonics occur for $K = 1.2, 1.9, 2.3, 2.6$, respectively.

The quantitative theory of undulator radiation has been worked out in detail and allows an exact prediction of the properties of a specific design. Many features are already easily derived from the model presented here:

1) There is no intensity in even harmonics along the axis ($\Theta = 0$). The reason is the antisymmetric emission within one «unit cell». The left wiggle emits radiation with opposite phase to the right wiggle which provides a N^2 dependence (N is the number of poles!) of the intensity in odd orders, but destructive interference for even orders.

2) Defining an in-plane angle ψ and an out-of-plane angle θ perpendicular to the plane of the linear wiggler such that $\Theta^2 = \psi^2 + \theta^2$, the antisymmetry of the unit cell mentioned under 1) is broken in the ψ-direction but not in the θ-direction. Even orders will appear only in the ψ-direction.

3) In analogy to other interference phenomena for a specific wavelength λ_m cancellation of intensity occurs if the phase shift of radiation from both ends of the undulator is 2π. For a fixed wavelength λ_m this occurs in analogy to eq. (2.16) at an angle $\Theta_0^2 L/2 = \lambda_m$, where L is the length of the undulator. Therefore,

(2.19)
$$\Theta_0 \approx \sqrt{2\lambda_m/L}$$

is roughly the width of the central maximum. For $\lambda_m = 20\,\text{Å}$ and $L = 4\,\text{m}$ we obtain $\Theta_0 \approx 3 \cdot 10^{-5}\,\text{rad}$, which demonstrates the need of high angular resolution to resolve the angular distribution of undulator radiation.

4) The same consideration as used in deriving eq. (2.19) at a fixed angle Θ yields a minimum in the intensity if $(\lambda - \lambda_m)\, m \cdot N/2 = \lambda_m$, $\Delta\lambda = \lambda - \lambda_m$. This gives for the width of the undulator lines

(2.20)
$$\Delta\lambda/\lambda = 2/N \cdot m = \lambda_0/L \cdot m.$$

This, *e.g.*, means that in first ($m = 1$) order the relative spectral resolution is the reciprocal of the number of periods.

5) Most of the present-day undulator structures are built from periodic arrangements of permanent magnets, which are glued to two bars mounted (mostly outside the vacuum chamber) on both sides of the electron beam. By varying the wiggler gap also B is varied and thus K. This is the way to tune the undulator harmonics to different wavelengths. In the new storage rings it will be attempted to vary K synchronously with the wavelength scan of the mono-chromators by microcomputer control. Thus the tuning to the maximum intensity is always maintained.

6) If the radiation pattern of an undulator is integrated over a wide angular acceptance in ψ and θ, or if the angular spread of the electron beam is very large, there is also intensity between the undulator peaks. This is due to the $\gamma^2 \theta^2$ term in eq. (2.18). Nevertheless, sharp structures in the spectrum persist since the θ^2 shift is always directed to longer wavelengths. Such a spectrum is shown in fig. 6.

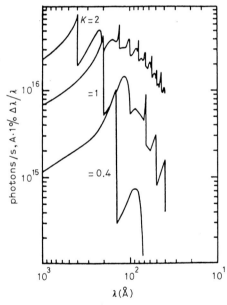

Fig. 6. – Undulator spectra integrated over all angles for different values of the undulator parameter K. (From ref. [26].)

7) For the design of the first optical element in a wiggler or undulator beam line the total emitted power is of great importance. This power P is given for wigglers and undulators likewise by

(2.21) $$P[\mathrm{kW}] = 0.633\, E^2[\mathrm{GeV}]\, B_0^2[\mathrm{T}]\, L[\mathrm{m}]\, j[\mathrm{A}].$$

8) The polarization of an undulator is linear since the circular polarization, which changes from left hand to right hand above and below the orbital plane of a bending magnet, disappears with the simultaneous presence of right-hand and left-hand bends in an undulator. The direction of linear polarization is horizontal in the undulator plane but changes direction outside the plane.

Several attempts have been made to produce circular polarization. I only want to mention one recent idea which was tested already successfully in the visible.

Figure 7 shows the arrangement originally proposed by ONUKI [27]. Later a prototype was built and tested [28]. The device consists of a crossed pair of linear undulators AA and BB. The gap of each pair can be tuned from 57 to 146 mm.

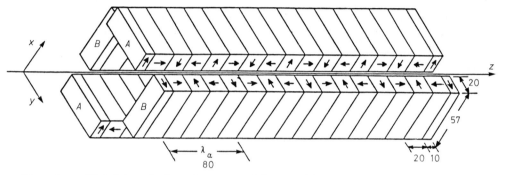

Fig. 7. – Undulator design for an undulator with variable left-hand/right-hand circular/elliptic polarization of any degree. The two sets of magnets AA and BB can be shifted with respect to each other by one period, the whole unit can be rotated around the beam. (From ref. [28].)

The relative phase α of the two pairs of magnet structures could be shifted mechanically between $-\pi/2 \leqslant \alpha \leqslant +\pi/2$, thus forcing the electron beam into a right-hand and a left-hand spiral at the extremes. With $\alpha = 0$ a linear undulator is realized. This undulator produces light with any degree of elliptical polarization. It may well be necessary to produce elliptical degrees of polarization in order to obtain circular polarization at the exit slit of the monochromator. The performance in the visible was in agreement with theory at the 600 MeV storage ring TERAS. The peak magnetic fields were 0.15 T in the helical field configuration and 0.21 T in the plane field configuration. The authors managed to modulate the polarization with a motor drive at a rate of 3 Hz. This arrangement appears to be a promising device.

There are other ideas under discussion for the production of circular polarization. In one case [29] the two perpendicular plane-field undulators are placed one behind the other and the phase shift is imposed on the electron beam in a single controllable wiggler between the two undulators. The other idea which is especially considered for wigglers in the hard-X-ray regime is to build a device with, say, high-field left-hand bends and low-field right-hand bends [30]. Of course, the field integral as in any other wiggler too has to be zero. Then, in the same way as with bending magnets, right- and left-hand elliptic polarization is obtained above and below the wiggler plane for the high-energy end of the spectrum.

Finally I want to point out that using a storage ring with a higher beam energy E (or γ) makes the design of wigglers and undulators much simpler. In order to achieve a certain short wavelength λ according to eq. (2.17), great efforts need to be made if γ is too small. If an undulator optimized for $m = 1$ is desired, $K = 1.2$ is fixed. As a consequence, λ_0 is determined and then B through eq. (2.15). The reduction in λ_0 requires increased fields B. Looking into the details of undulator design [31] it turns out that the magnet gap needs to decrease proportionally to λ_0 if the field B should be kept constant, but the gap needs to decrease faster than λ_0 if the field needs to grow proportionally to $1/\lambda_0$. From this requirement there follows a technical limit to the minimum λ_1 and maximum ε_1 achievable with a certain storage ring. Flux and brilliance of wigglers ($K \gg 1$) and undulators designed for BESSY II are shown in fig. 2 and 3.

3. – Monochromators and experimental stations.

3`1. – The main advantage of synchrotron radiation, namely its tunability over a very wide range of photon energies, is also the reason of one of its greatest problems in practice. Typically an energy band of 0.1 eV needs to be filtered out of a continuous spectrum $(1000 \div 50000)$ eV wide. This is done by diffracting elements, usually optical reflection gratings and crystals. In addition to the wanted radiation in the diffraction peak, unwanted radiation from surface scattering and higher-order harmonic diffraction penetrates the exit slit of the monochromator. While the problem of scattered radiation is somewhat relieved by using undulators in the first harmonic, tunable undulators which are scanned over a wide photon energy range enhance higher harmonics with increasing values of K more than the fundamental peak. At the beginning of experimental work with synchrotron radiation more than 25 years ago the problem of suppressing false light was considered to be so severe that many spectroscopists considered it to be not solvable. This is one of the reasons of the slow acceptance of synchrotron radiation in the community at that time. Even nowadays the problem is far from being solved satisfactorily and in many cases experimentalists are happy to obtain a $(90 \div 95)\%$ spectral purity with their monochromators. Indeed, in photoemission experiments the electron energy analysis and the tuning of the exciting radiation allow one to identify peaks in spite of this spectral impurity. Whenever intensity measurements are required on an absolute basis, serious problems arise.

The typical high-performance photoemission station at a bending magnet port of one the present-day high-performance storage rings like, *e.g.*, BESSY in Berlin or the NSLS in Brookhaven consists of one or several combined toroidal grating monochromators (TGMs) providing up to 10^{12} photons/s in the peak of the spectral range in a resolution interval of $\approx 0.2\,\text{eV}$. The spot on the sample is typically 1 mm wide or somewhat less. This spot size is matched to the acceptance of typical electron spectrometers (cylindrical mirror analyser, spherical mirror analyser or display-type analyser with wide-angle acceptance and multichannel read-out).

TGMs scan by rotating a toroidal grating with fixed entrance and exit slits (see fig. 8). They rapidly go out of focus and thus are restricted to a finite energy

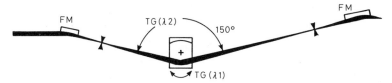

Fig. 8. – Layout of a toroidal grating monochromator with two interchangeable gratings TG(λ1), TG(λ2). FM are focusing mirrors. (From ref. [32].)

range. Nonetheless, the simplicity of the mounting, the nearly stigmatic imaging and the moderate requirements of resolution in solid-state spectroscopy gave the TGM a wide distribution after the Jobin-Yvon Company succeeded to put grating structures holographically on toroidal surfaces commercially. The first monochromator of this type was built in the late 70's [32] and installed at the ACO storage ring. It is this type of monochromator which has nowadays the widest distribution, but nevertheless appears not to be the instrument to fulfil the requirements at undulator lines in the future. It should be mentioned that there is recent progress with designs using only plane and spherical optical elements [33] and with plane gratings with variable grating constant [34] to design high-resolution monochromators which do not suffer from the imperfections of nonspherical optical elements any longer.

In the following I am not going to give a complete review of all the different monochromator mountings which have been designed and realized in this field, there exist good surveys and reviews in the literature [35-37]. I am also not reviewing the different types and arrangements of electron spectrometers which are in existence. These are altogether not typical for work with synchrotron radiation. They could be used or are actually used as well in combination with classical fixed photon energy sources. The rest of this section is devoted to the description of the FLIPPER station at the W1 undulator at HASYLAB. In this context those problems which have to be solved for the undulator/wiggler stations at the future dedicated storage rings like the ALS in Berkeley, the Sincrotrone Trieste and the BESSY II in Berlin are exemplified.

3˙2. *Monochromators at undulators.* – The FLIPPER monochromator at HASYLAB [38] is a prototype instrument covering almost the full range of interest in this context with the capability of suppressing higher-order light by taking advantage of the critical angle of total reflection. As a consequence of this special design the angular acceptance is fairly restricted, which is, however, no disadvantage in combination with an undulator.

The FLIPPER [39, 40] was successfully operated at a bending magnet beam line for almost a decade. Its optical design originates from the GLEISPIMO mounting invented at DESY in 1968 [41, 42] and was followed in the meantime by other similar designs, namely the SX 700 at BESSY [43] and the BUMBLE BEE at HASYLAB [44]. Six alternatively used plane mirrors $S_1 \ldots S_6$ (see fig. 9) cover

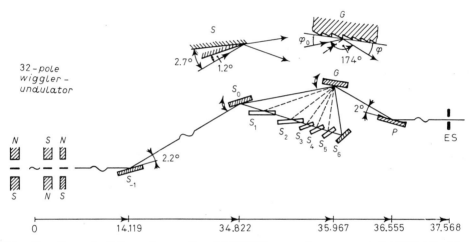

Fig. 9. – Layout of the order sorting FLIPPER monochromator at HASYLAB. The mirror S_{-1} can be inserted into the white beam which otherwise serves an X-ray wiggler station. The grating G is either illuminated directly by mirror S_0 or by one of the mirrors S_1-S_6. The paraboloid P focuses monochromatic radiation into the exit slit ES. (From ref. [38].) The scale at the bottom gives distances from the source in metres.

the energy range 15 to 450 eV. An additional mirror S_0 which is illuminated at 1.2° grazing angle was added for the set-up at the wiggler line and covers the energy range up to above 1000 eV. As a matter of fact the Al K-edge at 1560 eV can be observed in absorption. There the level of «good» light roughly equals that of stray light. A plane grating (1.5° blaze, 1200 lines/mm) disperses the prefiltered radiation and a paraboloid illuminated at now 2° grazing angle (formerly at 5.7°) focuses a beam of fixed direction into the exit slit. An experimental chamber suitable for photoemission and other surface-sensitive techniques is mounted behind the exit slit.

When the installation of a 32-pole wiggler [45] was planned at the 5.3 GeV storage ring DORIS, it was decided to split off a VUV beam with a 2.2° grazing-

TABLE I. – *Parameters of the* HASYLAB *wiggler/undulator* W1. Gap $= W$, maximum field on orbit $= B$, photon energy of first-order undulator radiation $= h\nu_1$, total emitted power $= P$.

Gap W (mm)	B (T)	K	3.7 GeV $h\nu_1$(eV)	5.3 GeV P(W)	3.7 GeV 100 mA P(W)	5.3 mA 50 mA P(W)
33.8	0.567	7.10	36.2	74.3	609	625
40.0	0.493	6.08	48.7	99.9	446	458
50.0	0.382	4.71	78.5	161	268	275
60.0	0.295	3.64	124	254	160	164
70.0	0.229	2.82	191	392	96	99
80.0	0.176	2.17	283	581	57	59
90.0	0.137	1.69	391	802	35	36
100	0.104	1.28	522	1071	20	21
120	0.063	0.777	729	1496	7.3	8
140	0.036	0.444	864	1773	2.4	2.5
160	0.022	0.271	916	1880	0.9	0.9
168	0.017	0.210	929	1906	0.5	0.6

incidence mirror alternatively to the X-ray beam. The FLIPPER mono-chromator with its angular acceptance of roughly 0.2 mrad vertically and 0.8 mrad horizontally almost perfectly matches the emission cone of this special wiggler W1. Furthermore, the undulator peak in first order of the HASYLAB wiggler can be tuned by the gap height over the whole energy range of the FLIPPER monochromator (see table I).

It is interesting to note that wigglers used in high-energy storage rings (from 2 GeV upwards) are not only excellent sources of X-rays, but as undulators also deliver more photons in a small solid angle (*i.e.* more brightness) in the vacuum ultraviolet than wigglers/undulators installed in low-energy storage rings. Moreover, much simpler technical solutions are possible and the influence on the storage ring is small. There are, however, some problems involved in using high-energy machines like DORIS in the VUV and SX regions which will not exists at smaller machines like, *e.g.*, BESSY II.

One of the problems is radiation safety due to the hard X-rays generated by nearby bending magnets in the storage ring tunnel. This problem was solved at the W1 station by inserting a first plane deflection mirror S_{-1} (see in fig. 9) and by using large beam paths, which are required by the FLIPPER mounting anyhow. There is a long-standing experience at DESY/HASYLAB in coping with such problems.

Another more severe problem was detected only during the continuous operation of this beam line over the past few years. The W1 wiggler is preceded and followed in the 10 m long straight section by the fringe field of two bending magnets and by several quadrupole magnets which focus the electron beam. A well-aligned quadrupole magnet has zero field on axis, but the wings of the beam extend already into regions of finite magnetic fields (in the order of 0.01 T). A beam which is not properly aligned, and this may occur frequently in actual storage ring operation, could be subjected to even larger fields. From eq. (2.5) it follows that $\varepsilon_c \approx 200\,\text{eV}$ at $E = 5.3\,\text{GeV}$ and $B = 0.01\,\text{T}$. As a consequence a background of not very well controlled radiation is superimposed on the undulator spectrum. With storage rings at lower energy this is not such a severe problem since quadrupoles are operated at magnetic-field gradients proportional to E. As a consequence $\varepsilon_c \propto E^3$ and ε_c thus decreases rapidly with energy.

A critical element for a successful use of a VUV undulator beam from a high-energy storage ring is the first mirror S_{-1}. It has to take a heat load of up to 600 W when the wiggler gap is completely closed (see table I). This power is concentrated in an angular interval of roughly 0.7 mrad horizontally and 0.25 mrad vertically taking into consideration an angular divergence of the electron beam of $0.5 \times 0.07\,\text{mrad}^2$ (FWHM) at the source point. Due to the grazing angle of 2.2° most of the power is distributed over a strip of $10 \times 100\,\text{mm}^2$ on the surface.

The mirror itself is a copper block with dimensions $200 \times 50 \times 50\,\text{mm}^3$ covered with Kanigen and polished to high accuracy (90% of reflected intensity in an angular interval of $\pm 2'' = \pm 10\,\mu\text{rad}$). It is pressed on a water-cooled support with a layer of indium in between providing good thermal contact. The rise in temperature of the mirror block is only some ten degrees measured about half-way between the irradiated surface and the cooled bottom. In order to keep the temperature of this mirror low, the Kanigen coating of the backside had to be removed partly for better thermal contact. Without this measure the temperature rose up to 200 °C. But even in the new configuration we observe large effects on the intensity of the monochromator output when the wiggler gap is small and thus the power high. We found that these losses originate from mirror S_{-1}. A drop in intensity of up to a factor of two is observed in the first 20 s after opening the beam stop. The performance of the monochromator appears to be otherwise unaffected.

From a series of investigations on test mirrors [46] it is understood now that the surface of the mirror reversibly develops a bump due to a large surface temperature decreasing rapidly with a high gradient into the inside of the copper block. Due to the elongated shape of the illuminated part of the surface this bump will act as a cylindrical convex horizontally defocusing mirror which deflects part of the incident radiation in such a way that it is lost at the horizontal apertures. Any additional horizontal divergence added this way to the beam does not pose serious problems to the monochromatization, since the spectral resolution of the monochromator depends mainly on vertical imaging. Such distortions of heavily loaded mirrors have been treated theoretically [47, 48] (see fig. 10).

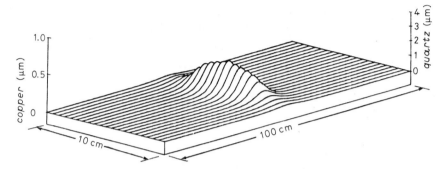

Fig. 10. – Calculated surface deformation by illuminating a copper and a quartz mirror with a wiggler beam at 2° grazing incidence. (From ref. [48].)

It is possible to avoid this effect by operating the wiggler at large gaps. The heat load on the mirror S_{-1} is then drastically reduced. The wiggler still provides a gain of roughly a factor of 40 compared to a bending magnet. The undulator peak in first order, however, is shifted to 929 eV at 3.7 GeV and $W = 168$ mm (see table I). With the completely closed gap, $W = 34$ mm, the undulator peak is lowered in energy to 36 eV at 3.7 GeV. As will be shown below (see fig. 11), the gain of intensity in the undulator peaks can be appreciable. It can, nevertheless, be advantageous for low photon energies at high currents in the storage ring to sacrifice this option in order to avoid the distortion of mirror S_{-1} and the subsequent losses.

This problem may not be unique to undulators operating at high-energy storage rings. In future rings dedicated to the VUV-SX region E will be lower than that of DORIS, but tunability over large spectral regions inevitably leads to large values of K, which together with the incresing length of undulators will lead to high power loads according to eq. (2.21). Moreover, with the much smaller divergences of the electron beam the power will be concentrated on a much smaller spot on the surface. The only advantage will be the absence of unwanted hard X-rays. It will become necessary to find ways for better cooling the mirrors, to find materials with high heat conductivity and low thermal expansion [48], and to distribute the heat load over larger areas on the mirror surface by going to larger distances or to apply even shallower angles of

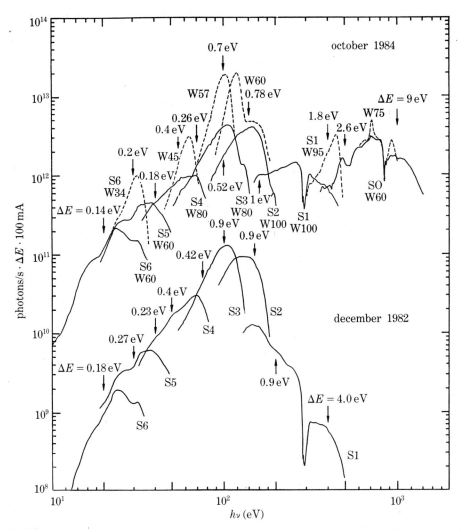

Fig. 11. – The output of the FLIPPER monochromator of fig. 9 with a maximum 200 μm slit width at a bending magnet (December 1982) and at the HASYLAB wiggler/undulator W1 (December 1984). The resolution ΔE as given is that for a 200 μm slit and can be reduced by a factor of roughly 1/3 by closing the slit. $S...$ = mirror number, $W...$ = wiggler gap (mm), 3.63 GeV. (From ref. [38].)

incidence for the first mirror in a beam line. The latter measure may well shift the problem to the following elements as was experienced also at the FLIPPER station to some extent.

For 3.7 GeV about 10% of the radiated power is reflected by mirror S_{-1}. This is incident on mirror S_0 which is made of $AlMg_5$ coated with Kanigen. This mirror has no water cooling. Due to a large reduction in the apparent aperture a maximum of about 2% of the total power hits the grating if S_0 is illuminating the grating directly and about 4% of the total power hits the mirrors S_1 to S_6 in the other mode. We have detected serious distortions of the mirrors S_1 to S_6 which originally were fabricated from glass (BK7). The heat load resulted in a convex curvature of the flat mirrors probably due to both a bump at the surface and a bending of the whole mirror. This was an especially serious problem for the less-grazing-incidence mirrors S_3 to S_6 for which most of the incident radiation is also absorbed. The distortion was observed as a defocusing stemming without doubt from these mirrors. It could be partly compensated by changing the focus of the paraboloid. Later on these mirrors were replaced by all-metal mirrors (AlMg$_5$ covered with Kanigen) and this problem is now solved. Small warm-up effects still occur on S_0 and on the grating which is etched on glass and coated with gold. We have attempted to measure the power incident on S_3 by directly measuring the rise in temperature of this mirror with a thermocouple. The first result, namely 1.5 W at 5.3 GeV storage ring energy and 30 mA, appears to be too low by roughly a factor of 3. A more detailed analysis is needed.

The main information on the performance of the FLIPPER monochromator is condensed in fig. 11. The old bending magnet version was actually operated until August 1984, but no characterization of the output was made immediately before the transfer to the wiggler line. Therefore, the status as of December 1982 is given.

The most striking feature is the appearance of the undulator peak in first order and also in higher orders (see ref. [45]). It shifts according to the theoretical calculation given in table I. The additional intensity gain in this peak amounts typically to 6. For the common photon energy ranges of the two FLIPPER versions a maximum gain of 200 over the December 1982 performance was observed at the undulator. This must not be completely attributed to the undulator, since all optical elements and the arrangement of the pre-mirrors have been changed in between.

An analysis of the shape of the undulator peak is given by GÜRTLER [45]. The analysis is based on the solid angle accepted by the FLIPPER optics and yields an excellent agreement between theory and experiment. By deflecting the beam with the help of the remote control of mirror S_{-1}, the sharp undulator peaks decay rapidly as soon as the central ray of the beam is outside of the entrance aperture of the monochromator. This also provides a simple means to align the monochromator to «look head-on» to the undulator. It has to be taken into account that the emittance of the electron beam in DORIS and subsequently its angular spread causes a spread of «forward directions». This distribution has the following dimensions at the entrance aperture of the FLIPPER, namely 18 mm horizontally and 2.5 mm vertically (FWHM). This perfectly matches the aperture of the FLIPPER. It becomes obvious from sect. **2** that the FLIPPER monochromator installed at one of the low-emittance storage rings could use smaller apertures without intensity loss in the undulator peak. But it should also be kept in mind that no additional flux would be obtained through the exit slit, since all «forward directions» of individual electrons within the beam envelope are collected by the monochromator already now. The real advantage would be an increase in brilliance at the exit slit position, provided perfect monochromator optics is available. Even with present-day state-of-the-art quality of aspherical optics, brilliance and thereby also spectral resolution could be improved, since the much smaller optical elements could be fabricated with a higher overall accuracy.

Finally, I should mention that the stability of the undulator beam depends very critically on the performance of the storage ring DORIS: If the monochromator is tuned to an undulator peak, very small motions and pulsations of the electron beam become perceptible as intensity fluctuation. A good monitor and good electronics are needed to eliminate this noise from the measurements. Long-term drifts change the shape of the whole spectral distribution and can cause serious troubles depending on the scope of the experiment. Definitely this problem will need to be solved in order to render the new dedicated storage rings useful, since they will be much more sensitive due to their smaller emittances.

4. – Microfocusing and microscopy.

4˙1. *General remarks.* – The general efforts to establish a soft-X-ray microscopy with synchrotron radiation are well documented in a series of conference proceedings [49-52]. A microscopy which involves photoelectrons as the signal which generates the image could follow in principle two quite different paths. One method uses an electron optical lens system designed to filter and image photoelectrons within a certain band of kinetic energies with large magnification onto a screen or a two-dimensional detector. With the other method a beam of monochromatized synchrotron radiation is focused down to a very small spot which is scanned across the sample. Then photoelectrons of selected energies, or the total emitted flux are providing the signal which generates the image. This type of microscope is already operating in the transmission mode for biological applications at several places [52].

The electron microscopy type of imaging is already in use with an ultraviolet xenon high-power lamp in the first existing LEED microscope built by BAUER and co-workers [53]. The operation of this microscope in connection with synchrotron radiation involves the incorporation of a filter element. There are also a few other projects of this type, *e.g.* the one by POLACK and co-workers in Paris [54] and the one by BETHGE and co-workers in Halle [55]. A resolution of 50 Å appears to be achievable. The main challenge for the SR part of this experiment is to maximize radiation density on the object within an area of about 10 μm in diameter. This is the typical field size for such microscopes. The optical system must be designed and incorporated in such a way that it does not interfere with the electrodes of the electron microscope.

A third type of microscope is only applicable in absorption spectroscopy on thin samples, the direct imaging light microscope. This type of microscopy has been developed by the group RUDOLPH, SCHMAHL, NIEMANN from Göttingen for more than ten years now (see, *e.g.*, [49-52]). They use specially designed Fresnel zone plates as lenses and have achieved a resolution in the order of 500 Å. They apply the microscopy exclusively to the investigation of biological objects in the so-called water window at 275 eV and at 500 eV photon energy. Additional contrast to the absorption contrast was recently obtained by means of the variations in the real part of the refractive index. This phase contrast needs a specially modified version of the microscope [56].

We turn now again to the scanning microscope which may not match the electron optical microscope with respect to ultimate resolution, but in many other respects has advantages over the other types of microscopy. The advantages lie in several places.

1) Radiation load on the sample can be minimized compared to the imaging microscope.

2) Imaging and electron analysis are well-separated measurements.

3) Electron spin analysis can be added.

4) Other secondary processes, as photostimulated desorption, reflectivity, scattering, can be analysed easily. Absorption can be measured in transmission and by means of the total photoelectric yield.

4'2. *Limitations.* – Figure 12 shows the general arrangement of a microprobe microscopy station. Starting with an undulator in one of the advanced new

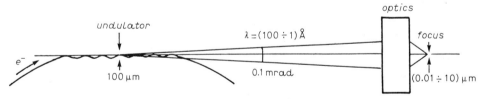

Fig. 12. – Typical layout and parameters of a microfocus arrangement at an undulator in one of the next-generation storage rings.

planned storage rings radiation originates from a source with an effective source size of roughly 100 μm and a divergence of considerably less than 0.1 mrad. The goal is monochromatization to 0.1 eV unless the natural monochromaticity of undulator radiation of $\Delta\varepsilon/\varepsilon \approx 1 : 50$ is sufficient for certain types of experiments. The further goal is focusing down the radiation to a diffraction-limited spot as small as possible. As focusing elements there are available quite generally mirrors with ordinary reflecting coatings, mirrors with specially tailored multilayer coatings, curved crystals and Fresnel zone plates. There are fundamental limitations to the minimum spot size available and technical limitations to the manufacturing of the optical elements. The technical limitations may be overcome step by step due to ingenuity or by investing enough effort and money.

We try now to obtain a rough estimate of the fundamental limitations. The final step of imaging onto the sample must be a large demagnification D of the source itself or of an intermediate image which is defined properly by a diaphragm. According to fig. 13 $D = d/d' = l/l'$. Then Liouville's theorem (in optics known as the Abbé condition)

(4.1) $$d \sin \Theta = d' \sin \Theta'$$

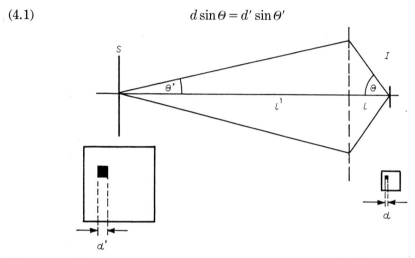

Fig. 13. – Illustration to explain Liouville's theorem in a microfocus arrangement, S = source, I = image, d' and d are corresponding elements of S and I, respectively.

provides a relation for the angular widths Θ, Θ' in the case of perfect imaging. Further, a focus of diameter d requires a minimum angular width due to the coherence condition

(4.2) $$d \sin \Theta \approx \lambda .$$

By d I mean in agreement with the usual definition of the diffraction limit of the classical light microscope the FWHM of the central diffraction peak which is roughly approximated for the purpose of this lecture to lie within 20% of the value given by eq. (4.2).

a) A Fresnel zone plate (see fig. 14) is a circular diffraction grating with the local grating constant Δr determined such that the beam is diffracted to the image point. The grating equation is

(4.3) $$\Delta r(\sin \Theta - \sin \Theta') = m\lambda ,$$

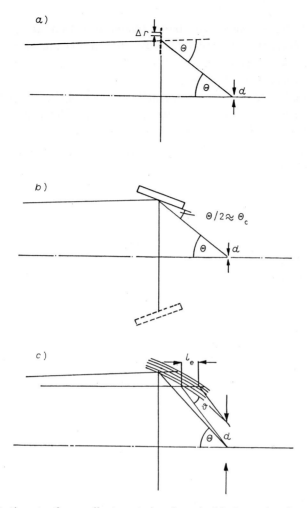

Fig. 14. – Limitations to the smallest spot size d attainable in a microfocus arrangement a) with a Fresnel zone plate with outermost zone separation Δr, b) with mirror optics with the critical angle of total reflection Θ_c, c) with a bent crystal with the extinction length l_e contributing to the spot size d, ϑ is the Bragg angle.

where m is the order of diffraction. With small D it follows $\Theta' \approx 0$ and eq. (4.3) reads $\Delta r \sin \Theta = m\lambda$. Inserting λ from eq. (4.2) and the local grating constant Δr_{min} of the outermost zone yields

$$(4.4) \qquad\qquad d \approx \Delta r_{min}/m \, .$$

If one remembers that all zones must be concentric and spherical to within a fraction of $\Delta r/m$, it is understandable that at present resolution of zone plates is limited to $d \approx 500$ Å. Nevertheless, producing zone plates with even lower dimensions of Δr_{min} appears to be a technical rather than a fundamental limitation. In this context it should be remembered that the «dark» zones need to have a certain thickness t either in order to block the beam or in order to shift the phase of the radiation by $\lambda/2$. The latter is achieved in the so-called «phase zone plates» taking advantage of the very small deviation of the real part of the index of refraction from 1 in the $(500 \div 1000)\,\mathrm{eV}$ range. The aspect ratio $2t/\Delta r_{min}$, however, becomes forbiddingly large with decreasing Δr_{min}.

b) Under the assumption that a single mirror constitutes the optical element in the final demagnification stage we come to an estimate for the grazing-incidence region (photon energies above 100 eV). From a Drude model for the index of refraction $n = 1 - \omega_p^2/2\omega^2 = 1 - \lambda^2/2\lambda_p^2$ with ω_p, λ_p being a plasma frequency and a plasma «wavelength», respectively. Θ_c the grazing critical angle of total reflection is defined by $n = \cos\Theta_c$. Θ_c is roughly the angle at which the reflectivity is 50%. $n \approx 1 - \Theta_c^2/2$ for small Θ_c yields $\Theta_c \approx \lambda/\lambda_p$ and with $\Theta = 2\Theta_c$ (see fig. 14) in combination with eq. (4.2)

$$(4.5) \qquad\qquad d \approx \lambda_p/2\,.$$

This effective plasma wavelength, of course, gives only a rough representation of the optical constants in a certain spectral range. In principle λ_p should be inversely proportional to the square root of the number of «free» electrons per unit volume. In addition to the conduction electrons, all those core electrons should be counted which can be excited at a given photon energy and which have exhausted their oscillator strength. Therefore, λ_p might decrease slowly with decreasing wavelength. In practice, however, reflectivities are well represented by a constant λ_p, which is, *e.g.*, for gold-coated mirrors $\lambda_p \approx 200\,\text{Å}$ over a wide energy range[57]. Taking this λ_p into eq. (4.5) gives an ultimate limit to the obtainable spot size of $d \approx 100\,\text{Å}$ with a single-mirror optics. With a double-mirror optics, *e.g.* the Wolter type-I arrangement[58, 59], the limit could become half this value. Other limitations lie, of course, in the aberrations and in the manufacturing accuracy of mirror elements with complicated shapes.

c) Bent-crystal optics and multilayer coated reflecting optics are limited to fixed photon energies, but with those elements grazing-reflection angles can be made large. There is, however, the problem that the grazing angle of reflection θ and the wavelength λ are linked by Bragg's law

$$(4.6) \qquad\qquad 2d_l \sin\theta = m\lambda\,,$$

where d_l is the lattice or multilayer spacing and m the order. Laterally graded lattice constants may be generated with multilayer coatings but not so easily with crystals. A near-normal-incidence optics like the Schwarzschild arrangement[60] (see fig. 15) is probably the optimum for such elements. Due to

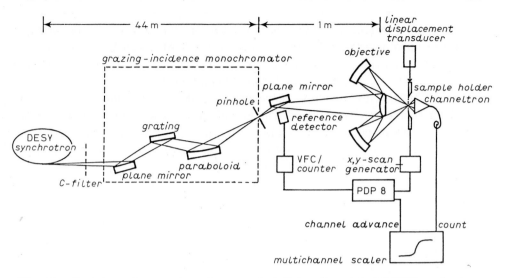

Fig. 15. – Prototype of a scanning microscope installed at the original DESY synchrotron using a multilayer coated Schwarzschild objective. (From ref. [60].)

the large roughness scattering at normal incidence by surface roughness there is a serious technical problem in this arrangement (see eq. (4.9) below).

Normal incidence will also minimize another limitation of such optics which is due to the finite depth in which radiation is reflected (see fig. 14). If l_e is the extinction length, the width of the focus due to this effect is

$$(4.7) \qquad d = l_e \frac{\sin \theta}{|\cos 2\theta|}.$$

Since l_e can be in the order of 5000 Å, this effect can seriously influence the size of the focal spot unless θ is less than a few degrees, or unless θ is equal to 90° within a few degrees (back reflection).

d) The simplest way to obtain a microfocus is by placing a pinhole in a parallel beam. The size of the light spot depends on the distance x between the pinhole of diameter d_0 and the sample. Then the size of the spot is given (see fig. 16) as

$$(4.8) \qquad d = \underset{\text{size}}{d'} + \underset{\text{divergence}}{2x \, \mathrm{tg}\,\theta} + \underset{\text{diffraction}}{x(\lambda/d')} \, .$$

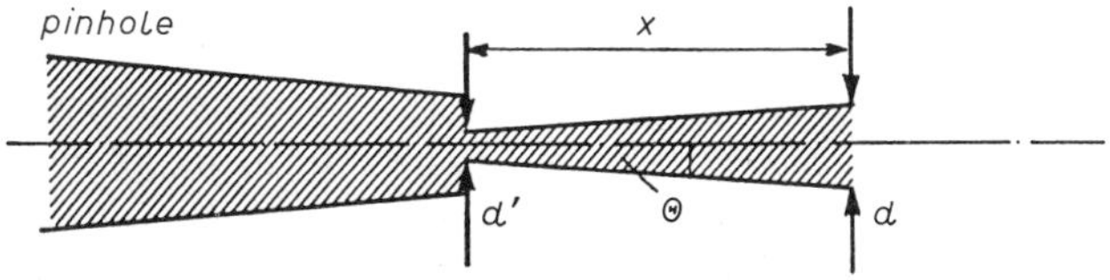

Fig. 16. – Definition of the parameters of a pinhole scanning microprobe.

We assume $x = 5\,\mathrm{mm}$, a distance barely large enough in order to be able to extract the photoelectrons from the sample surface. At $\lambda = 20\,\text{Å}$, neglecting the divergence term, the optimum pinhole size is then $d' = 3\,\mu\mathrm{m}$ yielding a spot size of $d = 6\,\mu\mathrm{m}$. The divergence of the beam illuminating the pinhole would have to be $\theta \leqslant 10^{-4}\,\mathrm{rad}$. This amounts, however, to a demagnification $D = 1$ of the undulator source onto the pinhole. The loss of intensity with an assumed source diameter of $100\,\mu\mathrm{m}$ corresponds to $(3\,\mu\mathrm{m}/100\,\mu\mathrm{m})^2 \approx 10^{-3}$. All this is just a simple demonstration of where the limitations for this simplest possible approach to spatially resolved photoemission lie. Nevertheless, in certain cases, *e.g.* photoemission on crystals which can only be grown in sizes up to $10\,\mu\mathrm{m}$, this approach might make sense. Also with photon-excited Auger analysis [61] which does not require monochromatization of the exciting radiation such an approach might be reasonable.

Let us finally consider emittance/acceptance matching between the source in the storage ring and the spot on the sample. If the source size of $100\,\mu\mathrm{m}$ is demagnified by $D = 10^{-3}$ to a spot of $1000\,\text{Å}$, the overall emission angle (electron beam divergence convoluted with the undulator angular distribution) which we assumed as $2\theta = 10^{-4}\,\mathrm{rad}$ transforms according to eq. (4.1) by a factor D^{-1} to $2\theta = 10^{-1}\,\mathrm{rad}$. This amounts to an angle $\theta = 3°$ or a grazing-incidence reflection angle $\theta/2 = 1.5°$. At $\lambda = 10\,\text{Å}$, $\theta_\mathrm{c} = \lambda/\lambda_\mathrm{p} = 3°$ for an Au coating is well above this angle. Even further demagnification is feasible without more loss of intensity than that due to the limited reflectivity of the optical elements. One important factor in this context is the loss of specular reflectivity due to roughness. The actual reflectivity is calculated in the simplest theoretical approach from the reflectivity of a perfectly smooth surface R_0 by

$$(4.9) \qquad R = R_0 \exp\left[-\left[(4\pi\,\sigma\sin\theta/2)/\lambda\right]^2\right],$$

where σ is the mean square roughness with Gaussian height distribution. The scattered intensity is unfortunately not lost but generates a spread-out background in the image of the sample which will reduce the contrast [62].

4˙3. *A ring mirror microscope.* – It can easily be shown that a rotationally symmetric (ring-shaped) single mirror has two fundamental problems. The most serious one is that such a mirror cannot image an extended area point by point onto another area. WOLTER has, therefore, designed arrangements involving two consecutive reflections which avoid this problem [57, 58]. With a small enough segment of a single mirror, however, imaging is feasible with a certain degree of accuracy. The second problem is due to the demagnification D, which is not constant for rays reflected at different points in the axial direction along the mirror. This effect limits the usable length w of such mirrors since $\Delta D/D \approx w/l$, where l is the distance to the image plane. In arrangements where only light is concentrated into a small spot $\Delta D/D \approx 0.3$ is still acceptable.

The simplest possible rotationally symmetric focusing element is a rotational ellipsoid which collects radiation originating from one focal spot into the other one. Figure 17 shows an arrangement which is presently set up at the HASYLAB laboratory by the University of Hamburg[63]. The elliptical ring mirror has a diameter of 5 mm, a length of 6 mm and distances $l' = 1000$ mm,

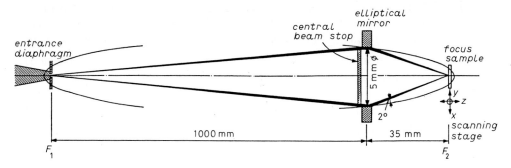

Fig. 17. – Principle and parameters of the scanning photoelectron microscope under construction at HASYLAB. F_1, F_2 are the foci of the elliptical mirror.

$l = 35$ mm, a demagnification $D = 1:3$, a grazing angle of reflection $\Theta/2 = 2°$. It will reflect radiation up to $\varepsilon \approx 2000$ eV. The imaging properties are such that a point in the object plane at a distance d from the axis generates a whole circle with a radius $d = Dd'$ in the image plane. This circle is concentric to the axis of the ellipsoid. Thus a concentric circle in the image plane is imaged on a concentric demagnified circle in the object plane and a concentric diaphragm of diameter d' has an image of diameter $d = Dd'$. For the purpose of forming a microfocus such a mirror fulfils all the necessary requirements. Also calculations involving diffraction show that the diffraction pattern is close, although not exactly equal, to that of an annular aperture of 5 mm diameter and 200 μm width. Therefore, the general considerations made above for the resolution limit due to diffraction apply. As a matter of fact due to the necessary apodization eq. (4.2) needs to be modified for a hollow cone yielding approximately

(4.10) $$2d \sin \Theta = \lambda.$$

As a consequence also eq. (4.5) is modified to

(4.11) $$d \approx \lambda_p/4.$$

In the case of the ellipsoid microscope, with $\Theta/2 = 0.035$, $\Theta/2 < \Theta_c$ for photon energies below 2000 eV.

The hollow-cone diffraction pattern is insofar disadvantageous that the central diffraction peak although very prominent contains only about 15% of the total intensity. The rest is smeared out into the diffraction rings of higher order and will generate a background in the image. With the phase-I microscope which is presently installed at HASYLAB a $10\,\mu$m diameter diaphragm will be demagnified to a $d = 0.33\,\mu$m diameter image. Because of the imperfections of this first mirror (manufactured by the Zeiss company with the best presently available technology) the final spot size is expected to be in the order of $(0.5 \div 1)\,\mu$m. This has to be compared with the diffraction limit according to eq. (4.10) with $\Theta = 0.07$ yielding, at $\lambda = 20$ Å, $d = 0.007\,\mu$m. Thus in this phase of the project there is no limitation to be expected due to diffraction. In a phase II which aims at a focus of $0.1\,\mu$m this mirror will reach the diffraction limit of $0.1\,\mu$m at $\lambda = 300$ Å corresponding to $\varepsilon = 40$ eV.

This microscope will be installed behind the FLIPPER station (see subsect. 3˙2) at the W1 undulator at HASYLAB. It can be operated with one single alignment in the whole range $\varepsilon = (15 \div 1500)$ eV, its angular acceptance is matched to that of the FLIPPER. The overall intensity loss is, at the exit slit of the monochromator, a factor of 10^{-3} ($10\,\mu$m diaphragm, $300 \times 250\,\mu$m^2 spot size), at the mirror a geometrical acceptance factor of $1.25 \cdot 10^{-2}$. The specular reflectivity of the mirror will depend on its roughness. Some initial tests on prototype mirrors were promising. Assuming an optimistic value of $R = 0.8$, we obtain a general loss factor of 10^{-5} by which the values given in fig. 11 need to be multiplied. At the oxygen K-edge (534 eV) approximately 10^7 photons are expected in the phase-I microscope.

Possible gains in intensity at storage rings of lower emittances with monochromators equipped with better optical elements are roughly estimated: A factor of 10 at the entrance aperture, a factor of 5 with a smaller horizontal divergence and a factor of 20 using longer undulators with more periods (conservative estimate). Thus three orders magnitude in the gain appear to be a reasonable estimate due to improvements of the storage rings. Using somewhat longer mirrors or mirrors with higher angles $\Theta/2$ could also gain some intensity but to a lesser extent. It should, however, be mentioned that using such a small angle as $\Theta/2 = 2°$ also for longer wavelengths enhances specular reflection and reduces roughness scattering according to eq. (4.9).

The mechanical installation is presently under construction in Hamburg. The most important factor is mechanical stability with respect to the axis of the ellipsoid and the scanning of the sample. Not much can be said about this set-up before the microscope is thoroughly tested.

* * *

I want to thank Dipl. Phys. S. CRAMM and Dipl. Phys. J. VOSS for a critical reading of the manuscript and A. SCHMIDT for her careful work in preparing the manuscript.

REFERENCES

[1]   E. E. KOCH, Editor: *Handbook on Synchrotron Radiation* (North Holland, Amsterdam, New York, N.Y., Oxford, 1983), Vol. 1*a*, 1*b*, 2.

[2]   C. KUNZ, Editor: *Synchrotron Radiation—Techniques and Applications* (Springer, Berlin, Heidelberg, New York, N.Y., 1979), *Topics in Current Physics*, Vol. 10.

[3]   H. WINICK and S. DONIACH, Editors: *Synchrotron Radiation Research* (Plenum Press, New York, N.Y., London, 1980).

[4]   D. VAUGHAN, Editor: X-*Ray Data Booklet* (Lawrence Berkeley Laboratory, Berkeley, Cal., 1986).

[5]   F. WUILLEUMIER and Y. FARGE, Editors: *Proceedings of the International Conference on Synchrotron Radiation Instrumentation and New Developments (Orsay, 12-14 September 1977), Nucl. Instrum. Methods*, 152 (1978).

[6]   D. L. EDERER and J. B. WEST, Editors: *Proceedings of the National Conference on Synchrotron Radiation Instrumentation (Gaithersburg, 4-6 June 1979), Nucl. Instrum. Methods*, 172 (1980).

[7]   M. R. HOWELLS, Editor: *Proceedings of the Japanese/USA Seminar on Synchrotron Radiation Facilities (Honolulu, 5-9 November 1979), Nucl. Instrum. Methods*, 177 (1980).

[8]   D. M. MILLS and B. W. BATTERMAN, Editors: *Proceedings of the Second National Conference on Synchrotron Radiation Instrumentation (Ithaca, 15-17 July 1981), Nucl. Instrum. Methods*, 195 (1982).

[9]   E. KOCH, Editor: *Proceedings of the International Conference on X-Ray and VUV Synchrotron Radiation Instrumentation (Hamburg, 9-13 August 1982), Nucl. Instrum. Methods*, 208 (1983).

[10] W. THOMLINSON and G. P. WILLIAMS, Editors: *Proceedings of the Third National Conference on Synchrotron Radiation Instrumentation (Upton, 12-14 September 1983), Nucl. Instrum. Methods*, 222 (1984).

[11] G. S. BROWN and I. LINDAU, Editors: *Proceedings of the International Conference on X-Ray and VUV Synchrotron Radiation Instrumentation (Stanford, 29 July-2 August 1985), Nucl. Instrum. Methods A*, 246 (1986).

[12] G. N. KULIPANOV, Editor: *Proceedings of the Seventh USSR National Conference on Synchrotron Radiation Utilization (SR86) (Novosibirsk, 3-5 June 1986), Nucl. Instrum. Methods A*, 261 (1987).

[13] M. A. GREEN, J. P. STOTT and P. R. WOODRUFF, Editors: *Proceedings of the Fifth* National Conference on Synchrotron Radiation Instrumentation (Madison, 21-25 June 1987), Nucl. Instrum. Methods A*, 266 (1988) (*actually the «Fourth»!).

[14] *Proceedings of the 3rd International Conference on Synchrotron Radiation Instrumentation: SRI-88 (Tsukuba 29 August - 2 September 1988) Rev. Sci. Instrum.*, to be published.

[15] M. R. HOWELLS, Editor: *Reflecting Optics for Synchrotron Radiation (Upton, 16-18 November 1981), Proc. SPIE*, 315 (1981).

[16] E. SPILLER Editor: *High Resolution Soft X-Ray Optics (Brookhaven, 18-20 November, 1981), Proc. SPIE*, 316 (1981).

[17] E. E. KOCH and G. SCHMAHL, Editors: *Soft X-Ray Optics and Technology (Berlin, 8-11 December 1986), Proc. SPIE*, 733 (1986).

[18] G. A. SCHOTT: *Electromagnetic Radiation* (Cambridge University Press, Cambridge, 1912).

[19] D. IVANENKO and J. POMERANCHUCK: *Phys. Rev.*, 65, 343 (1944).

[20] J. SCHWINGER: *Phys. Rev.*, 70, 798 (1946); 75, 1912 (1949).

[21] A. SOMMERFELD: *Elektrodynamik* (Akademische Verlagsgesellschaft, Leipzig, 1949).

[22] J. D. JACKSON: *Classical Electrodynamics* (Wiley and Sons, New York, N.Y., 1962), p. 481.

[23] D. H. TOMBOULIAN and P. L. HARTMAN: *Phys. Rev.*, **102**, 1423 (1956).

[24] Proposal: *BESSY II, Eine optimierte Undulator/Wiggler-Speicherring Lichtquelle für den VUV- und XUV-Spektralbereich* (Berliner Elektronenspeicherring-Gesellschaft für Synchrotronstrahlung mbH, November 1986).

[25] C. KUNZ: in *Synchrotron Radiation—Techniques and Applications*, edited by C. KUNZ (Springer, Berlin, Heidelberg, New York, N.Y., 1979), *Topics in Current Physics*, Vol. **10**, p. 1.

[26] S. KRINSKY, M. L. PERLMAN and R. E. WATSON: in *Handbook on Synchrotron Radiation*, edited by E. E. KOCH (North Holland, Amsterdam, New York, N.Y., Oxford, 1983), Vol. 1a, p. 65.

[27] H. ONUKI: in *Proceedings of the International Conference on X-Ray and VUV Synchrotron Radiation Instrumentation (Stanford, 29 July - 2 August 1985)*, edited by G. S. BROWN and I. LINDAU, *Nucl. Instrum. Methods A*, **246**, 94 (1986).

[28] H. ONUKI, N. SAITO and T. SAITO: *Appl. Phys. Lett.*, **52**, 173 (1988).

[29] K. J. KIM: *Nucl. Instrum. Methods*, **219**, 425 (1984).

[30] J. GOULON, P. ELLEAUME and D. RAUX: *Nucl. Instrum. Methods A*, **254**, 192 (1987).

[31] G. BROWN, K. HALBACH, J. HARRIS and H. WINICK: in *Proceedings of the International Conference on X-Ray and VUV Synchrotron Radiation Instrumentation (Hamburg, 9-13 August 1982)*, edited by E. E. KOCH, *Nucl. Instrum. Methods*, **208**, 65 (1983).

[32] C. DEPAUTEX, P. THIRY, R. PINCHAUX, Y. PETROFF, D. LEPÈRE, G. PASSERAU and J. FLAMMAND: in *Proceedings of the International Conference on Synchrotron Radiation Instrumentation and New Developments (Orsay, 12-14 September 1977)*, edited by F. WUILLEUMIER and Y. FARGE, *Nucl. Instrum. Methods*, **152**, 101 (1978).

[33] C. T. CHEN: *Nucl. Instrum. Methods Phys. Res. A*, **256**, 595 (1987).

[34] T. HARADA, T. KITA, M. ITOU, H. TAIRA and A. MIKUNI: in *Proceedings of the International Conference on X-Ray and VUV Synchrotron Radiation Instrumentation (Stanford, 29 July - 2 August 1985)*, edited by G. S. BROWN and I. LINDAU, *Nucl. Instrum. Methods A*, **246**, 272 (1986).

[35] W. GUDAT and C. KUNZ: in *Synchrotron Radiation—Techniques and Applications*, edited by C. KUNZ (Springer, Berlin, Heidelberg, New York, N.Y., 1979), *Topics in Current Physics*, Vol. **10**, p. 55.

[36] R. L. JOHNSON: in *Handbook on Synchrotron Radiation*, edited by E. E. KOCH (North Holland, Amsterdam, New York, N.Y., Oxford, 1983), Vol. 1a, p. 173.

[37] V. SAILE and J. B. WEST: in *Proceedings of the International Conference on X-Ray and VUV Synchrotron Radiation Instrumentation (Hamburg, 9-13 August 1982)*, edited by E. KOCH, *Nucl. Instrum. Methods*, **208**, 199 (1983).

[38] F. SENF, K. BERENS VON RAUTENFELDT, S. CRAMM, J. LAMP, J. SCHMIDT-MAY, J. VOSS, C. KUNZ and V. SAILE: in *Proceedings of the International Conference on X-Ray and VUV Synchrotron Radiation Instrumentation (Stanford, 29 July - 2 August 1985)*, edited by G. S. BROWN and I. LINDAU, *Nucl. Instrum. Methods A*, **246**, 314 (1986).

[39] W. EBERHARDT, G. KALKOFFEN and C. KUNZ: in *Proceedings of the International Conference on Synchrotron Radiation Instrumentation and New Developments*

(Orsay, 12-14 September 1977), edited by F. WUILLEUMIER and Y. FARGE, *Nucl. Instrum. Methods*, **152**, 81 (1978).

[40] J. BARTH, F. GERKEN, C. KUNZ and J. SCHMIDT-MAY: in *Proceedings of the International Conference on X-Ray and VUV Synchrotron Radiation Instrumentation (Hamburg, 9-13 August 1982)*, edited by E. KOCH, *Nucl. Instrum. Methods*, **208**, 307 (1983).

[41] C. KUNZ, R. HAENSEL and B. SONNTAG: *J. Opt. Soc. Am.*, **58**, 1415 (1968).

[42] H. DIETRICH and C. KUNZ: *Rev. Sci. Instrum.*, **43**, 434 (1972).

[43] H. PETERSEN and H. BAUMGÄRTEL: in *Proceedings of the National Conference on Synchrotron Radiation Instrumentation (Gaithersburg, 4-6 June 1979)*, edited by D. L. EDERER and J. B. WEST, *Nucl. Instrum. Methods*, **172**, 191 (1980).

[44] W. JARK and C. KUNZ: in *Proceedings of the International Conference on X-Ray and VUV Synchrotron Radiation Instrumentation (Stanford, 29 July - 2 August 1985)*, edited by G. S. BROWN and I. LINDAU, *Nucl. Instrum. Methods A*, **246**, 320 (1986).

[45] P. GÜRTLER: in *Proceedings of the International Conference on X-Ray and VUV Synchrotron Radiation Instrumentation (Stanford, 29 July - 2 August 1985)*, edited by G. S. BROWN and I. LINDAU, *Nucl. Instrum. Methods A*, **246**, 91 (1986).

[46] S. MOURIKIS, E. E. KOCH and V. SAILE: *Nucl. Instrum. Methods A*, **267**, 218 (1988).

[47] R. T. AVERY: in *Proceedings of the Third National Conference on Synchrotron Radiation Instrumentation (Upton, 12-14 September 1983)*, edited by W. THOMLINSON and G. P. WILLIAMS, *Nucl. Instrum. Methods*, **222**, 146 (1984).

[48] T. OHTA and T. FUJIKAWA: KEK Report 81-10 (August 1981).

[49] D. F. PARSONS, Editor: *Ultrasoft X-Ray Microscopy: Its Application to Biological and Physical Sciences (New York, 13-15 June 1979)*, *Ann. N.Y. Acad. Sci.*, **342** (1980).

[50] G. SCHMAHL and D. RUDOLPH, Editors: *Proceedings of the International Symposium, X-Ray Microscopy (Göttingen, 14-16 September 1983)*, *Springer Series in Optical Sciences*, Vol. **43** (Springer, Berlin, Heidelberg, 1984).

[51] PING-CHIN CHENG and GWO-JEN JAN, Editors: *X-Ray Microscopy—Instrumentation and Biological Applications (Proceedings Symposium, Taipei, August 1986)* (Springer, Berlin, Heidelberg, 1987).

[52] D. SAYRE, M. HOWELLS, J. KIRZ and H. RARBACK, Editors: *Proceedings of the International Symposium, X-Ray Microscopy II (Brookhaven 31 August - 4 September 1987)*, *Springer Series in Optical Sciences*, Vol. **56** (Springer, Berlin, Heidelberg, 1988).

[53] W. TELIEPS and E. BAUER: *Ultramicroscopy*, **17**, 57 (1985).

[54] F. POLACK, S. LOWENTHAL, D. PHALIPPOU and P. FOURNET: in *Proceedings of the International Symposium, X-Ray Microscopy II (Brookhaven 31 August - 4 September 1987)*, *Springer Series in Optical Sciences*, Vol. **56**, edited by D. SAYRE, M. HOWELLS, J. KIRZ and H. RARBACK (Springer, Berlin, Heidelberg, 1988), p. 220.

[55] H. BETHGE, T. KRAJEWSKI and O. LICHTENBERGER: *Ultramicroscopy*, **17**, 21 (1985).

[56] G. SCHMAHL, D. RUDOLPH and P. GUTHMANN: in *Proceedings of the International Symposium, X-Ray Microscopy II (Brookhaven 31 August - 4 September 1987)*, *Springer Series in Optical Sciences*, Vol. **56**, edited by D. SAYRE, M. HOWELLS, J. KIRZ and H. RARBACK (Springer, Berlin, Heidelberg, 1988), p. 228.

[57] B. L. HENKE, P. LEE, T. J. TANAKA, R. L. SHIMABUKURO and B. K. FUJIKAWA: *At. Data Nucl. Data Tables*, **27**, 123 (1982).

[58] M. WOLTER: *Ann. Phys.* (*N.Y.*), **10**, 94, 286 (1952).

[59] A. FRANKS and B. GALE: in *Proceedings of the International Symposium, X-Ray Microscopy (Göttingen, 14-16 September 1983), Springer Series in Optical Sciences*, Vol. **43**, edited by G. SCHMAHL and D. RUDOLPH (Springer, Berlin, Heidelberg, 1984), p. 129.

[60] R. P. HAELBICH, W. STAEHR and C. KUNZ: in *Ultrasolft* X-*Ray Microscopy: Its Application to Biological and Physical Sciences* (*New York, 13-15 June 1979*), edited by D. F. PARSONS, *Ann. N.Y. Acad. Sci.*, **342**, 148 (1980).

[61] E. UMBACH and Z. HUSSAIN: *Phys. Rev. Lett.*, **52**, 457 (1984), and private communication.

[62] H. HOGREFE and C. KUNZ: *Appl. Opt.*, **26**, 2851 (1987).

[63] C. KUNZ, A. MOEWES, G. ROY, H. SIEVERS, J. VOSS and H. WONGEL: in *HASYLAB Jahresbericht 1987*, p. 366, and to be published.

Core Level Spectroscopies and Synchrotron Radiation.

J. C. FUGGLE

Department of Physical Chemistry, Research Institute for Materials
University of Nijmegen, Toernooiveld - 6525 ED Nijmegen, The Netherlands

1. – Introduction.

This lecture is intended to give some of the basic information which is needed to appreciate the advantages of synchrotron radiation for core level spectroscopy of solids and to give a basis for further study and possible use of the techniques. It is a short account of a wide and active field. I have tried to make it a readable introduction, rather than a textbook. My name is on about 50% of the references quoted. That does not mean that we have done 50% of the work in the field, it means that it is much easier to illustrate a point with your own work.

2. – Basic concepts.

It is appropriate to start with a reminder of some of the different electron and X-ray spectroscopies and the transitions which lead to them. The figures illustrating these processes are drawn in the single-particle, or independent-electron approach. In photoelectron spectroscopy (PS, fig. 1a)), the sample is irradiated with monochromatic radiation of energy $h\nu$ and an electron is emitted with kinetic energy, KE, given by

$$(1) \qquad h\nu = E_B + \mathrm{KE},$$

where E_B is the one-electron binding energy. KE and E_B are referred to the Fermi level, to avoid long discussions about the role of the sample and instrument work functions. Conventional sources of radiation are gas discharge lamps emitting a radiation at discrete energy (*e.g.*, He I $= 21.2\,\mathrm{eV}$, He II $= 40.8\,\mathrm{eV}$) or characteristic X-ray line sources (*e.g.*, Al $K_\alpha = 1486.7$, Mg $K_\alpha = 1253.6\,\mathrm{eV}$). Part of this lecture will deal with some aspects of such investigations that can better be done using the continuum of X-ray radiation from a synchrotron, but much of the basic principles can also be illustrated with data taken using conventional sources and indeed for some purposes conventional sources are preferable.

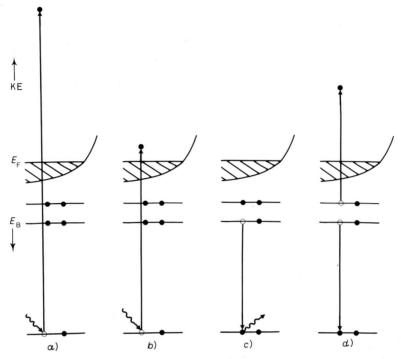

Fig. 1. – Schematic diagram of the transitions involved in various spectroscopies: *a*) photoelectron spectroscopy, PS; *b*) X-ray absorption spectroscopy, XAS; *c*) X-ray emission, XES; *d*) Auger electron emission, AES.

In X-ray absorption spectroscopy (XAS) one scans the radiation energy and observes the absorption. The resulting spectra show jumps in the absorption cross-section, at the thresholds when a new core level may be excited to the first unoccupied states. At threshold, for a metal,

$$(2) \qquad\qquad\qquad\qquad h\nu = E_B.$$

Discrete structure can be observed just above the thresholds, or «edges» in solids. It is divided into X-ray near-edge structure (XANES) and extended-absorption fine structure (EXAFS). In an atom one expects a few sharp lines from transitions to discrete, unoccupied levels, followed by a continuum.

The structure of XANES is traditionally used as a fingerprint of the local environment of an atom, or as a probe of the unoccupied density of electronic states. We have discussed elsewhere the limitations of this approach [1-3]. In this lecture other approaches will be explained.

Once a core hole has been created, it must decay. One decay mechanism is by transfer of an electron down from a higher level with simultaneous X-ray emission (XES, fig. 1*c*)). This process is the basis of X-ray fluorescence analysis.

The second relevant core hole decay process is by Auger electron emission (fig. 1d)). Here the hole is filled by an electron from a higher level and the excess energy is taken away by an electron. Note that the Auger process may be discussed as a stepwise process, but it is not stepwise at all. It is a single quantum-mechanical transition between two stationary eigenstates of the system which have different numbers of noncontinuum electrons. The combination of final-state holes is limited only by some (not very restrictive) selection rules[4] and the criterion that $E_{\mathrm{B1}} > E_{\mathrm{B2}} + E_{\mathrm{B3}}$. Note that the Auger electron energies are independent of the exciting radiation energy which created the core hole, except very close to threshold.

It is a consequence of the Heisenberg uncertainty relationship that a short core hole lifetime (rapid X-ray emission and/or Auger decay) is associated with a large uncertainty in energy in most cases. The lifetime broadening produces Lorentzian lineshapes in XPS and AES or X-ray emission with the form

$$(3) \qquad I(E) = I(E)_0 \cdot \Gamma_{\mathrm{L}}/[(E - E_0)^2 + \Gamma^2] \,.$$

In eq. (3) $I(E)$ is the intensity at energy E, E_0 is the peak centre and Γ_{L} is the core hole lifetime broadening and also one half of the full width at half maximum (FWHM), given by

$$(4) \qquad \tfrac{1}{2}\,\mathrm{FWHM} = \Gamma_{\mathrm{L}} = h/\tau = 6.58 \cdot 10^{-16}/\tau \ \mathrm{eV} \,,$$

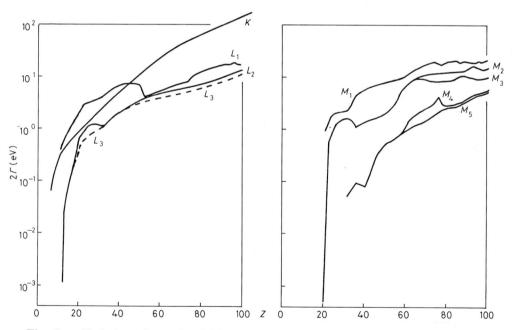

Fig. 2. – Variation of core level lifetimes with atomic number, Z. Data taken from McGuire's work, which has been partially superseded[5].

where τ is the lifetime involved. Lifetime broadening of core holes can range from about 0.1 to 100 eV, as shown in fig. 2.

Lifetime broadening is one of those things less well understood now than five years after its discovery. No energy gets lost or gained forever. If a core hole created in XPS has a short lifetime, the photoelectron may have more energy than E_0, but then the photon, or Auger electron resulting from decay of the core hole has less, or *vice versa*. In normal Auger spectra Γ_L is simply the sum of the lifetime broadenings of the single core hole in the initial state and the double-core-hole final state.

Decay of a core hole by a single X-ray emission channel has the rate

$$(5) \qquad W_{fi} = |M_{fi}|^2 = e^2 \Omega^3 / hc^3 2\pi \langle \Phi_f | t | \Phi_i \rangle^2 ,$$

where $\Omega = 2\pi c \nu$ is the radiation frequency and other symbols have their usual meaning. A given X-ray emission channel has a rate roughly proportional to Z^4, where Z is the atomic number. This dependence is caused by the contraction of the orbitals as Z is increased.

Decay of a core hole by Auger emission has the rate

$$(6) \qquad W_{fi} = 2\pi / h \langle \Phi_f | \sum_{i \neq j} e^2 / r_{ij} | \Phi_i \rangle^2 \rho E_f \text{ s}^{-1} .$$

For a given Auger decay channel this rate also increases for high Z, due to contraction of the orbitals, but the rate only increases roughly as Z^2.

As a consequence of these two relationships core hole lifetimes for a given core hole decrease dramatically with Z. Illustrations are given in fig. 2. Also the fluorescence yield from a given core hole tends to increase with Z, but it is only for the K-shell that the fluorescence yields ever exceed the Auger yield, as illustrated in fig. 3.

Auger transitions are classified as normal, Coster-Kronig (CK) and super Coster-Kronig (SCK). By definition, in CK transitions, one of the final-state holes has the same principal quantum number as the initial-state hole (e.g., $L_1 L_{2,3} M$). In SCK transitions, the principal quantum numbers of both final-state holes are the same as that of the initial-state hole (e.g., $L_1 L_{2,3} L_{2,3}$). CK and SCK processes are not possible if the angular-momentum quantum number of the initial-state hole is high. Because, in general, the rates of Auger decay follow the sequence Auger < CK < SCK, the core hole lifetime broadening tends to increase with decreasing angular-momentum quantum number because the number of possibilities for Coster-Kronig decay increases (see fig. 2). However, because CK and SCK processes are forbidden in some regions of the periodic table, plots of both fluorescence yield and core hole lifetime against Z show discontinuities.

If it is remembered that much of the industry associated with core level spectroscopies involves the measurements of small peak shifts and fine details of

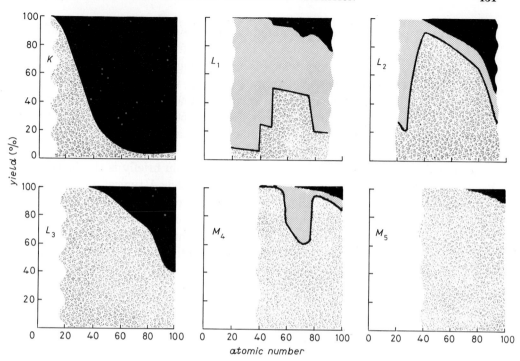

Fig. 3. – Variation of fluorescence yield (black), Auger yields (rough shading) and Coster-Kronig Auger yields (dots).

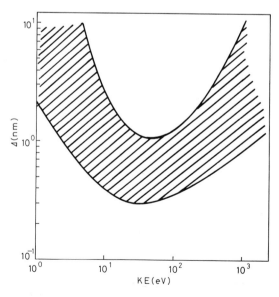

Fig. 4. – Schematic diagram of the variation of electron mean free path with electron kinetic energy. For more complete information see [6].

the spectra shapes, then it will be realized that narrow core levels are desirable. The lessons of fig. 2 and 3 are then that the most useful core levels are likely to be those with a binding energy which is not too high because for these the number and rate of the possible decay processes will be smaller and the consequent lifetime broadening will be smaller. I know of no core level with BE > 1 keV whose lifetime broadening is less than 500 meV. Levels with the highest angular-momentum quantum number are usually narrowest because of the absence of Coster-Kronig Auger decay.

A second major point which must be discussed again briefly is the inelastic mean free path of high-energy photoelectrons in a solid. A typical illustration, stripped of all details, is given in fig. 4. The mean free paths of electrons in solids do not follow a single universal curve, but they do tend to be minimum at about $(50 \div 100)$ eV. Thus a common use of synchrotron light sources involves tuning the radiation from the monochromator so that photoelectrons from the core levels under investigation are emitted with kinetic energies of $(50 \div 100)$ eV and the surface sensitivity is maximized.

3. – The relationship between spectroscopies and calculated eigenenergies.

This is a hot topic in 1988 and other authors in this book will also discuss the problem. Strictly speaking, there is no relationship. The individual eigen-energies and wave functions of, for instance, a density-of-states calculation have, strictly speaking, no physical meaning. It was KOOPMANS in 1933 [7] (yes, Koopmans is spelt with an «s») who proposed that, to a first approximation, the eigenenergies could be treated as ionization energies. In practice the other electrons of the system may be seen as changing when an electron is removed or added. Thus, for instance, when a core electron is removed in photoemission, we find a cluster of peaks in the photoelectron spectrum whose weighted mean energy is the one-electron eigenenergy, or Koopmans energy, in the sudden approximation. The lowest binding energy peak is one in which the wave functions of the $N-1$ electrons left over have the optimum, or lowest-energy distribution and they are regarded as having «relaxed» with respect to their distribution in the N-electron system. In general, it is purely coincidence if the ionization energies measured in photoelectron spectroscopy do correspond to *any* of the calculated one-electron eigenenergies.

The changes in distribution of the «spectator» electrons when one is added, removed, or excited have been discussed under various names. At the beginning of the 1970's the fashionable word was relaxation. Since then the words screening [8], exchange-correlation and self-energies have all been applied and there is a strong relationship to ideas concerning configuration interaction. It is not profitable to try to delineate between the words used, but it is sensible to try to illustrate the basic physics involved.

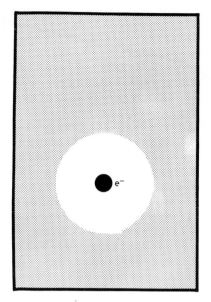 

Fig. 5. – Schematic representation of the exchange-correlation hole associated with an electron (left) and (right) the idea that high electron kinetic energies lead to decoupling of the electron and its exchange-correlation hole. This figure may help the casual reader remember the concept, but please remember that it is a very naive mixture of classical and quantum-mechanical ideas and cannot substitute a rigid formal description.

Consider the concept of an exchange-correlation hole, as illustrated in fig. 5. A (classical) electron in a sea of electrons will repel the other electrons due to exchange-correlation effects. The sea may be regarded as having developed a hole. The creation of such exchange-correlation holes reduces the energy of all the valence electrons and must be taken into account in any band structure calculations. It is of the order of 10 eV. However, for nearly free electrons the exchange-correlation term is not constant, as illustrated very schematically in fig. 5. At high electron kinetic energies the electron and its exchange-correlation hole are increasingly decoupled and the exchange-correlation stabilization is reduced. At infinite kinetic energies the entire exchange-correlation stabilization disappears.

An illustration of how the energy dependence of the exchange-correlation energy influences a spectrum is given in fig. 6, where the bremsstrahlung isochromatic spectrum (BIS, also known as inverse photoemission) is compared with a DOS spectrum [9]. Whilst there is a close correspondence of the peaks and valleys in the spectrum, the calculation produces peaks at too low energies because it does not recognize that the stabilization by the exchange-correlation term is larger at E_F than at higher energies.

This theoretical description of the energy dependence of the exchange-correlation energy is still incomplete. In the future one must expect fuller

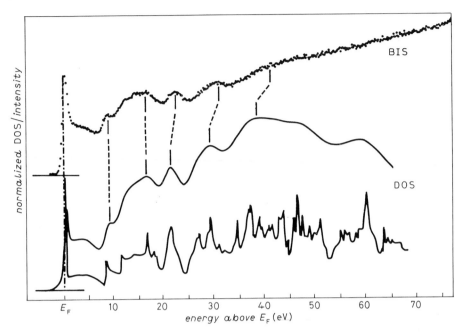

Fig. 6. – Bremsstrahlung isochromatic spectrum and calculated density of unoccupied states for Pd. The correspondence of structures in the measured spectrum (dots) with features in the broadened DOS (solid line) is indicated. The energy positions show, however, distinct differences for the high-energy states, which is due to the energy dependence of the exchange-correlation, or self-energy term. Figure taken from ref. [9].

treatments, taking into account the full details of the dielectric response of the solid, plasmons, etc. Also the above description should be modified for states like the rare-earth 4f states, where atomic correlations dominate.

We should now consider the concept of self-energies. The self-energy may simply be defined as the difference between the experimentally determined ionization energies and the eigenenergies of an *exact* calculation. For XPS/BIS of valence bands, the exchange-correlation potential is the leading term in the self-energies. For spectroscopies like X-ray absorption, or KLV Auger spectra there is an additional complication due to the core holes in the final states. Both spectroscopies have short-ranged matrix elements, so that the spectra are dominated by the local density of states for the atom where the core hole is placed. However, the core hole is a strong perturbation, so that XAS and KLV Auger reflect the local DOS on a core ionized site. The influence of the core hole was postulated many years ago[10], but there is now increasing quantitative evidence on its size[2-4, 11, 12]. As could have been predicted, it is clearly strongest when the valence electron wave functions are most compact. Note that it is experimentally far from trivial to separate the influence of core holes (relevant only to XAS and KLV Auger) from the energy dependence of the exchange-correlation potential. The latter was discussed in connection with XPS/BIS above, but is also relevant for XAS and KLV Auger.

4. – Experimental aspects.

Our purpose here is merely to pick out three aspects most salient to core level PS and XAS with synchrotron radiation.

The first point, that work at lower photon and electron energies eases the experimental difficulties associated with high resolution, will be illustrated at various points in this lecture. The second point will be less emphasized but should be remembered. It is a ramification of the law of conservation of misery that easing of experimental constraints on resolution is accompanied by the stiffening of constraints on surface cleanliness and vacuum at lower energy.

The third point concerns XAS. Traditional XAS spectra were measured in transmission mode, as illustrated in fig. 5. Here the spectrum is given by the ratio of the beam intensity before and after the absorber. In order to avoid saturation effects, it is desirable to use thin absorbers, so that the percentage of radiation absorbed is small, the spectrum is given as the ratio of two large numbers, and the shot noise is large. When the radiation energy is decreased, the absorption cross-section is in general higher and the task of preparing suitable, free-standing, pinhole-free, absorber foils is in most cases prohibitively difficult because of the extremely thin sample required (about $(20 \div 100)$ Å).

A second method of XAS measurement, known as photoyield (fig. 7), has largely replaced the traditional method for low-energy X-rays. In this method

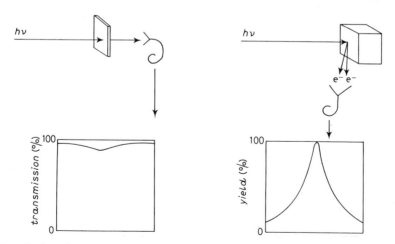

Fig. 7. – Left: schematic diagram of how XAS is measured in transmission, with a representation of the spectrum, in which the transmitted signal should not be excessively attenuated by the absorption. Right: schematic diagram of how XAS is measured in photoyield mode.

the radiation from the monochromator can be allowed to fall on a bulk sample, thus exciting a current of electrons and fluorescent radiation from the sample. The basis of the method is that if the mean absorption length is much longer than the mean escape depth of the electrons or fluorescent radiation, then the signal is proportional to the fraction of the incident radiation absorbed within the surface region. This condition is generally met. The use of XAS in photoelectron yield mode is much more sensitive than photoelectron spectroscopy because in XAS there is seldom any reason to energy-select the electrons. This also results in a greater probing depth, because one also accepts electrons that have suffered many inelastic-scattering events before escaping from the surface. The probing depth of XAS in the soft-X-ray region ((200 ÷ 2000) eV) is generally considered to be some hundreds of ångströms, rather than (5 ÷ 20) Å as in PS. For further ideas on the probing depth in photoyield XAS I refer to the beautiful article by ERBIL *et al.*[13].

5. – Uses of chemical shifts in PS.

The chemical shifts of core level peaks in photoelectron spectroscopy have their origin principally in the electrostatic interaction between the core electrons and the outer electrons. Thus the energy required to remove an electron from a core level is changed when the number of valence electrons is increased or decreased, as in a chemical compound in which the bonding is (partially) ionic[14]. Examples of this are given in fig. 8 and 9 (taken from Siegbahn's book[14]). In xenon fluorides the bonding has considerable ionic character because the fluorine atoms are highly electronegative. The charge transfer from

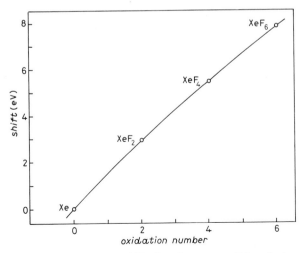

Fig. 8. – Measured binding energy shift for the xenon $3d_{5/2}$ electrons *vs.* oxidation number[14].

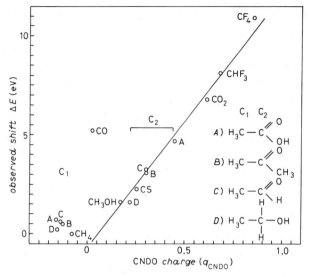

Fig. 9. – Plot of C 1s binding energy against the charge on the C atom, calculated by the CNDO method [14].

xenon increases as the number of fluorine ligands is increased, with consequent increase in the binding energy of the Xe core electrons, as typified in fig. 8 by the Xe $3d_{5/2}$ electrons. The increment of the binding energy per oxidation number increase is almost constant over the whole range and corresponds to approximately 0.3 electrons transferred per fluorine ligand [14].

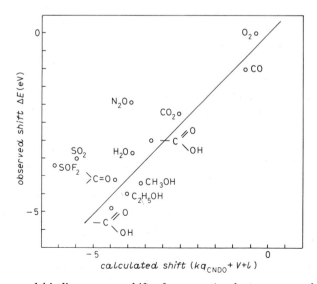

Fig. 10. – Measured binding energy shift of oxygen 1s electrons vs. calculated CNDO shift [14].

Figure 9 illustrates the shift of the C 1s level in a whole series of carbon compounds as a function of the computed charge on the carbon atom [14]. The correlation is rather good. It was figures like these that led to the belief that the core level shifts could be used to *deduce* charge distributions where calculations were difficult (big molecules, surfaces) or nearly impossible (amorphous solids and glasses). However, even Siegbahn's group realized very early [14] that the correlation was not always so good (fig. 10). The principal reason for this is the relaxation of the «spectator» electrons when a core electron is removed. The energy associated with relaxation is comparable to the chemical shift itself. Thus the observed ionization energy is not equal to the one-electron eigenenergy, or Koopmans energy, of the orbital. It is smaller. The main factors contributing to the reduction are the decrease in the electron-electron repulsions, a general decrease in the size of the electron wave functions of the «spectator electrons» when an electron is removed, and interatomic relaxation, or flow of valence electron charge from ligands to the core ionized atom when a core electron is removed. These relaxation processes are very complex and it is not surprising that they should not always be directly proportional to changes in the valence electron charge on the atom to be ionized. This is very relevant in the interpretation of core level shifts at surfaces, discussed below.

6. – Application of core level studies to surfaces.

Surface studies have been highly fashionable for nearly twenty years now, to such an extent that the question of whether a technique can be applied to surfaces is often (fallaciously) considered a prime test of its value. Nevertheless, surface phenomena are important and must be considered here. As the range of chemical forces between atoms is short, the most pertinent questions involve the first few monolayers. Obvious questions are what is the electronic structure of the first couple of monolayers of a clean material, and what changes occur in these layers as a result of a chemical reaction. In principle one could use photoemission studies of any core level as a probe of surface effects, but shallow core levels (BE typically $(10 \div 300)$ eV) provide the best probe because their natural linewidths are small. It is further desirable to choose the photon energy so that the kinetic energy of the outgoing electron is between about 30 and 100 eV. In this way the electron inelastic mean free path is minimized and the surface sensitivity is maximized. One may also choose p-polarized light to enhance surface sensitivity [15, 16]. The advantage of this choice of low photon and electron energies is that the constraints on resolution are less stringent: if one requires a resolution of better than 100 meV, and if $h\nu$ and electron KE are about 1 keV, one must have a resolving power of better than 10^4, which is very difficult to achieve. If we lower these energies to about 100 eV, the required resolving power is 10^3, which is well within the bounds of today's technology.

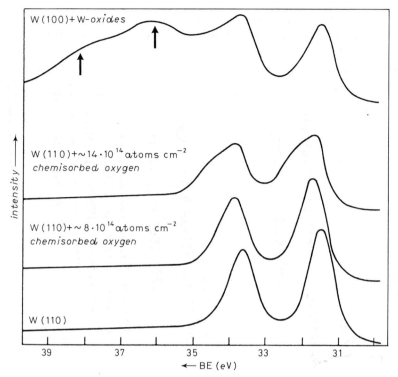

Fig. 11. – XPS study of the W $4f_{7/2}$ and W $4f_{5/2}$ peaks from W(110) and W(100) surfaces at various stages of the reaction with oxygen. The intensity scale of the W(100) and W(110) spectra are not normalized. For the W(110) spectra grazing take-off was used.

Figures 11-14 offer two examples of these ideas: first we consider an example of oxygen on tungsten studied with XPS[16, 17]. This work illustrated that shifts of the W $4f$ peaks at the surface layer, due to chemisorption, were much smaller than those due to oxidation (the oxide peaks are marked by arrows in fig. 11). In comparing the spectrum from clean W(110) with those from W(110) covered with chemisorbed oxygen one must realize that about $(60 \div 80)\%$ of the signal came from bulk W atoms which are almost unaffected by the chemisorption. The results in fig. 11 also illustrate occurrence of two phases in the chemisorption. In the second phase some sort of suboxide seems to be formed, in which the W $4f$ levels shift by about 600 meV to higher BE.

The advantage of synchrotron radiation can be seen when fig. 11 and 12 are compared. In fig. 12, taken with 70 eV p-polarized light, the bulk and surface peaks are clearly better resolved, not because of any fundamental physical law, but because of the better resolution of the experiment. After exposure to oxygen the peak from the surface atoms is seen to have shifted from the low to the high BE side of the bulk peak. The coverage at this point is probably about $(8 \div 9) \cdot 10^{14}$ atoms cm^{-2} [16, 17]. There is clearly an additional shoulder at about

0.6 eV to the high-BE side of the bulk peak at this coverage, and we may postulate that larger exposures to oxygen would have made this peak more prominent, as in the earlier XPS studies.

Chemical shifts of the core levels in surface atoms can be induced without

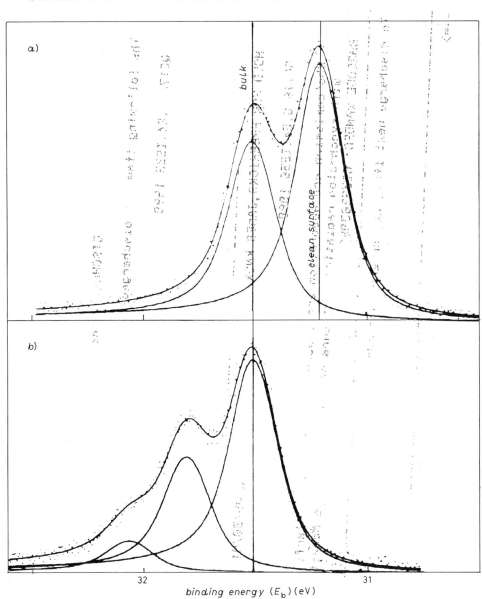

Fig. 12. – The PS spectrum of the W 4f level from W(110) recorded with 70 eV radiation [15, 18]. The spectra have been decomposed into many-body (Doniac-Sunjic [19, 20]) lineshapes a) for the clean surface and b) for the surface after exposure to 10 L of oxygen.

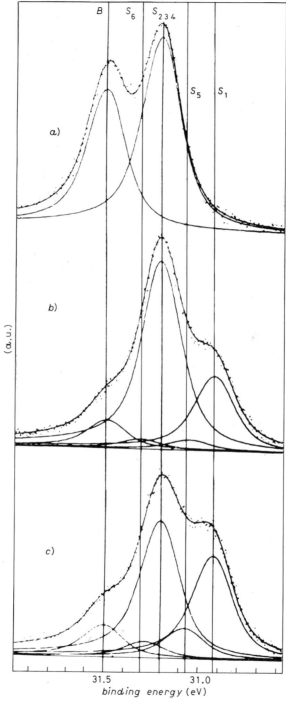

Fig. 13. – Experimental $4f_{7/2}$ core level spectra taken with 70 eV p-polarized light for a) a perfect W(110) surface, b) a stepped W[6(110) × (110)] surface at an azimuthal angle of $\varphi = 0°$ and c) as b) but with $\varphi = 90°$ [18, 21]; $h\nu = 70$ eV, $\theta = 0°$.

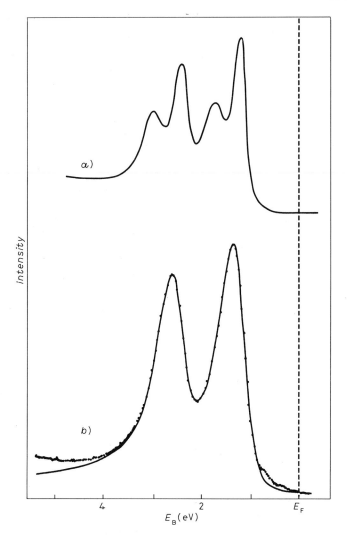

Fig. 14. – XPS spectra of the Yb 4f levels recorded with Al K_{α} (1486.7 eV) radiation and about 100 eV radiation: a) synchrotron PS, $h\nu = 0.1$ keV, $\Delta E = 0.3$ eV; b) XPS, $h\nu \sim 1.5$ keV, $\Delta E \sim 0.6$ eV; $2\Gamma = 0.16$, $\alpha = 0.15$ [22].

addition of gases, by surface reconstruction or steps in the surface. As an example we choose a study of the stepped W[6(110) × (110)] surface, illustrated in fig. 13. This surface has steps six atoms wide. In each step there is a row of atoms (S_1) at the outer edge of the step which protrude most from the surface. These are followed by five other rows (S_2-S_6), most of which are in an environment quite similar to that of the surface atoms on the perfect W(110) surface. By a geometric trick the bulk peak at about 31.5 eV has smaller weight on the stepped surface. Note the strong shoulder, S_1, at the low-BE side which is attributed to the peak from the row of surface atoms at the very edge of the step. The contributions from the other surface atoms are not well resolved.

In general the shifts due to reconstruction are less than those due to chemisorption or steps. It is interesting that they can be observed, but I am not quite sure what one can learn from the observations. Maybe angular dependence of the ratio of different surface peaks will be used as an indicator of surface geometry.

As a second example of core level shifts I consider the spectra of polycrystalline Yb metal in fig. 14 (adapted from ref. [22]). The lower spectrum was taken in using monochromated Al K_α radiation (1486.7 eV). The peaks are due to the Yb $4f$ electrons, and, as their binding energies are only a few volt, the electron KE is > 1480 eV and the electron mean free paths is about $(15 \div 20)$ Å. The contribution to the spectrum from the first atomic monolayer at the surface is thus only about 10% and at the experimental resolution of about 600 meV this cannot be resolved. The spectrum has been analysed in terms of two asymmetric lines and within this model a lifetime broadening of about 160 meV was found. (The asymmetry is a many-body effect; the lineshapes are normally described as Doniac-Sunjic because of the authors who gave the formula most widely used to fit data [19], although work by MAHAN was also important in this area [23]. A good account of these effects is given in ref. [20].)

The top spectrum in fig. 14 is a spectrum taken with 100 eV radiation [22] and the resolution is about 300 meV. This allows the surface and bulk contributions to be resolved. The surface contribution is at higher BE, and its weight is clearly comparable to that of the bulk contribution because of the lower electron escape depth of the about 10 eV electrons.

In accordance with our aim of giving a basis for appreciation of the role of synchrotron radiation in this field we should try to give a balanced account of what one can achieve with and without a synchrotron. State-of-the-art resolution using monochromated Al K_α radiation is about 250 meV, but this has been achieved in only two laboratories to my knowledge (Y. Baer's in Zürich/Neuchâtel and Siegbahn's at Uppsala). A resolution of about 600 meV is more common. The first studies of surface chemical shifts due to chemisorption were made using XPS [16, 17] and there has been some nice work done using the variation of the surface contribution with electron take-off angle to isolate the contribution from surface atoms [24]. However, there are at least half a dozen groups working at synchrotrons with a resolution of $(100 \div 150)$ meV, and $(200 \div 250)$ meV is considered almost trivial. By now the bulk of all work on surface core level shifts has been done with use of synchrotron radiation.

There are three quite orthogonal approaches to use and interpret core level shifts. The first use is simply to diagnose surface reconstructions and reactions such as chemisorption or oxidation. There can be little argument with that approach. The second uses a thermodynamic approach and Born-Haber cycles to relate the surface chemical shifts to surface energies and the energetics of surface alloy formation [25]. The relationship works well and allows unification of a lot of data [18, 25, 26], but some people ask whether we have learnt anything new with it about the surface. The third approach is to use the shifts as a measure of charge distributions at the surface, *i.e.* as a tool to answer questions like «What is the charge transfer between surface and bulk atoms?», «What is the charge transfer between substrate and chemisorbed atoms?» and «Is the charge transfer between substrate and adsorbate primarily from the first atomic monolayer of the substrate?». The role of relaxation effects in the observed chemical shifts at the surface is still not well understood, so that use of the core level shifts to diagnose charge distributions at surfaces is not very advanced. It seems that the charge transfer is always small (of the order of 0.1 electrons/atom?) because the shifts are small, but some specialists would be reluctant to regard even that as proven.

Finally we should consider the reliability of the curve resolutions (fitting procedures) that are done on core level PS spectra containing many component lines. Some of the curve resolutions in the literature are pure speculation. I know from our experience that such fits are seldom completely unambiguous, even with signal-to-noise ratios as high as 100:1. One always gives the fitting program a model. One can certainly say that, if a model of (say) X peaks *cannot* be made to fit the observed spectrum, then it is not a good model. However, if the spectrum has unresolved peaks, then just because a model with X peaks does fit, it does not prove that the model represents the real situation, merely that it is a feasible model.

7. – Atomic-multiplet effects.

I will illustrate here some uses of atomic-multiplet effects in synchrotron studies of solids. The uses frequently relate to the wider question of what is the relative importance of band structure and atomic correlation effects in solids. We know that the valence electrons of free-electron-like metals, such as aluminium, may be treated by band structure methods, and that the $4f$ electrons of most rare earths behave as if they were unfilled core levels and atomiclike interactions dominate their behaviour. In between these two extremes we have no clear theory of how to treat both interactions and some experimental guidance is necessary. XAS and XPS can provide some guidance to these questions.

Because the teaching of atomic-multiplet concepts is no longer universal, I give a short introduction. If an atom or ion has only closed shells (*e.g.*, Ne $1s^2 2s^2 2p^6$), then it only has one atomic energy level. The situation is more complex with an open shell because of the electron spin and orbit angular momentum. It is still more complex if there are holes in more than one shell, but this is precisely the situation found in our spectroscopies and which must be discussed.

The terminology associated with multiplet structure is not transparent because of its historical development. In the early 1920's spectroscopists working with independent electron diagrams, or similar tools, were faced with a forest of fine structure and discrete lines in atomic spectra. They realized that the spectra of atoms with one electron outside a closed shell were similar, and that those with two electrons outside a closed shell were basically similar, etc. They also realized that the lines in their spectra were due to transitions of electrons between the levels or electronic states of the atom and that these were grouped together. They had even found that they could associate different quantum numbers with these levels and talk about selection rules, without a clear idea of what their quantum numbers, L, S, J, really meant and at that stage they never could be sure whether they had found the last members in a series of transitions. (This is why they started the names for their series in the middle of the alphabet.) Their terminology could have been simplified on the light of later knowledge, but in fact never was.

Each electron has a spin quantum number, s, and an angular-momentum quantum number l. l and s must be regarded as a vector with a length ($s = 1/2$, $l = 0$ for s electrons, 1 for p electrons, 2 for d electrons, etc.). The vectors have certain allowed orientations such that the projection of s and l along the quantization axis may take the values $m_s = \pm 1/2$ and $m_l = +l, l-1, ..., -l$. The l and s of the different electrons must couple together and the way this is done gives the *term symbol* which is necessary for the full description of any atomic configuration. We first illustrate this in the Russell-Saunders scheme, developed for Ca in 1925 [27]. First consider a $1s^2 2s^2 2p^2$ configuration. There are two complete subshells, which do not interest us, and then two electrons in the six $2p$ orbitals which can be described by their angular momentum ($l = 1, 0, -1$) and spin ($+1/2, -1/2$) quantum numbers.

Consider first the arrangement with two electrons with $l = +1$, $s = +1/2$ and $-1/2$:

s	$l =$	1	0	-1
$+1/2$		↑		
$-1/2$		↓		

We want to find the term symbol for this state. It is described by three quantum numbers:

$$\text{angular momentum} \qquad L = l_1 + l_2 = 2;$$

$$\text{spin} \qquad S = s_1 + s_2 = 0,$$

$$J = L + S = 2.$$

It is conventional to use small letters for electron quantum numbers and capital letters for atomic configurations. The full term symbol is $^{2S+1}L_J$ with the added complication that an archaic nomenclature for L is used:

$$L = 0 \text{ is represented by } S,$$

$$L = 1 \text{ is represented by } P,$$

$$L = 2 \text{ is represented by } D,$$

$$L = 3 \text{ is represented by } F,$$

$$L = 4 \text{ is represented by } G,$$

$$L = 5 \text{ is represented by } H,$$

etc.

Thus our $2p^2$ configuration above would have the term symbol $^1D_2$.

There are apparently $6 \times 5 = 30$ ways of putting two electrons into these six boxes, but there are actually far less term symbols and energy levels because many of the arrangements are equivalent. This equivalence arises because the L, S and J quantum numbers should be regarded as vectors which have certain allowed directions. Thus the resultant value of L along the quantization axis can be

$$M_L = L, \quad L - 1, \quad L - 2, \quad \ldots, \quad -1, \quad [-L]$$

and for S

$$M_S = S, \quad S - 1, \quad \ldots, \quad -S,$$

and for J

$$M_J = L + S, \quad L = S - 1, \quad \ldots, \quad |L - S|.$$

There is no way we can give a *full* account of the physical meaning of the L, S and J quantum numbers. It is the subject of a large portion of Slater's classic two-volume book on the *Quantum Theory of Atomic Structure* [28] and it is also the subject of much of ESR and NMR theory. We will just give a short description of Russell-Saunders vector coupling ideas. They coupled the vectors one at a time. Let us describe the process in connection with the s's.

If the vectors are coupled one at a time then for 3 s's we start by coupling two of them to get $S = 1$ or 0. Then we couple an additional s of 1/2 to each of these S's. Thus, if we couple $s = 1/2$ to $S = 1$, we can have $S = 3/2$; if we couple $s = 1/2$ to $S = 0$, we can have 1/2. In other words, by adding $s = 1/2$ to any value of S, we get two new S's, one and half unit greater and one and half unit smaller. In this example the three electrons can have $S = 1/2$ or $3/2$ or doublets and quartets which will be split in a magnetic field. More generally we have what is called the branching diagram which is illustrated in fig. 15.

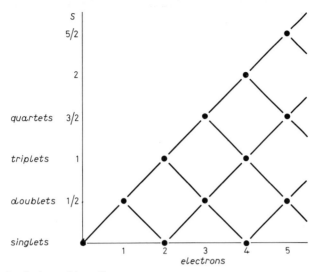

Fig. 15. – A simple branching diagram.

In a similar way RUSSELL and SAUNDERS coupled the l's one at a time.

There are many ways in which this scheme can become more complicated, *e.g.* if the atoms are heavy so that the j's couple first, or if there are more than two electrons outside the filled shell, but the basic ideas are similar. Worse still, it may be necessary to describe the coupling in the so-called intermediate-coupling scheme.

There are two ways of predicting the order and relative energies of the terms. The first is purely qualitative and uses Hund's rules:

1) The ground-state term always has the maximum spin multiplicity.

2) When comparing two states of the same spin multiplicity, the state with the higher value of L is usually more stable.

The second method is quantitative and uses tabulated matrices for multiplets (see ref. [28], p. 28 ff.) and tabulated integrals, when these are available.

We now come to consider observations of these multiplet effects in electron and X-ray spectra. The simplest is the case of spin-orbit coupling in XPS spectra. Here a single hole can be made in a closed, core, subshell. Let us take the 4f level of Yb. In the method of the boxes we could put the hole in a box with $L = +3$, $+2$, $+1$, 0, -1, -2, -3 and with $S = +1/2$, $-1/2$. However, these arrangements represent microstates of just two terms with $l = 3$, $s = +1/2$, $j = 7/2$ or

$l = 3$, $s = -1/2$, $j = 5/2$. There are thus just two $4f$ peaks in the XPS spectrum of Yb (with binding energy $\approx 2\,\text{eV}$, see fig. 14).

A second, and more complicated example is XPS with open valence shells in which the electrons are highly localized. Now we can get coupling between the core hole and the unfilled valence electrons and a very complex spectrum may ensue. As a simple example you can consider the Mn^{2+} ion with the configuration $1s^2 2s^2 2p^6 3s^2 3p^6 3d^5$. The ground-state configuration has the term symbol ^6S, with all the five $3d$ electrons having the same spin. If a hole is created in the $3s$ or $2s$ shells, then the spin of the other electron may be parallel or antiparallel to the 5 $3d$ electrons and either ^5S or ^7S final-state terms result. A more complicated example is shown for the $4f$ ionization of Eu in fig. 16. The ground-state term is

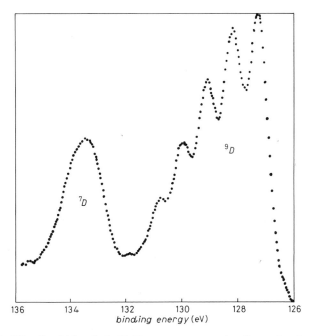

Fig. 16. – XPS of the Eu $4d$ levels in EuTe—as an example of core level multiplets. The coupling of the $4f^7$ and $3d^9$ multiplets means that $4d_{3/2}$ and $4d_{5/2}$ are no longer good descriptions. Picture adapted from Kowalczyk's Ph.D. thesis, Berkeley (1976).

presumably ^8S (Eu has 7 $4f$ electrons). If we now couple this to a $4d_{5/2}$ or $4d_{3/2}$ hole, we will get the ^9D or ^7D final-state terms. Note that I have not included the J subscripts here. One could label the regions in the spectra $4d_{5/2}$ and $4d_{3/2}$, but it is not really a good description. Note that in fig. 16 the intensity ratio of the two regions is quite different from the 6:4 expected of $4d_{5/2}$ and $4d_{3/2}$.

Next we should consider a complicated XAS spectrum: the $3d \rightarrow 4f$ spectra region of Nd. Nd has 3 $4f$ electrons in the ground state, so that X-ray absorption near the threshold probes the $3d^{10}4f^3 \rightarrow 3d^9 4f^4$ excitation. The $3d^9 4f^4$ multiplet is indeed rather complex and there are 1007 lines which overlap too strongly to be isolated, as seen in the top curve of fig. 17, which is the result of a calculation [29]. In practice the observed spectrum contains less lines because of the

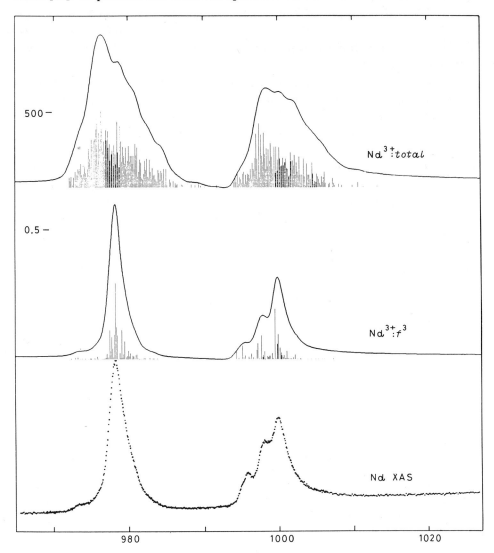

Fig. 17. – Spectra of Nd. The spectrum labelled XAS is the experimental curve. The label f^n denotes the theoretical dipole excitation spectrum from the Hund's rule ground level of the f^n configuration to the levels of $d^9 f^{n+1}$. The spectrum labelled «total» gives each level of $d^9 f^{n+1}$ with weight $2J + 1$. The horizontal axis gives the excitation energy in eV. The experimental spectrum has an unknown vertical scale. The scale of the theoretical one is indicated to the left. The dipole spectra give the absolute cross-section, σ, in Å², the «total» spectrum gives the density of states in numbers of atomic levels per eV for the curves. The vertical lines have bee normalized; close lying levels have been added. the figure has been adapted from ref. [29].

X-ray absorption selection rules. In fact, for any photon-induced transition the dipole selection rules are mostly good. These rules state that

$$\Delta L = \pm 1, \quad \Delta J = 0, \pm 1 \quad \text{(but not } J = 0 \rightarrow J = 0\text{)},$$

so that for a given ground-state term not all the terms of the final state can normally be reached.

It is not our purpose to point out all the potential uses of electron and X-ray absorption spectroscopies here, but a few comments are in order. The observation of structure due to the multiplet interactions between a core hole and the valence electrons can be diagnostic of localized valence electrons. Thus, for instance, one would not expect multiplet structure in the $2s$ XPS peaks of Al metal because the Al valence electrons are bandlike. But in the rare earths, where the $4f$ electrons are atomiclike, one does expect multiplet splittings in XPS, due to interaction between the core hole and the unpaired electrons. Unfortunately there are some complications which undermine the physical basis of such arguments in PS and I do not consider use of multiplet structure in PS any further here.

The use of multiplet effects in XAS as a probe of ground-state properties is less fraught with theoretical difficulties than XPS. A typical use concerns the situation when one needs to know which atomiclike configurations are important in the ground state of a material. This may be uncertain because of changes in

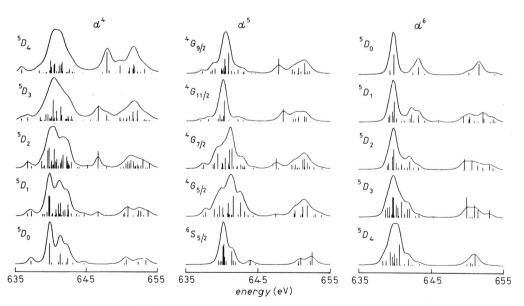

Fig. 18. – Calculated Mn $L_{2,3}$ EELS or XAS spectra for various possible ground-state terms. Only the lowest-energy configurations and terms for Mn are included. The figure is taken from ref. [30].

the order of low-energy configurations in a solid or to hybridization with the valence states which may lead to mixing of more than one configuration in the ground state. The low-energy configurations may only be some meV apart, but they can still lead to changes on a scale of volt in a spectrum centred at about 1 keV, due to the selection rules. Thus, for example, for Nd in fig. 16, the XAS spectrum would have looked very different if the ground-state term could have

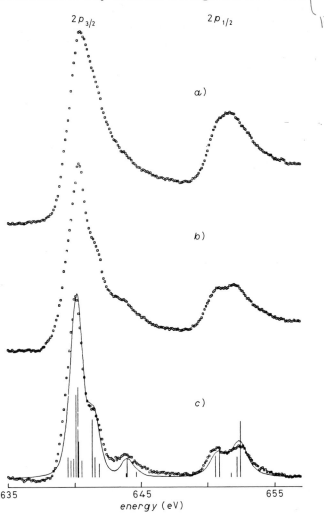

Fig. 19. – The Mn $L_{2,3} \rightarrow M_{4,5}$ XAS spectra of a) pure Mn metal, b) 3.5% Mn in Cu and c) 5% Mn in Ag. The theoretical lines were calculated for a d^5 (^6S) ground state as described in the text, with Slater integrals $F_{dd^2} = 8.25$ eV, $F_{dd^4} = 5.13$ eV and $\zeta_d = 0.04$ eV for the ground state and $F_{dd^2} = 8.92$ eV, $F_{dd^4} = 5.55$ eV, $F_{pd^2} = 5.06$ eV, $G_{pd^1} = 3.69$ eV, $G_{pd^3} = 2.09$ eV, $\zeta_p = 6.99$ eV and $\zeta_d = 0.05$ eV for the $2p^5d^6$ final states. The solid curve results from a convolution of a 0.4 and 1.0 eV full width at half maximum Lorentzian for the $2p_{3/2}$ and $2p_{1/2}$ regions, respectively, and a Gaussian of width 0.6 eV. The figure is taken from ref. [30].

been changed, so that transitions to a different subset of 1007 available $3d^9 4f^4$ states were allowed. For the rare earths this approach has been used for diagnosis of mixing of higher-energy terms into the ground state [31] and of mixed valence [31-33] in rare earths.

As another example of its uses we choose again a $2p \to 3d$ spectrum of a transition metal ion. Mn may be found in various configurations in its compounds. We illustrate in fig. 18 the multiplet structure calculated for $2p \to 3d$ XAS for various terms of the $3d^4$, $3d^5$ and $3d^6$ ground-state configurations [30]. There is clearly great sensitivity to the actual ground-state term. Note that the shape of the individual $2p_{3/2}$ and $2p_{1/2}$ groups are also, in general, quite different, which is important for distinguishing multiplet structure from intrinsic losses and various other effects. If we now examine the recorded Mn 2p XAS spectra for Mn as a dilute impurity in films of Ag or Cu (fig. 19), we find multiplet structure which is quite distinct from the bandlike structure of pure Mn metal. The multiplet structure observed is very similar to that calculated for excitation from a Mn ^6S $3d^5$ ground state. We conclude that the ^6S $3d^5$ term dominates the ground-state electronic structure for Mn dissolved in Ag or Cu [30] and that the hybridization of the Mn 3d levels with the Ag 4s valence band is too weak to result in mixing with higher-energy terms.

8. – Magnetic X-ray dichroism.

The last task of this lecture is to give a simple explanation of a new effect known as magnetic X-ray dichroism [34, 35]. It can be most clearly explained for the case of the $M_{4,5}$ absorption of Yb, which consists of just one line corresponding to the transition

$$\text{Yb } 3d^{10} 4f^{13}(^2F_{7/2}) \to d^9 4f^{14}(^2D_{5/2}).$$

As shown in fig. 20, the presence of a magnetic field H parallel to the x-axis splits both the initial and the final states into their Zeeman components with energy separation $g\mu H$. Such a splitting will remain unnoticed in the XAS spectrum as it is much smaller (only some meV) than the experimental linewidth ($\sim 300\,\text{meV}$). But now the degeneracy of the 18 ΔM lines is lifted as indicated by the arrows in the right-hand side of fig. 20. The numbers below these arrows are the relative amplitudes as given by the square of the 3-j symbol. In the diagram the lines are grouped according to the value of ΔM and also the polarization state of the light corresponding to each group is indicated.

The occupation of the levels of the initial state is governed by a Boltzmann distribution $\exp[-M/\Theta]$, where Θ is the reduced temperature $kT/g\mu H$. If $\Theta > 1$, the levels will be equally occupied. However, at lower Θ the upper n_j levels of the ground state J will become depleted and the absorption by these levels will

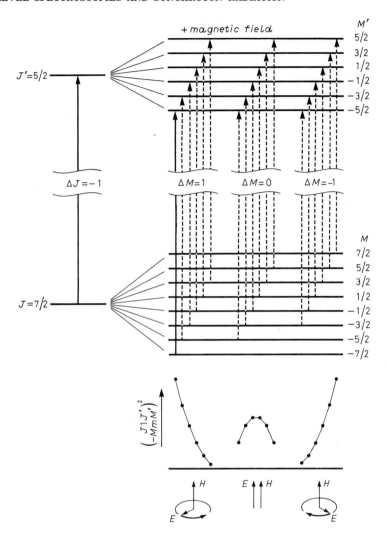

Fig. 20. – Schematic diagram of the Yb $3d^{10}4f^{13} \rightarrow 3d^{9}4f^{14}$ transition in the presence of a magnetic field. Figure taken from ref. [35].

diminish. As can be seen from the diagram, this causes the absorption strength of the different ΔM groups to become unequal and this means that the absorption coefficient is different for different polarization. This is most clearly demonstrated in the limiting case of $T = 0\,\mathrm{K}$ where only the $M = -7/2$ level is occupied, so that the solid arrow with $\Delta M = 1$ in fig. 20 is the only possible transition. In this situation the Yb absorption line is seen to be absent for clockwise circular polarization normal to H as well as for linear polarization parallel to H. This behaviour is typical for a $\Delta J = 1$ line. A similar mechanism acts on $\Delta J = 0$ and 1 lines, but leads to other dependences on θ.

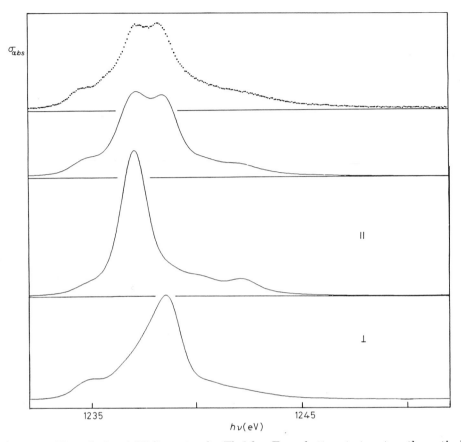

Fig. 21. – The calculated XAS spectra for Tb $3d_{5/2}$. From bottom to top: two theoretical spectra with $T = 0$ and with polarization perpendicular and parallel to the internal-magnetic-field direction, the theoretical spectrum for unpolarized light and the spectrum observed experimentally. Figure adapted from ref. [35].

In order to calculate the MXD effect in other rare-earth elements, it is necessary to follow the above procedure for all possible lines, taking into account the possible fractional-parentage effects. The result, for the M_5 groups of XAS lines, is shown in fig. 21. The theoretical result for $T > T_c$ strongly resembles the experimental spectrum taken with polarized light. However, the different ΔJ components of the spectrum have different weights in the different regions of the spectrum, and they have different temperature dependences. Thus the calculated spectra for a magnetically ordered system depend on the relative orientation of the polarization vectors of the light and the Dy ions. Strong effects are predicted for nine of the rare earths.

MXD is a new phenomenon and its potential uses must be investigated. However, there are few methods which can be used to investigate the orientation of the moments of individual ions. The most prominent method is neutron scattering, but this is not really applicable to thin magnetic layers. MXD and neutron scattering are thus complementary. In addition this MXD in XAS could be used to produce polarized X-ray beams.

* * *

I thank my many co-workers over the last few years, and in particular J. B. GOEDKOOP, G. A. SAWATZKY and G. VAN DER LAAN for letting me plagiarize abstracts from work as yet unpublished.

This work was supported in part by The Netherlands Foundation for Chemical Research (Stichting Scheikundig Onderzoek Nederland) with financial aid from The Netherlands Organization for Scientific Research (Nederlandse Organisatie voor Wetenschappelijk Onderzoek) and by the research program of the Stichting voor Fundamenteel Onderzoek der Materie (Foundation for Fundamental Research of Matter). We also had significant support from the Committee for the European Development of Science and Technology (CODEST) program. We thank the staff of LURE in Paris for their cooperation.

REFERENCES

[1] J. C. FUGGLE: *Phys. Scr.*, T17, 64 (1987).
[2] J. FINK, TH. MÜLLER-HEINZERLING, B. SCHEERER, W. SPEIER, F. U. HILLE-BRECHT, J. C. FUGGLE, J. ZAANEN and G. A. SAWATZKY: *Phys. Rev. B*, 32, 4899 (1985).
[3] J. ZAANEN, G. A. SAWATZKY, J. FINK, W. SPEIER and J. C. FUGGLE: *Phys. Rev. B*, 32, 4905 (1985).
[4] See, *e.g.*, J. C. FUGGLE: in *Electron Spectroscopy*, Vol. 4, edited by C. R. BRUNDLE and A. D. BAKER (Academic Press, London, New York, N.Y., 1981), p. 85 ff.
[5] J. C. FUGGLE and S. F. ALVARADO: *Phys. Rev. A*, 22, 1615 (1980), and references therein, especially those to E. J. McGUIRE.
[6] M. P. SEAH and W. DENCH: *J. Surf. Int. Anal.*, 1, 2 (1979).
[7] J. KOOPMANS: *Physica*, 1, 104 (1933).
[8] See, *e.g.*, J. G. FUGGLE: *Bull. Am. Phys. Soc. (AIP. Conf. Proc. Ser.)*, 94, 661 (1982).
[9] W. SPEIER, R. ZELLER and J. C. FUGGLE: *Phys. Rev. B*, 32, 3597 (1985).
[10] See, *e.g.*, L. G. PARRATT: *Rev. Mod. Phys.*, 31, 616 (1959).
[11] See, *e.g.*, R. LÄSSER and J. C. FUGGLE: *Phys. Rev. B*, 22, 2637 (1980).
[12] S. W. KORTBOYER, M. GRIONI, W. SPEIER, R. ZELLER, L. M. WATSON, B. T. GIBSON, F. SCHÄFERS and J. C. FUGGLE: *J. Phys. Condensed Matter*, 1, 5981 (1989).
[13] A. ERBIL, G. S. CARGIL III, R. FRAHM and R. F. BOEHME: *Phys. Rev. B*, 37, 2450 (1988).
[14] F. SIEGBAHN, C. NORDLING, G. JOHANSSON, J. HEDMAN, P. F.HEDÉN, K. HAMRIN, U. GELIUS, T. BERGMARK, L. O. WERME, R. MANNE and Y. BAER: *ESCA Applied to Free Molecules* (North Holland, Amsterdam, 1971).
[15] TRAN MINH DUC, C. GUILLOT, Y. LASAILLY, J. LECANTE, Y. YUGNET and J. C. VÉDRINE: *Phys. Rev. Lett.*, 43, 789 (1979).

[16] J. C. FUGGLE and D. MENZEL: *Surf. Sci.*, **53**, 21 (1975); *Chem. Phys. Lett.*, **33**, 37 (1975).

[17] J. C. FUGGLE: in *Handbook of X-ray and Ultra-Violet Photoelectron Spectroscopy*, edited D. BRIGGS (Heyden, London, 1977), p. 301.

[18] D. SPANJAARD, C. GUILLOT, M.-C. DESJONQUÈRES, G. TRÉGLIA and J. LECANTE: *Surf. Sci. Rep.*, **5**, 1 (1985).

[19] S. DONIAC and M. SUNJIC: *J. Phys. C*, **3**, 285 (1970).

[20] G. K. WERTHEIM and P. H. CITRIN: in *Photoemission in Solids I*, edited by M. CARDONA and L. LEY (Springer, Berlin, 1978), p. 197.

[21] D. CHAVEAU, P. ROUBIN, C. GUILLOT, J. LECANTE, G. TRÉGLIA, M. C. DESJONQUÈRES and D. SPANJAARD: *Solid State Commun.*, **52**, 635 (1984).

[22] S. F. ALVARADO, M. CAMPAGNA and W. GUDAT: *J. Electron Spectrosc. Relat. Phenom.*, **18**, 43 (1980); M. CAMPAGNA: private communication.

[23] See, *e.g.*, G. D. MAHAN: *Phys. Rev.*, **163**, 612, 1049 (1967); *Phys. Rev. B*, **11**, 4814 (1979), and references therein.

[24] P. H. CITRIN, G. K. WERTHEIM and Y. BAER: *Phys. Rev. Lett.*, **41**, 1425 (1978).

[25] B. JOHANSSON and N. MÅRTENSSON: *Phys. Rev. B*, **21**, 4427 (1980); A. ROSENGREN and B. JOHANSSON: *Phys. Rev. B*, **22**, 3706 (1980).

[26] G. K. WERTHEIM: *J. Electron Spectrosc.*, **47**, 271 (1988).

[27] H. N. RUSSELL and F. A. SAUNDERS: *Astrophys. J.*, **61**, 38 (1925).

[28] J. C. SLATER: *Quantum Theory of Atomic Structure*, Vol. I and II (McGraw-Hill, New York, N.Y., 1960).

[29] B. T. THOLE, G. VAN DER LAAN, J. C. FUGGLE, G. A. SAWATZKY, R. C. KARNATAK and J.-M. ESTEVA: *Phys. Rev. B*, **32**, 5107 (1985).

[30] B. T. THOLE, R. D. COWAN, G. A. SAWATZKY, J. FINK and J. C. FUGGLE: *Phys. Rev. B*, **31**, 6865 (1985).

[31] G. VAN DER LAAN, B. T. THOLE, G. A. SAWATZKY, J. C. FUGGLE, R. KARNATAK, J.-M. ESTEVA and B. LENGELER: *J. Phys. C*, **19**, 817 (1986).

[32] J. C. FUGGLE, F. U. HILLEBRECHT, J.-M. ESTEVA, R. C. KARNATAK, O. GUNNARSSON and K. SCHÖNHAMMER: *Phys. Rev. B*, **27**, 4637 (1983).

[33] G. KAINDL, G. KALKOWSKI, W. D. DREWER, B. PERSCHEID and F. HOLTZBERG: *J. Appl. Phys.*, **55**, 910 (1984).

[34] B. T. THOLE, G. VAN DER LAAN and G. A. SAWATZKY: *Phys. Rev. Lett.*, **55**, 2086 (1985); G. VAN DER LAAN, B. T. THOLE, G. A. SAWATZKY, J. B. GOEDKOOP, J. C. FUGGLE, J.-M. ESTEVA, R. C. KARATAK, J. P. REMEIKA and H. A. DABKOWSKA: *Phys. Rev. B*, **34**, 6529 (1986).

[35] J. B. GOEDKOOP, B. T. THOLE, G. VAN DER LAAN, G. A. SAWATZKY, F. M. F. DE GROOT and J. C. FUGGLE: *Phys. Rev. B*, **37**, 2086 (1988).

X-Ray Absorption and Reflection in the Hard-X-Ray Range.

B. LENGELER (*)

Siemens AG, Zentrale Forschung und Entwicklung - München, B.R.D.

1. – X-ray absorption spectroscopy (XAS).

1`1. *Introduction.* – When X-rays pass through matter, their intensity is attenuated. The relationship between the intensities $I_1(E)$ and $I_2(E)$ before and after passing a layer of thickness d reads

$$(1.1) \qquad I_2 = I_1 \exp[-\mu(E)\,d].$$

The linear absorption coefficient $\mu(E)$ depends on the photon energy E and on the absorbing material (fig. 1.1). $\mu(E)$ shows edges at certain energies, which are characteristic for the absorber. At these absorption edges the photon energy is high enough for strongly bound electrons to be excited into empty states. A new absorption channel opens up and μ increases abruptly with E. At a K-edge $1s$ electrons are excited, at L_1, L_2 and L_3 edges $2s$, $2p_{1/2}$ and $2p_{3/2}$ electrons are excited into empty states. Figure 1.2 shows the linear absorption coefficient for two Ti compounds (TiO and TiO$_2$) in the vicinity of the Ti K-edge. Two points are noteworthy in this figure. μd shows oscillations above the edge. The oscillating contribution in μ is called EXAFS (extended X-ray absorption fine structure) and contains information on the geometric structure (interatomic distances, coordination numbers) around the absorbing species [1-8]. Secondly, the form and the position of the edge depend on the chemical bonding of the absorbing species. Especially, the higher the valence of the absorber, the higher is the energy of the edge position. The chemical shift of the edge can be as high as 13 eV. The details in the structure of the edge reflect the details in the empty electronic density of states. References [1-8] are some recent review articles on the XAS.

The XAS in transmission mode is a bulk probe. On the other hand, there is an increasing interest in making X-ray techniques surface and interface sensitive. This can be achieved by total external reflection. This interesting possibility will be discussed in the second section of this lecture.

(*) Present address: Institut für Festkörperforschung, Forschungszentrum Jülich, Postfach 1913, D-5170 Jülich.

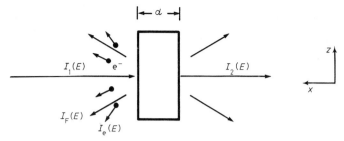

Fig. 1.1. – When passing through matter X-rays are attenuated in intensity. In addition, fluorescence photons and electrons are emitted.

1`2. *Experimental set-up in* XAS. – The most common procedure for measuring the X-ray absorption uses X-rays from an electron or positron storage ring. Electron energies range from 1 to 5 GeV. Electron currents between 50 and a few hundred mA are common. Figure 1.3 shows a schematic view of an absorption spectrometer. The X-rays are monochromatized by a Si double crystal (Si(111), (311) and (511)). The excellent vertical collimation of the SR allows an excellent monochromatization. At 10 keV an energy band 1.2 to 2.5 eV is transmitted by the monochromator depending on the crystal cut. At 3.7 GeV and 100 mA electron current about 10^{10} photons impinge per second on a sample

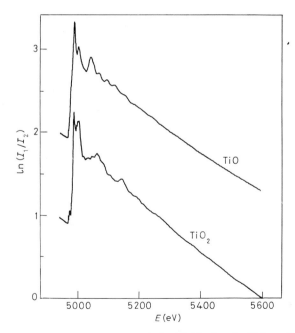

Fig. 1.2. – Absorption $\mu \cdot d$ in the vicinity of the Ti K-edge in TiO and TiO$_2$. There are EXAFS oscillations above the edge. The structure and position of the edge depends among others on the Ti valence.

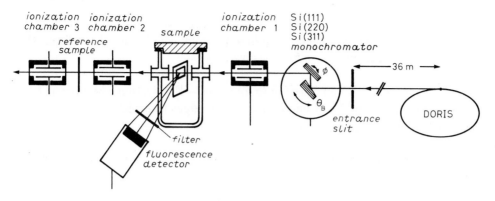

Fig. 1.3. – Schematic view of an X-ray absorption spectrometer.

spot 1×10 mm^2 without focusing. The photon flux can be increased by means of a focusing mirror or by using a wiggler instead of a bending magnet. The incoming intensity $I_1(E)$ and the transmitted intensity $I_2(E)$ are measured by ionization chambers (transmission mode).

In order to eliminate the influence of variations in the beam height on the energy calibration, it is common use to measure the absorption of a reference sample between the second and a third ionization chamber. In the case of the Ti compounds mentioned above this could be a Ti metal foil. By this method the position of an absorption edge can be determined within ± 0.1 eV.

The hole created in a $1s$, $2s$, $2p$... shell will be filled up by an electron from an upper shell. This process is accompanied by the emission of a fluorescence photon or by the emission of Auger electrons. At high ionization energies fluorescence is the dominant decay channel, whereas at lower energies the Auger channel is favoured. The intensities of the emitted photons and electrons show EXAFS oscillations just as the transmitted intensity does. Fluorescence photons are detected by NaI or Si detectors. When the intensities are high, ionization chambers can be used as well. The elastic radiation emitted by the sample can be separated by an appropriate filter through which the fluorescence photons are mostly transmitted[5]. The fluorescent radiation being characteristic for the absorbing species, the fluorescence mode is mainly used for dilute systems (from about 1 at. % down to about 10 at.p.p.m.). A great advantage of the transmission and fluorescence modes is the possibility to measure samples *in situ*. The relatively high penetration of X-rays through matter allows for a sample chamber with appropriate windows (Kapton, beryllium, Al...). The samples can be surrounded by a gas atmosphere or a liquid. This is especially interesting for high-pressure experiments and for the investigation of catalysts. On the other hand, the electrons emitted after X-ray absorption originate from a thin surface layer. This is a consequence of the short range of electrons in condensed matter when their energy is in the range of a few eV to about 1 keV[9]. XAS in the electron yield mode is, therefore, a surface-sensitive procedure (SEXAFS). Most electron detectors need ultra-high-vacuum conditions. Therefore, this detection mode excludes *in situ* measurements.

1`3. *Origin of the* EXAFS *and data analysis*. – For system which are not too strongly disturbed and for K and L_1 absorption edges the relative change $\chi(k)$ in the absorption coefficient μ can be expressed as [1-8]

(1.2) $\chi(k) \equiv [\mu(E) - \mu_0(E)]/\mu_0(E) =$

$$= \sum_j \frac{N_j}{kr_j^2} F_j(k) D_j(k) \exp\left[-2\sigma_j^2 k^2\right] \sin\left(2kr_j + \phi_j(k)\right).$$

$\mu_0(E)$ is the absorption coefficient that would be measured if the absorbing atoms had no neighbours by which the photoelectron could be scattered. The photoelectron emerges from the absorber as a spherical wave. Other atoms at a distance r_j from the absorber can scatter the photoelectron. The incoming and the scattered wave can interfere (fig. 1.4). When the interference is

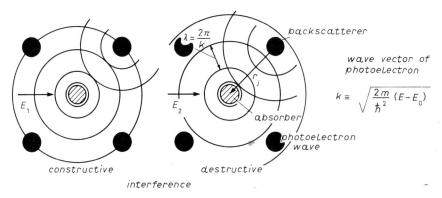

Fig. 1.4. – Origin of the fine structure (EXAFS) in the X-ray absorption coefficient μ.

constructive, the probability to find the photoelectron outside of the absorber is larger than in the case without neighbours, so that $\mu > \mu_0$. For destructive interference $\mu < \mu_0$. If we denote by E_0 the binding energy of the photoelectron, its wave vector k is given by

(1.3) $k = \sqrt{\frac{2m}{\hbar^2}(E - E_0)}.$

Therefore, one expects a periodicity of χ with k. One period is completed when the ratio of the double distance $2r_j$ to the wavelength of the photoelectron has increased by 2π. This is the meaning of the expression $\sin 2kr_j$ in eq. (1.2). The photoelectron suffers additional, energy-dependent phase shifts in the absorber atom and at the backscatterers. They are denoted by $\phi_j(k)$ in eq. (1.2). The amplitude of the EXAFS $\chi(k)$ is proportional to the coordination number N_j of each species in the j-th shell, which have a characteristic backscattering strength $F_j(k)$. When the atoms in the j-th shell are distributed around the average distance r_j (through thermal motion or by static disorder) and when the distribution is Gaussian with a standard deviation σ_j, then the amplitude is reduced by a Debye-Waller factor $\exp[-2\sigma_j^2 k^2]$. Inelastic-scattering events destroy the interferability of the photoelectron. This is the meaning of the loss factor $D_j(k)$ in eq. (1.2). The summation over j takes care of the different coordination shells with different atomic species in each shell. Similar although somewhat more complicated expressions for χ have been derived for the absorption at L_2 and L_3 edges [1-8].

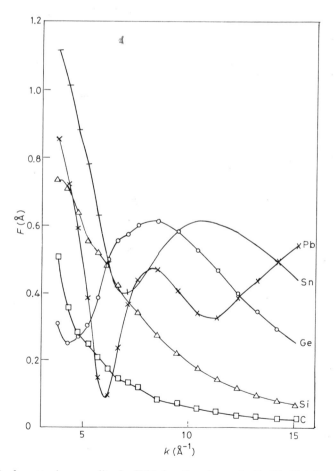

Fig. 1.5. – Backscattering amplitude $F(k)$ for the elements C, Si, Ge, Sn and Pb [10].

The information on the local structure around an absorber contains the distances to the nearest neighbours, the spread of the distribution around this mean distance and the corresponding coordination numbers. In order to deduce this information from the measured EXAFS χ, one needs the scattering phase shifts $\phi_j(k)$ and the amplitude factors $F_j(k)D_j$. There are detailed calculations of the k-dependent phases $\phi_a(k)$ and $\phi_b(k)$ suffered by the photoelectron in the absorber a and in the backscatterer b, as well as for the backscattering amplitude $F(k)$ [10]. Figure 1.5 shows $F(k)$ for the elements of the fourth column in the periodic table C, Si, Ge, Sn and Pb. $F(k)$ decreases for the light elements with increasing energy. The heavier elements show one or two pronounced minima which are due to Ramsauer-Townsend resonances. At certain energies the atoms are almost transparent for the electrons. Pb has two minima, at 6 and at 11 Å^{-1}. These resonances show also up in the backscatterer phase ϕ_b (fig. 1.6).

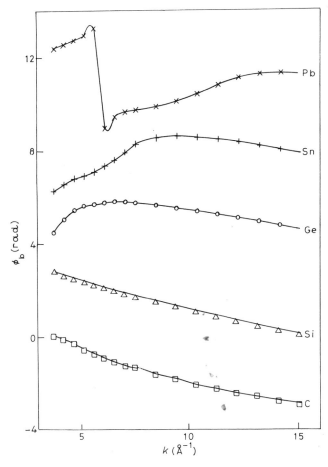

Fig. 1.6. – Phase shift $\phi_b(k)$ of a $1s$ electron scattered by the elements C, Si, Ge, Sn and Pb [10].

On the other hand, the phase of the absorber decreases always monotonously with k (fig. 1.7). The phases and amplitudes calculated by TEO and LEE[10] surely describe the behaviour of the phases and amplitudes qualitatively in a correct way. Nevertheless, there is an increasing tendency to use experimentally determined phases and amplitudes. For this purpose an appropriate model compound is chosen which is as similar in structure as possible to the system to be investigated. For instance, for the determination of the lattice distortion around Ti atoms in a dilute $\underline{Ni}Ti$ alloy the intermetallic compound $TiNi_3$ is an appropriate model. It has 12 Ni atoms around each Ti atom in a well-defined and known distance. The phases and amplitudes (including loss factor) determined from the EXAFS at the Ti K-edge in $TiNi_3$ are transferred to the system $\underline{Ni}Ti$ under investigation. The transferability of the phases is experimentally well established. That of the amplitudes depends more on the chemical bonding. For a

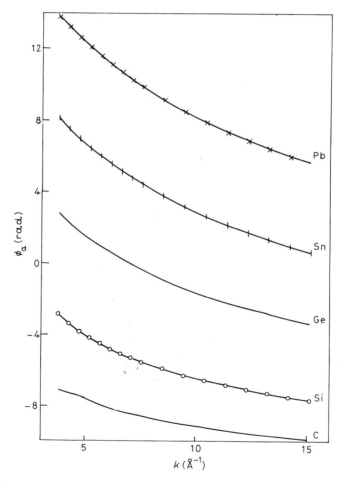

Fig. 1.7. – Phase shift $\phi_a(k)$ of a photoelectron in C, Si, Ge, Sn and Pb as absorber[10].

well-chosen model (as TiN_3 for NiTi) it is justified as well. For light backscatterers (like O, N, C, etc.) caution is needed [2]. Sometimes there is no
appropriate intermetallic compound to a given pair ab of absorber and
backscatterer. In this case a compound with neighbouring elements in the
periodic table can be used. The errors were shown to be small and linear in
ΔZ [11]. They can easily be corrected when ΔZ is only a few units.

The advantages of using experimental amplitudes and phases compared to
theoretical ones are the following. The EXAFS formula (1.2) has been derived
under a number of assumptions. The photoelectron is treated as a plane wave at
the position of the neighbouring shells. This is only an approximation which is
not well fulfilled for the nearest neighbours. The simple EXAFS formula (1.2)
does not contain multiple-scattering contributions, which are largest for small k
values and for collinear configurations of 3 atoms in a row. Furthermore, the
zero of energy E_0 is experimentally and theoretically not well defined. There is
good reason to assume that these effects are similar in a well-chosen model
compound and in the system to be investigated and that in first approximation
they cancel out.

An EXAFS analysis will now be demonstrated in the case of the lattice
distortion around Ti in a dilute NiTi alloy with 2 at. % Ti. First, the oscillating
contribution $\chi(k) \cdot k^2$ is extracted from the measured absorption (fig. 1.8 and 1.9).
Figure 1.10 shows the Fourier transform with at least 4 different coordination
shells. The contribution of the individual shells can be separated by means of a
filter function (dashed line in fig. 1.10) and back transformed into k-space

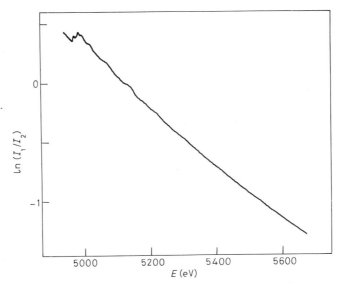

Fig. 1.8. – X-ray absorption at the Ti K-edge in a dilute NiTi alloy with 2 at. % Ti,
measured at 77 K.

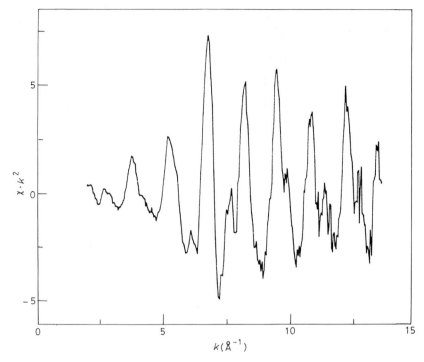

Fig. 1.9. EXAFS $\chi \cdot k^2$ at the Ti K-edge from fig. 1.8.

(fig. 1.11). This contribution is fitted with the phases and amplitudes taken from TiNi$_3$. In the present case the 12 nearest neighbours around a Ti atom are shifted outwards by (0.022 ± 0.005) Å. Within the accuracy of the data the higher shells are at the lattice sites in pure Ni.

A few more examples should show the possibilities and limitations of the EXAFS spectroscopy.

Example 1: evaporated amorphous Ge *films* [12]. Amorphous germanium (a-Ge) can be produced by sputtering, evaporation, dissociation of GeH$_4$, electrolytic deposition and by bombardment with high-energy particles. The most common model for the structure of a-Ge is the Polk model [13]. It can be best understood by considering the structure of crystalline Ge (c-Ge). c-Ge has the diamond structure (fig. 1.12). The elementary building block is a tetrahedron, in which each Ge atom is surrounded by 4 Ge atoms. The tetrahedral angle θ (fig. 1.12) has the value 109.5°. Two neighbouring tetrahedra form a dihedron with the basis atom in staggered configuration ($\varphi = 60°$). Amorphicity of Ge is due to a spread of θ and φ around their crystalline values. Nevertheless, almost each Ge atom is surrounded by 4 nearest neighbours and the mean nearest-neighbour distance is practically unchanged. Figure 1.13 shows this Polk model. Only one

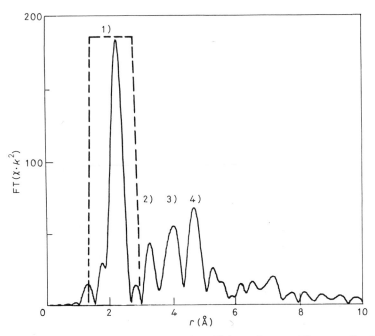

Fig. 1.10. – Fourier transform of the EXAFS $\chi \cdot k^2$ from fig. 1.9. The contribution of the first shell is separated by a filter function (dashed curve).

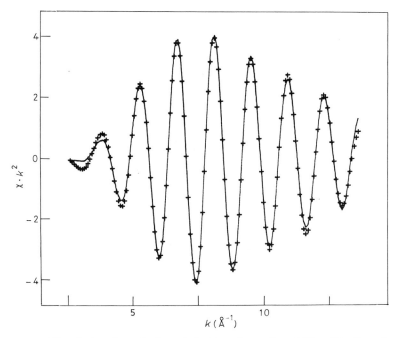

Fig. 1.11. – Fit of the first-shell contribution from fig. 1.10 with the phases and amplitudes taken from $TiNi_3$.

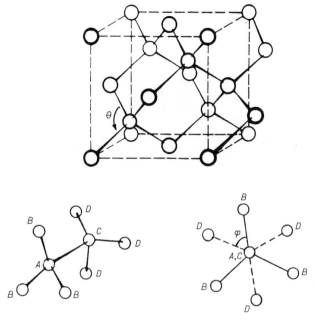

Fig. 1.12. – Diamond structure of c-Ge with tetrahedral angle $\theta_c = 109.5°$ and dihedral angle $\varphi_c = 60°$.

out of a thousand bonds is a dangling bond. It has turned out that the density of amorphous Ge films depends on the mode of production [12]. The densest films are those made by electrolytic deposition. Sputtered films without hydrogen are less dense by up to 10% compared to c-Ge. It seems that the reduced macroscopic

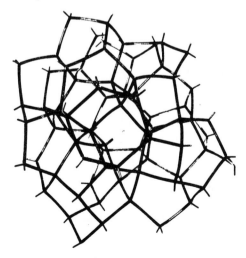

Fig. 1.13. – Polk model of a-Ge. One dangling bond is visible in the middle of the figure [14].

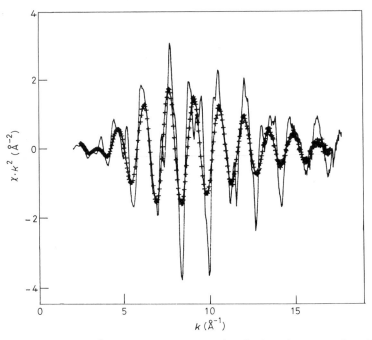

Fig. 1.14. – EXAFS $\chi \cdot k^2$ of a-Ge (–+–+–) and c-Ge (——) measured at 77 K.

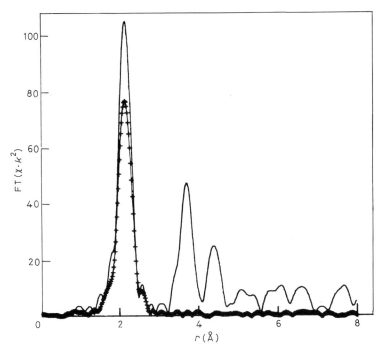

Fig. 1.15. – Fourier transforms of the EXAFS of a-Ge (–+–+–) and c-Ge (——) from fig. 1.14.

density is due to voids. But their form, size and size distribution are not well known[15]. The EXAFS spectroscopy has helped to elucidate this problem[12]. Amorphous Ge films 8 μm in thickness have been deposited by evaporation in high vacuum on oxidized Al foils. Figure 1.14 shows the EXAFS $\chi \cdot k^2$ of a-Ge and c-Ge. The corresponding Fourier transform is shown in fig. 1.15. In c-Ge 8 coordination shells are visible in the distance range up to 8 Å. The first three have been used in the analysis. In a-Ge only the first one is visible and that one with a reduced amplitude. This shell has been analysed with the phases and amplitudes from the first shell of c-Ge. Crystalline Ge is an excellent model for a-Ge. In a total 29 a-Ge samples have been analysed. The distribution of the interatomic distances r_1, coordination numbers N_1 and Debye-Waller factors

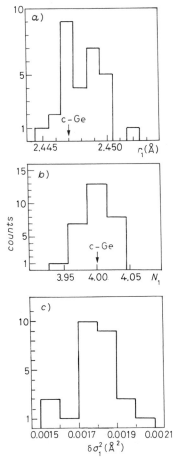

Fig. 1.16. – Distribution of the interatomic distance (*a*)) ($r_1 = (2.448 \pm 0.003)$ Å), coordination number (*b*)) ($N_1 = 4.00 \pm 0.04$) and Debye-Waller factor (*c*)) ($\delta\sigma_1^2 = (0.001\,80 \pm 0.000\,15)$ Å^2) for the first-neighbour shell in a-Ge measured at 77 K in 29 different amorphous Ge samples.

TABLE I. – *Comparison of the values* r_1, N_1, $\delta\sigma_1^2$ *for the first coordinatin shell in* a-Ge *with the corresponding values in* c-Ge.

	c-Ge	a-Ge
$r_1(\text{Å})$	2.447	2.448 ± 0.003
N_1	4	4.00 ± 0.04
$\delta\sigma_1^2(\text{Å}^2)$	0	$0.001\,80 \pm 0.000\,15$

$\delta\sigma_1^2 \equiv \sigma_{1a}^2 - \sigma_{1c}^2$, which were determined in the fits, are shown in fig. 1.16. Table I gives a comparison with the corresponding crystalline values. The accuracy reached is excellent.

The error in the distance is 0.1%, that of the coordination numbers is 1%, a value which compares very favourably with the usual error of about 10% in N_1. The excellent accuracy can only be reached when the samples are of high quality (homogeneous films of optimum thickness without pinholes), when the data are of high quality extending over a long k range and when the model is well chosen.

The following information can be deduced from the data on a-Ge:

1) In view of the fact that the first coordination number is 4 ± 0.04 and that the first interatomic distance is within 0.1% equal to the crystalline value, we must conclude that a-Ge contains at best 1% of vacancies or 0.5% of divacancies. Thus vacancies cannot be responsible for the reduced density, in contrast to an assumption made by CARGILL[16]. The reduced density must be due to voids many Å in diameter. Assuming spherical or needle-shaped voids and a reduced density of 9% and assuming the atoms at the surface of a pore to have 3 rather than 4 first neighbours, 3.96 first neighbours imply a pore size of at least 22 Å in diameter for a sphere and 14 Å diameter for a needle.

2) In view of the fact that the second and all higher coordination shells in a-Ge are not visible, the spread in the tetrahedral angle θ can be deduced. This spread generates a spread in the second interatomic distance r_2. In order to reduce the second-neighbour contribution by the amount shown in fig. 1.15, the spread in θ must be at least 9° r.m.s. X-ray scattering data give a spread of 10°[15].

Example 2: lattice site location of hydrogen in metals by EXAFS[17]. The backscattering strength of hydrogen with only one electron is too weak to be observable in EXAFS. Therefore, hydrogen shows up in XAS mainly by the lattice distortion it creates. This effect is not specific enough to determine the lattice site occupied by the hydrogen. Nevertheless, EXAFS can be used in many cases to determine the lattice site of hydrogen and other light atoms.

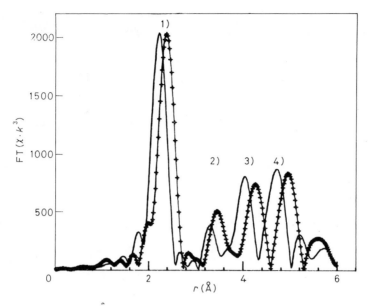

Fig. 1.17. – Fourier transforms of the EXAFS $\chi \cdot k^3$ of Ni (———) and $NiH_{0.85}$ (– + – + –). The hydrogen expands the Ni lattice by 6%.

When the hydrogen is located on the line joining the absorber and the backscatterer, then it changes the phase and the amplitude of the photoelectron on its way to the backscatterer and back, *i.e.* the hydrogen works as a lens for the photoelectron. This effect is demonstrated in fig. 1.17 for Ni and $NiH_{0.85}$. In both systems Ni forms a f.c.c. lattice, in which the hydrogen occupies octahedral sites. Those are the sites between an absorber and its second neighbour (fig. 1.18). The hydrogen acting as a lens increases the second-neighbour amplitude by 50% [17]. The reason for the hydrogen acting as a lens but not as a mirror for the photoelectron is found in the angular dependence of the scattering strength. The forward amplitude is about 10 times larger than the backward amplitude [3]. The lens effect is not limited to hydrogen. But hydrogen shows only the lens effect. When the hydrogen is replaced by a heavier element like oxygen in NiO, which also has the NaCl structure, then the second-neighbour contribution is enhanced (fig. 1.18). The oxygen shows up as a backscatterer as well. The lens effect has turned out to be very helpful in the lattice site location of other light elements (B in Pd [18]) and for investigating the internal oxidation of impurities in metals [19].

The interaction of hydrogen with impurities in Pd will be described in the following [20]. Yttrium in Pd generates a strong lattice expansion. Y pushes the first shell of 12 Pd neighbours outwards by 0.06 Å. When hydrogen is dissolved in Pd, the question arises whether the hydrogen prefers the vicinity of the Y atoms where the lattice has already been expanded, *i.e.* whether the Y traps the

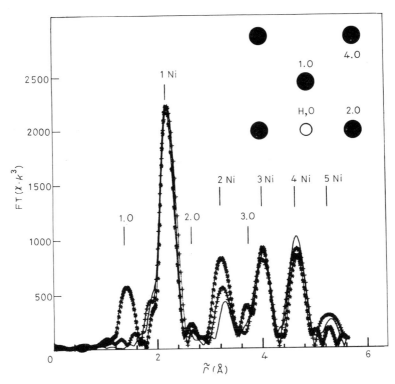

Fig. 1.18. – Fourier transforms of the EXAFS $\chi \cdot k^3$ of Ni (——), $NiH_{0.85}$ ($-+-+-$) and NiO ($-*-*-$). The abscissae for the hydride and the oxide have been rescaled in order to eliminate the influence of the lattice expansion. The inset shows the lattice position of H and O which shadow a second Ni neighbour.

hydrogen. Figure 1.19 shows the Fourier transform of the EXAFS at the Y K-edge in PdY 2 at. % and in PdY 2% H 80%. Pd having a f.c.c. lattice, it is again the second-neighbour shell that is of interest here. It is obvious that in the hydrogen-loaded sample the second shell is not enhanced by the hydrogen. This leads to the conclusion that Y does not trap the hydrogen. In view of the large hydrogen concentration one must even conclude that the Y expels the hydrogen from its neighbourhood. If the hydrogen were distributed statistically, 4.8 out of 6 nearest octahedral sites would be occupied with hydrogen for an alloy with 80% hydrogen. This is not the case, as shown in fig. 1.19. In other words, the hydrogen avoids the vicinity of the Y atoms. This conclusion is supported by an analysis of the lattice distortion around Y. Figure 1.20 shows the interatomic distances r_1 and r_3 of the first and third shells around Y in Pd. Y expands the Pd lattice. The hydrogen dissolved in pure Pd also expands the lattice (full line in fig. 1.20). The addition of hydrogen to the dilute alloy obviously expands the lattice less than what corresponds to the addition of both effects. If the hydrogen were trapped by Y, the points at 73 and 80% H should be above the dashed

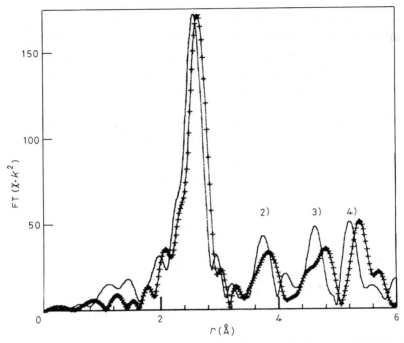

Fig. 1.19. – Fourier transform of the EXAFS $\chi \cdot k^2$ at the Y K-edge in $\underline{Pd}Y$ (2 at. % Y) (——) and in $\underline{Pd}YH_{0.80}$ (–+–+–).

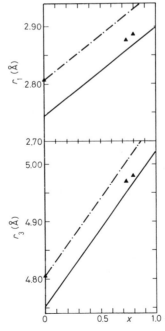

Fig. 1.20. – Interatomic distances r_1 and r_3 of the first and third neighbours around Y and Pd as a function of the hydrogen concentration: ▲ –·–·– $PdYH_x$, —— PdH_x.

curves. That they are below this line implies that the hydrogen is expelled from the vicinity of Y. It also implies that the high hydrogen concentration outside the Y vicinity acts as an external pressure shifting the Pd atoms in the nearest shells towards the Y [19]. The same observation was made for a great number of other oversized impurities in Pd (rare-earth atoms, Th...).

1`4. *Position and structure of the absorption edge.* – Figure 1.21 shows the K-edge of Cu in three Cu oxides (Cu_2O, CuO and $KCuO_2$) [21]. There are distinct differences in the form and in the position of the edges. It turned out that the position of the edge is shifted to higher energies with increasing valence of the absorber. This simple, experimentally well-established relation can be used to determine the valence of an atom in a compound. On the other hand, the form of the edge reflects the empty electronic densities of states. In fig. 1.22 is shown an energy diagram for the absorption process. Only those photons can be absorbed by, say, a 1s electron whose energy is larger than their binding energy. Due to the conservation of angular momentum only those final states are allowed which have p-character about the absorber atom. Before absorption the 1s electron has no angular momentum, whereas the photon carries (in the dipole approximation)

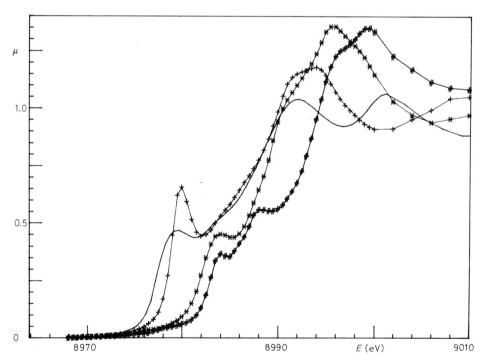

Fig. 1.21. – Copper K-edges in metallic copper (——), Cu_2O (– + – + –), CuO (– * – * –) and in $KCuO_2$ (– # – # –). The edge is shifted to higher energies with increasing Cu valence [21].

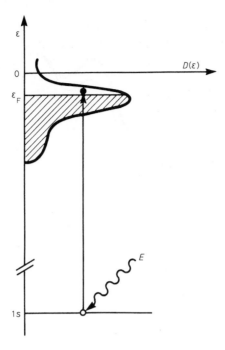

Fig. 1.22. – Energy diagram for the absorption of a photon with energy E by a $1s$ electron.

the momentum $\hbar$. After absorption the photon has disappeared, so that the electron has to carry the angular momentum $\hbar$. Therefore, the difference Δl in electron angular momentum has to be

$$(1.4) \qquad\qquad |\Delta l| = 1.$$

When the photons are adsorbed by p-electrons (*e.g.*, L_2 and L_3 edges) only electron transitions into empty s- and d-states are possible. However, calculations have shown that the cross-section for transition into empty d-states is about 50 times larger than that into empty s-states [10]. According to ref. [22] and [23] the near-edge structure $\mu(E)$ can be written as a product

$$(1.5) \qquad\qquad \mu(E) = M(E) \cdot \rho_l(E).$$

The matrix element $M(E)$ gives the probability for a transition from an occupied, say, $1s$ state into an empty p-state. The projected (l-dependent) density of states $\rho_l(E)$ gives the probability to find a final state of appropriate symmetry. Band structure calculations have shown that $M(E)$ is only weakly energy dependent [22, 23]. Thus the structure observed in the absorption edges must reflect the

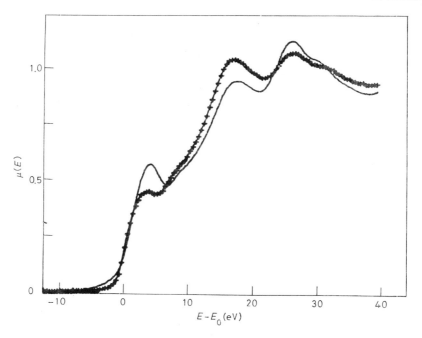

Fig. 1.23. – Mn K-edge in NiMn 2 at. % (experiment – + – + –, band structure calculation
——).

structure in the projected densities of states $\rho_l(E)$. There is a number of
calculations of $\mu(E)$ for elements and for stoichiometric compounds which confirm
this conclusion [22-26]. An example should show that this is also true for dilute
alloys [12, 27]. Since the X-ray absorption is specific to individual species, it is
possible to probe the projected densities of states locally around the impurity in a
dilute alloy. Figure 1.23 shows the Mn K-edge in a dilute NiMn 2 at. % alloy. A
self-consistent KKR band structure calculation with exchange and correlation in
the local density approximation has been performed for a Mn atom in a Ni
matrix [27]. The p-density of states in the Mn cell shows strong variations in the
energy range considered (up to 40 eV above the Fermi level). The matrix
element $M(E)$ is only weakly energy dependent. The calculated quantity $M \cdot \rho_l$
has to be convoluted with a Lorentzian that takes into account the finite lifetime
of the core hole, the finite lifetime of the end states and the energy resolution of
the spectrometer. It is obvious that this broadening smears out many details of
the calculation (fig. 1.24). The agreement between calculated and measured data
is satisfactory. The main results are the following: 1) The structure in the
absorption edges reflects the structure in the density of states. 2) The
measurement of absorption edges is a simple and stringent test of the quality of
band structure calculations.

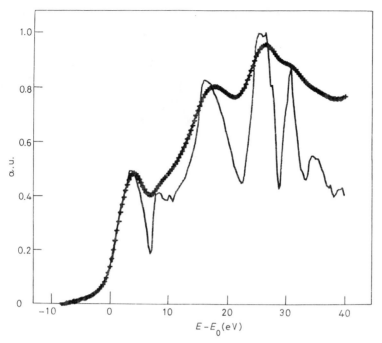

Fig. 1.24. – Effect of energy smearing on the calculated (——) local, empty DOS $\rho_1(\varepsilon)$ for the K-edge of Mn in Ni.

In another lecture of this course it will be shown that empty densities of electronic states can also be measured by bremsstrahlung-isochromatic spectroscopy (BIS). This spectroscopy gives the total density of states (summed over l). A disadvantage of this probe is that it is not a local one. Therefore, it does not allow one to determine the local density of states in the cell of a special atomic species in an alloy. Another difference between XAS and BIS is the fact that in XAS a hole is created in a core state, whereas in BIS no core hole is created. By comparing the results of these two spectroscopies the screening of a perturbation by the other electrons can be investigated.

It is rather tedious to perform band structure calculations for complicated crystal structures with many atoms per unit cell or for structures without translational lattice symmetry. On the other hand, the fingerprint procedure discussed above allowing one to determine the state of oxidation of individual atomic species is easy to apply. It will be demonstrated now for the internal oxidation of impurities in Pd [19]. P̲d̲Th alloys with 2.5 at. % of Th were oxidized in pure oxygen of 1 bar at 1037 °C. The oxygen uptake was found by weighing to be 5.7 at. %. The L_3 absorption edge of Th was measured in an oxidized and in a nonoxidized sample (fig. 1.25). There are distinct differences in the edges. The Th L_3-edge in ThO_2 is shown for comparison. The edges of ThO_2 and of the oxidized alloy are almost identical. This leads to the conclusion that the Th atoms

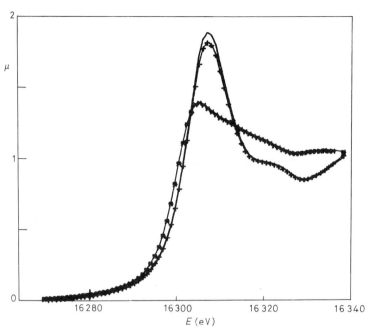

Fig. 1.25. – Th L_3-edge in P̲dTh 2.5 at. % (–*–*–), in oxidized P̲dTh (———) and in ThO$_2$ (–+–+–).

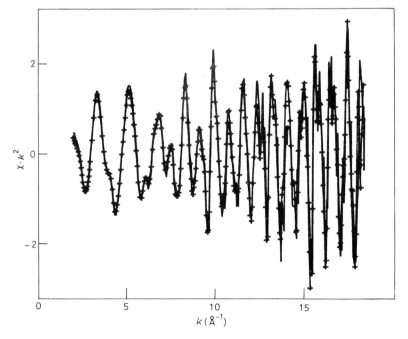

Fig. 1.26. – EXAFS $\chi \cdot k^2$ at the Th L_3-edge in oxidized P̲dTh (———) and in ThO$_2$ (–+–+–).

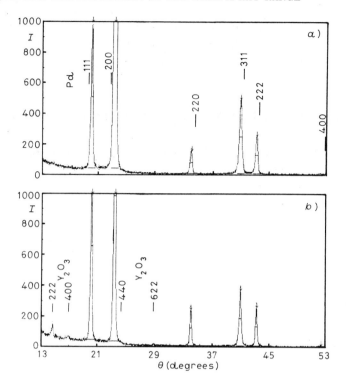

Fig. 1.27. – X-ray diffraction of an oxidized and a nonoxidized P̲dY alloy with 2 at. % Y.

in the oxidized alloy have precipitated as ThO_2. This conclusion is supported by the EXAFS shown in fig. 1.26. The local surrounding of Th atoms in the oxidized alloy is identical with that in the oxide ThO_2. In this way a number of impurities in Pd have been investigated[19]. Pt, Au, Pb and Bi turned out to be not oxidized. Y, Nb, Gd, Yb, Lu, Hf and Th precipitate at 1050 °C in 1 bar O_2 as oxides with the highest state of oxidation (Y_2O_3, Nb_2O_5, Gd_2O_3, Yb_2O_3, Lu_2O_3, HfO_2, ThO_2). Zr and Mo are also oxidized to the highest oxidation state. But in these two cases the precipitates are poorly crystallized. Only one shell of oxygen is visible around Zr and Mo. It is noteworthy that in the alloy (with 2 at. % of impurities) the precipitates were in all cases so small that they are barely visible in X-ray diffraction. Figure 1.27 shows the X-ray diffraction of an oxidized and nonoxidized P̲dY alloy. The Pd Bragg peaks dominate the diffraction pattern, whereas only one Y_2O_3 peak is barely visible. For the present problem XAS is superior to X-ray diffraction because absorption is specific to the atomic species of interest.

   In an absorption process by weakly bound electrons the emission of characteristic photons becomes less probable, whereas that of Auger electrons increases. The X-ray absorption by light elements (B, C, N, O, F…) is, therefore, mainly done in the electron yield mode. This detection mode is inherently surface sensitive. It is mostly applied to the investigation of surfaces and adsorbates. It will be treated in this course in a different lecture.

2. – Total reflection of hard X-rays and XAS under grazing incidence.

XAS in the transmission and fluorescence mode is a bulk probe. This is a consequence of the relatively large penetration depth of hard X-rays into matter. There is an increasing interest in science and technology for analytical tools to probe the geometric and electronic structure of surfaces and internal interfaces. The whole spectrum of X-ray techniques (X-ray absorption, diffraction, fluorescence analysis, topography ...) can be made surface sensitive by means of total reflection. This interesting new development will be described in the following and demonstrated in the case of XAS.

2'1. *Refraction of X-rays.* – The index of refraction n for a substance with elements j (atomic number Z_j, atomic mass A_j, density ρ_j) is given by [28-30]

$$(2.1) \qquad n = 1 - \frac{N_A}{2\pi} r_0 \lambda^2 \sum_j \frac{\rho_j}{A_j} f_j \, .$$

Here, λ is the photon wavelength ($E = hc/\lambda$). N_A is Avogadro's number and $r_0 = e^2/mc^2 = 2.818 \cdot 10^{-13}$ cm is the classical electron radius. Far above all absorption edges the form factor f_j is equal to Z_j. In the vicinity of absorption edges dispersive and absorptive corrections have to be taken into account:

$$(2.2) \qquad f_j = Z_j + f_j' + if_j'' \, .$$

The index of refraction is a complex quantity

$$(2.3) \qquad n = 1 - \delta - i\beta$$

with

$$(2.4) \qquad \delta = \frac{N_A}{2\pi} r_0 \lambda^2 \sum_j \frac{\rho_j}{A_j} (Z_j + f_j') \, ,$$

$$(2.5) \qquad \beta = \frac{N_A}{2\pi} r_0 \lambda^2 \sum_j \frac{\rho_j}{A_j} f_j'' = \mu \frac{\lambda}{4\pi} \, .$$

The absorptive correction β is proportional to the linear coefficient of absorption

μ. It turns out that for X-rays δ and β are small positive quantities. Typical values for Cu and Al are

	Cu	Al
δ (1.44 Å)	$23 \cdot 10^{-6}$	$7.3 \cdot 10^{-6}$
β (1.44 Å)	$0.4 \cdot 10^{-6}$	$0.1 \cdot 10^{-6}$
β (1.36 Å)	$2.8 \cdot 10^{-6}$	

Above the K absorption edge of Cu (1.38 Å, 8980 eV) β increases substantially but is still much smaller than 1. For hard X-rays condensed matter is optically thinner than the vacuum. Thus X-rays when entering into matter are refracted away from the normal, as indicated in fig. 2.1. The fact that n is near 1 and a little

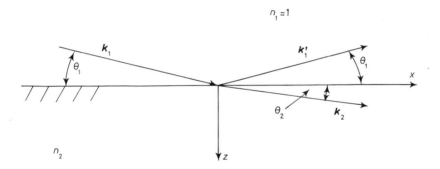

Fig. 2.1. – Refraction and reflection of X-rays at the transition from vacuum ($n_1 = 1$) into matter with index of refraction n_2.

bit smaller than 1 can be understood by considering the phase and amplitude of a forced oscillation. The binding energies of most electrons in an atom are below 10 keV. The E vector of the X-rays excites the electrons to forced oscillations with a frequency which is in general much higher than the eigenfrequencies of the oscillator. Therefore, the amplitude of the forced oscillations is small and matter is only weakly polarized. Thus the index of refraction is almost 1. In addition, the polarization and the exciting electric field of the photon are out of phase. Thus the polarization enhances the excitation. In other words, the phase velocity is larger than the velocity of light in vacuum, or the index of refraction is smaller than 1. In the vicinity of an absorption edge the contribution of the corresponding electrons to δ may be negative. But the contribution of the many electrons whose eigenfrequencies are smaller than the photon frequency will make δ, in general, positive in the whole hard-X-ray range.

Snell's law has the same form as in the visible range (fig. 2.1)

(2.6) $n_1 \cos \theta_1 = n_2 \cos \theta_2$.

The beam entering into matter being refracted away from the normal, there is a finite angle θ_1 for which $\theta_2 = 0$. This critical angle θ_{1c} is the angle of total external reflection ($n_1 = 1$)

(2.7) $\cos \theta_{1c} = n_2$.

When the absorption can be neglected, θ_{1c} is given by

(2.8) $\theta_{1c} = \sqrt{2\theta_2} = \sqrt{\dfrac{N_A r_0}{\pi} \sum_j \dfrac{\rho_j}{A_j}(Z_j + f_j') \cdot \lambda}$

The critical angle is proportional to the photon wavelength λ and to the square root of the mass density. The density dependence of θ_{1c} in condensed matter is rather weak. The increase in θ_{1c} between water ($1\,\mathrm{g/cm^3}$) and, say, platinum ($21\,\mathrm{g/cm^2}$) is by a factor of 4.6. Typical values of θ_{1c} are $0.39°$ (6.8 mrad) for Cu and $0.22°$ (3.8 mrad) for Al at 1.44 Å. Total external reflection of X-rays in the 10 keV range occurs only at grazing incidence below about $0.5°$.

2`2. *Electric and magnetic field at grazing incidence.* – For s-polarization ($\boldsymbol{E}$ vector perpendicular to the plane of reflection (fig. 2.1)) the electric-field vectors are [29]

(2.9)

$$\begin{cases} \boldsymbol{E}_1(\boldsymbol{r}, t) = \begin{pmatrix} 0 \\ A_1 \\ 0 \end{pmatrix} \exp\left[i(\boldsymbol{k}_1 \cdot \boldsymbol{r} - \omega t)\right], \\[20pt] \boldsymbol{E}_1'(\boldsymbol{r}, t) = \begin{pmatrix} 0 \\ A_1' \\ 0 \end{pmatrix} \exp\left[i(\boldsymbol{k}_1' \cdot \boldsymbol{r} - \omega t)\right], \\[20pt] \boldsymbol{E}_2(\boldsymbol{r}, t) = \begin{pmatrix} 0 \\ A_2 \\ 0 \end{pmatrix} \exp\left[i(\boldsymbol{k}_2 \cdot \boldsymbol{r} - \omega t)\right]. \end{cases}$$

The reflected beam is denoted by a prime. According to Maxwell's equations the corresponding magnetic fields are given by

(2.10) $\boldsymbol{H} = n\hat{\boldsymbol{k}} \times \boldsymbol{E}$.

Here $\hat{k}$ is a unit vector in the direction of the photon k vector. At the boundary $z = 0$ of the two media the tangential components of E and H must be continuous for all x, y. The invariance in time t for elastic processes has already been taken care of by using the same frequency ω in all three expressions (2.9).

Continuity gives

(2.11)
$$\begin{cases} k_{1x} = k'_{1x} = k_{2x} \quad \text{and} \quad k_2 = n_2 k_1, \\ A_1 + A'_1 = A_2, \\ (A_1 - A'_1) \sin \theta_1 = A_1 n_2 \sin \theta_2. \end{cases}$$

This leads to the Fresnel equations

(2.12)
$$\begin{cases} \dfrac{A'_1}{A_1} = \dfrac{\sin \theta_1 - n_2 \sin \theta_2}{\sin\theta_1 + n_2 \sin \theta_2} = \dfrac{\sin (\theta_1 - \theta_2)}{\sin (\theta_1 + \theta_2)}, \\ \dfrac{A_2}{A_1} = \dfrac{2\sin \theta_1}{\sin\theta_1 + n_2 \sin \theta_2} = \dfrac{2\sin \theta_1 \cos \theta_2}{\sin (\theta_1 + \theta_2)}. \end{cases}$$

Since total reflection occurs only at grazing incidence, θ_1 and θ_2 can be treated as small compared to 1:

(2.13)
$$k_1 = k_1 \begin{pmatrix} 1 \\ 0 \\ \theta_1 \end{pmatrix}, \qquad k'_1 = k_1 \begin{pmatrix} 1 \\ 0 \\ -\theta_1 \end{pmatrix}, \qquad k_2 = k_1 \begin{pmatrix} 1 \\ 0 \\ \theta_2 \end{pmatrix},$$

(2.14)
$$\frac{A'_1}{A_1} = \frac{\theta_1 - \theta_2}{\theta_1 + \theta_2} \quad \text{and} \quad \frac{A_2}{A_1} = \frac{2\theta_1}{\theta_1 + \theta_2}.$$

At grazing incidence, expressions (2.14) for the reflected and transmitted amplitudes are valid for s-polarization (E perpendicular to the plane of reflection) and for p-polarization (E in the plane of reflection) [28, 29].

At grazing incidence and according to Snell's law θ_2 can be expressed as a function of θ_1

(2.15)
$$\theta_2 = \sqrt{\theta_1^2 - 2\delta_2 - i2\beta_2} = p_2 + iq_2.$$

When θ_1 is below the angle of total reflection, θ_2 is complex even for vanishing absorption:

(2.16)
$$\begin{cases} p_2^2 = \dfrac{1}{2}\left[\sqrt{(\theta_1^2 - 2\delta_2)^2 + 4\beta_2^2} + (\theta_1^2 - 2\delta_2)\right], \\[3mm] q_2^2 = \dfrac{1}{2}\left[\sqrt{(\theta_1^2 - 2\delta_2)^2 + 4\beta_2^2} - (\theta_1^2 - 2\delta_2)\right]. \end{cases}$$

Without absorption it is

(2.17)
$$\theta_2 = \begin{cases} i\sqrt{\theta_{1c}^2 - \theta_1^2}, & \theta_1 \lesssim \theta_{1c}, \\[3mm] \sqrt{\theta_1^2 - \theta_{1c}^2}, & \theta_1 \gtrsim \theta_{1c}. \end{cases}$$

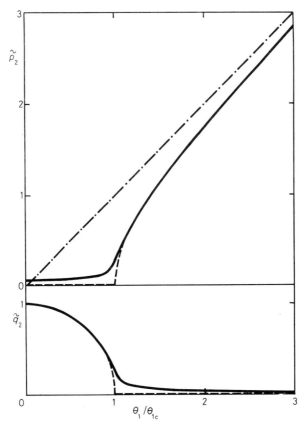

Fig. 2.2. – Real and imaginary parts of $\theta_2 = p_2 + i q_2$ for $\beta_2 = 2.8 \cdot 10^{-6}$ and $\delta_2 = 23 \cdot 10^{-6}$ (Cu metal at 1.36 Å, 9100 eV). Without absorption p_2 vanishes below and q_2 above θ_{1c} ($\tilde{p}_2 = p_2/\theta_{1c}$ and $\tilde{q}_2 = q_2/\theta_{1c}$).

At $\theta_{1c} = \sqrt{2\delta_2}$, θ_2 has the value

(2.18) $$\theta_2 = (1 + i) \sqrt{\beta_2} = (1 + i) \sqrt{\frac{\mu_2 \lambda}{4\pi}}.$$

Figure 2.2 shows the dependence of p_2 and q_2 on the angle of incidence for a typical case (refraction of 9.1 keV photons (1.36 Å) at the interface vacuum-metallic copper with $\delta_2 = 23 \cdot 10^{-6}$ and $\beta_2 = 2.8 \cdot 10^{-6}$). p_2 and q_2 are positive quantities. At θ_{1c} and below it θ_2 is substantially different from θ_1.

2`3. *Transmitted amplitude at grazing incidence.* – The transmitted amplitude is given by (s-polarization)

(2.19) $$E_{2y}(r) = A_2 \exp\left[i(k_1 x + k_1 p_2 z - \omega t)\right] \cdot \exp\left[-k_1 q_2 z\right].$$

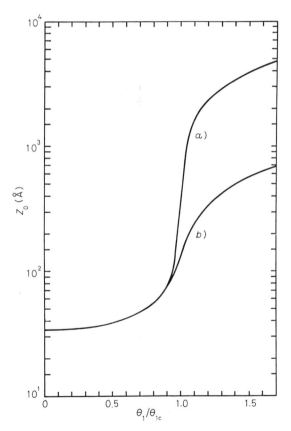

Fig. 2.3. – Decay constant z_0 of the transmitted amplitude $\boldsymbol{E}_2(\boldsymbol{r}, t)$ for 8.6 (curve a), $\theta_{1c} = 6.8$ mrad) and 9.1 keV (curve b), $\theta_{1c} = 6.4$ mrad) photons in metallic copper. The copper K-edge is at 8980 eV. Note the small penetration depth of the X-rays below θ_{1c} which is the basis of making all X-ray techniques surface sensitive at total reflection.

The amplitude is damped exponentially in the z direction with a decay constant

(2.20) $\qquad z_0(\theta_1) = (k_1 q_2)^{-1} = \dfrac{\lambda\sqrt{2}}{2\pi}[\sqrt{(\theta_1^2 - 2\delta_2)^2 + 4\beta_2^2} - (\theta_1^2 - 2\delta_2)]^{-1/2}\,.$

Figure 2.3 shows the decay constant z_0 as a function of the angle θ_1 for metallic copper. Three special cases are of interest

(2.21) $\qquad \begin{cases} z_0 = 2\theta_1/\mu_2\,, & \theta_1 \gg \theta_{1c}\,, \\[2mm] z_0 = (\lambda/\pi\mu_2)^{1/2}\,, & \theta = \theta_{1c}\,, \\[2mm] z_{00} = \dfrac{1}{2\pi}\left(\dfrac{N_A r_0}{\pi}\sum_j \dfrac{\rho_j(z_j + f'_j)}{A_j}\right)^{-1/2}\,, & \theta_1 \to 0\,. \end{cases}$

Above the critical angle θ_{1c} the absorption determines the penetration depth z_0. For $\theta_1 \to 0$, z_0 tends towards a value that is independent of the absorption. In this regime z_0 depends mainly on the mass density. Its energy dependence is that of the dispersive correction f'_j. The absolute value of z_{00} varies between about 20 Å for a dense material like Pt and about 70 Å for water. The small value of z_{00} means that all X-ray techniques become inherently surface sensitive when they are performed under total reflection, in the sense that the signal originates in a 20 to 70 Å thick layer below the surface. A second interesting feature of fig. 2.3 is the possibility to determine depth profiles in X-ray techniques when the signal is measured as a function of the grazing angle θ_1. The probing depth of the X-ray techniques increases gradually from 20 to 70 Å at $\theta_1 \to 0$ towards values up to about $1\,\mu$m when θ_1 has reached a few times the value of θ_{1c}.

The amplitude A_2 of the transmitted wave is given by eqs. (2.14), (2.15):

(2.22) $\qquad \dfrac{A_2}{A_1} = \dfrac{2\theta_1}{\sqrt{(\theta_1 + p_2)^2 + q_2^2}}\exp\left[-i\,\mathrm{arctg}\,\dfrac{q_2}{\theta_1 + p_2}\right]\,.$

The modulus $|A_2/A_1|$ is shown in fig. 2.4. When there is no absorption, the amplitude at the interface is twice the value of the incoming wave at θ_{1c}. So, the intensity is four times that of the incoming wave. This can be understood in the following way. In front of the interface, a standing wave field builds up which for s-polarization has only a y-component:

(2.23) $\qquad \begin{cases} E_{1y}(\mathbf{r}) + E'_{1y}(\mathbf{r}) = A_1 \exp\left[i(k_1 x - \omega t)\right]A_{\mathrm{st}}\,, \\[3mm] A_{\mathrm{st}} = \exp\left[ik_1\theta_1 z\right] + \exp\left[-ik_1\theta_1 z\right]\dfrac{\theta_1 - \theta_2}{\theta_1 + \theta_2}\,. \end{cases}$

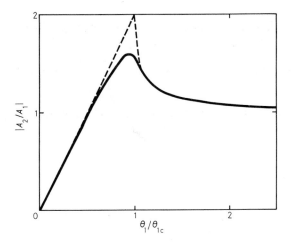

Fig. 2.4. – Modulus of the transmitted amplitude A_2. At θ_{1c} the intensity of the X-rays (without absorption) is 4 times as large as the incoming intensity (broken line). Absorption drastically reduces this factor (full line: Cu metal at 9.1 keV with $\hat{\delta}_2 = 23 \cdot 10^{-6}$ and $\beta_2 = 2.8 \cdot 10^{-6}$).

When the absorption can be neglected

$$(2.24) \qquad A_{\mathrm{st}} = \begin{cases} 2 \cos k_1 \theta_1 z \,, & \theta_1 = \theta_{1c} \,, \\[2mm] 2i \sin k_1 \theta_1 z \,, & \theta_1 = 0 \,. \end{cases}$$

The standing wave field changes its phase by 90° between $\theta_1 = 0$ and θ_{1c}. At θ_{1c} it has an antinode at the interface with twice the amplitude of the incoming field. The next node is at $\pi z_{00}/2$ or about 30 to 100 Å in front of the interface. The absorption reduces the amplitude of the standing wave field considerably at θ_{1c} (fig. 2.4).

The wave field penetrating into the medium is able to excite fluorescence photons if the energy of the photons is properly chosen. The intensity of fluorescence photons emitted by a given atomic species in medium 2 can be calculated in a straightforward way as a function of the angle θ_1. The result will be given in the following. If we denote by $g_{\mathrm{F}}(z)$ the mass density of these atoms in a homogeneous matrix, the intensity of fluorescence photons registered per second by a detector above the illuminated surface is

$$(2.25a) \qquad I_{\mathrm{F}} = I_1 \cdot \varepsilon \frac{\Omega}{4\pi} \frac{8\theta_1 p_2}{|\theta_1 + \theta_2|^2} k_1 q_2 \frac{1}{\mu_2} \left(\frac{\mu}{\rho} \right)_{\mathrm{F}} \cdot J(\theta_1) \,,$$

$$(2.25b) \qquad J(\theta_1) = \int\limits_0^\infty \mathrm{d}z \, g_{\mathrm{F}}(z) \exp\left[-2k_1 q_2 + \mu_2^{\mathrm{F}} \right] z \right] \,.$$

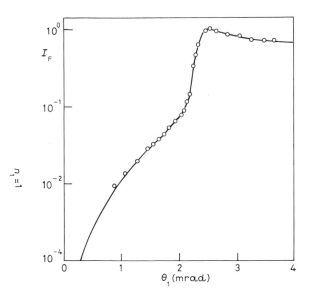

Fig. 2.5. – X-ray fluorescence intensity I_F emitted by As implanted in a Si wafer[31].

Here I_1 is the number of photons per second impinging on medium 2. ε is the fluorescence yield, Ω the solid angle seen by the detector. μ_2 and μ_2^F are the absorption coefficients for the incoming and fluorescence photons in medium 2. The integral (2.25b) is the Laplace transform of the mass density distribution $g_F(z)$. The Laplace variable $2k_1 q_2 + \mu_2^F$ is the reciprocal of the probing depth. The density distribution $g_F(z)$ can be obtained from eq. (2.25) by taking the Laplace backtransformation.

Figure 2.5 shows the angular dependence of the fluorescence photons for a Si wafer implanted with As[31]. The depth profile has been determined by SIMS and the intensity $I_F(\theta_1)$ has been calculated according to eq. (2.25). The agreement with the measured data is excellent.

Equation (2.25) is only valid for a homogeneous matrix in which the variations in $g_F(z)$ do not change substantially the index of refraction. For layered substrates or substantial variations in the index of refraction induced by g_F the fluorescence intensity $I_F(\theta_1)$ is a more complicated function (see below).

2'4. *Reflected intensity at grazing incidence.* – The Fresnel reflectivity R_F is given by eq. (2.14):

$$(2.26) \qquad R_F = \left| \frac{A_1'}{A_1} \right|^2 = \left| \frac{\theta_1 - \theta_2}{\theta_1 + \theta_2} \right|^2 = \frac{(\theta_1 - p_2)^2 + q_2^2}{(\theta_1 + \theta_2)^2 + q_2^2}.$$

This expression is shown in fig. 2.6 for a flat copper mirror. When the absorption

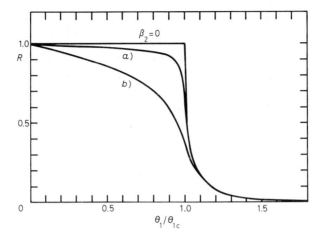

Fig. 2.6. – Reflection of X-rays from a flat Cu surface at grazing incidence: curve a) $E = 8.60$ keV, $\theta_{1c} = 6.78$ mrad; curve b) $E = 9.10$ keV, $\theta_{1c} = 6.40$ mrad. The K-edge of Cu is at 8980 eV. Absorption reduces the reflectability substantially at and below θ_{1c}.

can be neglected ($\beta_2 = 0$), $R_F = 1$ below θ_{1c}. Above θ_{1c} the reflectance decreases gradually towards zero. It is

$$(2.26') \qquad\qquad R_F = \frac{\delta_2^2}{4\theta_1^4} \qquad\qquad \text{for } \theta_1 \geq 1.5\,\theta_{1c}.$$

Above the angle of total reflection the reflectivity becomes more and more independent of the absorption. The absorption has a drastic influence on the reflectivity below $\sqrt{2\theta_2}$. For strong absorption the critical angle of total reflection can no longer be defined (fig. 2.6). Besides the surface sensitivity which can be achieved by total reflection, grazing incidence of X-rays offers more interesting possibilities. X-rays can be focused by curved mirrors. This feature is very appealing, the more there are no lenses for X-rays. It is extensively used in synchrotron radiation research. Total reflection can be applied for determining the index of refraction in the X-ray regime. In particular, the dispersive correction f' to the atomic form factor can be determined. It can also be used to deduce the mass density in a surface layer, which often differs from the bulk value. Finally, surface roughness reduces the reflectivity. Deviations from the Fresnel reflectivity, eq. (2.26), can be used to determine the roughness of interfaces, as will be demonstrated in the following. We denote by $h(x)$ the position of the interface as a function of x with a distribution $w(z)$ around the average value z_0 (fig. 2.7). We now replace the rough interface by an ensemble of flat surfaces with the distribution $w(z)$ around z_0 (fig. 2.7). If we assume

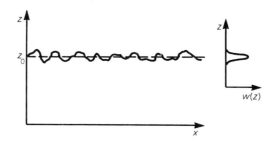

Fig. 2.7. – Model for the roughness of an interface. The rough surface is replaced by an ensemble of flat surfaces with a distribution function $w(z)$, or the density increases gradually from zero to the bulk value in such a way that its variation is also given by $w(z)$. This model contains no diffuse scattering.

furthermore the distribution to be Gaussian

$$(2.27) \qquad w(z) = \frac{1}{\sqrt{2\pi\sigma^2}} \exp\left[-(z-z_0)^2/2\sigma^2\right]$$

with a standard deviation σ, then the reflectivity is reduced by an exponential damping factor[32]

$$(2.28) \qquad R = R_{\mathrm{F}} \exp\left[-\left(\frac{4\pi}{\lambda}\theta_1\sigma\right)^2\right],$$

where R_{F} is the Fresnel reflectivity eq. (2.26). The more grazing the incidence, the smaller is the effect of roughness. For very flat surfaces with σ about 5 Å (liquids, float glass, polished Si wafers...) X-rays are well suited to detect surface roughness because the X-ray wavelength is of the same order of magnitude as σ. The model leading to expression (2.28) does not contain diffuse scattering because the rough surface is replaced by an ensemble of flat surfaces which all reflect the beam in the same direction. This is a drawback of the present model. Equation (2.28) can also be understood in a different way[33]. We replace the rough surface by an interface in which the density increases gradually from zero (in vacuum) to the bulk density (in medium 2) in such a way that the change in density has the Gaussian form (2.27) (fig. 2.7). Then the reflectivity is again given by eq. (2.28). This model, although containing no diffuse scattering, describes well the reflectivity of X-rays from a clean water surface[33]. Figure 2.8 shows the reflectivity over 7 orders of magnitude. For a photon energy $E = 8109$ eV ($\lambda = 1.529$ Å) the angle of total reflection is at 2.64 mrad. The angular range extends to 18 times θ_{1c}. The Fresnel reflectivity

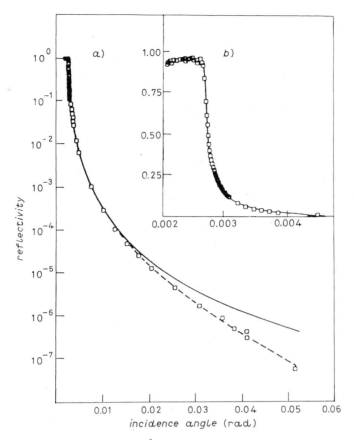

Fig. 2.8. – Reflection of X-rays (1.529 Å) from a clean water surface. Deviations from the Fresnel reflectivity (full line) are due to surface roughness[33].

$(\sigma = 0)$ is shown as a full line. Substantial differences between R and R_F show up only above about 20 mrad. The fit of the measured data with eq. (2.28) gives $\sigma = 3.2$ Å. The authors have assumed the water surface to be roughened by the thermal excitation of capillary waves. The restoring force for these waves is the surface tension. This model gives a value for σ of (2.9 ± 0.1) Å in good agreement with the experimental result. In tetrachloromethane (CCl_4) the same authors have found a surface roughness 3 times as large as that of water in agreement with a surface tension in CCl_4 3 times smaller than that of water.

2˙5. *Reflection from layers on a substrate.* – Next, the reflection of X-rays from a layer on top of a flat substrate will be considered (fig. 2.9). Since the reflectivity is independent of the polarization at grazing incidence, we consider the case of s-polarization (E vector in y-direction). The amplitudes of the electric and magnetic fields are referred to the middle of the layers. The reflected waves

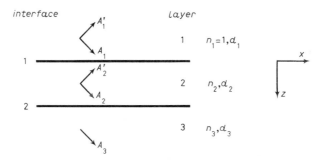

Fig. 2.9. – Amplitude of the s-polarized X-ray wave fields for reflection from a layer (n_2, d_2) on a substrate (n_3, d_3). For a thick substrate there is no reflected amplitude A_3'.

are again denoted by a prime:

(2.29)
$$
\begin{cases}
\boldsymbol{E}_j(\boldsymbol{r}, t) = \begin{pmatrix} 0 \\ A_j \\ 0 \end{pmatrix} \exp\left[i(k_1 x + k_1 \theta_j z_j - \omega t)\right], \\[20pt]
\boldsymbol{E}_j'(\boldsymbol{r}, t) = \begin{pmatrix} 0 \\ A_j' \\ 0 \end{pmatrix} \exp\left[i(k_1 x - k_1 \theta_j z_j - \omega t)\right].
\end{cases}
$$

z_j is limited to the j-th layer and $z_j = 0$ is in the middle of this layer. The corresponding magnetic fields are given by eq. (2.10). The continuity of the tangential components of $\boldsymbol{E}$ and $\boldsymbol{H}$ at each interface relates the amplitudes A_j, A_j' and A_{j+1}, A_{j+1}'. Since there is no wave coming back from the substrate ($A_3' \equiv 0$), A_1' can be expressed by A_1. A straightforward calculation leads to [34]

(2.30)
$$
R_{\mathrm{F}} = \left| \frac{A_1'}{A_1} \right|^2 = \left| \frac{r_1 + r_2 \exp\left[2ik_1 \theta_2 d_2\right]}{1 + r_1 r_2 \exp\left[2ik_1 \theta_2 d_2\right]} \right|^2
$$

with

(2.31)
$$
r_1 = \frac{\theta_1 - \theta_2}{\theta_1 + \theta_2} \quad \text{and} \quad r_2 = \frac{\theta_2 - \theta_3}{\theta_2 + \theta_3},
$$

the amplitudes reflected from the interfaces 1 and 2. The phase factor $\exp\left[2ik_1 \theta_2 d_2\right]$ gives rise to oscillations in the reflected intensity [35]. The origin of these oscillations is seen in fig. 2.10. When the angle θ_1 of total reflection for the layer is exceeded, part of the beam will penetrate into the layer. The

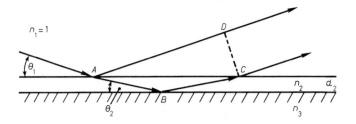

Fig. 2.10. – Reflection of X-rays under grazing incidence from a layer (n_2, d_2) on a substrate (n_3).

amplitudes 1 and 2 reflected from the interfaces can interfere. The difference in optical path is

$$(2.32) \qquad \Delta = (\overline{AB} + \overline{BC}) \, n_2 - \overline{AD} = 2d_2 \, \theta_2 \approx 2d_2 \sqrt{\theta_1^2 - \theta_{1c}^2} \, .$$

The reflectivity shows maxima and minima as a function of θ_1. The maxima are at

$$(2.33a) \qquad \nu\lambda = 2d_2 \sqrt{\theta_{1\nu}^2 - \theta_{1c}^2} \qquad \text{for } n_2 > n_3 \, ,$$

$$(2.33b) \qquad (\nu + 1/2)\,\lambda = 2d_2 \sqrt{\theta_{1\nu}^2 - \theta_{1c}^2} \qquad \text{for } n_2 < n_3 \, .$$

The factor 1/2 in eq. (2.33b) is due to a phase shift by π when $\delta_2 > \delta_3$. Note that the denominator in eq. (2.30) describes multiple reflections in the layer d_2. Figure 2.10 describes the leading term (numerator in eq. (2.30)) of only one reflection at the interface 1 and one reflection at the interface 2. In general, the multiple reflections can be neglected at least if r_1 or r_2 are below 10%. In this case a plot of the position $\theta_{1\nu}$ of the ν-th maximum vs. ν, according to the equations

$$(2.34a) \qquad \theta_{1\nu}^2 = \theta_{1c}^2 + \frac{\lambda^2}{4d_2^2} \, \nu^2 \qquad \text{for } n_2 > n_3 \, ,$$

$$(2.34b) \qquad \theta_{1\nu}^2 = \theta_{1c}^2 + \frac{\lambda^2}{4d_2^2} \, (\nu + 1/2)^2 \qquad \text{for } n_2 < n_3 \, ,$$

gives a linear plot $\theta_{1\nu}^2$ vs. ν^2 or $(\nu + 1/2)^2$. The slope gives the thickness d_2 of the layer. Since the wavelength of the photons is known with an accuracy of $1 \cdot 10^{-4}$, the thickness can be determined with a similar accuracy. It is noteworthy that this method for determining the thickness of a layer is contact-free, nondestructive and independent of optical constants. The last point is especially interesting since the optical constants of thin films (needed in ellipsometry with

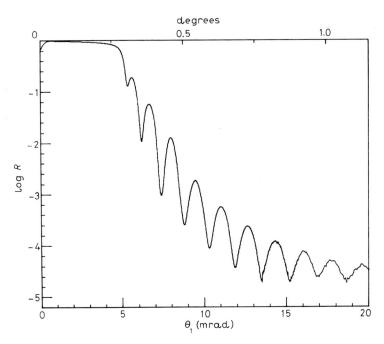

Fig. 2.11. – Reflectivity of 9800 eV photons from a Ta oxide layer on Si deposited at 410 °C by CVD.

visible light) are often not known very well, due to missing knowledge at the composition and density of the layers. The second piece of information in the plot $\theta_{1\nu}^2$ vs. ν^2 or $(\nu + 1/2)^2$ is the intercept, which gives the angle θ_{1c} of total reflection and thus the density of the layer (eq. (2.8)). Figure 2.11 shows the reflectivity from a tantalum oxide layer deposited by CVD at 410 °C on a 4 inch silicon wafer. At 5.2 mrad the photon beam (9800 eV) penetrates into the oxide. Above this angle 10 oscillations due to the interference of the amplitudes reflected at the oxide and silicon interfaces are observed. A plot $\theta_{1\nu}^2$ vs. $(\nu + 1/2)^2$ according to eq. (2.34b) shows the predicted linear relationship (fig. 2.12). The slope gives a Ta oxide layer thickness of (352 ± 3) Å. The intercept gives a value θ_{1c} of 4.90 mrad. Assuming the layer to be Ta_2O_5 (see below), we deduce a density of 94% in the layer compared to that of crystalline Ta_2O_5.

The amplitudes of the oscillations in fig. 2.11 are damped by roughness at the two interfaces which destroys the interferability of the reflected amplitudes. When the rough surfaces are replaced by Gaussian ensembles with standard deviations σ_1 and σ_2 (as in subsect. 2·4), then the reflectivity reads[36]

$$(2.35) \quad R = \left| \frac{A_1'}{A_1} \right|^2 = \left| \frac{r_1 \exp\left[-2(k_1\,\theta_1\,\sigma_1)^2\right] + r_2 \exp\left[2ik_1\,\theta_2\,d_2\right] \exp\left[-2(k_1\,\theta_2\,\sigma_2)^2\right]}{1 + r_1 r_2 \exp\left[2ik_1\,\theta_2\,d_2\right] \exp\left[-2(k_1\,\theta_1\,\sigma_1)^2\right] \exp\left[-2(k_1\,\theta_2\,\sigma_2)^2\right]} \right|^2 .$$

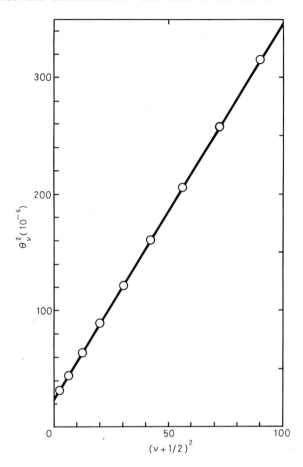

Fig. 2.12. – Plot of the position of the maxima in fig. 2.11 according to eq. (2.34b). The slope of the line gives the thickness and the intercept the density of the Ta oxide layer.

In principle, reflectivity measurements can determine the roughness of interfaces a few thousand Å deep below the external surface. But it has turned out that eq. (2.35) is not a satisfactory model for most systems (sputtered and evaporated films on glass or Si wafers). The reason for the discrepancy is not yet completely clear. Equation (2.35) does not contain diffuse-scattering contributions, nor is the warp of the substrates included in this equation. We have found the bow of pure 4 inch Si wafers to be always at least a few micrometres. It increases, in general, when a layer is deposited on it and can reach easily 50 μm and more when the metal layer reacts with the Si (silicide formation). The influence of diffuse scattering and of wafer bow on the reflectivity has not yet been treated in a satisfactory way.

The next point to be discussed is the reflectivity and the fluorescence intensity from two layers on a substrate as shown in fig. 2.13. A Ta layer 150 Å

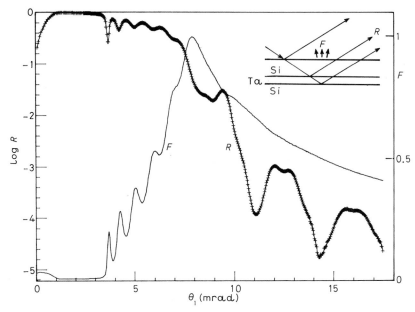

Fig. 2.13. – Reflectivity R and Ta fluorescence intesity F of 10 000 eV photons reflected from a Ta (150 Å)-Si (500 Å) double layer on a 4 inch Si wafer.

thick is deposited by sputtering on a 4 inch Si wafer and a 500 Å thick Si layer covers the Ta layer. Photons, 10 000 eV in energy, are reflected from this package and the fluorescence photons emitted by the excitation of Ta L_3 electrons have been measured as a function of the angle θ_1. Below 3.6 mrad the beam is totally reflected from the Si top layer. No Ta characteristic photons are emitted. Between 3.6 and 7.9 mrad the beam is totally reflected by the Ta layer with typical oscillations, the periodicity of which gives the Si layer thickness. Above 8 mrad the beam enters into the Ta layer and there are now 3 reflected beams that can interfere. The long-periodicity oscillations above 8 mrad reflect the Ta layer thickness. It is obvious that the simple formula (2.25) no longer describes the fluorescence intensity in a correct way. The layered structure produces oscillations in F in such a way that minima in F correspond to maxima in R and *vice versa*. The position of the main maximum in F (fig. 2.13) corresponds to the angle of total reflection for the Ta layer. When the Ta layer is very thick, then F levels of into a plateau above 8 mrad [37].

As a last example in this section the reflectivity from a multiple-layer system on a substrate will be treated. In addition to the total reflection observed at small angles, Bragg reflection can now be expected. Figure 2.14 shows the reflectivity from a system of 75 double layers of 20 Å Si and 10 Å Ta, each. The layers have been deposited by sputtering on a 4 inch Si wafer. Total reflectivity for the 9800 eV photons is observed below 5 mrad. Above this angle the

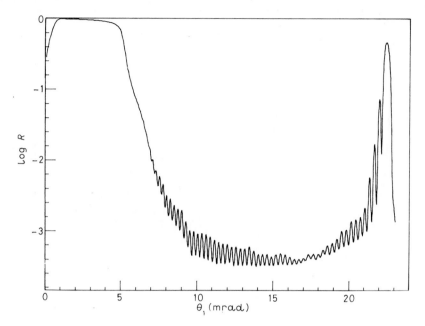

Fig. 2.14. – Reflectivity of 9800 eV photons from 75 double layers of 20 Å Si and 10 Å Ta deposited by sputtering on a 4 inch silicon wafer. At 22.5 mrad a Bragg peak is observed with a reflectivity of 50%.

reflectivity drops by 3 orders of magnitude and increases again at 22.5 mrad to about 50%. This reflectivity, limited to a small angular range, is a Bragg reflection from the 75 double layers. Correcting Bragg law for the refraction of the X-rays, the position of the Bragg reflection gives a periodicity D of 29.1 Å compared to 30 Å as given by the deposition parameters. The oscillations in the angular range from 7 to 22 mrad are a consequence of the finite number of periods. Their period $\Delta\theta$ is related to the thickness $N \cdot D$ of the package by

$$(2.36) \qquad\qquad N \cdot D = \frac{\lambda}{2\,\Delta\theta}\,.$$

Using the data from fig. 2.13 this equation leads to a number of periods N of 75.2 in good agreement with the 75 layers known to have been deposited. These multiple layers are excellent monochromator crystals for soft X-rays. In addition they can be used as mirrors for soft X-rays working even at normal incidence [38-42].

2˙6. XAS *at grazing incidence*. – Measuring the fluorescence signal at fixed angle θ_1 as a function of energy E gives the EXAFS and edge structure of the absorbing species and thus the geometric and electronic structure around the absorber. This version of XAS is surface sensitive and the probing depth is

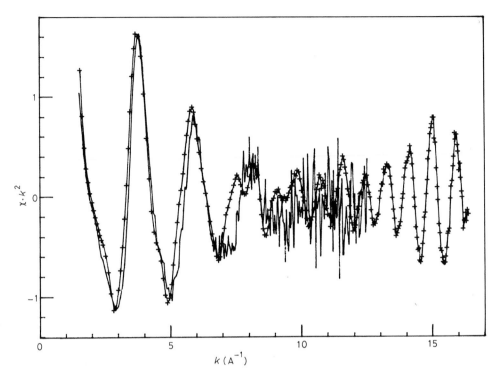

Fig. 2.15. – EXAFS at the Ta L_3-edge in crystalline Ta$_2$O$_5$ (–+–+–) and in a Ta oxide layer deposited on Si by CVD at 410 °C (———).

varied by varying the angle θ_1. In the case of the Ta oxide deposited by CVD on Si at 410 °C (see above) we have measured the EXAFS at different angles. Figure 2.15 shows the EXAFS at 6 mrad where the bulk of the oxide layer is probed. For comparison the EXAFS of crystalline Ta$_2$O$_5$ is shown as well. The CVD layer turned out to be X-ray amorphous. X-ray absorption also shows that the layer is poorly crystallized. But the first oxygen coordination shell is visible and it is obviously identical to that of crystalline Ta$_2$O$_5$. No difference was found in the contribution to the oxide at the surface (first 30 Å) compared to that of the bulk.

An interesting feature of the XAS at grazing incidence is the possibility to investigate the structure of interfaces buried up to about 1 μm deep below the surface. Possible applications are the adhesion of a film on a substrate, the metal-semiconductor contact, the interdiffusion of layers and the formation of intermetallic compounds. An example is shown in fig. 2.16 [43]. A copper film 2000 Å thick was deposited by evaporation on a float glass and covered by a 1000 Å thick Al film. The Cu-Al interface was studied at grazing incidence by measuring the Cu EXAFS. The angle of incidence was chosen in such a way that the beam was totally reflected by the Cu layer. The EXAFS were measured

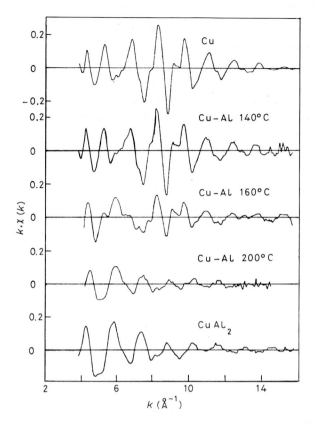

Fig. 2.16. – EXAFS at the Cu K-edge measured at grazing incidence on an Al-Cu layer on glass. The beam passes through the Al layer and is totally reflected by Cu. The double layer has been sequentially annealed for 5 min at 140 °C, 160 °C and 200 °C. The EXAFS of metallic Cu and CuAl$_2$ are shown for comparison [43].

after sequentially annealing for 5 min at 140 °C, 160 °C and 200 °C. The EXAFS of pure Cu and of the intermetallic compound CuAl$_2$ are shown for comparison. The result of the investigation can be summarized as follows. At 140 °C no substantial reaction has taken place between Cu and Al. At 160 °C the reaction has started and at 200 °C the compound CuAl$_2$ has formed at the interface.

3. – Conclusions.

The XAS is a local probe for the geometric and electronic structure of specific atomic species in condensed matter. The information obtained includes interatomic distances, coordination numbers, static and dynamic Debye-Waller factors, projected empty electronic DOS and the valence of the absorbing species. In general, the probe is a bulk probe, but it can be made surface sensitive by means of total external reflection with a probing depth as low as 20 to 70 Å. An especially appealing feature is the possibility to investigate internal interfaces which can be up to 1 μm below a flat surface. The reflectivity of X-rays can be used to determine the thickness of layers, indices of refraction, the mass density and the roughness of interfaces. The measurements can easily be combined with a chemical analysis by X-ray fluorescence and depth profiles in chemical composition can be made by measuring the angular dependence of the fluorescence.

XAS has a wide range of applications. It can be applied to crystalline, microcrystalline, amorphous and liquid phases which contain the atomic species of interest in concentrated and in diluted form. The sample preparation in XAS as bulk probe is simple. In general, no single crystals are needed. On the other hand, in the grazing-angle version large and flat substrates are needed. The sample size should exceed 1 cm². Silicon wafers and float glass substrates have turned out to be excellently suited. XAS is a nondestructive probe and many investigations can be made *in situ*. This is especially interesting for the investigation of liquid-solid and gas-solid interfaces. Since XAS is a local probe, it does not give information on the long-range order in condensed matter. Other techniques (like X-ray, neutron or electron diffraction) have to be used for obtaining this information. Since XAS needs the tunability of the X-ray energy, the experiments are done in general with synchrotron radiation. The limited availability of synchrotron radiation is a serious drawback of XAS especially for applications in industry. Some examples of special interest in materials sciences where XAS has been applied successfully are: structure of alloys, precipitations, internal and external oxidation, corrosion, catalysis, trapping of light elements by impurities, crystallization of amorphous systems, structure of surfaces and internal interfaces, reactions at interfaces, glass surfaces, coatings on glass and on Si wafers, interface electrode-electrolyte.

* * *

The author would like to thank Drs. H. FÖLL, H. GÖBEL and M. SCHUSTER from Siemens AG for many helpful discussions and Mr. R. BAUR and W. RUDNIK from Siemens AG for preparing the Ta oxide and Ta-Si films used as illustrating examples in this lecture. The author would also like to thank Mr. U. DEDEK and F. STANGLMEIER from the Kernforschungsanlage Jülich for their invaluable help during the measurements at the Hamburger Synchrotronstrahlungslabor.

REFERENCES

[1] E. A. STERN: *Contemp. Phys.*, **19**, 289 (1978).
[2] P. A. LEE, P. H. CITRIN, P. EISENBERGER and B. M. KINCAID: *Rev. Mod. Phys.*, **53**, 796 (1981).

[3] B. K. TEO and D. C. JOY, Editors: *EXAFS Spectroscopy* (Plenum Press, New York, N.Y., London, 1981).

[4] T. M. HAYES and J. B. BOYCE: *Solid State Phys.*, **37**, 173 (1982).

[5] E. A. STERN and S. M. HEALD: in *Handbook of Synchrotron Radiation*, Vol. 1*B*, edited by D. E. EASTMAN and Y. FARGE (North Holland Co., Amsterdam, New York, N.Y., Oxford, 1983).

[6] B. LENGELER: *Z. Phys. B*, **61**, 421, 439 (1985).

[7] B. K. TEO: *EXAFS: Basic Principles and Data Analysis* (Springer-Verlag, Berlin, Heidelberg, New York, N.Y., Tokyo, 1986).

[8] R. PRINZ and D. KONIGSBERGER, Editors: X-*ray Absorption: Principles and Techniques of EXAFS, SEXAFS and XANES* (J. Wiley, New York, N.Y., 1968).

[9] M. P. SEAH and W. A. DENCH: *Surf. Interf. Anal.*, **1**, 2 (1979).

[10] B. K. TEO and P. A. LEE: *J. Am. Chem. Soc.*, **101**, 2815 (1979).

[11] B. LENGELER: in *Proceedings of the 4th International Conference EXAFS and Near Edge Structure, J. Phys. (Paris)*, C8, **47**, 75 (1986).

[12] G. STEGEMANN: *Röntgenabsorptionsmessungen an* Fe, Ni, Cu *Legierungen und an amorphem* Ge, Jülich Report JUL-2075 (1986).

[13] D. E. POLK and D. S. BOUDREAUX: *Phys. Rev. Lett.*, **31**, 92 (1973).

[14] N. F. MOTT and E. A. DAVIS: *Electronic Processes in Non-Crystalline Materials* (Clarendon Press, Oxford, 1979).

[15] W. PAUL, G. A. N. CONNELL and K. J. TEMKIN: *Adv. Phys.*, **22**, 529 (1979).

[16] G. S. CARGILL: *Phys. Rev. Lett.*, **28**, 1372 (1972).

[17] B. LENGELER: *Phys. Rev. Lett.*, **53**, 74 (1984).

[18] B. LENGELER: *Solid State Commun.*, **55**, 679 (1985).

[19] B. LENGELER: *Ber. Bunsenges. Phys. Chem.*, **90**, 649 (1986).

[20] B. LENGELER: *J. Phys. (Paris)*, C8, **47**, 1015 (1986).

[21] B. LENGELER, M. WILHELM, B. JOBST, B. SEEBACHER and W. SCHWÄN: *Solid State Commun.*, **65**, 1545 (1988).

[22] J. E. MÜLLER, O. JEPSEN, O. K. ANDERSON and J. W. WILKINS: *Phys. Rev. Lett.*, **40**, 11, 720 (1978).

[23] J. E. MÜLLER, O. JEPSEN and J. W. WILKINS: *Solid State Commun.*, **42**, 365 (1982).

[24] B. LENGELER and R. ZELLER: *Solid State Commun.*, **51**, 889 (1984).

[25] L. A. GRUNES: *Phys. Rev. B*, **27**, 2111 (1983).

[26] G. MATERLIK, J. E. MÜLLER and J. W. WILKINS: *Phys. Rev. Lett.*, **50**, 267 (1983).

[27] R. ZELLER, G. STEGEMANN and B. LENGELER: *J. Phys. (Paris)*, C8, **47**, 1101 (1986).

[28] R. W. JAMES: *The Optical Principles of the Diffraction of X-rays* (Cornell University Press, Ithaca, N.Y., 1967).

[29] M. BORN and E. WOLF: *Principles of Optics* (Pergamon Press, London, 1975).

[30] G. H. VINEYARD: *Phys. Rev. B*, **26**, 4146 (1982).

[31] A. IIDA, K. SAKURAI, Y. YOSHINAGA and Y. GOHSHI: *Photon factory activity report VI*, 29 (1984-85).

[32] P. BECKMANN and A. SPIZZICHINO: *The Scattering of Electromagnetic Waves from Rough Surfaces* (Pergamon Press, London, 1963).

[33] A. BRASLAU, M. DEUTSCH, P. S. PERSHAN, A. H. WEISS, J. ALS-NIELSEN and J. BOHR: *Phys. Rev. Lett.*, **54**, 114 (1985).

[34] L. G. PARRATT: *Phys. Rev.*, **95**, 359 (1954).

[35] H. KIESSIG: *Ann. Phys. (N.Y.)*, **10**, 715 (1931).

[36] E. SPILLER and A. E. ROSENBLUTH: *SPIE J.*, **563**, 221 (1985).

[37] B. LENGELER: *Mikrochim. Acta*, **1**, 455 (1987).

[38] R. P. HAELBICH and C. KUNZ: *Opt. Commun.*, **17**, 287 (1976).

[39] E. SPILLER, A. SEGMÜLLER, J. RITE and R. P. HAELBICH: *Appl. Phys. Lett.*, **37**, 1048 (1981).

[40] L. GOLUB and E. SPILLER: *Appl. Opt.*, **23**, 3529 (1984).

[41] P. DHEZ: *J. Microscopy*, **138**, 267 (1985).

[42] T. W. BARBEE and J. H. UNDERWOOD: *Opt. Commun.*, **48**, 161 (1983).

[43] S. M. HEALD, J. M. TRANQUADA, H. CHEN and D. O. WELCH: *J. Vac. Sci. Technol. A*, **3**, 2432 (1985).

Core Level Spectroscopy at Silicon Surfaces and Interfaces.

F. J. HIMPSEL, B. S. MEYERSON, F. R. McFEELY, J. F. MORAR,
A. TALEB-IBRAHIMI and J. A. YARMOFF (*)

IBM Research Division, T. J. Watson Research Center
Box 218, Yorktown Heights, NY 10598

1. – Introduction.

Silicon is the dominant material in today's semiconductor technology. Therefore, there has been a lot of activity in the study of silicon surfaces and interfaces [1-5]. Silicon surfaces need to be characterized at an atomic level in order to properly understand the methods for processing Si wafers, *e.g.*, reactive ion etching (RIE), wet chemical etching, chemical vapor deposition (CVD), molecular-beam epitaxy (MBE), oxidation and many other techniques. Apart from these somewhat messy surfaces occurring in technology, there are the clean Si surface structures [1], which exhibit beautiful long-range order such as the Si(111)7 × 7 surface. In many cases, one is also able to put down overlayers of adsorbates on a clean Si surface in a well-ordered fashion such as a $\sqrt{3} \times \sqrt{3}$ arrangement or, even simpler, a 1 × 1 structure. There are also a few lattice-matched solids that can be deposited epitaxially on Si without intermixing at the interface. In this lecture, we will make extensive use of such simple model systems in order to demonstrate the principles of core level spectroscopy at Si surfaces and interfaces. It turns out that core level measurements are easy to interpret [6, 7]; they give more direct information than valence band data or other types of spectroscopies. Therefore, it is straightforward to apply the same principles to complex, technologically important surfaces. The methods developed here for silicon can easily be transferred to other semiconductors.

The use of synchrotron radiation has been a great boon to the field. The main reason is that surface sensitivity can be achieved by tuning the photon energy to (130 ÷ 150) eV where the photoelectrons come out with the minimum escape depth. Under these conditions the escape depth can be as short as 3 Å (two silicon

(*) Present address: National Bureau of Standards, Surface Science Division, Gaithersburg, MD 20899.

layers) compared to about 20 Å (more than 10 layers) in conventional X-ray photoelectron spectroscopy (XPS). This short escape depth makes it possible to detect a monolayer of chemically altered Si atoms at a surface or an interface.

2. – Characteristics of the Si2p core level.

In order to obtain the optimum results, one uses the sharpest core level of Si, the $2p$ level, with a binding energy of about 100 eV. The fundamental resolution limit is given by Auger lifetime broadening [7, 8]. (The sharpest core levels of all elements are tabulated in ref. [5].) For the Si2p level a lifetime broadening of 0.08 eV (Lorentzian full width half maximum) has been determined from high-resolution XPS (see fig. 1 and ref. [9]).

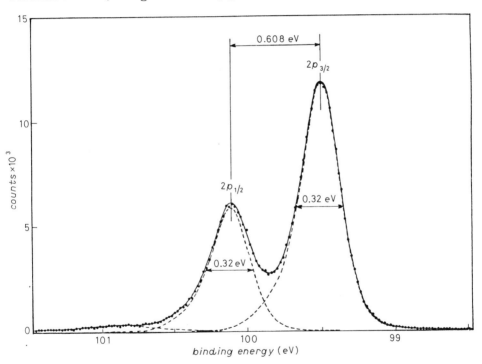

Fig. 1. – The Si2p core level of bulk Si measured by X-ray photoelectron spectroscopy (from ref. [9]); Si(111), $T = 300$ K, $\theta = 45°$.

The Si2p level is split into the $2p_{1/2}$ and $2p_{3/2}$ spin-orbit partners (split by $\delta_{so} = 0.608$ eV [9]) which exhibit roughly the statistical intensity ratio $I_{so} = I_{1/2}/I_{3/2} = 0.5$. For interpreting the spectra it is helpful to subtract out the weaker Si2p$_{1/2}$ component (fig. 2). This can be achieved easily in a two-step process. First, the secondary-electron background has to be subtracted in order to have zero intensity at the high-kinetic-energy end of the spectrum. This background can be measured by reducing the photon energy by about 10 eV such that the Si2p core level drops out of the kinetic-energy window of the spectrum. In the second step, the intensity of the 2$p_{1/2}$ spin-orbit partner is subtracted

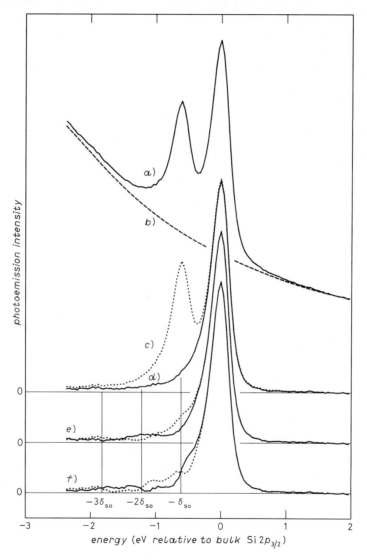

Fig. 2. – Background subtraction and spin-orbit decomposition of the Si2p core level spectrum, Si(111) + HF 1 × 1, $h\nu$ = 108 eV. Curve a) represents the raw data. The secondary-electron background b) is measured at a lower photon energy using the same kinetic-energy range. After background subtraction c) the Si2$p_{1/2}$ contribution is subtracted d) using eq. (1), δ_{so} = 0.608 eV, I_{so} = 0.52. Curves e) and f) are generated with a spin-orbit intensity ratio I_{so} and splitting δ_{so} which deviate from the ideal values: e) I_{so} ± 0.03, f) δ_{so} ± 0.015 eV.

according to the formula

(1) $I_{3/2}(E) = I_{tot}(E) - I_{so} \cdot I_{3/2}(E + \delta_{so})$,

where $I_{tot}(E)$ is the measured total $2p$ intensity, $I_{3/2}(E)$ the $2p_{3/2}$ component, and E the kinetic energy. In order to get the subtraction process started, one has to begin at the high-kinetic-energy end of the spectrum and set $I_{3/2}(E) = I_{tot}(E)$ from the maximum kinetic energy E_{max} to $E_{max} - \delta_{so}$. (This implies $I_{3/2}(E_{max}) = 0$, which is a good approximation if one starts far above the center of the core level emission.) Usually, this procedure requires some fine-tuning of I_{so} and δ_{so}, which is achieved by minimizing residual oscillations at the low-kinetic-energy end of the spectrum (see fig. 2): If the intensity I_{so} is off, one observes maxima (minima) at intervals of $-\delta_{so}$, $-2\delta_{so}$, etc. from the Si2$p_{3/2}$ line, if the splitting δ_{so} is off, one observes turning points. The spin-orbit splitting is mainly determined by the atomic properties of Si and varies very little in different chemical environments (the largest deviation is found for clean Si(100)2 × 1 where δ_{so} is smaller by 0.05 eV, probably due to surface effects [10]). The intensity ratio I_{so} tends towards the statistical 0.5 for high photon energies ($h\nu \gtrsim 130$ eV), but varies near threshold due to the photon energy dependence of the cross-section (see fig. 2 and sect. 5).

For a detailed line shape analysis of core level spectra one has to take the intrinsic secondary-electron background into account. These electrons are losses coming from the core line itself, and are not contained in the extrinsic background subtraction described above. The intrinsic losses depend on the electronic structure of the surface, particularly on the band gap. For oxide and fluoride overlayers one has a large band gap, and the energy losses are too high in energy to interfere with the core level spectrum. Only the losses in the Si substrate have to be taken into account which exhibit a weak peak above the 1.1 eV band gap of silicon (see fig. 1). For clean Si surfaces the situation is more complicated, since there exist strong loss features across the surface band gap. The surface band gap of Si(111)7 × 7 is close to zero, thereby giving rise to infinitesimally small losses and an asymmetric core line like in a metal. In practice one often approximates the intrinsic losses by a step function, which is equivalent to taking the integral curve of the spectrum as the loss function.

Apart from measuring photoemission spectra from the Si2p core level, there are several other techniques involving the Si2p level. In fig. 3 the corresponding LVV Auger spectrum is shown. Its high-energy part exhibits a strong surface contribution [11], which disappears upon adsorption of H_2O, CaF_2, etc. It is difficult to analyze, though, due to the self-convolution of valence band features.

The Si2p absorption edge (fig. 4) exhibits a steplike onset that is caused by excitonic effects [12-14]. In analogy to the C1s edge in diamond, where the first exciton is resolvable [15], one would expect an exciton line about 0.03 eV below the onset of transitions into the conduction band minimum [13, 15]. In addition,

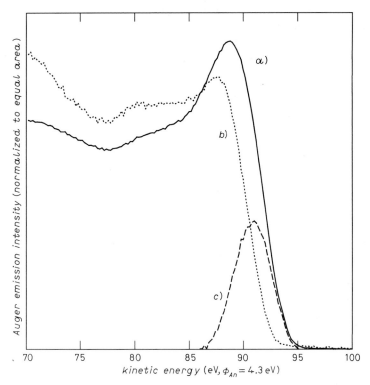

Fig. 3. – The Si2p (LVV) Auger spectrum from a) clean and b) adsorbate-covered (+ 100 L H$_2$O) Si(100). The difference spectrum c) emphasizes emission from surface states.

the transition probability into the conduction band is enhanced by excitonic effects[14]. Due to lifetime broadening, all the excitonic features are smeared out in Si, and the actual edge (*i.e.* the turning point of the steplike absorption curve) occurs at (0.1 ÷ 0.2) eV below the onset of transitions into the conduction band. The Si2p absorption can also be measured in a surface-sensitive mode[16] by collecting secondary electrons with kinetic energy near the escape depth minimum (typically 30 eV, see sect. 5). Under such conditions, transitions from the Si2p level into unoccupied surface states have been observed below the bulk absorption edge. Due to dipole selection rules, the surface states with s-like (or d-like) character are projected out. The surface states have mainly p-like character in Si and would be observable only using the (much broader) Si1s core level. Adsorbates change the energy and the shape of the Si2p edge due to chemical shifts of the Si2p level and due to changes in the density of states near the conduction band minimum[17]. Such variations may be used as fingerprints for the state of the surface. In particular, photoelectron microscopy experiments[18] employing high-brightness undulator sources look very promising in this mode. In such experiments the emitted photoelectrons are sent through

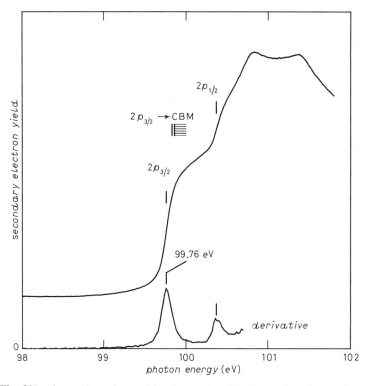

Fig. 4. – The Si2p absorption edge and its derivative. The intensity of secondary electrons is detected, which is proportional to the absorption coefficient.

magnifying electron optics and imaged onto a channel plate multiplier with phosphor screen. For example, oxidized portions of a silicon surface are easily distinguished from clean portions by an upwards shift of the Si2p edge [19]. This shift may be detected by taking pictures at photon energies above and below the Si and the SiO$_2$ absorption edges. Thereby an electron energy analysis is not necessary, since all secondary electrons may be collected, such as in fig. 4. Avoiding energy analysis of the electrons makes it easier to achieve high spatial resolution.

3. – Fermi level pinning.

The general potential diagram at a semiconductor surface (or interface) is given in fig. 5. Charges in surface/interface states cause an electrostatic potential that bends the bands near the surface. The characteristic length scale of this band bending is the Debye screening length, which decreases for increasing doping levels. Typical values are thousands of Å for intrinsic material and tens of Å for the highest achievable doping levels. Therefore, the probing

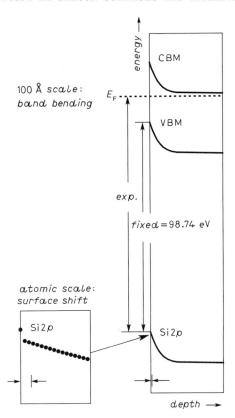

Fig. 5. – The two contributions to the Si2p core level energy from Fermi level pinning and from chemical effects at the surface. The former affects the position of the whole core level spectrum (bulk plus surface), the latter introduces extra surface core level peaks originating from the outermost Si layer. Examples for the two effects are given in fig. 6a), b).

depth of a photoemission experiment (typically 5 Å, see sect. 5) is smaller than the width of the band bending region, and one probes the bands (and the Si2p level) in their position at the surface. An important quantity for the electrical performance of an electronic device is the Schottky barrier, *i.e.* the potential barrier that charge carriers have to overcome to cross the surface (interface). For an *n*-type sample (see fig. 5) this barrier is given by the difference between the conduction band minimum (CBM) and the Fermi level (E_F) at the surface. For *p*-type samples, the barrier is $E_F - E_{VBM}$ (VBM = valence band maximum). This barrier can be measured accurately (with a reproducibility of $< 20\,\mathrm{meV}$) using the Si2p core level [20-22]. As shown in fig. 5, one measures the energy of the Si2p core level relative to E_F, and has only to subtract the fixed energy of the Si2p level relative to E_{VBM} to get the *p*-type barrier. For determining this fixed reference energy one has to obtain a sample with flat bands, where the quantity $E_F - E_{VBM}$ is known from bulk doping. It is difficult to obtain a flat-band situation at Si surfaces as it requires to reduce the number of surface/interface states in the gap to about 1/1000 of a monolayer. There have been successful attempts using overlayers of dopands [20, 22]. In fig. 6a) we achieve a flat-band condition for a highly p-doped and a highly n-doped Si sample by hydrogen termination [23]. The fact that there is no remaining band bending (within 0.04 eV) is

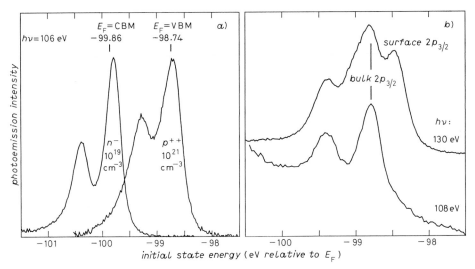

Fig. 6. – Si2p core level shifts induced by Fermi level movement and by chemical shift: a) Effect of the Fermi level position on the bulk Si2p energy. For well-passivated surfaces (see ref. [23]) the Fermi level can be moved from the valence band maximum (p^{++} sample) to the conduction band minimum (n^- sample). Intermediate Fermi level positions are determined using these two spectra as a reference. A photon energy near threshold is chosen to keep the escape depth large and thereby avoid contributions from surface chemical shifts. b) Identification of a surface chemical shift CaF$_2$/Si(111) by tuning into the minimum of the escape depth. At $h\nu = 130$ eV one has about two layers probing depth, and the surface core level emission is enhanced compared to the spectrum taken near threshold.

evident from the energy difference of the n- and p-type spectra (1.08 eV), which nearly equals the difference of 1.12 eV in the bulk Fermi levels (the Fermi level lies 0.05 eV below the conduction band minimum for the n-type 10^{19} cm^{-3} sample and 0.05 eV below the valence band maximum for the p-type 10^{21} cm^{-3} sample).

For practical purposes, it is convenient to measure the Si2p level relative to some reference surface with a known value of $E_F - E_{VBM}$. Clean Si surfaces can be prepared in a reproducible manner, which makes them a good reference. One has to be careful, though, that the Fermi level is pinned at a stable position and does not move within a band gap between occupied and empty surface states. The cleaved Si(111)2 × 1 surface [25] and the Si(100)2 × 1 surface [26] have such a surface band gap, the Si(111)7 × 7 surface is metallic [27-30]. We have indeed observed that E_F moves relative to E_{VBM} for the former two surfaces when the doping is changed from p-type to n-type (see table I). From the range of Fermi level positions *vs.* doping for Si(111)2 × 1 and Si(100)2 × 1, one may obtain an estimate for the extension of the surface band gap. The Fermi level shift *vs.* doping occurs only for nearly perfect surfaces, since the Fermi level tends to be pinned near the top of the surface band gap due to defect states. (This makes n-type samples more attractive for photoemission measurements since the Fermi level position is homogeneous even in the presence of defects. In addition, there is a problem with surface segregation of boron when heating B-doped p-type samples above 900 °C.) Si(111)7 × 7 is a good reference surface, since the Fermi level is stable at 0.65 eV above E_{VBM}, as expected from its metallic behavior. Results on $E_{\text{F}} - E_{\text{VBM}}$ are listed in table I for various reference surfaces.

For semiconductor-semiconductor interfaces, the band offset is the analog to the Schottky barrier for metal-semiconductor interfaces. It can be measured in a

TABLE I. – *Fermi level position* E_{F} *(relative to the valence band maximum* E_{VBM}*), work function* ϕ*, surface core level shifts* $E_{\text{S}} - E_{\text{B}}$ *and intensities* $I_{\text{S}}/(I_{\text{S}} + I_{\text{B}})$ *for various silicon surfaces and interfaces.*

Surface	$E_{\text{F}} - E_{\text{VBM}}$ (eV)	ϕ (eV)	$E_{\text{S}} - E_{\text{B}}$	$I_{\text{S}}/(I_{\text{S}} + I_{\text{B}})$ ($h\nu = 130$ eV)
Si(111) 2 × 1	0.39 (intr.) 0.55 (n^-)	4.85	+ 0.40, (− 0.12)	0.19, (0.19)
Si(111) 7 × 7	0.65 (n, p) (^{f})	4.55	+ 0.8	0.05
Si(100) 2 × 1	0.19 (p) 0.45 (n)	4.9	+ 0.49, (− 0.23)	0.17, (0.17)
Si(110) 4 × 5 (^{a})	0.7 (p)	4.55	+ 0.7	0.07
Si(311) 3 × 2 (^{a})	0.6 (p)	4.65	+ 0.55, − 0.65	0.15, 0.21
Si(111) 1 × 1 H	0.5 ÷ 1.1 (^{b})		− 0.26 (^{b})	
Si(111) 1 × 1 As	0.7 ÷ 0.9 (^{c})		− 0.75 (^{c})	0.40 (^{c})
Si(111) 1 × 1 CaF$_2$ (^{d})	0.05 (p)		+ 0.36	0.43
Si(100) 1 × 1 H$_2$O (^{e})			− 0.9, (− 0.3)	0.17, (0.17)
CoSi$_2$ (111) 1 × 1 (^{g})			− 0.75	0.33

(a) See. ref. [31] for the reconstruction.
(b) Ref. [16] and [32] for H/Si(111) 2 × 1 and H/Si(111) 7 × 7, respectively.
(c) Ref. [33].
(d) Compare ref. [34, 35].
(e) Compare ref. [36, 37].
(f) This value is more accurate than the value of (0.63 ± 0.05) eV given in ref. [20].
(g) Ref. [38].

fashion similar to the Schottky barrier. Essentially, measurement of $E_F - E_{VBM}$ on both sides of the heterojunction gives automatically the valence band offset, and adding the respective band gaps gives the conduction band offset. Such measurements have been performed for Si/Ge$_x$Si$_{1-x}$ heterojunctions [22].

4. – Core level shifts.

The key ingredient in Si$2p$ core level measurements is the energy shift that occurs for surface and interface atoms. As we will demonstrate in the following, this shift gives very direct chemical and bonding information. Looking back at fig. 5 (inset), we leave Fermi level pinning effects aside and concentrate on the core level shift for the outermost atomic layer. Fermi level pinning determines the position of the «bulk» peak in the spectra, whereas we are now looking at the shift between the «bulk» and the «surface» peak. Bulk and surface effects can easily be separated by taking spectra with different photon energies $h\nu$ (fig. 6b)), corresponding to different electron escape dephts. Near threshold one has a probing depth of about ten layers, whereas at $h\nu = 130$ eV one probes only two layers (see sect. 5). Figure 7 demonstrates surface core level shifts induced by various adsorbates. For oxygen- and fluorine-covered Si surfaces one can clearly distinguish four discrete surface core levels, which are assigned to Si bonding to one, two, three and four oxygen [36, 39] or fluorine [40] atoms. The SiF$_4$ species is weak for monolayer coverage, but becomes stronger at higher exposure, when the surface begins to be etched [41]. On the other side of the electronegativity spectrum we display results from various Ca-Si configurations, which exhibit a surface core level shift opposite to O and F. These shifts are plotted in fig. 8 $vs.$ the number of ligands, and a simple linear relation is found. Exploiting a connection between core level shift and electronegativity difference, we compare the core level shift per ligand [33, 36, 39-46] with the electronegativity [47] difference between ligand and silicon in fig. 9. (The corresponding data are tabulated in table II.) Empirically, there seems to be a correlation although exceptions exist [48]. A similar S-shaped curve has been found for a variety of metals on Si by DEL GIUDICE et al. [49]. For small electronegativity difference $\Delta\chi$ the core level shift ΔE is nearly equal to $\Delta\chi$, at larger $\Delta\chi$ the core level shift saturates. The behavior for positive and negative $\Delta\chi$ is symmetric. From such an empirical curve one may predict core level shifts for new ligands, or find out from a measured core level shift which ligand is involved. An example is Ca-Si $vs.$ F-Si bonding at the CaF$_2$/Si(111) interface (see sect. 8).

For a quantitative understanding of core level shifts, one has to take a variety of mechanisms into account [4, 6, 7]. These can be classified into initial-state and final-state effects. The former are due to a change in the chemical environment of a surface atom and comprise charge transfer, Madelung energy and rehybridization. The latter occur after the core level is ionized and can be viewed as a

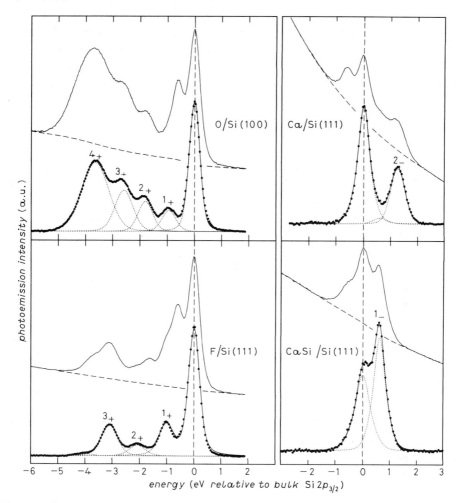

Fig. 7. – Si2p core level spectra for a variety of reference surfaces. Discrete chemical shifts are seen at the surface that correspond to the number of ligands. Nominal oxidation states (1 +, 1 −, etc.) are used as labels for the number of ligands. (Note that the charge transfer is smaller than this oxidation state, e.g. only half an electron for a Si-O bond [47].) Electronegative ligands (F, O) shift the Si2p level down since the Si atom becomes more positive. The reverse is true for electropositive elements (Ca). The raw data are shown (top curves) together with the Si2p$_{3/2}$ component (dots).

change in dielectric screening at the surface. One of the advantages of Si surfaces is that the dielectric constant of Si is so large ($\varepsilon \approx 12$) that the screening is nearly complete, like in a metal (the screening charge is $((\varepsilon - 1)/(\varepsilon + 1)) \cdot e \approx 0.85 \, e$ for Si). Therefore, even with a reduced number of Si neighbours at the surface, there is still nearly complete screening. Consequently, the final-state effect is minimized, and long-range fields in the initial state, like the Madelung potential, are screened out. If one proceeds from monolayer adsorbates to bulk Si

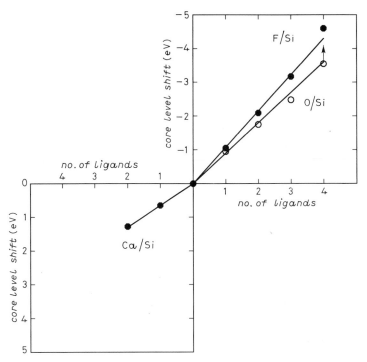

Fig. 8. – Core level shift *vs.* number of ligands for the data in fig. 7. The relationship is nearly linear as long as one stays in the monolayer regime, where the strong dielectric screening by the Si substrate is effective [50].

compounds with low dielectric constant (*e.g.*, SiO_2), one runs into irregularities caused by lack of screening [50]. The importance of initial-state shift ΔE^i *vs.* final-state shift ΔE^f can be assessed by comparing the core level shift ΔE_c with the shift ΔE_A of an Auger peak. In a somewhat simplified picture [4] one finds

$$(2) \qquad \Delta E_c = \Delta E^i + \Delta E^f, \qquad \Delta E_A = \Delta E^i + 3\,\Delta E^f.$$

The factor of 3 in the final-state shift for the Auger peak comes from the fact that a doubly charged ion is left after Auger decay giving 4 times as large a relaxation energy. One unit of relaxation energy needs to be subtracted since it has already been taken into account for the relaxation around the core hole before Auger decay takes place. Solving eq. (2) with respect to ΔE^i and ΔE^f one obtains

$$(3) \qquad \Delta E^i = \frac{1}{2}\,(3\,\Delta E_c - \Delta E_A), \qquad \Delta E^f = \frac{1}{2}\,(\Delta E_A - \Delta E_c).$$

Unfortunately, the Si2p LVV Auger spectrum (see fig. 3) cannot be used to determine the Auger shift since it involves the valence band. By using the KLL

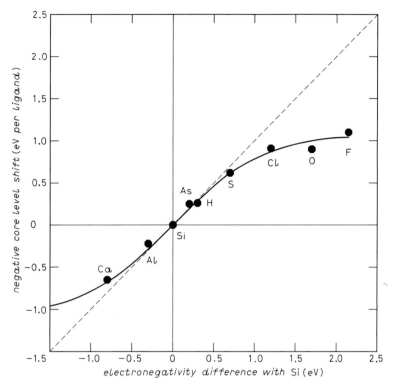

Fig. 9. – Core level shift per ligand, corresponding to the slope of the curves in fig. 8. A few well-defined adsorbate structures have been chosen (see table II for a listing). A symmetric curve interpolates the data, which can be used to empirically predict core level shifts from electronegativity differences (and *vice versa*). The saturation of the core level shift at high electronegativity difference is due to a saturation of the charge transfer[47].

Auger spectrum instead, one loses all surface sensitivity due to the high kinetic energy of the Auger electrons (1.6 keV). A shift of $\Delta E_A \approx -7$ eV has been observed[51] for bulk SiO_2 *vs.* Si, compared with $\Delta E_c \approx -4.5$ eV. Inserted in our simplified formula, this gives $\Delta E^i \approx -3.25$ eV, $\Delta E^f \approx -1.25$ eV. The negative final-state shift shows that SiO_2 has less screening than Si, as expected[50] for the lower dielectric constant of SiO_2. Analogous experiments at the surface have yet to succeed.

The initial-state shift can be analyzed in terms of charge transfer in a systematic way[4, 6, 7]. As fig. 10 shows, there are two contributions, one from the charge transferred towards the Si atom (this charge is located in a valence orbital of Si), the other from the atom that lost the charge (this gives the Madelung energy). The two charges give rise to opposite electrostatic potentials at the Si core, thereby shifting the Si2p core level in opposite directions. However, the charge sitting in the Si valence orbital dominates since it is closer than the charge on the neighbour atom. The charge transfer can be correlated

TABLE II. – *Core level shift per ligand ΔE and electronegativity difference $\Delta\chi$ for various adsorbates on silicon.*

Surface-adsorbate	ΔE (eV)	$\Delta\chi$ (eV)	References
Si-F	1.10	2.15	[40, 41]
Si-O	0.90	1.7	[36, 39]
Si-Cl	0.91	1.2	[43]
Si-S	0.62	0.7	[44]
Si-H	0.26	0.3	[16, 32, 45]
Si-As	0.25	0.2	[33]
Si-Al	− 0.22	− 0.3	[46]
Si-Ca	− 0.65	− 0.8	(compare [42])

with the electronegativity difference $\Delta\chi$ according to PAULING [47]. This correlation explains the empirical success of relating core level shift to electronegativity difference in fig. 9. Furthermore, the saturation of the core level shift for large $\Delta\chi$ can be understood. It is due to the nonlinear relation [47] between charge transfer and $\Delta\chi$. At large $\Delta\chi$ the charge transfer saturates, like the core level shift. If we were to plot core level shift *vs.* charge transfer, we would obtain a nearly linear relationship [6]. However, the concept of charge transfer is somewhat ambiguous (it depends on how the size of an atom is defined), and we prefer the better-defined electronegativity. In order to give an idea of the magnitude of the charge transfer we quote a few values from Pauling's book [47]: Si-F ($0.6e$), Si-O ($0.5e$), Si-Cl ($0.3e$). The corresponding core level shifts (per ligand) are Si-F (1.1 eV), Si-O (0.9 eV), Si-Cl (0.9 eV). Thus one

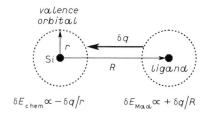

Fig. 10. – Core level shift induced by charge transfer. The charge δq transferred from a ligand to a valence orbital of silicon induces an electrostatic potential $\delta q/r$ at the Si core. The complementary charge $-\delta q$ at the ligand site induces an opposite, but smaller potential $-\delta q/R$ at the Si core, which corresponds to the Madelung energy.

has about 2 eV core level shift per unit charge. This holds for Si. For smaller atoms the core level shift is larger since the charge sits closer to the core. Consequently, C has about twice the core level shift per ligand relative to Si (*e.g.*, 2.8 eV per C-F bond [3]).

The Madelung contribution has been calculated for various bulk compounds of Ge (see ref. [52]). It compensates about three quarters of the on-site core level shift. At a surface the Madelung energy involves a summation over the half space of the substrate, plus a possible overlayer. It has been worked out for a few semiconductor surfaces [53]. The influence of overlayer Madelung energies on a Si surface atom has not been estimated yet, but the nearly complete screening in Si will strongly reduce this effect relative to its strength within the overlayer itself. The variable screening condition in overlayers has made it impossible so far to analyze core level shifts of adsorbates in a manner similar to the Si substrate.

There is a different way to explain core level shifts: Using a Born-Haber cycle, one can replace a core-ionized atom with atomic number Z (*e.g.*, silicon) by a neutral atom with atomic number $Z + 1$ (*e.g.*, phosphorous) assuming the core level is fully screened [7, 49, 54]. A core hole on a bulk Si atom corresponds to a P impurity in Si, a surface core hole next to an adsorbate A corresponds to a P atom bonding to A. The difference in bond energy between Si-P and A-P is then related to the core level shift. At first glance, this method appears to be a re-writing of the problem in terms of new variables, *i.e.* bond energies. However, there are empirical models for bond energies and heats of solution [55] which allow systematic treatment of core level shifts. Such an approach has been used for metal-silicon systems [49].

With all these semi-empirical descriptions of surface core level shifts one may ask the question whether there exist first-principles methods that can handle the problem for semiconductor surfaces. Although steps have been taken [56] in this direction, there exists no complete *ab initio* calculation of initial- and final-state effects yet. The computational problems are enormous since a core excitation destroys the periodicity of a crystal, making Bloch's theorem inapplicable. In addition, a proper calculation of the screening involves the energy- and momentum-dependent dielectric tensor, which is a complicated quantity to obtain. For using core level spectroscopy to characterize surfaces and interfaces one does not require a first-principles theory, as long as there exist simple empirical rules.

5. – Cross-section and escape depth.

In addition to the information contained in the magnitude of the Si2p core level shifts one may use the intensity of the shifted core levels as stringent test of models of the surface structure. To do so requires not only that each peak be identified with a specified local coordination structure, but also that the intensity of each such peak be related to the number density of Si atoms predicted by the model.

To perform such an optimal analysis of the photoemission data, the relative photoemission cross-section of each chemically distinct species and the photo-electron escape depth of each chemically distinct region (substrate, interface, overlayer) in the structure must be known. The effect of cross-section and escape depth can be assessed qualitatively from fig. 11. The bulk Si2p doublet plus a

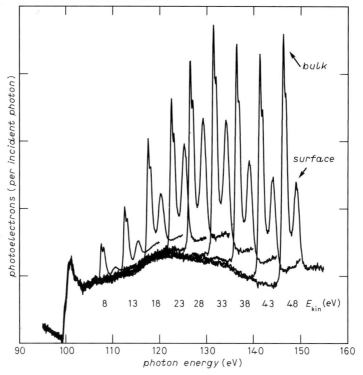

Fig. 11. – Constant-final-state spectra for oxygen on Si(100) exhibiting energy-dependent variations in the photoionization cross-section and in the escape depth. A maximum in the surface-to-bulk intensity ratio is reached at about 130 eV photon energy.

shifted surface peak are measured for a range of kinetic energies. In this case the constant-final-state (CFS) detection mode is used where the kinetic energy of the photoelectrons is fixed and the photon energy is varied in order to scan through the core level. This type of measurement provides a constant density of final states and constant escape depth across the whole spectrum, whereas in a conventional photoelectron spectrum one has a variation in escape depth and final-state energy between surface and bulk core levels and between the spin-orbit partners. The overall core level intensity increases in fig. 11 when going from threshold to kinetic energies of about 30 eV due to the increasing photoionization cross-section. The ratio of the surface to the bulk intensity exhibits a maximum around 30 eV kinetic energy. This maximum is due to two effects: The escape depth is shortest near 30 eV kinetic energy, and the cross-section for the shifted surface core level exhibits a resonant enhancement relative to the bulk cross-section at this energy. The surface sensitivity thus reaches a maximum near a photon energy of 130 eV. A monolayer of surface atoms has an intensity comparable to the whole bulk emission.

The high surface sensitivity of synchrotron sources complicates the analysis, though. In the region of photoelectron kinetic energies where the surface sensitivity is largest, chemical effects on the cross-section can be large, and escape depths vary rapidly. Such effects may lead to errors in the analysis of surface and interface structure. To the best of our knowledge, there is no case in which all of the information necessary for a complete analysis is available. However, reasonable conclusions can nonetheless be made. To illustrate the process we consider a concrete case, the interface of thermally grown SiO_2 on Si(111).

The Si$2p$ photoemission spectrum of the SiO_2/Si interface consists of five resolvable peaks (see fig. 7). Those at lowest and highest binding energy correspond to the bulk Si and the fully formed SiO_2, respectively. The other three arise from Si^{1+}, Si^{2+} and Si^{3+} species at the interface. The basic information to be determined from such spectra is the width of the interface region and the relative amounts of the various oxidation states. This requires a knowledge of the cross-sections of each state relative to the bulk Si.

Figure 12 shows the results of an experiment designed to look for cross-section variations in this system. A thin layer of SiO_2 was grown on Si(111) and a series of constant-final-state (CFS) spectra were taken at various kinetic energies. The CFS mode was used to insure there was no variation in the escape depths within a spectrum. The top panel of fig. 12 shows the raw intensities as measured for the three intermediate oxides. The ordinate scale is in electrons per incident photon, the photon flux being measured by monitoring the photocurrent from the mirror which focuses the output of the monochromator onto the sample. Each curve thus reflects a product of the cross-section and the photon-energy-dependent response of the electron spectrometer. The bottom panel shows the intensities of the 2 + and the 3 + moieties divided by the 1 + intensity. This cancels the instrumental-response term and gives curves which are the product of the photoemission cross-section (relative to Si^{1+}) and the number of emitting species, assuming that inelastic scattering within the thin interfacial layer is the same for all species. Two different types of behavior are observed. The emission intensity ratio for the Si^{2+} peak is, within experimental error, a constant above about 20 eV kinetic energy, indicating that above this energy the relative cross-sections of Si^{2+} and Si^{1+} are equal. The Si^{3+} curve, however, shows a large, broad peak centered at about 23 eV, suggestive of a

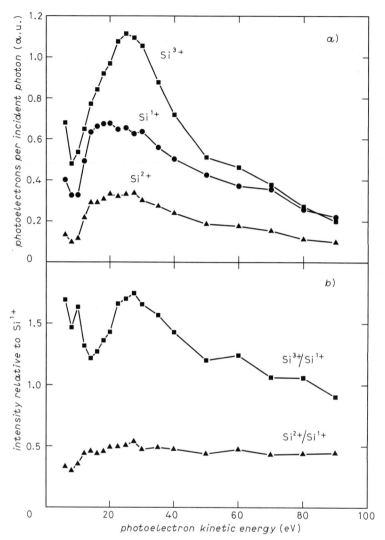

Fig. 12. – Photoionization cross-section for intermediate oxidation states of Si, measured at monolayer oxygen coverage on Si(111): a) raw intensities, b) relative intensities.

shape resonance in the cross-section. Only at higher energies do the 1 + and 3 + features settle down to nearly equal intensity. We take the constancy of the intensity ratios obtained at the highest energies as an empirical indication of equal relative cross-sections. With these data we can, therefore, divide out the enhancement of the 3 + oxidation state found at more surface-sensitive kinetic energies.

In making this assessment, we are assuming that escape depth effects within the oxidized region do not distort the spectra significantly. This would be rigorously true only if the suboxides had identical spatial distributions, which is almost certainly not the case. Although we do not know the attenuation function for the surface region (which would be depth- as well as energy-dependent for a region of nonuniform composition), the total thickness of this is only about 1/3 of SiO_2 escape depth. The errors introduced by this assumption are, therefore, probably small.

For the SiO_2 overlayer, the photoelectron escape depth $\lambda(E)$ may be determined by a method which is free of any assumptions concerning the photoemission cross-sections [57]. This is done by measuring CFS photoemission

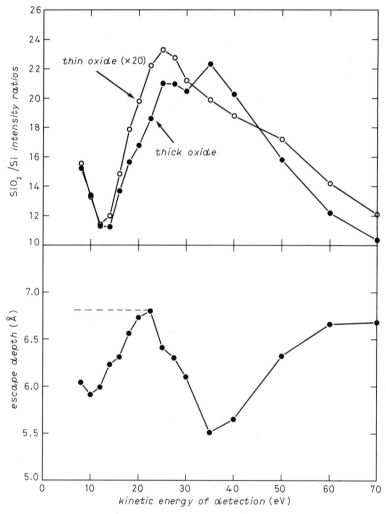

Fig. 13. – Determination of the electron escape depth in SiO_2 from the overlayer-to-substrate intensity ratio for two SiO_2 films of different thickness. The raw data are given in panel a). The escape depth shown in b) is obtained by solving eq. (4). A more complete analysis [57], treating elastic scattering explicitly, would show the escape depth becoming constant below 20 eV for sufficiently thick SiO_2 films (dashed line).

spectra for two samples with different thicknesses of SiO_2. Then, for each sample, the ratio R_1 (R_2) of fully formed SiO_2 emission to Si bulk emission is measured. From these data one then calculates the ratio of these ratios, the value of which is given by

$$(4) \qquad R(E) = \frac{R_2}{R_1} = \frac{(1 - \exp[-d_2/\lambda(E)]) \exp[-d_1/\lambda(E)]}{(1 - \exp[-d_1/\lambda(E)]) \exp[-d_2/\lambda(E)]},$$

where $\lambda(E)$ is the escape depth in the SiO_2, and d_1 and d_2 are the thicknesses of the two SiO_2 films. Knowledge of the latter two quantities, either from ellipsometric measurements or from published growth kinetic data, allows this equation to be solved numerically for $\lambda(E)$. Note that the use of intensity ratios causes all photoemission cross-section terms to cancel exactly, and, if the interfacial suboxide region has the same thickness and composition in the two samples (which can be verified from the spectra themselves), its contribution also vanishes. The only limitation of this method generally is given by the necessity to maximize the differences in the oxide thickness in the two films, in order to achieve sufficient variation in R.

The results of such a determination are shown in fig. 13. The top panel shows the ratios of the SiO_2 to bulk Si emission for the two films. The bottom panel

TABLE III. – *Escape depths for* Si2p *photoelectrons in* Si *and* SiO_2 *at various photon energies. The kinetic energy is* 104 eV (108 eV) *smaller than the photon energy for* Si (SiO_2).

	$h\nu$ (eV)	λ_{SiO_2} (Å)	λ_{Si} (Å)
Threshold:	104	—	12 [a], 25 [b]
	120	8.5 [c], 5.9 [f]	4.0 [c]
	130	7.1 [c], 6.8 [f]	3.3 [c]
	145	6.3 [c], 5.5 [f]	3.3 [c]
	200	6.5 [c]	5 [c]
	400	11 [c]	10 [c]
	1254	21 [d], 25 [e]	13 [d], 23 [e]
	1487	26 [d], 27 [e]	16 [d], 26 [e]

(a) C. SEBENNE, D. BOLMONT, G. GUICHAR and M. BALKANSKI: *Phys. Rev. B*, 12, 3280 (1975).
(b) F. G. ALLEN and G. W. GOBELI: *Phys. Rev.*, 127, 150 (1962).
(c) Ref. [36].
(d) M. F. HOCHELLA jr. and A. H. CARIM: *Surf. Sci.*, 197, L260 (1988).
(e) R. FLITSCH and S. I. RAIDER: *J. Vac. Sci. Technol.*, 12, 305 (1975).
(f) Ref. [57]; see fig. 13.

shows the escape depths derived from these data and eq. (4). As is evident, the escape depth curve for SiO_2 deviates strikingly from the «universal curve» form. In addition to the expected escape depth minimum at about 35 eV, there is also a region of short escape depth at very low energies [57]. This latter effect is due to a combination of strong inelastic scattering by optical phonons and by large-angle elastic scattering, which strongly increases the effective path length of the photoelectrons. The phonon losses also influence the lineshape of the bulk peak, causing strong asymmetric tailing in the thick oxide spectra. (The effect is negligible in sufficiently thin samples.)

The escape depth in pure Si can be determined from the surface-to-bulk intensity ratio at clean Si surfaces [16, 36, 59, 60]. Cross-section changes are expected to be small in this case, since the oxidation state does not change significantly. This is reflected in the small core level shifts. By collecting photoelectrons over a wide acceptance angle one avoids diffraction effects in crystalline silicon [61]. Results from such an analysis [36] are compiled in table III.

6. – Clean silicon surfaces.

Core level shifts occur already at clean Si surfaces [16]. These shifts originate from the charge transfer that takes place between surface atoms while they are rearranging themselves to minimize the density of broken bonds. An additional, but less-explored mechanism is the rehybridization (*e.g.*, sp^3 to sp^2) that accompanies π-bonding [62] at Si surfaces. In fig. 14, the Si2p spectra are shown for the common low-index surfaces, plus the (311) surface. The latter represents a surface with extremely high step density, where Si(100) and Si(111) terraces alternate. The interpretation of these surface core level spectra is not completely settled yet [32, 58-60]. We believe, however, that the following assignments give a picture that agrees best with other information on these surfaces, *e.g.* LEED (low-energy electron diffraction), STM (scanning tunneling microscopy) and first-principles band calculations. The Si(111) surface comes in two varieties, the metastable (2×1) structure obtained by cleaving and the stable (7×7) structure. In addition, there are (1×1) structures (probably disordered 2×2) obtained at temperatures above 870 °C (ref. [63]) and by laser annealing [64]. Spectra for the latter are shown in fig. 15.

The Si(111)2 × 1 structure consists of π-bonded chains [62] with two inequivalent surface atoms. According to LEED analyses [65] the two inequivalent chain atoms move up and down, respectively. Such a movement causes them to rehybridize and acquire negative and positive charge, respectively. Therefore, one expects surface core levels on either side of the bulk line with equal intensities corresponding to half a layer of Si atoms each. The upper surface core level at 0.4 eV above the bulk line is seen in the raw data as a shoulder (fig. 14). It corresponds to the up-atoms in this assignment. The lower surface core level causes a shift of the main peak below that of the bulk core level. (The latter is obtained at a lower photon energy of 106 eV, where the escape depth is large.) By fixing the intensity of the lower surface core level to be equal to that of the upper and by freezing the position of the bulk line one obtains the dashed fit line given in fig. 14. This fit of the lower surface core level is rather uncertain and

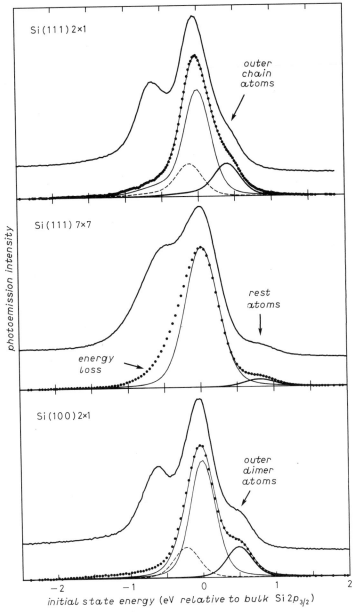

Fig. 14. – Core level shifts at clean Si surfaces and their tentative assignment to special surface atoms. All spectra are taken in the surface-sensitive mode at $h\nu = 130$ eV. The raw data are shown (top curves) together with the Si$2p_{3/2}$ component (dots).

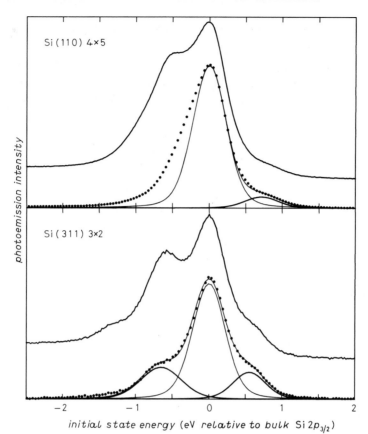

Fig. 14. – Core level shifts at clean Si surfaces and their tentative assignment to special surface atoms. All spectra are taken in the surface-sensitive mode at $h\nu = 130$ eV. The raw data are shown (top curves) together with the Si$2p_{3/2}$ component (dots).

shown for completeness only. In order to explain the low-energy tail of the spectrum it is necessary to take energy loss features into account. A prominent loss feature is indeed seen at 0.5 eV for Si(111)2×1. It is caused by transitions across the surface band gap [25]. For our fit we add this loss to each line with 15% of the core level intensity.

The Si(111)7×7 surface is described by a model [66-68] containing dimers, adatoms and a stacking fault. There are at least three types of special surface atoms to be considered, $i.e.$ 12 adatoms, $6 + 1 = 7$ rest atoms (counting the atom at the corner of the unit cell) and $3 \times 12 = 36$ atoms bonding to the adatoms. (These numbers are referred to the 7×7 unit cell containing 49 atoms per layer.) The dimer atoms may also be considered, but they are fully coordinated and less likely to exhibit a chemical shift. We assign the most-pronounced surface core level at 0.8 eV above the bulk line to the rest atoms. These atoms correspond to isolated broken bonds of the original truncated bulk structure. They pick up an extra electron from the adatoms to fill their broken-bond orbital. It is seen to be

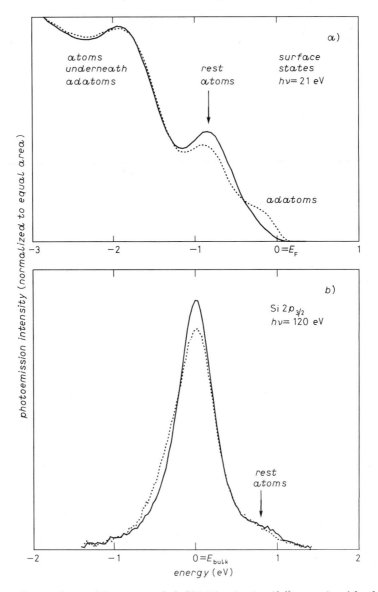

Fig. 15. – Comparison of laser-annealed Si(111) «1 × 1» (full curve) with thermally annealed Si(111)7 × 7 (dotted curve) (from ref. [69]). The laser-annealed surface has more rest atoms (= atoms in truncated bulk sites) than Si(111)7 × 7 (compare [69]).

completely filled by photoelectron spectroscopy (see fig. 15 top) and by STM [70]. The adatoms tie up three broken bonds of the truncated bulk structure. They are positively charged, since they lose electrons to the rest atoms. Using this charge transfer picture we expect the adatom core levels at an energy below the bulk line in a region that is obscured by energy loss features. Adatoms exhibit empty broken-bond orbitals as evidenced by STM[70] and inverse photoemission[28]. A portion of the adatom orbitals is still filled for the Si(111)7 × 7 surface since there are twice as many adatoms as rest atoms, *i.e.* only half an electron needs to be taken from each adatom to fill the bond orbitals of the rest atoms. These states show up near the Fermi level in photoemission (fig. 15). First-principles band calculations[71] are consistent with this picture. The third kind of surface atoms, *i.e.* the 3 original surface atoms underneath an adatom, can also be seen in the valence band photoelectron spectrum (fig. 15). They are difficult to identify in the core level spectrum of Si(111)7 × 7, but they can be seen[69, 72-74] on Ge(111)c(2 × 8) at 0.3 eV above the bulk line. On Si(111)7 × 7 they are probably located in the upper part of the asymmetric «bulk» line. There is a significant contribution from energy loss features to the low-energy tail of the core level spectrum. The 0.5 eV loss is absent in Si(111)7 × 7, but metallic low-energy losses are observed[29]. These are left out in the fit in fig. 14. They could be simulated by an asymmetric Doniach-Sunjic line shape[58].

The laser-annealed Si(111) «1 × 1» surface may be viewed as a disordered 2 × 2 adatom structure (compare LEED[64] and STM[75] results). It differs from Si(111)7 × 7 by the lack of a stacking fault. A stacking fault in half of the 7 × 7 unit cell causes the long-range 7 × 7 reconstruction by reducing the density of broken bonds[66, 76]. In the 2 × 2 structure the number of broken bonds at rest atom sites increases and becomes equal to the number of adatoms. One adatom ties up three broken bonds; the fourth broken bond in the 2 × 2 unit cell remains as a rest atom. Using the same charge transfer idea[71] as for Si(111)7 × 7 one finds that the broken-bond orbitals of the adatoms should be completely empty for the 2 × 2 structure. These changes show up in the photoemission spectrum. The occupied adatom states near the Fermi level are gone, and there is stronger emission from rest atoms. The corresponding core level spectra are also consistent with an increase in the intensity from rest atoms. The sharpening of the bulk line can be explained by the fewer number of inequivalent surface atoms and by the absence of low-energy losses (the metallic Fermi edge disappeared, thereby making low-energy transitions impossible).

The Si(100)2 × 1 (or c(4 × 2)) structure can be described by dimers[77]. We adopt the asymmetric dimer model[2,78], where one of the dimer atoms moves up and becomes negatively charged, and the other moves down and becomes positive. This model is consistent with first-principles total-energy calculations[2] and with the observation of a c(4 × 2) reconstruction. In STM pictures both symmetric and asymmetric dimers have been identified[79]. The symmetric dimers observed with STM are viewed as oscillating asymmetric dimers. The time scale of photoemission is faster than the oscillation frequency, and only asymmetric dimers are seen frozen in time. Correspondingly, we assign the pronounced surface core level at 0.5 eV above the bulk line in Si(100)2 × 1 to half a monolayer of outer dimer atoms. The core level of the inner dimer atoms can be inferred like that of the inner chain atoms on Si(111)2 × 1 (dashed line in fig. 14).

With these assignments one makes the areas of all the surface core levels on the various low-index Si surfaces consistent with a common escape depth.

The Si(110)4×5 and the Si(311)$\times 2$ surfaces [31, 80] have unknown structures. We do see surface core level emission, particularly for Si(311), where surface features are observed on either side of the bulk line. Their intensity corresponds to about half a monolayer each. They are very sensitive to contamination. In this regard the Si(311) surface resembles the Si(100) surface, which has a high sticking coefficient of H_2O. Models have been made [80] for Si(311) which include dimers, like on Si(100). However, the pronounced core level emission below the bulk line has no counterpart on any other Si surface. The Si(110) spectrum is reminiscent of that for Si(111)7×7. Both the surface core level emission above the bulk line and the asymmetry of the bulk line are very similar for the two surfaces. Also, the work function and Fermi level pinning are similar (table I). It could be that the Si(110) reconstruction exhibits locally the adatom/rest atom structure of Si(111)7×7.

7. – Model adsorbates.

Most overlayers on silicon surfaces exhibit complex reconstructions like the clean Si surfaces. Metals tend to react and form (often disordered) layers with variable stoichiometry. There are a few adsorbates, though, that exhibit ordered structures with a small unit cell (1×1 or 2×1). We will discuss representative examples, which will not only be interesting per se, but also provide a reference for calibrating the core level intensities and the core level shifts.

One of the simplest and best-studied [33] overlayer structures on silicon is Si(111)1×1-As. The high degree of order on this surface shows up in a sharp

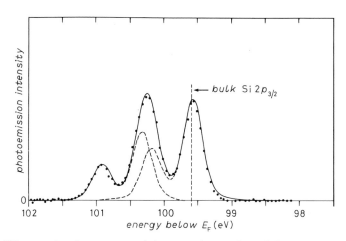

Fig. 16. – Si$2p$ core level spectrum of the arsenic-terminated Si(111) surface showing a shifted surface core level doublet below the bulk line, $h\nu = 130\,\text{eV}$ (from [33]).

surface core level spectrum (fig. 16). The As atoms replace the outermost Si layer. Arsenic has one extra valence electron compared with Si. Therefore, it can form a stable lone pair to terminate the surface instead of an unstable broken bond at the Si(111) surface. Each Si atom in the second layer bonds to three As in the first layer. The corresponding core level shift of $0.75\,eV/3 = 0.25\,eV$ per As ligand serves as one of the data points in fig. 9.

Another 1×1 structure can be formed by adsorbing monovalent atoms (*e.g.*, H, F, Cl, Br) on top of a broken bond. In this case, the adsorbate occupies a site on the inner layer of the characteristic double-layer structure of Si(111), whereas for the As structure the adsorbate occupied the outer layer. The H-, F- and Cl-terminated Si surfaces can be prepared by exposing cleaved Si(111)2×1 to the atomic species [16, 40, 43]. To reach full monolayer coverage is difficult, since etching starts before the first layer is complete [81]. On the Si(111)7×7 surface higher hydrides, fluorides, etc. are formed. H-terminated surfaces can also be established [23, 24] by dipping a wafer in aqueous HF (high purity of the water and short exposure to ambient atmosphere is important). These surfaces are quite stable with respect to oxidation in air and provide a good starting point for obtaining highly ordered Si surfaces.

With a doubling of the unit cell comes a lot more complexity in the (2×1) structures. These occur mostly on Si(100) where the pairing of atoms into dimers forms a natural template. A variety of adsorbates (As, Cl, H_2O) form such structures, where the adsorbate either replaces a dimer atom [33] (As) or bonds to it via the remaining broken bond [37, 43, 82] (Cl, H_2O). These structures are useful for calibrating the surface core level intensity on Si(100), because the uncertainty about symmetric *vs.* asymmetric dimers is removed. They are either

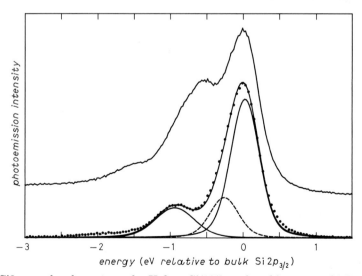

Fig. 17. – Si$2p$ core level spectrum for H_2O on Si(100) analyzed in terms of 0.5 monolayer of Si-OH bonds plus 0.5 monolayer of Si-H bonds.

all symmetric (As, Cl) or all asymmetric (H$_2$O). In the case of H$_2$O adsorption, it has been established by vibrational spectroscopy [82] that the water molecule dissociates into OH and H. These two fragments occupy the two dimer sites and give rise to a large shift (-0.9 eV for Si-OH) and to a small shift (-0.3 eV for Si-H), each corresponding to half a monolayer. The Si2p spectrum and the corresponding assignment are shown in fig. 17. For quantitative measurements it is important to establish full coverage, which occurs at about 20 langmuir (1 L $= 10^{-6}$ Torr $\cdot$ s). The surface peak is sharper at lower exposure, but less intense. Additional surface core levels appear at higher exposure corresponding to higher oxidation states. We find that the intensity under the Si-OH peak equals that of the clean Si(100)2 $\times$ 1 surface core level (fig. 14). The cross-section of Si^{1+} in Si-OH is equal to that of pure Si according to the findings in sect. 5. Therefore, we must have half a monolayer of shifted surface core levels in both cases, in agreement with our assignment for the clean Si(100)2 $\times$ 1 spectrum in terms of asymmetric dimers.

The next larger unit cells of adsorbates comprise $\sqrt{3} \times \sqrt{3} R \, 30°$ and 2×2 structures, which are quite common on Si(111). They are most likely associated with adatom structures, where the adsorbate (*e.g.*, Al, In, Ga, etc.) ties up three broken bonds. In the 2×2 unit cell, there is an extra single broken bond. Such structures are becoming accessible to first-principles band calculations, a helpful step for a complete understanding.

8. – Interfaces.

So far we were mainly concerned with calibrating core level spectroscopy using a set of well-defined reference surfaces. When it comes to interfaces, we can put this knowledge to use and learn something new about the bonding at interfaces. First, we will discuss an ordered interface [34, 35] (CaF$_2$/Si(111)), then we will venture onto the disordered SiO$_2$/Si(100) interface [36, 39], which will bring us into contact with the real world of silicon devices.

At the CaF$_2$/Si(111) interface, the obvious question to ask is whether Ca or F bonds to Si. It can be decided simply by looking at the sign of the Si core level shift. Figure 7 shows that Ca and F induce opposite shifts due to their opposite electronegativities. The Si2p spectrum of CaF$_2$/Si(111) in fig. 18 clearly shows that most of the interface bonds [83] are between Si and Ca since the interface core level lies above the bulk line. The shift of 0.36 eV compares with 0.65 eV for CaSi$_2$. In CaSi$_2$ each Si atom is bonded to three other Si in a double-layer structure like Si(111). (CaSi$_2$ actually forms a well-ordered epitaxial interface [42] with Si(111).) The remaining bond is connected to a layer of Ca with the Ca atoms in hollow sites. A very similar structure has been determined for the a CaF$_2$/Si(111) overlayer using ion scattering [84]. The smaller core level shift for CaF$_2$/Si(111) relative to CaSi$_2$ may be explained by different Madelung energies

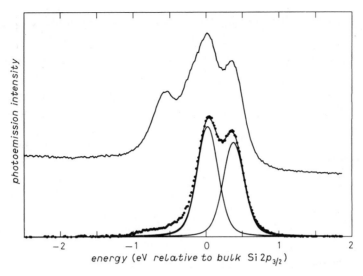

Fig. 18. – Si2p core level spectrum of the CaF$_2$/Si(111) interface. The bonding is mainly[83] from Si to Ca as evidenced by the interface core level above the bulk line.

or by a smaller ionicity of the interface bond compared with CaSi$_2$. It has been found from Ca core level spectroscopy[34] that Ca is close to the 1 + oxidation state at the interface. An extra electron is needed to convert Ca^{2+} in bulk CaF$_2$ to Ca^{1+} at the interface. It can be provided by removing a neutral F^0 from the F$^-$Ca^{2+}F$^-$ slab at the interface and leaving an electron behind[35]. Part of this electron is transferred to the Si at the interface and causes the upwards Si2p core level shift.

The SiO$_2$/Si(100) interface represents the crucial interface in today's silicon technology. It has been studied extensively[36], but a clear picture of the bonding at the interface is still lacking. Since the SiO$_2$ layer is amorphous, many structural techniques are inapplicable. Core level spectroscopy measures short-range order (the four immediate ligands) and can tell us about the local bonding at the interface. The simplest model for the SiO$_2$/Si(100) interface is a truncated bulk structure connected to an amorphous SiO$_2$ lattice. Apart from SiO$_2$ (*i.e.* Si^{4+}), an interface should exhibit only Si^{2+} (*i.e.* Si bonded to two oxygens) since the truncated Si(100) surfaces has two broken bonds per surface atom. As the Si2p core level spectrum in fig. 7 shows, there is a substantial amount of Si^{1+} and Si^{3+}, thus ruling out a simple abrupt interface. A variety of more sophisticated models can be tested this way, by just comparing the expected distribution of oxidation states with the core level spectrum (see ref. [36]). The models comparing favorably with the core level spectra consist of an extended interface (about 2 layers wide), and exhibit protrusions of Si^{3+} into the SiO$_2$ overlayer. Thereby the intensity of the interface core levels has also been taken into account. The reason for an extended interface is a huge density mismatch between Si and SiO_2. The density of Si atoms in Si is a factor of 2.2 higher than in SiO_2. Strain energy calculations [85] show that the elastic energy can be reduced by going from an abrupt to an extended interface.

9. – Summary.

To summarize, let us recall some of the key features of core level spectroscopy at silicon surfaces and interfaces: core level shifts can be used to measure electrical parameters (Schottky barrier, band offset) as well as structural parameters (number and electronegativity of ligands, number of Si atoms bonding to an adlayer). The interpretation of core level spectra is simpler at surfaces than for bulk compounds because of the nearly metallic screening by the underlying Si. Using synchrotron radiation makes it possible to achieve a depth resolution such that monolayers can be analyzed easily.

REFERENCES

[1] G. LE LAY: *Atomic structure of semiconductor surfaces*, in *Semiconductor Interfaces: Formation and Properties*, edited by G. LE LAY, J. DERRIEN and N. BOCCARA, *Springer Proceedings in Physics*, Vol. 22 (Springer-Verlag, Berlin, 1987), p. 48.
[2] J. POLLMANN, R. KALLA, P. KRÜGER, A. MAZUR and G. WOLFGARTEN: *Appl. Phys. A*, 41, 21 (1986).
[3] F. J. HIMPSEL: *Appl. Phys. A*, 38, 205 (1985).
[4] G. HOLLINGER: *Photoelectron spectroscopies: probes of chemical bonding and electronic properties at semiconductor interfaces*, in *Semiconductor Interfaces: Formation and Properties*, edited by G. LE LAY, J. DERRIEN and N. BOCCARA, *Springer Proceedings in Physics*, Vol. 22 (Springer-Verlag, Berlin, 1987), p. 210.
[5] F. J. HIMPSEL: *Photoemission and inverse photoemission from semiconductor interfaces*, in *Semiconductor Interfaces: Formation and Properties*, edited by G. LE LAY, J. DERRIEN and N. BOCCARA, *Springer Proceedings in Physics*, Vol. 22 (Springer-Verlag, Berlin, 1987), p. 196.
[6] K. SIEGBAHN: *J. Electron Spectrosc.*, 5, 3 (1974).
[7] M. CARDONA and L. LEY: *Photoemission in Solids I* (Springer, Berlin, 1978), Chapt. 1.5, p. 60; W. F. EGELHOFF jr.: *Surf. Sci. Rep.*, 6, 253 (1986); J. C. FUGGLE: this volume, p. 127; G. K. WERTHEIM: this volume, p. 237.
[8] O. KESKI-RAHKONEN and M. O. KRAUSE: *At. Data Nucl. Data Tables*, 14, 139 (1974); for very shallow core levels other broadening mechanics take over, *e.g.* band dispersion.
[9] U. GELIUS, L. ASPLUND, E. BASILIER, S. HEDMAN, K. HELENELUND and K. SIEGBAHN: *Nucl. Instrum. Methods B*, 1, 85 (1984); K. SIEGBAHN: *Philos. Trans. R. Soc. London Ser. A*, 318, 3 (1986).
[10] D. E. EASTMAN, F. J. HIMPSEL and J. F. VAN DER VEEN: *J. Vac. Sci. Technol*, 20, 609 (1982).

[11] P. J. FEIBELMAN, E. J. McGUIRE and K. C. PANDEY: *Phys. Rev. B*, **15**, 2202 (1977); M. C. MUNOZ, V. MARTINEZ, J. A. TAGLE and J. L. SACEDON: *Phys. Rev. Lett.*, **44**, 814 (1980); D. E. RAMAKER, F. L. HUTSON, N. H. TURNER and W. N. MEI: *Phys. Rev. B*, **33**, 2574 (1986).

[12] F. C. BROWN and O. P. RUSTGI: *Phys. Rev. Lett.*, **28**, 497 (1972); W. EBERHARDT, G. KALKOFFEN, C. KUNZ, D. ASPNES and M. CARDONA: *Phys. Status Solidi B*, **88**, 135 (1978); R. S. BAUER, R. S. BACHRACH, J. C. McMENAMIN and D. E. ASPNES: *Nuovo Cimento B*, **39**, 409 (1977).

[13] M. ALTARELLI and D. L. DEXTER: *Phys. Rev. Lett.*, **29**, 1100 (1972).

[14] R. J. ELLIOTT: *Phys. Rev.*, **108**, 1384 (1957).

[15] J. F. MORAR, F. J. HIMPSEL, G. HOLLINGER, G. HUGHES and J. L. JORDAN: *Phys. Rev. Lett.*, **54**, 1960 (1985).

[16] F. J. HIMPSEL, P. HEIMANN, T.-C. CHIANG and D. E. EASTMAN: *Phys. Rev. Lett.*, **45**, 1112 (1980).

[17] J. A. YARMOFF, A. TALEB-IBRAHIMI, F. R. McFEELY and PH. AVOURIS: *Phys. Rev. Lett.*, **60**, 960 (1988).

[18] W. TELIEPS and E. BAUER: *Ultramicroscopy*, **17**, 57 (1985); E. BAUER, W. TELIEPS and G. TURNER: *J. Vac. Sci. Technol. A*, **6**, 573 (1988); B. P. TONNER and G. R. HARP: *J. Vac. Sci. Technol. A*, **7**, 1 (1989).

[19] F. C. BROWN, R. Z. BACHRACH and M. SKIBOWSKI: *Phys. Rev. B*, **15**, 4781 (1977); A. BIANCONI: *Surf. Sci*, **89**, 41 (1979).

[20] F. J. HIMPSEL, G. HOLLINGER and R. A. POLLAK: *Phys. Rev. B*, **28**, 7014 (1983), for early work see J. HEDMAN, Y. BAER, A. BERNDTSSON, M. KLASSON, G. LEONHARDT, R. NILSSON and C. NORDLING: *J. Electron Spectrosc.*, **1**, 101 (1972/73).

[21] For Ge and GaAs see E. A. KRAUT, R. W. GRANT, J. R. WALDROP and S. P. KOWALCZYK: *Phys. Rev. B*, **28**, 1965 (1983).

[22] W. X. NI, J. KNALL and G. V. HANSSON: *Phys. Rev. B*, **36**, 7744 (1987); G. P. SCHWARTZ, M. S. HYBERTSEN, J. BEVK, R. G. NUZZO, J. P. MANNAERTS and G. J. GUALTIERI: *Phys. Rev. B*, **39**, 1235 (1989).

[23] The sample is dipped in aqueous HF just before insertion into the vacuum system. This treatment gives a H-terminated surface (see ref. [24]) with very little band bending for n-type samples. For p-type samples, the Fermi level is still pinned by donorlike impurities, but the band bending can be removed by extremely high doping levels (10^{21} cm^{-3}) and by mild heating (until the work function reaches a maximum).

[24] E. YABLONOVITCH, D. L. ALLARA, C. C. CHANG, T. GMITTER and T. B. BRIGHT: *Phys. Rev. Lett.*, **57**, 149 (1986); B. S. MEYERSON, F. J. HIMPSEL and J. A. YARMOFF: unpublished.

[25] M. OLMSTEAD: *Surf. Sci. Rep.*, **6**, 159 (1986).

[26] F. J. HIMPSEL and D. E. EASTMAN: *J. Vac. Sci. Technol.*, **16**, 1297 (1979); P. MARTENSSON, A. CRICENTI and G. HANSSON: *Phys. Rev. B*, **33**, 8855 (1986); R. D. BRINGANS, R. I. G. UHRBERG, M. A. OLMSTEAD and R. Z. BACHRACH: *Phys. Rev. B*, **34**, 7447 (1986).

[27] The Si(111)7 × 7 surface is metallic at room temperature as evidenced by photoemission and in inverse photoemission [28]. Electron energy loss spectroscopy gives a metallic, asymmetric line shape at room temperature [29] and a half-filled localized state in a 0.1 eV gap of delocalized states [30] at 15 K. The odd number of electrons in the 7 × 7 unit cell requires a half-filled band.

[28] F. J. HIMPSEL and TH. FAUSTER: *J. Vac. Sci. Technol. A*, **2**, 815 (1984).

[29] U. BACKES and H. IBACH: *Solid State Commun.*, **40**, 575 (1981).

[30] J. E. DEMUTH, B. N. J. PERSSON and A. J. SCHELL-SOROKIN: *Phys. Rev. Lett.*, **51**, 2214 (1983).

[31] The reconstruction of high-index Si surfaces is inferred from B. Z. OLSHANETSKY and V. I. MASHANOV: *Surf. Sci.*, **111**, 414 (1981). For Si(110), a more complex reconstruction has been observed more recently, see H. AMPO, S. MIURA, K. KATO, Y. OHKAWA and A. TAMURA: *Phys. Rev. B*, **34**, 2329 (1986) and H. NEDDERMEYER and ST. TOSCH: *Phys. Rev. B*, **38**, 5784 (1988).

[32] C. J. KARLSSON, E. LANDEMARK, L. S. O. JOHANSSON, U. O. KARLSSON and R. I. G. URBERG: *Phys. Rev. B*, submitted.

[33] M. A. OLMSTEAD, R. D. BRINGANS, R. I. G. UHRBERG and R. Z. BACHRACH: *Phys. Rev. B*, **34**, 6041 (1986); for As/Si(100) see R. D. BRINGANS, R. I. G. UHRBERG, M. A. OLMSTEAD, R. Z. BACHRACH and J. E. NORTHRUP: *Phys. Scr.*, T **17**, 7 (1987).

[34] F. J. HIMPSEL, F. U. HILLEBRECHT, G. HUGHES, J. L. JORDAN, U. O. KARLSSON, F. R. MCFEELY, J. F. MORAR and D. RIEGER: *Appl. Phys. Lett.*, **48**, 596 (1986); U. O. KARLSSON, F. J. HIMPSEL, J. F. MORAR, D. RIEGER and J. A. YARMOFF: *J. Vac. Sci. Technol. B*, **4**, 1117 (1986); D. RIEGER, F. J. HIMPSEL, U. O. KARLSSON, F. R. MCFEELY, J. F. MORAR and J. A. YARMOFF: *Phys. Rev. B*, **34**, 7295 (1986); F. J. HIMPSEL, U. O. KARLSSON, J. F. MORAR, D. RIEGER and J. A. YARMOFF: *Phys. Rev. Lett.*, **56**, 1497 (1986); *Mater. Res. Soc. Symp. Proc.*, **94**, 181 (1987).

[35] M. A. OLMSTEAD, R. I. G. UHRBERG, R. D. BRINGANS and R. Z. BACHRACH: *Phys. Rev. B*, **35**, 7526 (1986); *Mater. Res. Soc. Symp. Proc.*, **94**, 195 (1987).

[36] F. J. HIMPSEL, F. R. MCFEELY, A. TALEB-IBRAHIMI, J. A. YARMOFF and G. HOLLINGER: *Phys. Rev. B*, **38**, 6084 (1988).

[37] D. SCHMEISSER, F. J. HIMPSEL and G. HOLLINGER: *Phys. Rev. B*, **27**, 7813 (1983).

[38] J. E. ROWE, G. K. WERTHEIM and R. T. TUNG: *J. Vac. Sci. Technol. A*, **7**, 2454 (1989).

[39] G. HOLLINGER and F. J. HIMPSEL: *J. Vac. Sci. Technol. A*, **1**, 640 (1983); *Appl. Phys. Lett.*, **44**, 93 (1984); W. BRAUN and H. KUHLENBECK: *Surf. Sci.*, **180**, 279 (1987).

[40] J. F. MORAR, F. E. MCFEELY, N. D. SHINN, G. LANDGREN and F. J. HIMPSEL: *Appl. Phys. Lett.*, **45**, 174 (1984); F. R. MCFEELY, J. F. MORAR, N. D. SHINN, G. LANDGREN and F. J. HIMPSEL: *Phys. Rev. B*, **30**, 764 (1984). Compare also K. J. GRUNTZ, L. LEY and R. L. JOHNSON: *Phys. Rev. B*, **24**, 2069 (1981). A core level shift of -1.2 eV has been calculated for Si-F by M. SEEL and P. S. BAGUS: *Phys. Rev. B*, **28**, 2023 (1983).

[41] F. R. MCFEELY, J. F. MORAR and F. J. HIMPSEL: *Surf. Sci.*, **165**, 277 (1986).

[42] J. F. MORAR and M. WITTMER: *Phys. Rev. B*, **37**, 2618 (1988); *J. Vac. Sci. Technol. A*, **6**, 1340 (1988).

[43] R. D. SCHNELL, D. RIEGER, A. BOGEN, F. J. HIMPSEL, K. WANDELT and W. STEINMANN: *Phys. Rev. B*, **32**, 8057 (1985).

[44] TH. WESER, A. BOGEN, B. KONRAD, R. D. SCHNELL, C. A. SCHUG and W. STEINMANN: in *Proceedings of the 18th International Conference on the Physics of Semiconductors, Stockholm, 1986*, edited by O. ENGSTRÖM (World Scientific, Singapore, 1987), p. 97.

[45] L. LEY, J. REICHARDT and R. L. JOHNSON: *Phys. Rev. Lett.*, **49**, 1664 (1982).

[46] L. J. BRILLSON, A. D. KATNANI, M. KELLY and G. MARGARITONDO: *J. Vac. Sci. Technol. A*, **2**, 551 (1984). A coordination of Si with three Al has been assumed, relying on the assignment by the authors as dissociated Si at the free Al surface.

[47] L. PAULING: *The Nature of the Chemical Bond* (Cornell University Press, Ithaca, N.Y., 1948).

[48] Alkali metals do not cause any observable core level shift despite their large electronegativity difference with Si, see H. I. STARNBERG, P. SOUKIASSIAN, M. H. BAKSHI and Z. HURYCH: *Phys. Rev. B*, **37**, 1315 (1988).

[49] M. DEL GIUDICE, J. J. JOYCE and J. H. WEAVER: *Phys. Rev. B*, **36**, 4761 (1987).

[50] The arrow above the SiO_2 data point in fig. 8 indicates that for thicker oxides the core level shift increases due to a decrease in screening. The electronic part of the dielectric constant of SiO_2 that is relevant for screening is $\varepsilon_{el} \approx n^2 = 2.1$. Therefore, the screening charge is only $((\varepsilon - 1)/(\varepsilon + 1)) \cdot e = 0.35\ e$ in SiO_2. A similar irregularity for the core level shift of thick oxides has been found for Ta and W. The fully oxidized material exhibits a larger shift than expected from the regular sequence of core level shifts for the lower oxidation states: F. J. HIMPSEL, J. F. MORAR, F. R. McFEELY, R. A. POLLAK and G. HOLLINGER: *Phys. Rev. B*, **30**, 7236 (1984).

[51] C. D. WAGNER: *Faraday Discuss. Chem. Soc.*, **60**, 291 (1975). J. S. JOHANNESSEN, W. E. SPICER and Y. E. STRAUSSER: *Appl. Phys. Lett.*, **27**, 452 (1975). See also J. C. WOICIK, B. B. PATE and P. PIANETTA, *Phys. Rev. B*, **39**, 8593 (1989).

[52] G. HOLLINGER, P. KUMURDJIAN, J. M. MACKOWSKI, P. PERTOSA, L. PORTE and TRAN MINH DUC: *J. Electron Spetrosc.*, **5**, 237 (1974); J. Q. BROUGHTON and P. S. BAGUS: *J. Electron Spectrosc.*, **20**, 261 (1980).

[53] J. W. DAVENPORT, R. E. WATSON, M. L. PERLMAN and T. K. SHAM: *Solid State Commun.*, **40**, 999 (1981); R. E. WATSON, J. W. DAVENPORT, M. L. PERLMAN and T. K. SHAM: *Phys. Rev. B*, **24**, 1791 (1981).

[54] B. JOHANSSON and N. MARTENSSON: *Phys. Rev. B*, **21**, 4427 (1980); P. CITRIN and G. K. WERTHEIM: *Phys. Rev. B*, **27**, 3176 (1983).

[55] A. R. MIEDEMA: *Z. Metallkd.*, **70**, 345 (1979); A. R. MIEDEMA, P. F. DE CHATEL and F. R. DeBOER: *Physica B*, **100**, 1 (1980).

[56] M. SEEL and P. S. BAGUS: *Phys. Rev. B*, **28**, 2023 (1983); C. PRIESTER, G. ALLAN and M. LANOO: *Phys. Rev. Lett.*, **58**, 1989 (1987).

[57] F. R. McFEELY, E. CARTIER, J. A. YARMOTT, and S. JOYCE: *Phys. Rev. B*, submitted.

[58] D. H. RICH, T. MILLER and T.-C. CHIANG: *Phys. Rev. B*, **37**, 3124 (1988).

[59] D. H. RICH, T. MILLER and T.-C. CHIANG: *Phys. Rev. Lett.*, **60**, 357 (1988).

[60] The assignment given in ref. [58, 59] doubles the number of surface atoms corresponding to a given core level intensity relative to our assignment. For Si(100) 2×1 both dimer atoms are assumed to contribute, giving a full monolayer for the shifted core level. For Si(111)7×7 the surface core level emission is assigned to adatoms, not rest atoms, giving almost twice as many surfaces atoms (a factor of 12/7). Essentially, such an assignment is not compatible with the surface core level intensities on other model surfaces (the As/Si(111)1×1 and CaF$_2$/Si(111)1×1 structures would give about two layers of Si with shifted core levels), and it is opposite to our knowledge about charge transfer at the Si(111)7×7 surface and the laser-annealed Si(111) «1×1» surface, as discussed in the text. Reference [32] comes to a similar conclusion based on Si(111) 7×7 H data.

[61] J. M. HILL, D. G. ROYCE, C. S. FADLEY, L. F. WAGNER and F. J. GRUNTHANER: *Chem. Phys. Lett.*, **44**, 225 (1976).

[62] K. C. PANDEY: *Phys. Rev. Lett.*, **47**, 1913 (1981).

[63] P. A. BENNETT and M. B. WEBB: *Surf. Sci.*, **104**, 74 (1981); R. J. PHANEUF and E. D. WILLIAMS: *Phys. Rev. B*, **35**, 4155 (1987).

[64] D. M. ZEHNER, C. W. WHITE and G. W. OWNBY: *Surf. Sci.*, **92**, L67 (1980); *Appl. Phys. Lett.*, **37**, 456 (1980). See also R. M. TROMP, L. SMIT and J. F. VAN DER VEEN: *Pyhs. Rev. B*, **30**, 6235 (1984).

[65] R. FEDER and W. MÖNCH: *Solid State Commun.*, **50**, 311 (1984); F. J. HIMPSEL, P. M. MARCUS, R. TROMP, I. P. BATRA, M. R. COOK, F. JONA and H. LIU: *Phys. Rev. B*, **30**, 2257 (1984); H. SAKAMA, A. KAWAZU and K. UEDA: *Phys. Rev. B*, **34**, 1367 (1986); S. Y. TONG *et al.*: to be published.

[66] K. TAKAYANAGI, Y. TANISHIRO, M. TAKAHASHI and S. TAKAHASHI: *J. Vac. Sci. Technol. A*, **3**, 1502 (1984); K. TAKAYANAGI, Y. TANISHIRO, S. TAKAHASHI and M. TAKAHASHI: *Surf. Sci.*, **164**, 367 (1985).

[67] I. K. ROBINSON, W. K. WASKIEWICZ, P. H. FUOSS and L. J. NORTON: *Phys. Rev. B*, **37**, 4325 (1988).

[68] S. Y. TONG, H. HUANG, C. M. WEI, W. E. PACKARD, F. K. MEN and M. B. WEBB: *J. Vac. Sci. Technol. A*, **6**, 615 (1988).

[69] F. J. HIMPSEL, D. E. EASTMANN, P. HEIMANN, B. REIHL, C. W. WHITE and D. M. ZEHNER: *Phys. Rev. B*, **24**, 1120 (1981).

[70] G. BINNIG, H. ROHRER, CH. GERBER and E. WEIBEL: *Phys. Rev. Lett.*, **50**, 120 (1983); R. J. HAMERS, R. M. TROMP and J. E. DEMUTH: *Phys. Rev. Lett.*, **56**, 1972 (1986).

[71] J. E. NORTHRUP: *Phys. Rev. Lett.*, **57**, 154 (1986).

[72] T. MILLER, E. ROSENWINKEL and T.-C. CHIANG: *Solid State Commun.*, **47**, 935 (1983).

[73] S. B. DICENZO, P. A. BENNETT, D. TRIBULA, P. THIRY, G. K. WERTHEIM and J. E. ROWE: *Phys. Rev. B*, **31**, 2330 (1985)

[74] R. D. SCHNELL, F. J. HIMPSEL, A. BOGEN, D. RIEGER and W. STEINMANN: *Phys. Rev. B*, **32**, 8052 (1985)

[75] R. S. BECKER, J. A. GOLOVCHENKO, G. S. HIGASHI and B. S. SWARTZENTRUBER: *Phys. Rev. Lett.*, **57**, 1020 (1986)

[76] F. J. HIMPSEL: *Phys. Rev. B*, **27**, 7782 (1983); D. VANDERBILT: *Phys. Rev. B*, **36**, 6209 (1987).

[77] J. A. APPELBAUM and D. R. HAMANN: *Surf. Sci.*, **74**, 21 (1978)

[78] D. J. CHADI: *Phys. Rev. Lett.*, **43**, 43 (1979); *J. Vac. Sci. Technol.*, **16**, 1290 (1979).

[79] R. J. HAMERS, R. M. TROMP and J. E. DEMUTH: *Phys. Rev. B*, **34**, 5343 (1986).

[80] J. E. ROWE *et al.*: to be published.

[81] G. SCHULZE and M. HENZLER: in *Proceedings of the Fourth International Conference on Solid Surfaces* and *The Third European Conference on Surface Science*, edited by D. A. DEGRAS and M. COSTA, Supplement a la Revue Le Vide, Les Couches Minces, No. 201, p. 967.

[82] H. IBACH, W. WAGNER and D. BRUCHMANN: *Solid State Commun.*, **42**, 457 (1982); E. M. OELLIG, R. BUTZ, H. WAGNER and H. IBACH: *Solid State Commun.*, **51**, 7 (1984); Y. CHABAL: *Phys. Rev. B*, **29**, 3677 (1984).

[83] The Ca-terminated interface is obtained by annealing a CaF_2 film close to its desorption point. Films grown at lower temperature exhibit significant Si-F bonding [34, 35], particularly on stepped surfaces.

[84] R. M. TROMP and M. C. REUTER: *Phys. Rev. Lett.*, **61**, 1756 (1988).

[85] I. OHDOMARI, H. AKATSU, Y. YAMAKOSHI and K. KISHIMOTO: *J. Appl. Phys.*, **62**, 3751 (1987).

Core Level Spectroscopy of Metals, Alloys and Intermediate-Valence Compounds.

G. K. WERTHEIM

AT&T Bell Laboratories - Murray Hill, N.J. 07974

1. – Core electron spectroscopy.

Core electron photoemission spectroscopy became a viable technique through the efforts of Kai SIEGBAHN [1], who eventually received a Noble Prize for this work. It has replaced X-ray spectroscopy as the major source of information about the core electrons. SIEGBAHN also demonstrated that X-ray excited photoemission spectroscopy from core levels can provide information about the outer electronic states of atoms, molecules and solids. Although initially conceived as a technique for obtaining chemical information, and called electron spectroscopy for chemical analysis (ESCA), it rapidly transcended that limited domain. Today these experiments are carried out not only with the (monochromatized) characteristic X-rays used by SIEGBAHN, but also with synchrotron radiation, closing the gap between classical photoemission and ESCA.

We will here restrict ourselves to metallic systems, and investigate the information which emerges from photoemission experiments on core levels. The salient feature that distinguishes core electron spectroscopy from valence band spectroscopy is the failure of the one-electron approximation, which is used with reasonable success for shallow electron states. The many-electron aspects of core electron spectroscopy are usually described in terms of specific final-state effects, many of which we will consider below. To avoid confusion, it seems worthwhile pointing out that the term «final state» is used to denote two distinct aspects of the state created by the photoelectric transition. In angle-resolved photoemission one is concerned with the *final state of the excited electron* in the empty band structure, and pays little attention to final-state effects on the hole state that is left behind. In core electron spectroscopy the opposite is true. The photoelectron is excited to states well above the Fermi level, where there are many broad, overlapping bands without structure of interest. Major interest here focusses on the processes which modify the *final state of the hole state atom* and its environment. This is justified by the fact that a localized core hole perturbs its environment much more strongly than a delocalized valence or conduction band hole.

1'1. *Three-step model.* – The photoemission process has conventionally been broken down into three steps. In the first step, photoionization, an electron is excited from a bound state to an empty state accessible by dipole selection rules. In the second step, transport, this electron propagates through the lattice, exciting other bound electrons, individually or collectively, provided it has sufficient kinetic energy. The coupling to collective excitations, called plasmon, is strong and usually determines the mean free path of the electron. This length, often called the escape depth, has a broad minimum of $\sim 5\,\text{Å}$ in the interval from 15 to 75 eV and then increases as $E_{\text{kin}}^{1/2}$. It also increases rapidly for E_{kin} less than 10 eV. In the third step, emission, the electron emerges from the solid into free space with a reduced kinetic energy, now measured from the vacuum level. The physics of core electron photoemission is largely contained in the first step, and resides in the properties of the final state resulting from the photoejection of a localized electron.

1'2. *Intra-atomic relaxation.* – Even in the free atom a core electron photoemission causes profound change. Outer orbitals relax because they experience a stronger nuclear potential when an inner electron is removed. As a result of this intra-atomic relaxation, measured binding energies are smaller by many electronvolt than the eigenenergies of the unperturbed atom. The corresponding theoretical values are obtained by taking the difference of the total energies calculated for the hole state and unperturbed atoms. It is worth noting that relaxation need not go to the ground state of the hole state atom (or solid). Outer electrons may be excited into low-lying, empty orbitals, resulting in higher-energy final states. These are usually called shake-up satellites. In the sudden approximation a sum rule states that the centroid of the line and its satellites lies at the eigenenergy of the electron in the initial state. In photoelectron spectroscopy from free atoms these and other satellites have been studied in great detail, and the sum rule verified. In solids, where the intrinsic satellites are superimposed on the extrinsic losses mentioned in the discussion of the three-step model, this sum rule is of little practical value. At the excitation threshold, where the sudden approximation is no longer valid, the satellite intensities should descrease.

Intra-atomic relaxation is largely independent of the environment of the atom, and is included in the free-atom core electron ionization potential, which will be used as the starting point for further analysis. The intra-atomic relaxation makes the outer electronic orbitals assume the properties of a singly ionized atom with next higher atomic number. This comes about because the nuclear potential is screened by one electron less in the hole state atom. This is a very useful point of view which goes under the name of «equivalent-core approximation». It is valid provided the radial extent of the hole state is significantly smaller than that of the valence orbitals.

1'3. *Experimental techniques.* – The short escape depth of the photoelectrons makes photoemission a surface spectroscopy. The most important experimental task is the preparation of a suitable surface, *i.e.* one which is clean and detect free. Commonly used techniques include cleaving, scraping or fracturing in vacuum, sputtering with rare-gas ions followed by annealing, and vacuum evaporation onto a clean substrate. Surface chemistry is usually monitored by Auger spectroscopy, and surface order by low-energy electron diffraction (LEED). Most of this work must be carried out in a vacuum better than 10^{-9} Torr, unless the sample has a very small sticking coefficient for the residual gases of the vacuum system. The measurement of the electron kinetic energy is done routinely with commerical cylindrical-mirror or hemispherical electrostatic-deflection analyzers.

2. – Born-Haber cycles and core electron binding energy.

It may well come as a surprise to find that one can relate the core electron binding energy in a metal to the core electron ionization potential of the free atom, without recourse to detailed electronic considerations.

2'1. *From free atom to metal.* – The Born-Haber cycle shown in fig. 1 provides a methodology that accomplishes this task, one which can be extended to deal with many interesting related problems[3]. It contains only macroscopic

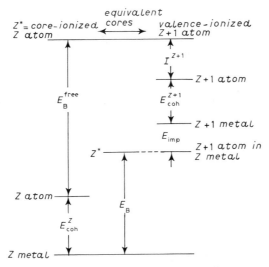

Fig. 1. – Born-Haber cycle relating the core electron binding energy in a metal to the corresponding core electron ionization potential of the free atom. (From ref. [2a].)

thermodynamic quantities, but requires use of the equivalent-core approximation. Starting at the lower left: 1) An atom of atomic number Z is desorbed from
the metal. 2) The resulting free atom is core-ionized and 3) replaced by a valence-
ionized $Z + 1$ atom, invoking the equivalent-core approximation. 4) This ion is
neutralized, using the electron left at the vacuum level in the second step, and 5)
adsorbed into a $Z + 1$ metal. 6) In the final step this $Z + 1$ atom is transferred
into the Z metal. The final result is a fully screened, core-ionized Z atom in its
native Z host. These steps lead to the equation

$$(1) \qquad E_B^{\text{metal}} = E_B^{\text{atom}} - I^{Z+1} + E_{\text{coh}}^Z - E_{\text{coh}}^{Z+1} - E_{\text{impl}},$$

where coh denotes a cohesive energy, and impl the implanation energy for
moving a $Z + 1$ atom from its native $Z + 1$ metal into the Z metal. For β-Sn,
ignoring the implantation term, we have $31.99 - 8.64 + 3.13 - 2.74 = 23.74\,\text{eV}$,
which agrees reasonably well with the measured value of $23.95\,\text{eV}$. The
discrepancy suggests that the implantation term is itself negative, confirming
the tendency of Sb to segregate to the surface of Sn predicted by CHELIKOW-
SKY [4]. This approach is of considerable value in that it relates core electron
properties to macroscopic physical quantities that are independently
measurable.

It is worth noting that the Born-Haber cycle yields the core electron binding
energy relative to the Fermi level without recourse to the work function of the
metal. This comes about because the atomic core electron ionization potential of
the Z atom and the first ionization potential of the $Z + 1$ atom, which appear with
opposite sign, are both measured from the vacuum level. The binding energy is,
in essence, the energy required to excite a core electron to the Fermi level,
where it becomes the screening charge. It consequently depends directly on the
location of the Fermi level. This becomes an important consideration when we
consider binding-energy shifts.

2˙2. *Surface atom core level shift*. – Because of the small escape depth, a
significant fraction of the PE signal comes from atoms in the first monolayer.
These atoms have a different environment from that of the bulk atoms, *e.g.*,
fewer near neighbors and lower symmetry. As a result they also have distinct
core electron binding energies. This emerges readily from a Born-Haber
analysis, as carried out by JOHANSSON and MARTENSSON [3].

Taking the bulk cycle of fig. 1 as a starting point, it is sufficient to consider the
modification of the two cohesive energies at the surface, to obtain a good
estimate of the surface atom core level shift. Guided by the difference in
coordination number at the surface, a 20 percent reduction in cohesive energy
has generally been assumed. The surface atom core level shift then is given by

$$(2) \qquad \Delta E_{\text{surf}} = 0.2\,(E_{\text{coh}}^{Z+1} - E_{\text{coh}}^Z) - \Delta E_{\text{impl}},$$

where the contribution from the change in the implantation term should be small. For Au, where the cohesive energy of the $Z + 1$ element Hg is much smaller than that of Au itself, a negative shift is obtained, in agreement with experiment. Quantitative agreement is, however, often no better than 50 percent, because of the overly simple characterization of the surface.

The shift can also be interpreted in terms of the changes in the band structure at the surface. Narrowing of the surface bands, combined with the requirement that the Fermi levels of bulk and surface coincide, results in centroid shifts that are comparable to the core level shifts [5].

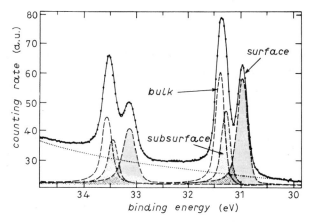

Fig. 2. – Surface atom core level shifts at the W(111) surface. Emission from surface and subsurface layers is indicated. (From ref. [6c].)

The surface atom core level shift has been determined in many metallic and semiconducting systems. In close-packed structures it is generally limited to the first atomic layer, but in open structures shifts have also ben observed for other layers as well (see fig. 2) [6]. It has also been found that the shifts are different for different crystallographic planes, as one might anticipate simply on the basis of their stabilities.

2˙3. *Adsorbate atoms on metals.* – The same approach can also be used to deal with adsorbate atoms on metals. It is particularly simple for the isolated adsorbate atom, and can be readily extended to treat the coverage dependence of the binding energy [2].

3. – Extra-atomic relaxation processes.

A core electron binding energy in a metal can be related to that of the free atom, by considering the changes which take place as the metal is formed and the response of the metal to the presence of a core hole. During the formation of the metal the outer wave functions form bands and the electronic charge redistributes among the outer orbitals. In a metal one or more of the resulting bands will be partially occupied. In the final state the resulting conduction electrons will screen the core hole, reducing the energy of the final state. Since the atomic core electron ionization potential is measured relative to the vacuum level, the work function ϕ of the metal is required to obtain the core electron binding energy, E_B, relative to the Fermi level. These events are summarized by

$$(3) \qquad\qquad E_B = E_B^{atom} - E_{conf} - E_{relax} - e\phi ,$$

where we have lumped the configurational and other initial-state changes into E_{conf} and the final-state changes into E_{relax}. This formulation makes sense from a theoretical point of view, in which initial- and final-state effects are treated separately. It is, however, not especially helpful to the experimentalist, since the effects of band formation in the initial state and relaxation in the final state cannot be separately determined by photoemission alone. Under favorable circumstances one can, however, obtain an experimental value for the relaxation energy by combining the photoemission binding-energy shift with an Auger electron kinetic-energy shift for a transition between core levels[7]. The formulation of eq. (3) is often employed in qualitative discussions of environment-dependent changes in binding energy. We will now consider some specific aspects of the final-state response of a metal to core electron photoemission.

3‘1. *Conduction electron screening.* – In simple metals the dominant response to a core hole is provided by the conduction electrons, which are free to screen the localized unit positive charge associated with the core-ionized atom. The screened final state closely resembles that associated with a chemical impurity, so that the formalism can be couched in terms of the Friedel phase shifts. The photoemission experiment is unique in providing an image of the spectrum of the electron-hole pair excitations which are required to produce the screening configuration of the conduction electrons. Theoretical considerations show that this spectrum contains an integrable infinity of infinitesimal excitations, *i.e.* an infrared catastrophe. In the usual mathematical representation (see below) it has a nonintegrable tail toward higher binding energies, but in real systems this tail is cut off by the finite width of the conduction band[8, 9].

It is worth considering the effect of the screening process on the measured core electron binding energy. On the one hand, the energy of the optimally screened final state is reduced, on the other, energy is consumed by the creation of the electron-hole pairs. The net result is to shift the peak of the spectrum to smaller binding energy, while adding a long tail toward greater binding energy. In order to distinguish between these two effects, it is essential to define the binding energy as the energy of an optimally screened state, *i.e.* in terms of the position of the singularity, rather than in terms of the centroid of the spectrum. This in accord with conventional practice which uses the peak position to define the binding energy. This does, however, introduce a small error, since the singularity always lies at a slightly smaller value. A more accurate determination is obtained by fitting the data with a mathematical function incorporating the singularity and other broadening mechanisms (see below).

In principle, such a spectrum contains sufficient information to obtain the relaxation energy associated with the conduction electron screening. According to the usual sum rule, the centroid falls at the energy of the unscreened state, and the singularity at the energy of the optimally screened state. The difference between the two is the relaxation energy. Unfortunately, the centroid is difficult to define experimentally, because the long tail merges into the extrinsic satellites and the cut-off is not well defined.

For a simple metal the spectrum of excitations is well approximated by a one-sided power law of the form $1/\varepsilon^{1-\alpha}$, where ε is the excitation energy and α is called the singularity index. It may be expressed in terms of the Friedel phase shifts, δ_l,

(4) $$\alpha = 2 \sum_{l=0} (2l+1)(\delta_l/\pi)^2.$$

By using the Friedel sum rule it can be shown that $0 < \alpha \leqslant 1/2$. The same phase shifts describe the X-ray absorption edge singularity.

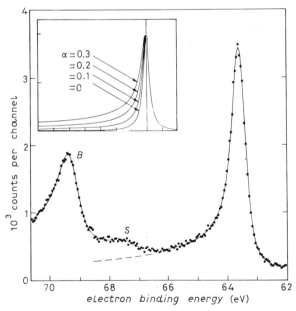

Fig. 3. – The effect of the conduction electron screening response on the 2s core electron line shape of Na. (From WERTHEIM (see bibliography).)

The convolution of the power law with a Lorentzian is known as the Doniach-
Šunjić [10] function:

$$
(5) \qquad I(\varepsilon) = \frac{\Gamma(1-\alpha)\cos\left[\pi\alpha/2 + (1-\alpha)\arctan(\varepsilon/\gamma)\right]}{(\varepsilon^2 + \gamma^2)^{(1-\alpha)/2}},
$$

where γ is the half width of the Lorentzian corresponding to the core hole lifetime. Data are analyzed by fitting them with a convolution of this function and a
Gaussian, adjusting the parameters to minimize the sum of the squares of the
deviations. The three parameters have quite different effects on the line shape,
so that the Gaussian and Lorentzian widths are readily separated, and the
singularity index obtained without ambiguity. It is important, however, to avoid
assigning the long many-body tail to the inelastic background.

The screening response in the simple metals is compatible with this
description (see fig. 3). It has also been shown that a single set of phase shifts
successfully accounts not only for the XPS core electron line shape, but also for
the X-ray absorption and emission edges [11].

3`2. *Neighbor polarization and charge transfer.* – In insulators the most
common response to the creation of a core hole is the polarization of the
neighboring atoms, displacing the negative charge towards the hole state atom.
If the core hole has pulled a normally empty orbital down below the Fermi level,
charge may transfer from the ligands, *i.e.* from the valence band, into this
orbital. The transition probability depends on the covalent mixing between the
wave functions of the valence band and the empty orbital in the initial state. This
charge transfer process is not restricted to insulators, and will be encountered
below in the discussion of the lighter rare-earth metals.

3`3. *Phonon excitation.* – Photoionization of an atom in a solid results not
only in electronic but also in vibrational excitation of the host. In metals the two
main mechanisms for this excitation are the recoil of the emitted photoelectron
and the size change of the hole state atom [12]. In polar materials the Coulomb
forces between the hole state atom and its neighbors are also important [13].
Theoretical calculations are based on the Frank-Condon approximation. The
excitation energy is generally much larger than the phonon energy, resulting in a
many-phonon process, so that the resulting Poisson distribution is well
approximated by a Gaussian.

In VUV photoemission from metals the size change term is dominant and
results in line widths between 30 and 300 meV. Few measurement of phonon
widths have been reported, because the phonon width adds in quadrature to the
·instrumental resolution function, which is often not known with sufficient
precision.

4. – Alloys and intermetallic compounds.

4˙1. *Resonant photoemission.* – One of the interesting problems in the study of alloys is the determination of the contributions of the various component elements to the density of states. This has historically been accomplished with soft-X-ray spectroscopy, by observing the emission spectrum arising from valence-band-to-core-hole radiative transitions in each of the constituent elements. Photoemission provides two new approaches to this problem.

At high photon energy, say above 100 eV, the photoemission spectrum begins to resemble the cross-section-weighted density of states. For materials which contain one of the 4d and 5d elements it may then be possible to obtain information about the contributions of the individual constituents to the density of states by making use of the Cooper minimum in the photoelectric cross-section. These elements have one or more nodes in their d wave functions which cause a deep minimum in their photoelectric cross-section in the 100 to 200 eV photon energy range [14]. The drop is typically as large as two orders of magnitude, making it possible to «turn off» the emission from that element.

For many elements the valence band emission can be enhanced by making use of a core electron resonance. This enhancement is obtained at a photon energy just large enough to excite a core electron into an empty state. If the resulting excited atom decays by autoionization and leaves a valence band hole, then the final state is indistinguishable from that obtained by direct photoionization of the valence band. Since core electron cross-sections are usually larger than those of the valence band, this may result in a substantial enhancement of the valence band signal. This technique has found application in the study of transition metal and rare-earth compounds, notably those of Ni and Ce [15].

4˙2. *Charge transfer in alloys.* – In a binary alloy the difference between the electronegativities of the constituents is likely to result in a flow of charge to the more electronegative element. In band structure calculations the charge transfer can be estimated by summing the charges in the Wigner-Seitz cells. This is, however, an arbitrary partitioning. Photoelectron binding-energy shifts provide a natural way of defining the charge transfer in terms of the Coulomb potential of the charge surrounding each atom. This should work well provided there is no significant redistribution of charge among the orbitals of the constituent atoms as a result of alloy formation.

With that restriction in mind, it would be reasonable to expect that the binding energies of one of the constituents will increase while that of the other decreases. Surprisingly this is seldom the case. In order to understand the origin of this unexpected behavior, it is necessary to examine the implications of using the Fermi level as a reference level for these measurements. It should be emphasized at the outset that the following should not be construed as an objection to the use of this reference level in measurements. It is clearly the best choice. However, it must be recognized that the Fermi levels of two different materials will not, in general, coincide; there is always a contact potential difference. Consequently we cannot compare binding energies measured in two different materials relative to their individual Fermi levels without taking account of this difference in reference level. When the indicated correction is applied to the data, the anomaly disappears[16].

5. – Incomplete shells and mixed valence.

The electrons in the incomplete shells of metals may have very different properties. They range from the sp bands of metals like Na, Mg or Al, which are free-electron-like, to the $4f$ shells of the rare earths, which are strongly localized and corelike in radial extent. The d-group transition metals have intermediate properties. Quantitative discussion is based on the relative sizes of the band width W, the Coulomb correlation energy U of the electrons and the core hole potential V. Until recently it was widely believed that the one-electron approximation is adequate to describe valence band photoemission from the simple metals, where $W \gg U$. However, work on Na has now shown that correlations have significant effects even in these systems[17]. Nevertheless, photoemission does yield information about the initial-state bands, but with relatively minor distortions. When $W \ll U$, as in the rare earths, an entirely new phenomenology appears, one that is dominated by properties of the final state.

Compounds with noninteger valence are familiar in inorganic chemistry. They include systems like magnetite, Fe_3O_4, in which the valence of the iron on the B site is half-integer, the sodium tungsten bronzes, Na_xWO_3, in which the valence of the tungsten can be continuously adjusted between 5 and 6 by changing the sodium content, and rare-earth compounds like SmB_6, in which the $4f$ level lies at the Fermi level and exchanges charge with the conduction band. Photoemission has made major contributions to the study of these rare-earth compounds.

5`1. *Ionization of incomplete shells.* – Consider a $4f$ shell containing n electrons, coupled according to Hund's rule. To place another electron in the $4f$ shell requires an energy very large compared to the width of the $4f^n$ state. Removal of an electron produces one of many possible $4f^{n-1}$ states, which include the ground and many of the optical excited states of that configuration[18, 19]. Theoretical spectra are readily derived by the fractional-parentage method, by expanding the initial $4f^n$ state in terms of the final f^{n-1} states and the outgoing photoelectron[20]. Examples are shown in fig. 4. The utility of these spectra rests on the fact that the $4f$ spectra of the rare earths are largely insensitive to the environment of the atom, whether it be metal[21] or insulator. For a given value of n we will obtain a recognizable spectrum of final-state multiplets.

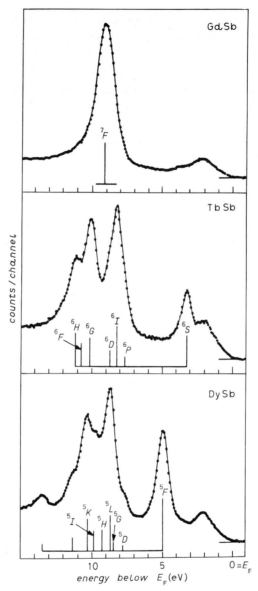

Fig. 4. – Final-state 4f multiplets in some rare-earth antimonides. (From ref. [22].)

Transition metal ions do not have this useful property. They are much more sensitive to their environment, because the d-electrons interact with their ligand and with each other. Insulating compounds may have narrow d-band widths and behave somewhat like the rare earths, but spectra depend on the strength of the crystal field. A strong crystal field can even change the coupling in the ground state. Metallic transition metal compounds tend to have larger band widths, and yield bandlike spectra, but with strong correlation effects.

5˙2. *Rare-earth mixed valence.* – The fact that the 4*f* spectra of the rare earths unambiguously identify the initial-state occupancy of the 4*f* shell has been useful in the study of mixed-valence systems [19, 22-24]. In these systems one obtains two 4*f* multiplet spectra corresponding to the two valences which participate in the charge fluctuation process, never a spectrum corresponding to an average valence (see fig. 5). The reason is that photoionization is a large perturbation, which projects the ion into one of two stable final states. The configuration of the ion in the final state is then no longer appropriate for charge fluctuations. The intensities of the two spectra provide a good measurement of

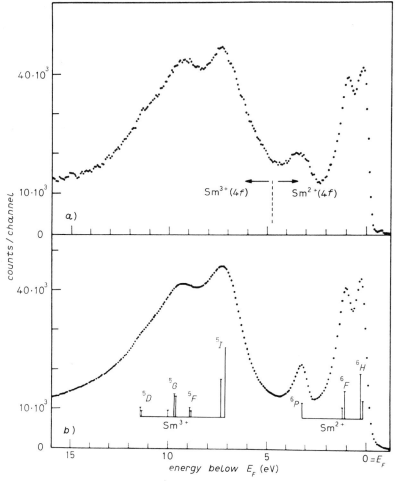

Fig. 5. – The final-state 4*f* multiplets in the mixed-valence compound SmB_6 obtained with monochromatized Al K_α radiation (*a*)), compared with theory (*b*)). (From ref. [23].)

the initial-state average valence. The spectra also make it possible to distinguish between homogeneous mixed-valence compounds which contain ions with fractional valence in a single lattice site and inhomogeneous systems with two stable valences in different lattice sites in the initial state. In the former the lowest-energy multiplet of the lower-valence final state always lies at the Fermi level, and may be thought of as an image of the fluctuating 4f level.

It would be a great advantage if core electron spectra could be used to characterize the valence of rare-earth compounds. The deeper core electrons have much larger cross-sections and weaker multiplet structure, and can provide data more rapidly. Such spectra are, however, more difficult to interpret, because the occupancy of the 4f shell may change during core photoionization. The equivalent-core approximation shows that this is almost always energetically favorable. Removal of a core electron (by photoionization) makes the core of an atom with atomic number Z look like that of a $Z + 1$ atom to the valence electrons. In the rare-earth series each increase in Z increases the 4f shell occupancy by one. Consequently in the final state of photoemission there will generally be an empty f-state below the Fermi level, ready to accept an electron from the valence or conduction band. Such charge transfer leads to core electron satellite structure at *smaller* binding energy, because of the Coulomb potential of the added charge in the 4f level. This process is well documented for the lighter rare-earth metals[25], as well as in many compounds with stable valence. This charge transfer process is often referred to as «shake-down» to distinguish it from the shake-up process in which outer electrons are excited to higher-lying empty orbitals, creating satellite structure at *greater* binding energy. The degree to which the empty f orbital is filled depends not only on its energy relative to valence band edge but also on the hybridization between the 4f's and the 5d band in the initial state. In mixed-valence compounds this process can result in three final states, corresponding to three distinct f-shell occupancies. The initial-state valence cannot then be deduced without the help of a complete theoretical calculation.

5˙3. *Time scale.* – The discussion of mixed valence inevitably raises the question of the time scale of the photoemission process. There is no time scale associated with the photoelectric effect itself, but the photoelectron, once excited, requires a finite time to traverse its atom. This interval determines the nature of the relaxation processes. If it could be made arbitrarily long, the final state would be created in its ground state, but in practical photoemission one is closer to the sudden limit. For a kinetic energy of 10 eV this time is of the order of 10^{-16} s, clearly too short to average out thermally driven fluctuations.

The lifetime of the hole state provides another constraint. For most of the levels that are useful in PES it ranges from 10^{-14} to 10^{-16} s. These times are short compared to lattice vibration periods, which are also representative of charge fluctuation times, so that there is no prospect of seeing thermal average properties.

But what if the inital state is a quantum-mechanically mixed state, not a fluctuating state? From the equivalent-core point of view it is clear that the removal of a core electron from an atom in such a system is a strong perturbation which will destroy the mixing in the final state. It will also prevent fluctuations in the final state, so that one should always see the properties of two distinct final states.

For valence ionization the result depends on the width of the fluctuating state and the strength of the perturbation. For systems like the sodium tungsten bronzes, in which the W $5d$ band has a sizable width, direct examination of the $5d$ states yields a fractionally occupied conduction band. Tungsten $4f$ core electron spectra, on the other hand, exhibit final states corresponding to two different screening configurations. In one the $5d$ state of the Re-like, hole state atom is filled, in the other it is empty and screening is provided by the ligands. In this case the core hole provides a perturbation comparable to the band width. The intensity of these two states does not provide a good measure of the initial-state valence.

For mixed-valence rare-earth compounds fluctuating between $4f^n$ and $4f^{n-1}$, in which the $4f$ band is narrow, one sees the final-state multiplets corresponding to $4f^{n-1}$ and $4f^{n-2}$. Band widths, Coulomb correlation energies and core hole potentials provide a much better way of understanding the observations than do attempts to appeal to a time scale of photoemission.

6. – Core electron multiplet structure.

6`1. *Core electron multiplets*. – We will now consider in greater detail the information obtained from core electrons in systems with paramagnetic shells. The essential point is that in the final state all the spins and angular momenta of a given atom are coupled into one overall symmetry. Any of the symmetries that can be formed by coupling the core and outer shells are available. When both the core hole and paramagnetic shells have orbital angular momentum, this leads to complex spectra, with possibly hundreds of lines.

For simplicity we will use Mn^{2+} with ^6S configuration as an illustrative example [26]. If a core s shell is photoionized, final states with two different total spins are formed by coupling the spin of the core hole state to that of the half-filled $3d$ shell. A ^7S state in which the core spin is parallel to that of the d shell and a ^5S in which it is antiparallel. The resulting splitting of the $3s$ photoemission spectrum of MnF_2 is readily resolved (see fig. 6). The $2s$ shell exhibits a somewhat smaller splitting [27]. For core s-shell ionization the number of final states is also limited even if the d shell has orbital angular momentum. Consequently the core electron spectrum obtained for Fe^{2+} in FeF_2 is similar but

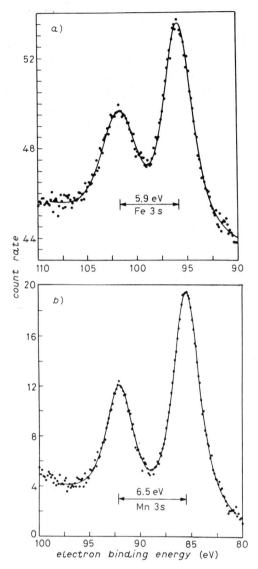

Fig. 6. – The multiplet splitting of the 3s state in a) FeF_2 and b) MnF_2. (From ref. [27, 28].)

has a smaller splitting [28]. It should not come as a surprise that the 3s spectrum of the divalent iron in the diamagnetic, strong-crystal-field compound $K_4Fe(CN)_6$ in unsplit. In rare-earth compounds the core s electrons are similarly split by coupling to the spin of the 4f shell [29] (see fig. 7).

So far we have ignored the possibility that many-electron processes might complicate these spectra. One indication that something is amiss lies in the intensities of the two components in fig. 6. They are not in the ratio of the

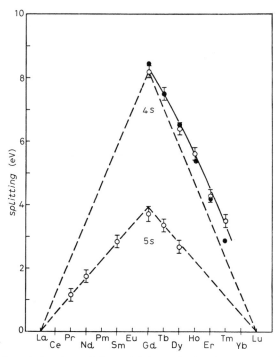

Fig. 7. – Systematics of the 4s and 5s multiplet splitting in rare-earth trifluorides: ● theoretical (Slater model), ○ experimental. (From ref. [29].)

degeneracies of the two states; the ^5S state is too weak. This is most readily understood in terms of the configuration interaction formalism, which states that excited states with same symmetry as those produced by coupling the core hole to the initial-state configuration of the outer shell will also be populated by photoemission. These states lie at a significantly greater binding energy and are relatively broad. For $3s$ photoionization of Mn^{2+}, an S-state transition metal ion with half-filled d shell, the final states produced by simply coupling the two shells have the configuration $3s^1 3p^6 3d^5$, ^5S or ^7S. There are, however, other quintet S-states which can be obtained by rearranging the electrons with in the M shell, e.g., $3s^2 3p^4 3d^6$, ^5S, which have significantly greater energy. There are no similar septet final states. As a result of configuration interaction the intensity associated with the quintet transition is now shared by all of these states, reducing the intensity of the main line. The other quintet states have now been observed experimentally [30].

6˙2. *Shake-down.* – A further complication arises from the possibility that the population of the partially occupied shell has changed as the result of core electron photoemission. There are basically two modes for the core hole screening process in these ionic solids. Either the cation core hole is screened simply by the polarization of the neighboring anions, or an electron may transfer from the anion-derived valence band to a normally empty cation orbital. The former process is benign as far as the interpretation of the splitting is concerned, the latter is not because it changes the final-state configuration of the ion. This problem is potentially just as severe for transition metal ions where the low-lying empty states are d-states as in the rare earths.

The simple test for the presence of shake-down into d- or f-states lies in the systematics of the splitting with increasing atomic number. In these weak-crystal-field, Hund's-rule-coupled outer shells the spin reaches its maximum value at the half-filled shell. If the splitting of the core s-electron is also largest at that point, then there is no shake-down. The data for the $5s$ electrons of the rare earths in a variety of compounds[29] (fig. 6) rule out shake-down. This is intuitively reasonable because the $5s$ shell has a radial extent greater than that of the $4f$ shell, and so should not pull down the empty $4f$ level. However, data for the $4s$ electrons[31] lead to the same conclusion. The data for the divalent transition metal fluorides unfortunately do not extend below the half-filled shell. Failing that, there is convincing evidence that a $3s$ hole in MnF_2 is screened by polarization, in that the spectrum calculated without charge transfer agrees in detail with the experimental one[32]. From an equivalent-core point of view one might well have predicted the opposite, since the equilibrium charge state of an Fe-like core-ionized Mn atom in the MnF_2 lattice is $2+$. The $3s$ hole is, however, so shallow that this approximation may be inappropriate. The systematics of the $3s$ splittings in the other divalent fluorides[28] then indicates that there is no shake-down in any of these compounds in response to $3s$ ionization. Either the hole is too shallow or else the mixing between metal d-states and anion p-states is too weak. It follows that these shallow core s-levels can be used with confidence to study the initial-state occupancy of the magnetic shells. However, metal ion $2p$ electron spectra for the transition metal difluorides, when properly interpreted, clearly demonstrate that deeper states do induce shake-down transitions in these materials[33]. Shake-down effects are also well established for most Ce compounds, including CeF_4[34]. A full interpretation of these effects is usually based on the theory developed by KOTANI and TOYOZAWA[35-37].

BYBLIOGRAPHY

Photoelectron and Auger Spectroscopy, edited by T. A. CARLSON (Plenum, New York, N.Y., and London, 1975).

Handbook of X-Ray and Ultra-Violet Photoelectron Spectroscopy, edited by D. BRIGGS (Heyden, London, 1977).

Electron Spectroscopy for Surface Analysis, edited by H. IBACH, *Topics in Current Physics*, Vol. 4 (Springer, Berlin and Heidelberg, 1977).

Electron Spectroscopy, Theory, Techniques and Applications, Vol. 1, edited by C. R. BRUNDLE and A. D. BAKER (Academic, London and New York, N.Y., 1977); and subsequent volumes in this series.

Photoemission in Solids, edited by M. CARDONA and L. LEY, *Topics in Applied Physics*, Vol. **26** and **27** (Springer, Berlin and Heidelberg, 1978 and 1979).

G. WENDIN: *Breakdown of the one-electron picture in photoelectron spectra*, in *Structure and Bonding*, Vol. **45** (Springer, Berlin and Heidelberg, 1981), p. 1.

G. K. WERTHEIM: X-*ray photoelectron spectroscopy*, in *Microscopic Methods in Metals*, edited by U. GONSER, *Topics in Current Physics*, Vol. **40** (Springer, Berlin and Heidelberg, 1986), p. 193.

REFERENCES

[1] K. SIEGBAHN, C. NORDLING, A. FAHLMAN, R. NORDBERG, K. HAMRIN, J. HEDMAN, G. JOHANSSON, T. BERGMARK, S. E. KARLSSON, I. LINDGREN and B. LINDBERG: *ESCA, Atomic, Molecular, and Solid State Structure Studied by Means of Electron Spectroscopy*, Nova Regia Societas Scientiarum Upsaliensis, Ser. IV, **20** (1967).

[2] K. GÜRTLER and K. JACOBI: *Surf. Sci.*, **152/153**, 272 (1985); G. K. WERTHEIM: *Appl. Phys. A*, **41**, 75 (1986); G. K. WERTHEIM and D. N. E. BUCHANAN: *Phys. Rev. B*, **33**, 914 (1986).

[3] B. JOHANNSON and N. MARTENSSON: *Phys. Rev. B*, **21**, 4427 (1980).

[4] J. R. CHELIKOWSKY: *Surf. Sci.*, **139**, L197 (1984), and private communication; see also A. R. MIEDEMA: *Z. Metallkd.*, **69**, 455 (1978).

[5] P. H. CITRIN, G. K. WERTHEIM and Y. BAER: *Phys. Rev. Lett.*, **41**, 1425 (1978); *Phys. Rev. B*, **27**, 3160 (1983); P. H. CITRIN and G. K. WERTHEIM: *Phys. Rev. B*, **27**, 3176 (1983).

[6] T. M. DUC, C. GUILLOT, Y. LASAILLY, J. LECANTE, Y. JUGNET and J. C. VEDRINE: *Phys. Rev. Lett.*, **43**, 789 (1979); J. F. VAN DER VEEN, F. J. HIMPSEL and D. E. EASTMAN: *Phys. Rev. Lett.*, **44**, 89 (1980); *Phys. Rev. B*, **25**, 7388 (1982); *Solid State Commun.*, **37**, 555 (1981); **40**, 57 (1981); G. K. WERTHEIM, P. H. CITRIN and J. F. VAN DER VEEN: *Phys. Rev. B*, **30**, 4343 (1984).

[7] C. D. WAGNER: *The role of Auger lines in photoelectron spectroscopy*, in *Handbook of X-Ray and Ultraviolet Photoelectron Spectroscopy*, edited by D. BRIGGS (Heyden, London, 1977), p. 249.

[8] P. H. CITRIN, G. K. WERTHEIM and Y. BAER: *Phys. Rev. Lett.*, **35**, 885 (1975); *Phys. Rev. B*, **16**, 4256 (1977).

[9] G. K. WERTHEIM and P. H. CITRIN: *Fermi surface excitations in XPS line shapes from metals*, in *Photoemission in Solids*, Vol. I, edited by M. CARDONA and L. LEY, *Topics in Applied Physics*, Vol. **26** (Springer, Berlin and Heidelberg, 1978), p. 197.

[10] S. DONIACH and M. ŠUNJIĆ: *J. Phys. C*, **3**, 285 (1970).

[11] P. H. CITRIN, G. K. WERTHEIM and M. SCHLÜTER: *Phys. Rev. B*, **20**, 3067 (1979).

[12] C. P. FLYNN: *Phys. Rev. Lett.*, **37**, 1445 (1976).

[13] P. H. CITRIN, P. EISENBERGER and D. R. HAMANN: *Phys. Rev. Lett.*, **33**, 965 (1974).

[14] J. J. YEH and I. LINDAU: *At. Data Nucl. Data Tables*, **32**, 1 (1985).

[15] C. GUILLOT, Y. BALLU, J. PAIGUE, J. LECANTE, K. P. JAIN, P. THIRY, R. PINCHAUX, Y. PETROFF and L. M. FALIKOV: *Phys. Rev. Lett.*, **39**, 1642 (1977); L. I. JOHANSSON, J. W. ALLEN, I. LINDAU, M. H. HECHT and S. B. M. HAGSTROM: *Phys. Rev. B*, **21**, 1408 (1980); S.-J. OH and S. DONIACH: *Phys. Rev. B*, **26**, 1859 (1982).

[16] G. K. WERTHEIM, R. L. COHEN, G. CRECELIUS, K. W. WEST and J. H. WERNICK: *Phys. Rev. B*, **20**, 860 (1979).

[17] E. JENSEN and E. W. PLUMMER: *Phys. Rev. Lett.*, **55**, 1912 (1985).

[18] G. K. WERTHEIM, A. ROSENCWAIG, R. L. COHEN and H. J. GUGGENHEIM: *Phys. Rev. Lett.*, **27**, 505 (1971).

[19] M. CAMPAGNA, G. K. WERTHEIM and Y. BAER: *Unfilled inner shells: rare earths and their compounds*, in *Photoemission in Solids II*, edited by L. LEY and M. CARDONA, *Topics in Applied Physics*, Vol. **27** (Springer, Berlin and Heidelberg, 1979), p. 217.

[20] P. A. COX: *Struct. Bonding*, **24**, 59 (1975).

[21] J. K. LANG, Y. BAER and P. A. COX: *J. Phys. F*, **11**, 121 (1981); P. A. COX, J. K. LANG and Y. BAER: *J. Phys, F*, **11**, 113 (1981).

[22] M. CAMPAGNA, G. K. WERTHEIM and E. BUCHER: *Struct. Bonding*, **30**, 99 (1976).

[23] J.-N. CHAZALVIEL, M. CAMPAGNA, G. K. WERTHEIM and P. H. SCHMIDT: *Solid State Commun.*, **19**, 725 (1976) (SmB$_6$).

[24] See *Valence Fluctuations in Solids*, edited by L. M. FALICOV, W. HANKE and M. P. MAPLE (North Holland, Amsterdam, 1981).

[25] G. CRECELIUS, G. K. WERTHEIM and D. N. E. BUCHANAN: *Phys. Rev. B*, **18**, 6519 (1978).

[26] C. S. FADLEY, D. A. SHIRLEY, A. J. FREEMEN, P. S. BAGUS and J. V. MALLOW: *Phys. Rev. Lett.*, **23**, 1397 (1969).

[27] S. HÜFNER and G. K. WERTHEIM: *Phys. Rev. B*, **7**, 2333 (1973).

[28] G. K. WERTHEIM, S. HÜFNER and H. J. GUGGENHEIM: *Phys. Rev. B*, **7**, 556 (1973).

[29] R. L. COHEN, G. K. WERTHEIM, A. ROSENCWAIG and H. J. GUGGENHEIM: *Phys. Rev. B*, **5**, 1037 (1972).

[30] S. P. KOWALCZYK, L. LEY, R. A. POLLAK, F. R. MCFEELY and D. A. SHIRLEY: *Phys. Rev. B*, **7**, 4009 (1973).

[31] F. R. MCFEELY, S. P. KOWALCZYK, L. LEY and D. A. SHIRLEY: *Phys. Lett. A*, **49**, 301 (1974).

[32] B. HERMSMEIER, C. S. FADLEY, M. O. KRAUSE, J. JIMENEZ-MIER, P. GERARD and S. T. MANSON: *Bull. Am. Phys. Soc.*, **33**, 291 (1988). See, however, B. W. VEAL and A. P. PAULIKAS: *Phys. Rev. Lett.*, **51**, 1995 (1983).

[33] A. ROSENCWAIG, G. K. WERTHEIM and H. J. GUGGENHEIM: *Phys. Rev. Lett.*, **8**, 479 (1971).

[34] G. KAINDL, G. K. WERTHEIM, G. SCHMIESTER and E. V. SAMPATHKUMARAN: *Phys. Rev. Lett.*, **58**, 606 (1987).

[35] A. KOTANI and Y. TOYOZAWA: *J. Phys. Soc. Jpn.*, **35**, 1073, 1082 (1973); **37**, 912 (1974).

[36] S. ASADA and S. SUGANO: *J. Phys. Soc. Jpn.*, **41**, 1291 (1976).

[37] G. VAN DER LAAN, C. WESTRA, C. HAAS and G. A. SAWATZKY: *Phys. Rev. B*, **23**, 4369 (1981).

Inverse Photoemission Spectroscopy.

V. Dose

Max-Planck-Institut für Plasmaphysik, EURATOM Association
D-8046 Garching/München, B.R.D.

1. – Introduction.

Ultraviolet inverse photoemission spectroscopy has emerged as the principal spectroscopy technique for the investigation of the energy *vs.* momentum dispersion of unoccupied electronic states in the bulk and at clean and adsorbate-covered surfaces of single crystalline solids[1-10]. The basic phenomenon, namely emission of radiation from solids under electron bombardment, has now been known since over ninety years as X-ray emission[11]. This process has only recently been recognized to be intimately related to the well-matured photoelectron spectroscopy (PES)[12, 13]. The relation will be elucidated by reference to fig. 1. The left-hand panel shows an energy diagram for conventional photoemission. In ultraviolet photoelectron spectroscopy a quantum $\hbar\omega$ of monochromatic radiation is absorbed by a valence electron of the solid at initial energy E_i. The electron is raised by absorption of the photon to a previously unoccupied electronic band of energy

$$E_f = E_i + \hbar\omega.\tag{1}$$

If the final state E_f lies above the vacuum level of the solid, the electron may be transmitted into the vacuum and can be subjected to energy and momentum analysis using some suitable electron spectrometer. From the measured final-state energy E_f the binding energy of the electron in the initial state E_i can be inferred from the known photon energy $\hbar\omega$ by energy conservation. From conservation of momentum we have

$$\boldsymbol{k}_f - \boldsymbol{k}_i = \boldsymbol{q} + \boldsymbol{G},\tag{2}$$

where $\boldsymbol{k}_f$, $\boldsymbol{k}_i$ are the electron momenta in the final and initial state, respectively, $\boldsymbol{q}$ is the momentum of the emitted photon and $\boldsymbol{G}$ a reciprocal-lattice vector

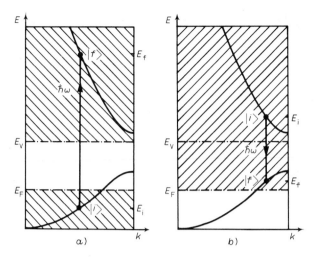

Fig. 1. – Schematic of the a) photoemission and b) inverse photoemission process.

originating from the potential inducing the radiative transition. Since $\boldsymbol{G}$ is of the order of the Brillouin zone size (2π in atomic units) and $q = \hbar\omega/c$ typically of the order 10^{-2} in the ultraviolet spectral range, q can be neglected in the momentum balance and we end up with $\boldsymbol{k}_i = \boldsymbol{k}_f$ in a reduced zone scheme. This is the famous direct-transition concept.

It is immediately obvious from the band panel of fig. 1 that electrons with final-state energy between the Fermi level E_F and the vacuum level E_V remain trapped in the solid. On the other hand, electronic states in this range are empty and, therefore, cannot serve as initial states in PES either. The energy range between E_F and E_V remains, therefore, inaccessible to ordinary photoemission. This apparent gap of information is closed by inverse photoemission (IPE). A schematic representation of this transition is depicted in the right-hand panel of fig. 1. Electrons of well-defined initial energy E_i and momentum $\boldsymbol{k}_i$ occupy an empty electronic band above E_V. This excited state may decay via emission of radiation $\hbar\omega$ to a lower-lying also previously empty state. If we measure the energy of the emitted photon, we can deduce from the known initial energy E_i the energy E_f of the band occupied by the electron after the radiative transition. If the energy of the emitted photon is in the ultraviolet range, we can further make use of the direct-transition concept and infer from the known momentum of the electron in the initial state the momentum $\boldsymbol{k}_f$. The validity of neglecting the photon momentum in the momentum balance is of utmost importance in inverse photoemission beyond the simplicity of interpreting spectra within the direct-transition scheme, because it allows one to employ photon detectors accepting as large solid angles as desired. The success of the experiment relies very heavily on this possibility, since PES and IPE, being governed by the same dipole transition matrix elements, differ in «brightness» as a consequence of phase space factors entering the golden rule by roughly a factor of α^2, where α is Sommerfeld's fine-structure constant[14].

Having established the full equivalence of PES and IPE, a final inspection of fig. 1 reveals that the common range of application for both methods lies above the vacuum level E_V. In this range lie the initial states for IPE and the final states for PES. Information on electronic states below E_F can only be obtained from PES. Conversely, electronic states between E_F and E_V are exclusively accessible by IPE. Considering the long-standing and continuing efforts in PES we dare predict quite a prosperous future for IPE.

2. – Experimental.

The experimental requirements for the two spectroscopies, PES and IPE, result in a straightforward manner from our introductory discussion of their basic physics. PES requires a monochromatic light source which is either a traditional noble-gas resonance lamp or a monochromator coupled to a white-light synchrotron radiation source for excitation and an electron spectrometer for detection of the photoemitted electron[12, 13]. Conversely, IPE needs a source of monochromatic electrons and an energy-selective radiation detector. The latter can be either some sort of band pass detector with fixed mean energy in analogy to the noble-gas resonance lamp, or a spectrograph offering tunability of the photon energy. Since the special impetus to IPE in the ultraviolet range came from the introduction of a photon-energy-selective Geiger-Müller counter[15], we shall discuss the principles of band pass detectors first. The previously mentioned Geiger-Müller counter employs an alkaline-earth fluoride single-crystal entrance window and a filling of iodine with some noble buffer gas. The threshold for molecular photoionization of iodine limits the lower response energy and is approximately 9.2 eV. Suitable candidates for the entrance window are CaF_2[15, 16] and SrF_2[17] with approximate transmission cut-offs at 10.2 eV and 9.7 eV, respectively. The resulting band passes are characterized by mean energies of 9.57 eV and 9.43 eV with mean square deviations of $((240 \pm 18)\,meV)^2$ and $((113 \pm 8)\,meV)^2$, respectively[18]. The design of an actual IPE spectrometer based on a photon band pass counter is shown in fig. 2[19]. The counter is mounted in a standard UHV system equipped with the usual facilities for sample preparation and analysis. The sample can be rotated through 360° in the figure plane. It is shown in the IPE position opposite to a small three-electrode electron gun mounted coaxially to a vacuum ultraviolet mirror. The gun delivers currents in the $(10 \div 20)\,\mu A$ range at 10 eV with an angular spread below 5°. Radiation produced by electrons hitting the sample at a polar angle θ is

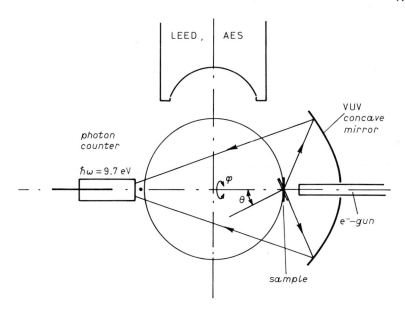

Fig. 2. – Spectrometer for inverse-photoemission experiments employing a VUV band pass photon detector.

imaged into the band pass counter by the concave mirror. The mirror is made from a glass blank by evaporation of aluminium and subsequent coating with MgF_2.

A variation of the band pass principle uses the rapidly decaying quantum efficiency of a channeltron combined with a crystal window to define the high-energy sensitivity limit [20]. This type of detector exhibits considerable tailing of its sensitivity vs. long wavelengths. This deficiency can be largely corrected for by sensitizing the channeltron with a KBr layer [21].

Band pass detectors are simple, cheap and easy to build, but suffer from their fixed energy in exactly the same way as resonance lamps in conventional PES. Tunable photon detectors in IPE experiments abandon this restriction [22]. Figure 3 shows the set-up used in our laboratory. It consists of a sample preparation and analysis chamber and an UV spectrograph. The spectrograph is equipped with two holographic gratings with f acceptance. It operates at normal incidence with equal source and image distances for optimum brightness.

Photons are detected with a position-sensitive detector. The energy ranges are $(8 \div 18)\,eV$ and $(16 \div 40)\,eV$, respectively. Several entrance slits between 0.1 mm and 0.8 mm are available for compromising between resolution and sensitivity.

An alternative approach to tunable photon energy has been reported by ROYER and SMITH [23]. They exploit the strong chromatic aberration of a LiF

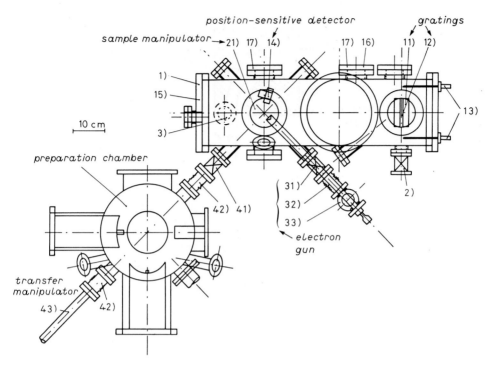

Fig. 3. – Spectrometer for inverse-photoemission studies based on a VUV normal-incidence spectrograph.

lens in the vicinity of the LiF transmission cut-off. The operating range of such an instrument is limited to a narrow spectral range between 8 and 12 eV.

The overall resolution of an IPE experiment is, of course, a composite from contributions of the optical resolution and the energy resolution of the electron source. Let σ_{opt} be the root mean square width of the photon detector and σ_{el} the r.m.s. width of the electron source, then

$$\sigma_{tot}^2 = \sigma_{opt}^2 + \sigma_{el}^2 \,. \tag{3}$$

An optimum compromise between resolution and intensity is obtained when σ_{opt} equals approximately σ_{el}. Optimum combinations in this sense are a tungsten cathode in the electron source combined with an I_2/CaF_2 band pass counter and a BaO cathode in the electron gun combined with an I_2/SrF_2 counter [18].

Better resolution in the electron source can at least in principle be obtained with the GaAs negative electron affinity (NEA) photoemitter [24, 25]. The importance of this source lies even more in its capability of producing spin-polarized electrons. Its principal applications are in the study of the electronic structure of ferromagnetic materials [26, 27]. A schematic of an apparatus for

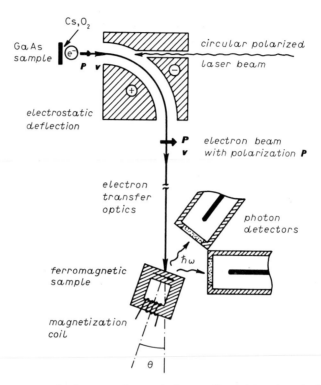

Fig. 4. – Spectrometer for inverse-photoemission studies with spin-polarized electrons. The two photon counters allow also to derive information on the polarization of the emitted light.

spin-polarized inverse-photoemission studies is depicted in fig. 4 [28]. Photoelectrons excited by circularly polarized light are emitted with longitudinal spin polarization from the GaAs crystal. Transformation to transverse spin polarization is accomplished by means of a spherical electrostatic deflector. The electrons impinge then onto a picture frame target crystal magnetized parallel to the sample surface. The radiation emitted from the target crystal is detected by two photon counters viewing the target under different angles of observation. Intensity differences in the two counters allow in certain cases to derive information on the polarization of the emitted light [29].

3. – Bulk electronic bands.

Momentum-resolved inverse photoemission is well suited to measure the energy and momentum of electrons injected into previously empty electronic states. Thus it is possible to determine the dispersion $E(k)$, which describes an energy band, experimentally. Since the band structure of even simple crystals is quite complicated at general points in momentum space, experimental work is frequently restricted to measurements in mirror planes. Working in mirror planes has several advantages. Bands may become degenerate there due to symmetry and this will reduce the complexity of the observed spectra. Because of symmetry the bands cross mirror planes with zero slope. This effect has been explored for absolute k-space location of observed transition in some cases. Further, special polarization rules apply for transition in mirror planes. They can be used to determine whether the final band wave function involved in the transition has even or odd parity with respect to the mirror plane once the polarization of the emitted radiation has been analysed.

Figure 5 shows experimental inverse-photoemission spectra[17] obtained with the spectrometer displayed in fig. 2. The only variable in work with fixed quantum energy is the energy of the electrons incident on the solid and the only parameter the angle of incidence with respect to the surface normal. The electron momentum results from the definition of these two quantities. The electron momentum is, of course, a three-component vector. We shall split this

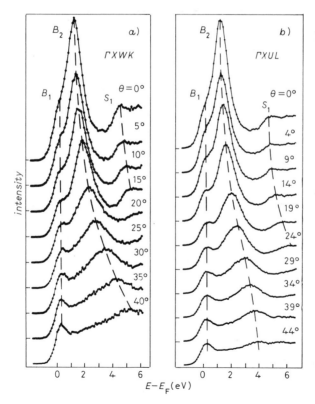

Fig. 5. – Inverse-photoemission spectra from Ni(001) at various polar angles θ in the two bulk mirror planes; $\hbar\omega = 9.7$ eV. The dashed lines indicate the dispersion of two bulk and one surface state.

vector into $K_\parallel$, the two-component vector of momentum parallel to the crystal surface, and $K_\perp$, the component of momentum perpendicular to the crystal surface. The spectra in fig. 5 have been obtained for $K_\parallel$ in ΓXWK and ΓXUL mirror planes of the Ni(001) crystal, respectively. The data represent the intensity of the emitted radiation as a function of *final*-state energy of the electrons with respect to E_F for a variety of incidence angles. Three different emission features can be distinguished. Peak B_1 right at the Fermi energy remains stationary upon variation of the polar angle. It will turn out to result from transitions into weakly dispersing empty d-bands. Peak B_2, starting at 1.3 eV above E_F, shows rapid dispersion and will be shown to result from transitions into empty sp-like bands. The step S_1 finally will turn out to be a surface feature and will be discussed later. It is a common property of all electron spectroscopies that they reveal bulk and surface features simultaneously. Whether a particular transition is of bulk or surface origin requires additional analysis.

The two series of spectra in the ΓXWK and ΓXUL mirror planes exhibit nearly the same overall behaviour except that peak B_2 shows a somewhat slower dispersion at high polar angles.

Given such a set of data, the goal is to derive energy *vs.* momentum dispersion for the involved electronic bands. This encounters some difficulty, because the electron wave vector K_i (or momentum) is only known in vacuum before the electron has penetrated into the solid. The wave vector in the solid k has changed, however, because the surface potential has exerted a force on the electron. This force shall again be decomposed into a two-component vector parallel to the surface and a component normal to the surface. The force parallel to the surface originates from a periodic potential and leads to diffraction such that

$$(4) \qquad\qquad k_{i\parallel} = K_{i\parallel} + g_\parallel,$$

where $g_\parallel$ is a surface reciprocal-lattice vector. At sufficiently low quantum energy and hence initial energy the requirement that

$$(5) \qquad\qquad |K_i| \geqslant |k_{i\parallel} - g_\parallel|$$

leads to a preference for $g_\parallel = 0$. $K_{i\parallel}$ is, on the other hand, given by

$$(6) \qquad\qquad |K_{i\parallel}| = [2m(E_f + h\omega - \Phi_p)/\hbar^2]^{1/2} \sin\theta,$$

where E_f is the final-state energy with respect to E_F and Φ_p is the sample work function.

The force normal to the surface results from the potential change at the surface and introduces a change in $K_\perp$ of so far unknown magnitude.

The possibility to condense data as those given in fig. 5 into dispersion relations is consequently in principle limited to plots of final-state energy as function of $k_\parallel$. This is shown in fig. 6 [17]. Solid dots labelled B_1, B_2 and S_1 refer to the corresponding emission features in fig. 5. Emissions S_2, B_3 and B_4 are outside

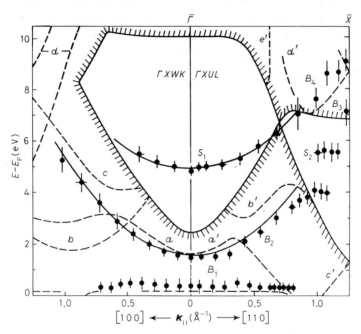

Fig. 6. – Final-state energies $(E - E_F)$ as a function of $k_\parallel$ for the emission peaks shown in fig. 5. Dashed lines indicate energetically allowed bulk direct transitions. Unshaded areas mark gaps of the projected bulk band structure.

the data range displayed in fig. 5. The areas surrounded by shaded boundaries are the gaps of the projected bulk band structure from a theoretical calculation. For $(E, k_\parallel)$ combinations inside these gaps no bulk electronic states exist and observed emissions inside these regions must, therefore, be of surface origin. A comparison between experiment and theory is also given in fig. 6 in terms of optical curves for a transition energy of 9.6 eV. These curves show the final-state energy as a function of $k_\parallel$ for any two empty bands 9.6 eV apart. They are shown as dashed lines. The experimental data fall close to such curves, but only part of the possible final states is observed experimentally. This is a consequence of matrix element effects which suppress certain transitions in course of selection rules. Alternatively, transitions that are not strictly forbidden may remain undetectably weak due to inefficient coupling of the incident plane wave to the initial-state bulk Bloch wave. This coupling is strongest for bulk bands with nearly free-electron character. This is the case for the bulk transition B_2; the solid line in fig. 6 shows the dispersion derived from a nearly free-electron two-state approximation and is seen to reproduce the data quite well.

The analysis of the data in fig. 5 terminates at this point and we are left with the unpleasant situation that a single experiment does not allow to derive the three-dimensional $E(k)$ information we were hunting for. An approximation to the three-dimensional $E(k)$ can be obtained if we assume a free-electron dispersion of the initial band. In this approximation the bulk solid is represented by a structureless potential trough of depth $V_0 + \Phi_p$. The electron kinetic energy inside the solid $E_{kin,i}$ is then given in terms of the initial-state energy in vacuum E_i by

$$(7) \qquad\qquad E_i = E_{kin,i} - |V_0|\,,$$

$$(8) \qquad\qquad E_{kin,i} = \hbar\omega + E_f + |V_0| = \hbar^2(k_\parallel^2 + k_\perp^2)/2m\,.$$

If V_0 is known from some independent source as, for example, a theoretical band structure calculation, $k_\perp$ can be calculated from (8). Experience from both photoemission and inverse photoemission has shown that this approximation meets with considerable success [17, 30].

An alternative to assuming free-electron behaviour in order to determine $k_\perp$ is to recover it from a second independent measurement. The procedure has become known as triangulation or energy coincidence method and was proposed already quite early [31]. If we happen to observe a particular bulk transition at the same fixed energy from two different crystal faces, we know from the experimental conditions the associated two values of $k_\parallel$ for the two surfaces. These values define in turn two straight lines in momentum space parallel to the respective surface normal direction. Their intersection in momentum space specifies then the third unknown momentum component. An example for such a procedure is given in fig. 7 [17].

So far our interpretation of experimental IPE spectra has concentrated on final-state energies. We shall now refine the analysis and turn to polarization effects. As already mentioned, special selection rules apply for mirror planes. The solution of the Schrödinger equation must have definite (odd or even) parity under reflection in a mirror plane. The wave function of the electron impinging on the crystal surface is a plane wave and has, therefore, even parity. It can couple only to those bulk electronic wave functions which have also even parity. Consider now the dipole transition matrix element in the form

$$(9) \qquad\qquad A\langle f| \nabla V |i\rangle = A_\parallel \langle f| \nabla V_\parallel |i\rangle + A_\perp \langle f| \nabla V_\perp |i\rangle\,,$$

where $\parallel$ and $\perp$ refer to the mirror plane. Clearly the matrix element vanishes if the intergrand is odd under reflection. Since $\nabla V_\parallel$ is even and $\nabla V_\perp$ odd under reflection, it follows that radiation with polarization parallel to the mirror plane

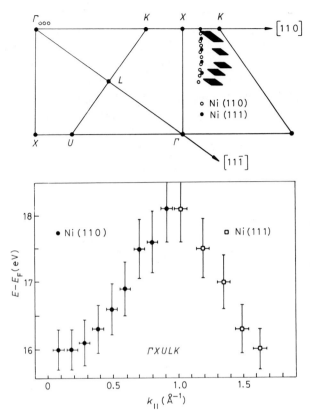

Fig. 7. – Triangulation of a bulk direct transition observed from the (110) and (111) faces of nickel.

results from transitions into even final states. For radiation polarized perpendicular to the mirror plane the final state must be odd. In the absence of efficient polarizers for ultraviolet radiation, information on the polarization of the emitted light must be derived from its angular distribution. Sample data obtained with the apparatus depicted in fig. 4 are shown in fig. 8. Two observation directions are indicated in fig. 4. A third direction can be obtained if the crystal is rotated counterclockwise to the equivalent electron incidence angle since one photon counter will then look under an inequivalent photon collection angle onto the surface. The electron angle of incidence is $\theta = 43°$ in fig. 8. The photon collection angle is taken positive if electron incidence is on the same side of the surface normal as photon collection. From the intensity variations of the bulk emission B_2 we conclude immediately that the associated final states are of even parity. We can even get more detailed information on the direction of the transition dipole. It has an intensity maximum at glancing exit $\alpha_p = 80°$ and vanishes nearly at normal exit $\alpha_p = 5°$. The conclusion that the dipole is oriented along the surface normal is supported from the intensity ratio under $\alpha_p = 80°$ and

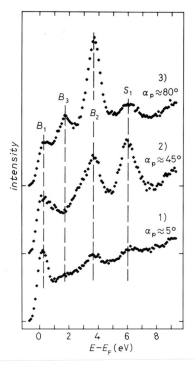

Fig. 8. – Variation of the intensity of emitted photons at three different photon collection angles. The angle of incidence of the electrons is kept constant; Ni(110), $\overline{\Gamma X}$, $\theta = 43°$, $\hbar\omega = 9.6\,\text{eV}$.

$\alpha_p = -45°$ if an undistorted $\sin^2$ distribution is assumed. Another clear-cut case is the surface transition S_1. The prominent intensity at $\alpha_p = -45°$ corresponding to 90° with respect to the incident electron beam and the roughly equal and by a factor of 3 weaker emissions at an observation angle of 40° with respect to the electron beam direction lead us to conclude that for surface transition S_1 the dipole moment is oriented along the direction of the incident electrons. The discussion of B_1 and B_3 is more difficult due to their small intensities and will be skipped here.

We shall turn finally to the spin resolution capability of the apparatus in fig. 4. In a strong ferromagnet such as nickel, the empty $3d$ states have a magnetic moment opposite to the direction of sample magnetization. Radiative transitions into the empty d-bands should, therefore, be only possible for electrons of the appropriate spin orientation. For incompletely polarized electrons of polarization degree P_0 the spin-dependent asymmetry in photon emission A is

$$(10) \qquad A = [(n_\uparrow - n_\downarrow)/(n_\uparrow + n_\downarrow)]/P_0 \cos\theta = (N_\uparrow - N_\downarrow)/(N_\uparrow + N_\downarrow),$$

θ is the angle of incidence, $n_\uparrow$, $n_\downarrow$ are the registered counts corresponding to

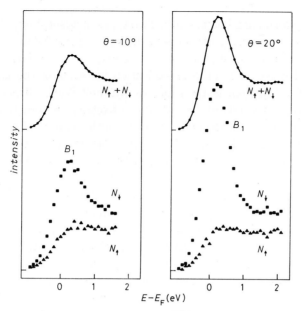

Fig. 9. – Spin-resolved spectra from Ni(110), $\overline{\Gamma X}$, $\hbar\omega = 9.6\,\mathrm{eV}$.

either spin direction and $N_\uparrow$, $N_\downarrow$ the photon fluxes for a hypothetical 100% polarized incident beam. Figure 9 shows experimental data in terms of the corrected quantities $N_\uparrow$ and $N_\downarrow$ which confirm the initial expectation. A strong asymmetry is observed in the d-band transition, however also the structureless background shows a finite spin dependence as expected from Stoner excitations preceding the radiative transition.

4. – Surface states.

Electronic states which are strongly localized in the surface region are called surface states. In this section we shall consider both, surface states intrinsic to the clean surface or adsorbate-induced states.

The termination of a solid by a surface breaks its three-dimensional symmetry, changes the coordination number of the atoms in the surface layer and may, therefore, give rise to various modifications of the electronic and geometric structure. Electronic wave functions need no longer be periodic in the direction of the surface normal. This introduces electron states forbidden in an infinite crystal [32, 33].

New electronic states with strong localization in the surface region can (and will usually) also arise from the formation of an adsorbate layer. Adsorption of atoms or molecules to a solid surface occurs when the total energy of the system can be lowered by forming a bound state. Two types of adsorption, physisorption and chemisorption, are usually distinguished. We shall not follow the usual way to classify a system as being physisorbed or chemisorbed according to its binding energy. We shall rather rely on a classification in terms of the nature of the binding force. Physisorption is then due to van der Waals dispersion forces, while chemisorption proceeds via an attractive exchange interaction. Since the size of the exchange integrals depends critically on the amount of mutual overlap between adsorbate and substrate, we expect the chemisorption force to be directed and specific not only for the particular adsorbate combination but also for the individual crystal face and adsorption site.

The common property of both types of surface states, intrinsic or adsorbate induced, is that their wave functions are no longer periodic in the direction normal to the surface. Unlike bulk electronic states we expect in an attempt to map the energy *vs.* momentum dispersion of electronic surface states only dispersion of E as a function of the electrons momentum parallel to the surface, $k_\parallel$. An example of what might happen to an IPE spectrum upon formation of an ordered adsorbate overlayer is shown in the left-hand panel of fig. 10 [34]. Sulfur has been adsorbed in an ordered $c(2 \times 2)$ overlayer on Ni(110) here. Unlike all previously shown spectra the data in fig. 10 have been obtained with the apparatus depicted in fig. 3, that is with tunable photon energy. The electrons impinge normal to the surface with initial-state energy E_i as given in the figure.

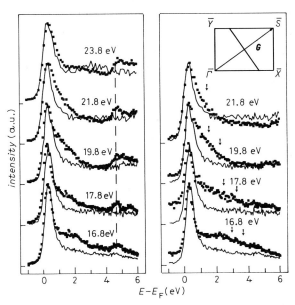

Fig. 10. – Constant initial-state inverse-photoemission spectra from Ni(110). Solid traces refer to normal incidence on the clean sample. Solid squares show features observed upon adsorption of a $c(2 \times 2)$ sulfur overlayer (left-hand panel) and spectra taken at $\overline{S}$ for the clean sample (right-hand panel).

The continuous traces represent the spectra from the clean nickel crystal, while the solid squares indicate the data for Ni-$c(2 \times 2)$ S. Two spectral features are introduced by the sulfur overlayer. The peak at 4.6 eV remains stationary upon variation of the initial-state energy, while a shoulder on the d-band peak at $E = 23.8$ eV disperses and develops into a separated peak at 16.8 eV. Normal incidence to the surface means $k_{\parallel} = 0$ irrespective of the initial-state energy. $k_{\perp}$, however, varies with E_i. The sulfur-induced feature at 4.6 eV is, therefore, immediately identified as an adsorbate-induced surface state, while the other structure must be of different origin. Its nature can be identified with the data shown in the right-hand panel of fig. 10. The adsorbate does not only change the electronic surface structure but leads also to a reduced surface Brillouin zone. The surface Brillouin zone of clean Ni(110) is shown in the upper corner of the right-hand panel in fig. 10. Formation of the sulfur layer halves the original zone as indicated by the dashed line. The former $\bar{S}$ point of the Brillouin zone of the clean surface becomes a $\bar{\Gamma}$ point for the adsorbate-covered surface. Diffraction of the normally incident electrons by exchange of reciprocal surface lattice vector G of the adsorbate-covered surface brings the beam to an equivalent point of the surface Brillouin zone but to an unequivalent point of the momentum space for the bulk of the substrate. The proof that the dispersing shoulder is in fact a substrate bulk transition can be obtained in the following way. By appropriate adjustment of the angle of incidence one may choose the momentum conditions for each initial-state energy such that $k_{\parallel}$ is always equal to G. Variation of E_i leads then to a variation of k along the L-W line of the bulk Brillouin zone. The spectra obtained by this procedure are shown as solid squares in the right-hand panel of fig. 10, while the continuous traces are the spectra at normal incidence. The arrows indicate the theoretical energetic positions of bulk direct transitions. Our analysis shows that considerable care must be exercised in interpreting adsorbate-induced spectral features and the importance of two-dimensional *vs.* three-dimensional dispersion.

Having identified the peak at 4.6 eV in fig. 10 as a sulfur-induced state, we are left with the problem of what can be learned from that. The answer is not very satisfactory. From a theoretical point of view chemisorption is an intermediate-coupling problem in the sense that there is no obviously small expansion parameter in powers of which the solution could be constructed from a simple limiting case. Interpretation of observed adsorbate-induced electronic states continues, therefore, to rely and depend on the availability of appropriate theoretical calculations. This strict statement can be slightly relaxed in the case of molecular chemisorption which shall not be discussed here.

We finally return to the so far unexplained spectral feature S_1 in fig. 5. Its surface nature was already suggested from a comparison of its $E(k_{\parallel})$ dispersion to the projected bulk band structure (see fig. 6). The dispersion of this state has also been proved to be purely two-dimensional in just the same manner as discussed above. Features like S_1 in fig. 5 have been observed on different crystal faces of various materials [33, 35-37]. The most important property common to all those observations is that the structure is pinned to the vacuum level of the sample rather than to the Fermi level. Further it shows up always in a gap of the projected bulk band structure at roughly $(500 \div 800)$ meV below the vacuum level for normal electron incidence. Mainly these two observations have led to the interpretation that this emission results from radiative transitions into bound states of the image potential an electron experiences upon approaching a metal surface. The essential characteristic of these states can be derived from a grossly simplified model which replaces the band gap by the rigidly reflecting plane $V(Z) = \infty$ for $Z > 0$ and the surface potential by its asymptote for $V = 1/4 Z$ for all $Z < 0$. The problem is then that of a one-dimensional hydrogen atom with fractional charge 1/4. Consequently, the discrete energy eigenvalues of this model are (in atomic units)

(11) $$E_n = - 1/32 \, n^2 \, .$$

This Rydberg series spans a range of 850 meV. In view of the instrumental resolution employed in the measurements displayed in fig. 5 individual members of the series cannot be resolved, but the whole series is broadened into a step.

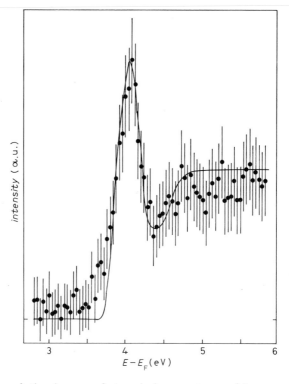

Fig. 11. – High-resolution inverse-photoemission spectrum of image potential states.

More recent IPE experiments with considerably higher resolution have been able to clearly separate the $n = 1$ state from the rest of the series (fig. 11 [18]). Two-photon photoemission with its even higher resolving power has succeeded in resolving even the three lowest-lying states [38].

The lowest-order one-dimensional hydrogen atom model for the image potential bound states can be modified in a few respects to fully account for the experimental observations. The first correction is the saturation of the Coulomb potential at some finite value V_0. The entire wave function is then composed of the solution of the Coulomb potential, the solution in a constant potential and the exponentially decaying solution in the band gap with continuous logarithmic derivatives at the two boundaries. Such a more realistic potential will, depending on the parameters, frequently accommodate a further bound state below the Rydberg series. Such situations prevail, for example, on the (111) surfaces of Cu and Ag [10]. Wave functions from a model calculation for Ag(111) are shown in fig. 12. They are classified in terms of their nodes and exhibit quite

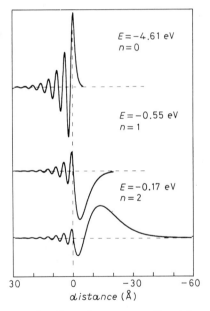

Fig. 12. – Surface state wave functions for a conventional ($n = 0$) and image potential surface states ($n = 1, 2$).

different shapes. The $n = 0$ state describes the from PES well-known occupied surface state on Ag(111). Its wave function is clearly much more concentrated in the surface plane than those of the image potential $n = 1$ and $n = 2$ states. This explains the frequent observation that strongly bound surface states are often quenched by adsorbate contamination, while the image potential states usually show only an energetic shift equal to the amount of work function change introduced by the adsorbate [37].

Precise knowledge of the binding energy of a few surface states is of considerable importance. In view of the difficulties to obtain the correct surface potential barrier from first-principles calculations such data have been used to determine a parametrized ansatz for the surface potential barrier with considerable precision [39].

* * *

I wish to thank my co-workers over the last years for their contributions to this lecture. I am particularly indebted to M. DONATH for permission to include figures which have not yet been published elsewhere. M. DONATH has also been very helpful with the references, Mrs. M.-L. HIRSCHINGER and G. DAUBE have done expert jobs with figures and typing.

REFERENCES

[1] V. DOSE: *Prog. Surf. Sci.*, **13**, 225 (1983); *J. Phys. Chem.*, **88**, 1681 (1984); *Appl. Surf. Sci.*, **22/23**, 338 (1985); *Festkörperprobleme*, **25**, 555 (1985); *Surf. Sci. Rep.*, **8**, 337 (1985); *J. Vac. Sci. Technol. A*, **5**, 2032 (1987).

[2] D. P. WOODRUFF, P. D. JOHNSON and N. V. SMITH: *J. Vac. Sci. Technol. A*, **1**, 1104 (1983).

[3] N. V. SMITH: *Vacuum*, **33**, 803 (1983); *Appl. Surf. Sci.*, **22/23**, 349 (1985).

[4] F. J. HIMPSEL and TH. FAUSTER: *J. Vac. Sci. Technol. A*, **2**, 815 (1984).

[5] F. J. HIMPSEL: *Comments Cond. Mat. Phys.*, **12**, 199 (1986).

[6] TH. FAUSTER and V. DOSE: in *Chemistry and Physics of Solid Surfaces VI*, edited by R. VANSELOW and R. HOWE (Springer, Berlin, 1986), p. 483.

[7] P. O. NILSSON and A. KOVACS: *Phys. Scr.*, T4, 61 (1983).

[8] B. REIHL: *Surf. Sci.*, **162**, 1 (1985).

[9] G. BORSTEL: *Appl. Phys. A*, **38**, 193 (1985).

[10] G. BORSTEL and G. THÖRNER: *Surf. Sci. Rep.*, **8**, 1 (1988).

[11] W. C. RÖNTGEN: *Sitz. Ber. Med. Phys. Ges. Würzburg*, 137 (1895).

[12] E. W. PLUMMER and W. EBERHARDT: *Adv. Chem. Phys.*, **49**, 533 (1982).

[13] F. J. HIMPSEL: *Adv. Phys.*, **32**, 1 (1983).

[14] J. B. PENDRY: *Phys. Rev. Lett.*, **45**, 1356 (1980); *J. Phys. C*, **14**, 1381 (1981).

[15] V. DOSE: *Appl. Phys.*, **14**, 117 (1977).

[16] G. DENNINGER, V. DOSE and H. SCHEIDT: *Appl. Phys.*, **18**, 375 (1979).

[17] A. GOLDMANN, M. DONATH, W. ALTMANN and V. DOSE: *Phys. Rev. B*, **32**, 837 (1985).

[18] V. DOSE, TH. FAUSTER and R. SCHNEIDER: *Appl. Phys. A*, **40**, 203 (1986).

[19] K. DESINGER, V. DOSE, M. GLÖBL and H. SCHEIDT: *Solid State Commun.*, **49**, 479 (1984).

[20] N. BABBE, W. DRUBE, I. SCHÄFER and M. SKIBOWSKI: *J. Phys. E*, **18**, 158 (1985).

[21] I. SCHÄFER, W. DRUBE, M. SCHLÜTER, G. PLAGEMANN and M. SKIBOWSKI: *Rev. Sci. Instrum.*, **58**, 710 (1987).

[22] TH. FAUSTER, F. J. HIMPSEL, J. J. DONELON and A. MARX: *Rev. Sci. Instrum.*, **54**, 68 (1983).

[23] W. A. ROYER and N. V. SMITH: *Rev. Sci. Instrum.*, **59**, 737 (1988); see also T. T. CHILDS, W. A. ROYER and N. V. SMITH: *Rev. Sci. Instrum.*, **55**, 812 (1984).

[24] D. T. PIERCE and F. MEIER: *Phys. Rev. B*, **13**, 5484 (1976).

[25] D. T. PIERCE, R. J. CELOTTA, G.-C. WANG, W. N. UNERTL, A. GALEJS, C. E. KUYATT and S. R. MIELCZAREK: *Rev. Sci. Intrum.*, **51**, 478 (1980).

[26] U. KOLAC, M. DONATH, K. ERTL, H. LIEBL and V. DOSE: *Rev. Sci. Instrum.*, **54**, 1933 (1988).

[27] J. UNGURIS, A. SEILER, R. J. CELOTTA, D. T. PIERCE, P. D. JOHNSON and N. V. SMITH: *Phys. Rev. Lett.*, **49**, 1047 (1982).

[28] H. SCHEIDT, M. GLÖBL, V. DOSE and J. KIRSCHNER: *Phys. Rev. Lett.*, **51**, 1688 (1983).

[29] M. DONATH, M. GLÖBL, B. SENFTINGER and V. DOSE: *Solid State Commun.*, **60**, 237 (1986).

[30] F. J. HIMPSEL: *Appl. Opt.*, **19**, 3964 (1980).

[31] E. O. KANE: *Phys. Rev. Lett.*, **12**, 97 (1964).

[32] A. GOLDMANN, V. DOSE and G. BORSTEL: *Phys. Rev. B*, **32**, 1971 (1985).

[33] A. GOLDMANN: *Surf. Sci.*, **178**, 210 (1986).

[34] K. DESINGER, W. ALTMANN and V. DOSE: *Surf. Sci.*, **201**, L 491 (1988).

[35] V. DOSE, W. ALTMANN, A. GOLDMANN, U. KOLAC and J. ROGOZIK: *Phys. Rev. Lett.*, **52**, 1919 (1984).

[36] D. STRAUB and F. J. HIMPSEL: *Phys. Rev. Lett.*, **52**, 1922 (1984); *Phys. Rev. B*, **33**, 2256 (1986).

[37] V. DOSE: *Phys. Scr.*, **36**, 669 (1987).

[38] K. GIESEN, F. HAGE, F. J. HIMPSEL, H. J. RIESS and W. STEINMANN: *Phys. Rev. Lett.*, **55**, 300 (1985); *Phys. Rev. B*, **33**, 5241 (2986).

[39] G. THÖRNER and G. BORSTEL: *Appl. Phys. A*, **41**, 99 (1986).

X-Ray Absorption Fine Structure of Adsorbates: SEXAFS and NEXAFS.

J. HAASE

Fritz-Haber-Institut der Max-Planck-Gesellschaft - Faradayweg 4-6, D-1000 Berlin 33

1. – Introduction.

In a photoabsorption experiment the absorption coefficient above the absorption edge of a core level is measured as a function of the photon energy. Above the edge the absorption coefficient does not monotonically decrease with energy but exhibits a fine structure. The X-ray absorption fine structure (XAFS) is generally divided into two parts: into a near-edge region within some ten eV above an absorption edge and into an extended region well above the edge. The near-edge X-ray absorption fine structure (NEXAFS), also known as X-ray absorption near-edge structure (XANES), exhibits strong and sharp features, whereas the extended X-ray absorption fine structure (EXAFS) consists of relatively weak and broad wiggles (see fig. 1). In a scattering picture these two regions can be distinguished by the effectiveness of multiple-scattering events: single scattering of the outgoing photoelectrons dominates well above the edge (*i.e.* for high kinetic energies of the photoelectrons), whereas multiple scattering cannot be neglected close to the edge.

Surface EXAFS (SEXAFS) as an experimental technique is the surface version of (bulk) EXAFS. In a SEXAFS experiment the signal from less than a monolayer of surface (adsorbate) atoms has to be detected and discriminated from the bulk signal. These demands can only be met by using synchrotron radiation and surface-sensitive detection methods. To date mainly electron yield detection (see subsect. 2˙6) has been applied in SEXAFS studies. Surface EXAFS studies on atoms and molecules adsorbed on single crystal surfaces began in 1978 at SSRL with the pioneering work of Citrin *et al.* [1] and Stöhr *et al.* [2]. In less then a decade the technique has become an established ˙and powerful tool for the determination of adsorbate structure. Two parameters can be obtained from a SEXAFS measurement in a straightforward manner independent of theory: adsorbate-substrate bond lengths and adsorption sites. The accuracy of SEXAFS derived bond lengths (roughly ± 0.01 Å) and the general reliability of adsorption site identification indicate that some time in the

277

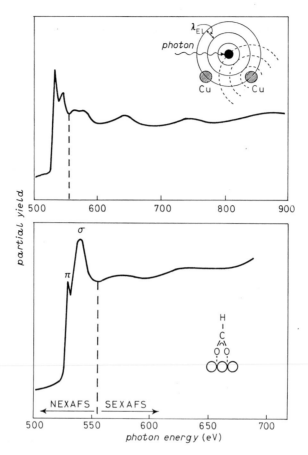

Fig. 1. – Photoabsorption spectra of O (above) and HCOO (below) adsorbed on Cu{110} taken above the oxygen K-edge.

future enough information will be available, at least for atomic adsorption systems, to establish a comprehensive picture of structure and bonding at surfaces[3] similar to that already available in three-dimensional chemistry[4]. These developments have been accelerated by the more stable and more intense synchrotron radiation facilities of the second generation which, combined with better monochromators, have recently improved the quality of SEXAFS data and allowed new structural problems to be tackled, particularly in the area of low-Z elements. Thus quite recently adsorbate-induced reconstruction has been investigated[5, 6] and intramolecular bond lengths have been indirectly determined[7] with SEXAFS. Improved data quality has also stimulated the study of problems related to SEXAFS analysis, $e.g.$ the anisotropies in the surface mean free path[6] and in the surface Debye-Waller factor[6, 8-10], as well as the anharmonicity of the Debye-Waller factor[10, 11] and the possibility of multiple scattering[12]. Features deriving from multiple scattering might in the future be used to determine interatomic distances within the substrate. Even higher beam intensities are desirable for low-Z SEXAFS studies, particularly at the carbon K-edge [13] if fluorescence detection is to be applied [14].

To date the majority of SEXAFS studies (for an overwiew see ref. [3, 15]) has been on atomic adsorbates. Molecules are generally more complicated because, even at high-symmetry adsorption sites, there is the possibility of several similar bond lengths and the analysis requires great care. In general, multishell analysis (curve fitting) should then be applied (see subsect. 2˙3). The examples given (subsect. 2˙7) have been chosen not only to demonstrate the determination of bond lengths, adsorption sites and adsorbate-induced reconstructions but also to present the different SEXAFS analysis procedures (single-shell and multishell analysis). They also demonstrate the effectiveness of three different methods for adsorption site determination in a single-shell analysis: higher-neighbour adsorbate-substrate distances, relative polarization-dependent surface amplitudes and surface/bulk absolute amplitudes.

The first NEXAFS spectra for molecules adsorbed on single crystal surfaces were reported in 1981 by STÖHR *et al.* [16]. The development of the NEXAFS technique and its rapid application to chemisorbed molecules and surface intermediates have been even more spectacular than that of SEXAFS. This is due to the fact that both the experiment and its interpretation—at least on a superficial level—are very straightforward [17]. NEXAFS is in the first instance a structural tool: Since the X-ray absorption spectrum of an adsorbed molecule is dominated by resonances corresponding to transitions into unoccupied orbitals, information can be obtained on molecular orientation from the polarization dependence of the dipole matrix element, *i.e.* without the necessity of performing complicated multiple-scattering calculations. It has even been suggested [18] that intramolecular bond lengths can be estimated from the spectra, although this probably applies only to simple, isostructural molecules [19]. Thus for adsorbed molecular species the structural information extracted from NEXAFS and SEXAFS is complementary making the combination of the two techniques quite powerful. By probing the unoccupied states of the adsorbate/substrate system, NEXAFS also provides information on the chemisorption bond.

NEXAFS studies are now being performed on increasingly more complex molecules and reaction products. The polarization dependence of the NEXAFS is always determined by the effective point group of the adsorbed species and by the symmetry of the orbitals involved [20]. There is no way of deciding *a priori* whether the point group is that of the free molecule or that of the adsorbate plus substrate. However, even for relatively strongly bound adsorbates, symmetry lowering on adsorption often appears not to occur [21]. The second problem concerns the assignment of the resonances which, of course, belong to the irreducible representations of the point group. This is necessary in order to apply the simple formulae for the polarization dependence [17]. It is now clear that the approach in which polyatomic molecules are described in terms of diatomic- and ring-like «building blocks» [17] is inadequate and, in many cases, incorrect. The accuracy of an orientation determination with NEXAFS—usually between 5° and 15°—is often limited by the uncertain background under the resonances as well as by the presence of «substrate scattering» resonances, Rydberg transitions and multielectron excitations. Multiple-scattering cluster calculations have shown that the NEXAFS of atomic adsorbates (the «substrate scattering» in the molecular case) is sensitive to the adsorbate-substrate separation and to the adsorption site [22-24]. However, these structural parameters are in general determined more easily and reliably with SEXAFS [24]. NEXAFS of atomic adsorbates cannot be used as a fingerprint technique to determine adsorption geometries but will remain a valuable structural tool especially for those cases where SEXAFS fails.

The examples in sect. 3 have been chosen to demonstrate the potential of NEXAFS and, in particular, its interplay with SEXAFS in the determination of adsorption geometry. They also point out the difficulties encountered in analysing the NEXAFS from simple and more complex adsorbed molecules.

2. – SEXAFS.

2‘1. *Single-scattering formalism.* – All EXAFS theories are based on Fermi's golden rule for the absorption cross-section σ in the dipole approximation:

$$(1) \qquad \sigma(E_f) \propto |\langle f| \boldsymbol{EM} |i\rangle|^2 \rho(E_f).$$

Here $|i\rangle$ and $|f\rangle$ are the initial and final state, respectively, E is the electric-field vector, M is the momentum operator and $\rho(E_f)$ is the density of states at the energy of the final state E_f.

EXAFS is an interference effect involving the final-state photoelectron wave function which is modified due to scattering by neighbours surrounding the absorbing atom. It is the modulation in the matrix element that describes EXAFS, and not the density of states. Phenomenological approaches [25] as well as full one-electron single-scattering theories [26-28] in a plane-wave approximation yield the following expression for the modulation of the absorption coefficient μ above K- and L_1-edges (s initial state) which could also be derived intuitively by considering the scattering of the photoelectron waves from neighbouring atoms back towards the excited atom at the origin [29]:

$$(2) \qquad \chi(k) = \frac{\mu - \mu_0}{\mu_0} = - \sum_i A_i(k) \sin[2kR_i + \psi_i(k)],$$

where μ_0 is the absorption coefficient of the free atom, and the summation extends over all neighbouring shells i of atoms separated from the absorbing atom by the same distance R_i. The photoelectron wave number k is defined as

$$(3) \qquad\qquad k = \hbar^{-1}[2m\,(E - E_0)]^{1/2},$$

where E is the photon energy and E_0 is the threshold energy (see subsect. 2'1.1). The total phase shift $\psi_i(k)$ is given by

$$(4) \qquad\qquad \psi_i(k) = 2\delta_c(k) + \phi_i(k),$$

where $\delta_c(k)$ is the phase shift due to the central atom and $\phi_i(k)$ is the phase shift due to the backscattering atom. The EXAFS amplitude is polarization dependent:

$$(5) \qquad\qquad A_i(k) = \frac{N_i^*}{kR_i^2} F_i(k) \cdot \exp\left[-2\sigma_i^2 k^2\right]\exp\left[-2R_i/\lambda_i(k)\right]$$

according to the polarization-dependent effective coordination number N_i^*:

$$(6) \qquad\qquad N_i^* = 3\sum_j^{N_i} \cos^2 \alpha_{ij}, \qquad\qquad \alpha_{ij} = \measuredangle\,(\boldsymbol{E}, \boldsymbol{r}_{ij}),$$

where α_{ij} is the angle between the E vector at the absorbing-atom site and the vector r_{ij} from the absorbing atom to the j-th atom in the i-th shell. For polycrystalline samples, cubic crystals and amorphous samples N_i^* equals the number of atoms in the i-th shell, N_i. $F_i(k)$ in eq. (5) is the backscattering amplitude of the neighbouring atoms and the exponential terms are due to Debye-Waller-like effects and to inelastic scattering of the photoelectrons with a mean free path $\lambda_i(k)$.

For L_2- and L_3-edges with an initial state of p symmetry and final states of either s or d symmetry formally the same eqs. (2)-(5) hold [15]. The polarization dependence, however, is weakened.

Recent developments in EXAFS theory include its formulation with spherical waves [30, 31] and the treatment of multiple scattering [32]. The curved-wave formalism may be recast into the same form as eqs. (2)-(6) with backscattering amplitudes and phase shifts that are functions of the interatomic separation R_i [33] (see subsect. 2'1.1 and 2'1.2). The extent to which multiple scattering has to be taken into account in SEXAFS is discussed in subsect. 2'2.

From the simple sine-shaped phase term in eq. (2) it is obvious that a Fourier transform of the measured function $\chi(k)$ directly yields bond lengths R_i if the phase shifts $\psi_i(k)$ are known [34]. This outstanding feature of a SEXAFS analysis contrasts with the analysis in two other surface-sensitive structural techniques: LEED and photoelectron diffraction (PhD), where experimental curves have to be compared with model calculations which are often very time consuming. A more detailed comparison of LEED, PhD and SEXAFS is made in sect. 4. If first-neighbour and second-neighbour shells are separated by more than roughly 1 Å, the first-neighbour bond length can be determined to an accuracy of better than ± 0.02 Å using the simple Fourier transform method (single-shell analysis). Higher-neighbour bond lengths normally determine the second structural parameter provided by a SEXAFS experiment: the adsorption site. Compared with bulk EXAFS measurements SEXAFS data suffer from low signal-to-background and signal-to-noise ratios (see subsect. 2·6), so that very often only the nearest-neighbour shell shows up as a peak in the Fourier transform. In this case the adsorption site can be determined using either relative polarization-dependent surface amplitudes or surface and bulk absolute amplitudes (see subsect. 2·1.3). If there are overlapping bond lengths, a curve-fitting procedure (multishell analysis) should be applied, in which the experimental $\chi(k)$ (and/or its Fourier transform) is compared with a calculated $\chi(k)$ (and/or its Fourier transform). The multishell analysis provides the same accuracy for the nearest-neighbour bond lengths as the single-shell analysis. Single-shell and multishell analyses are discussed in more detail in subsect. 2·3.

2·1.1. Phase shifts. The total phase shift $\psi_i(k)$ in eq. (2), which is made up of the central atom and the backscattering phase shifts according to eq. (4), can be either calculated or determined experimentally. Until recently theoretical phase shifts had been calculated mainly in the plane-wave approximation and were tabulated down to k-values of about $3.8 \, \text{Å}^{-1}$ [35]. Quite recently the k-range could be extended to $2 \, \text{Å}^{-1}$ by using a full curved-wave formalism [33], thus providing the capability to analyse experimental data much closer to the absorption edge than was previously possible. In general the effects of the curved-wave formalism are more prominent at low k and involve a weak dependence of the calculated phase shifts (and backscattering amplitudes) on the interatomic distance.

The common way of reliable SEXAFS bond length determination makes use of model compound (bulk) EXAFS phase shifts and is based on the concept of phase shift transferability [36]. Phase shifts are insensitive to the chemical environment because at sufficiently high photoelectron kinetic energies (typically $\geqslant 50$ eV) the scattering process is dominated by the core electrons. Therefore, phase shifts derived from bulk EXAFS measurements on samples with the same atom pair absorber/backscatterer, for which the interatomic distances are known, can be used as input for the SEXAFS analysis. In cases where the absorbing or backscattering atoms differ in atomic number Z by $\leqslant \pm 2$ from those in a model compound, the phase shifts of the latter can be modified with differences of theoretical phase shifts [37]. Care should be taken, however, that surface system and reference sample are measured at the same temperature, preferentially at low temperature (see subsect. 2·1.4).

In a single-shell analysis (see subsect. 2'3) peaks in the (complex) Fourier transform are back transformed to yield amplitude and phase functions[29]. The measured phase function $2kR + \psi$ is then compared to the reference phase shift ψ_{ref} to yield the bond length R:

$$(7) \qquad R(k) = \frac{2kR + \psi - \psi_{ref}}{2k}.$$

In this analysis procedure the threshold energy E_0 (cf. eq. (3)) is an adjustable parameter fitted in the back transformation in a way to get constant R-values over a large region in k-space[38].

Numerous comparisons between EXAFS derived bulk bond lengths and those determined from X-ray diffraction have shown that the accuracy is as good as ± 0.01 Å. Unless there are limitations due to the data length and quality (see subsect. 2'6) the same holds for SEXAFS.

2'1.2. Backscattering amplitudes.   By use of a full curved-wave formalism backscattering amplitudes $F_i(k)$ have been calculated and tabulated quite recently[33] for k-values above 2 Å^{-1}. They mainly differ from back-scattering amplitudes calculated in a plane-wave approximation[35] at low k and are weakly dependent on the interatomic distance. This behaviour is similar to that of the phases shifts (see subsect. 2'1.1) as backscattering amplitudes and phase shifts are part of the same complex scattering function[33].

In fig. 2 backscattering amplitudes $F_i(k)$ for $Z = 9$, 47 and 82 are plotted. For light atoms $F_i(k)$ peaks at small k and decreases for higher k, whereas for

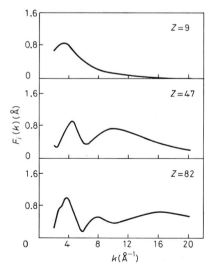

Fig. 2. – Backscattering amplitudes $F(k)$ as a function of k for F ($Z = 9$), Ag ($Z = 47$) and Pb ($Z = 82$) calculated with the curved-wave formalism at $R = 2.5$ Å. After[33].

heavier atoms there is a peak at intermediate k which has the same physical origin as the Ramsauer-Townsend effect [39, 40]. $F_i(k)$ of heavy elements shows a two-peak structure. Thus from the shape of the backscattering amplitude of a well-separated peak we can differentiate between low-Z, intermediate-Z and high-Z backscatterers. Due to the Ramsauer-Townsend-like resonances in the backscattering amplitude of sufficiently high-Z atoms there is an additional peak at lower k-values in the Fourier transform (see subsect. 2·7.3) which can compli- cate the single-shell SEXAFS analysis. In the SEXAFS region well above the edge backscattering from high-Z atoms dominates over backscattering from low- Z atoms. This is one reason why in a SEXAFS experiment on a low-Z adsorbate we measure the absorption coefficient above the absorption edge of an adsorbate atom. And this explains also the fact that in a SEXAFS study on adsorbed low-Z molecules we measure adsorbate-substrate bond lengths and do not directly «see» intramolecular bond lengths with the present-day instrumentation.

2·1.3. Effective coordination numbers. According to the polarization- dependent effective coordination number N_i^* (eq. (6)) the electric-field vector E can be regarded as a searchlight sorting out all neighbours in a given direction for a bond length determination. In favourable cases it is thus possible to measure nearest and second-nearest neighbour bond lengths independently even if the absolute values are similar or the same (see subsect. 2·7.2). This is important because in a SEXAFS measurement the resolution is only some 10th of an Å depending on the k-range, which is much more limited compared to that of a low-noise bulk EXAFS experiment.

In the surface geometry there are in general two angular dependences: 1) on polar angle θ, which measures the deviation of the E vector from the surface normal, and 2) on azimuthal angle ϕ, which measures the orientation of the E vector in the surface plane. The latter dependence only exists if the rotational symmetry about the absorbing atom is twofold or less. Effective coordination numbers N_i^* expressed as functions of θ and ϕ are easily obtained for atop, hollow and bridge sites [15].

In all cases where only the nearest-neighbour bond length shows up in the Fourier transform, the measurement of N_i^* is essential for an adsorption site determination. The concept of amplitude transferability, however, does not work as well as the phase shift transferability mainly because the exponential terms in eq. (5) are strongly dependent on the chemical environent [41]. Therefore, a normalization of the surface absolute amplitude with the absolute amplitude for a bulk reference sample of known coordination number, N_i, yields an effective coordination number which by comparison with calculated N_i^* values for model geometries can often only give a hint for the correct adsorption site. Much more reliable is a comparison of relative SEXAFS amplitudes obtained in the same sample at different polarizations with calculated relative N_i^* values. In this case both the Debye-Waller-like and the electron loss terms in eq. (5)

practically cancel:

(8)
$$\frac{A_i(\theta_i)}{A_i(\theta_2)} = \frac{N_i^*(\theta_1)}{N_i^*(\theta_2)}.$$

This experimentally determined amplitude ratio is in general sufficient to distinguish between various adsorption sites.

2'1.4. Debye-Waller factors.   Atoms are vibrating aroung their equilibrium lattice positions. These vibrations are temperature dependent and result in a damping and broadening of the SEXAFS oscillations. It is, therefore, advisable to take data at low temperatures to minimize this effect. If the interatomic potential is harmonic, the exponential damping term $\exp[-2\sigma_i^2 k^2]$ in eq. (5) holds with $\sigma_i^2(T)$ being the mean square average of the difference of displacements [29]. When the potential is nonharmonic, phase and amplitude corrections arise [42]. Phase corrections lead to systematic errors in bond lengths [11] (see subsect. 2'7.3), especially when surface system and bulk reference sample are measured at different temperatures. Therefore, for systems where anharmonicity is important, all measurements should be performed at low temperatures.

The EXAFS Debye-Waller factor differs from that for X-ray diffraction. In the harmonic approximation $\sigma_i^2(T)$ in eq. (5) is given by [43]

(9)
$$\sigma_i^2(T) = \langle u_0^2(T) \rangle + \langle u_i^2(T) \rangle - 2 \langle u_0(T) u_i(T) \rangle,$$

where $\langle u_0^2(T) \rangle$ and $\langle u_i^2(T) \rangle$ are the mean square displacements of the central absorbing atom and of the backscattering atom, respectively, measured by X-ray diffraction. In general, the correlation term $2 \langle u_0(T) u_i(T) \rangle$ is most effective for nearest neighbours because of their direct bonding and less important for neighbours at larger distances. Only recently anisotropies in the mean square relative displacements in directions parallel and perpendicular to the surface have been reported [6, 8-10] which show the importance of the correlation term.

2'1.5. Mean free paths.   The exponential mean-free-path term $\exp[-2R_i/\lambda_i(k)]$ in eq. (5) is only an approximation for an inelastic loss factor associated with the medium between the central atom and the i-th neighbour [29]. Working with the generally accepted mean-free-path concept one should bear in mind that the EXAFS mean free path is a convolution of the mean free path of the photoelectron and the lifetime of the core hole. For low-Z elements, however, the EXAFS λ is dominated by the photoelectron [44]. Its k-dependence should then follow the behaviour expected for the electron mean free path as determined from Auger and photoemission data, which is shown in fig. 3. In the SEXAFS energy range (energy from the edge) from some ten eV to some

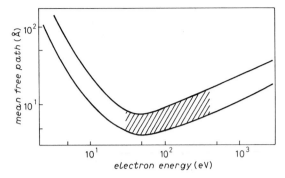

Fig. 3. – Electron mean free paths in solids as a function of the kinetic energy of the electron. The shaded area characterizes electron energies in a typical SEXAFS experiment. After [15].

hundred eV there is only a weak k-dependence (if there is any at all) and the mean-free-path values centre around 5 Å [44]. Recently anisotropies in the mean free path in directions parallel and perpendicular to the surface have been reported [6] which can be explained by a jellium-type model of the surface.

2.2. Multiple scattering. – In deriving eq. (2) for the EXAFS oscillations only single scattering was taken into account. The single-scattering approximation works quite well for energies well above the absorption edge with only one important exception. This can be envisaged from fig. 4, where the electron scattering amplitude as a function of the scattering angle is plotted with the

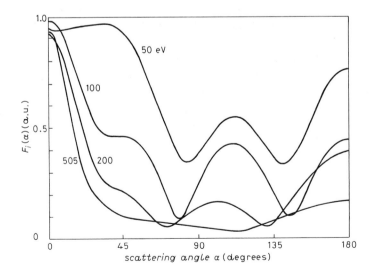

Fig. 4. – Scattering amplitudes $F(\alpha)$ for Ni as a function of the scattering angle α with the electron kinetic energy as parameter. After [45].

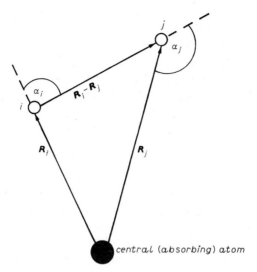

Fig. 5. – 3-atom system with a central (absorbing) atom and two scattering atoms i and j.

electron kinetic energy as parameter. The scattering amplitude is high in the foward direction ($\alpha = 0$), then decreases with increasing scattering angle, has some structure for intermediate angles α between 90° and 135° and finally rises again for backscattering ($\alpha = 180°$). When taking multiple-scattering paths into account—e.g., in fig. 5 the path from the absorbing atom via the scattering atom i to the scattering atom j and back to the central atom—then the product of two (double scattering) or more small values $F(\alpha)$ has to be compared with $F(\pi)$. Moreover, multiple-scattering paths are longer than the nearest-neighbour backscattering path, so that they show up farther out in the Fourier transform and have no influence on the nearest-neighbour distance and they are more damped according to eq. (5). The serious failure of the single-scattering approach occurs for cases where atoms of a near-neighbour shell are directly in the line of sight of atoms in a higher-neighbour shell [26] (e.g., Cu EXAFS). The outgoing electron is strongly forward scattered at the near-neighbour atoms (cf. fig. 4) and thereby enhances the electron amplitude in the higher-neighbour shell. Amplitude and phase shift changes can be quantitatively explained by a multiple-scattering calculation for the three-atom system of fig. 5 ignoring the exponential damping terms and neglecting higher than double scattering [46]

$$(10) \qquad \chi(k) = -\frac{3(\boldsymbol{ER}_i)^2}{kR_i^2} F_i(\pi, k) \sin\left[2kR_i + 2\delta(k) + \phi_i(k)\right] -$$

$$-\frac{3(\boldsymbol{ER}_j)^2}{kR_j^2} F_j(\pi, k) \sin\left[2kR_j + 2\delta(k) + \phi_j(k)\right] -$$

$$-\frac{6(\boldsymbol{ER}_i)(\boldsymbol{ER}_j)}{kR_iR_jR_{ij}} F_i(\alpha_i, k) F_j(\alpha_j, k) \sin\left[k(R_i + R_j + R_{ij}) + 2\delta(k) + \phi_i(k) + \phi_j(k)\right].$$

The third term in eq. (10) corresponds to double scattering and contains information on the interatomic distance $R_{ij} = |R_i - R_j|$.

Due the improved data quality multiple-scattering features have recently been observed in SEXAFS studies [12, 47]. They open up new possibilities to determine surface reconstruction with SEXAFS via the interatomic distance R_{ij}.

2`3. Single- and multi-shell analysis.

2`3.1. Single-shell analysis. The first step in a SEXAFS analysis is the background subtraction. Normally a smooth spline polynomial is fitted through the oscillating structure which mimics μ_0 in eq. (2) as is seen in fig. 6a). The normalization is often performed with the «edge jump», which is the difference of the total signal and the background signal at the threshold energy E_0. The resulting $\chi(E)$ is then converted to a k-scale (cf. fig. 6b)) and weighted with k^n ($n = 0, 1, 2, ...$) before Fourier transforming. To minimize noise both the lower and upper limits of the complex Fourier transformation should coincide with a zero in the χ function. Single peaks of the Fourier transform are then backtransformed by choosing an appropriate Fourier window function (dashed line in fig. 6c)). The backtransform $\chi'(k)$ can be separated into an amplitude $A(k)$,

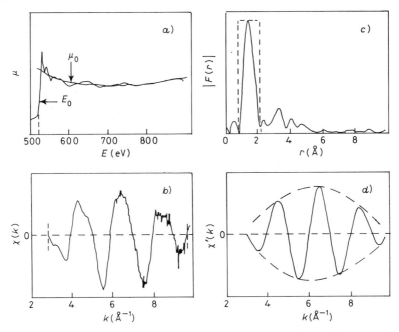

Fig. 6. – Fourier analysis of a SEXAFS spectrum: a) raw data with a smooth spline polynomial fitted through the oscillations, b) SEXAFS oscillations, c) Fourier transform of b) with a Fourier window (dashed line) for the backtransformation, d) backtransform of the dominating peak in c); the envelope (dashed line) is the SEXAFS amplitude.

which is the envelope of the oscillations (dashed line in fig. 6d)), and into a phase shift $\psi(k)$. For bond length determination the latter is compared with an empirical phase shift (cf. subsect. 2'1.1).

2'3.2. Multishell analysis (curve fitting). In a multishell analysis a theoretical SEXAFS spectrum according to eqs. (2)-(5) is calculated and fitted to that of the experiment using an iterative least-square procedure. Input parameters are $F_i(k)$, $\psi_i(k)$, σ_i^2 and λ_i, whereby in the existing programs[48] anisotropies in σ_i^2 and λ_i (cf. subsect. 2'1.4 and 2'1.5) are neglected. Due to the fact that amplitude transferability does not hold, absolute values of effective coordination numbers, especially beyond the first, should be used with caution[48]. Shell distances, however, can be determined with good precision; nearest-neighbour distances are accurate to within ± 0.02 Å. As already mentioned in subsect. 2'1, a multishell analysis must be applied in case of overlapping bond lengths. Close-lying shells can then be distinguished, which in general is impossible in a single-shell analysis. Moreover, the problem of Ramsauer-Townsend-like resonances is solved automatically as its physics is included in the input parameters.

2'4. *Low-Z molecular adsorbates*. – SEXAFS studies on low-Z molecules adsorbed on surfaces need particular attention and care because of several reasons. In principle both adsorbate-substrate and intramolecular bond lengths can be determined. However, due to the relatively low backscattering amplitude of low-Z elements (cf. fig. 2) backscattering from the substrate, which in general is a medium- or high-Z element, dominates. Intramolecular bond lengths, on the other hand, are small compared to adsorbate-substrate bond lengths, so that their periodicity in k-space is large and might be mistaken as a slow variation of the background intensity. Therefore, it is advisable to resort to indirect methods for an intramolecular bond length determination as are, *e.g.*, curve-fitting procedures[7] and measurements at different absorption edges when the atoms constituting the bond are different[49]. But also the determination of adsorbate-substrate bond lengths bears some problems. The close proximity of the K-edges for C, N and O results in small k-ranges for molecules containing two of these elements. The experimental errors for the bond lengths are then large. And even for hydrocarbons the k-range above the C K-edge is small when using the partial-yield technique of detection (see subsect. 2'6) which generally is the method of choice in the soft-X-ray region. Finally, because of overlapping bond lengths a multishell analysis normally has to be performed.

2'5. *Clean surfaces*. – SEXAFS is not ideally suited to study clean surfaces and to tackle the problem of multilayer relaxation (see subsect. 2'6). Even if only electrons of minimum escape depth are detected, the information is an average over at least two lattice spacings (cf. fig. 3). Therefore, only one such investigation has been published [50] in which high and low surface sensitivity spectra in a limited energy range have been compared. Using the same principle by measuring Auger electrons of different escape depths (from different Auger transitions in the same sample), however, questions related to local ordering and crystallization in the outer layers of an amorphous solid can be successfully addressed [51].

2'6. *Detection techniques.* – Any signal which is proportional to the filling of the core hole created by the absorption of radiation can be used as a measure of XAFS. XAFS can, therefore, be monitored by the yield of Auger electrons or fluorescent photons produced in nonradiative and radiative transitions, respectively. SEXAFS (and NEXAFS) detectors, however, have to be surface sensitive (see the introduction). Because of their inherent surface sensitivity and UHV compatibility electron detectors have been used for most of the SEXAFS measurements reported to date. Due to shortcomings of electron yield detection, especially in the soft-X-ray region, and changing demands in surface science (*e.g.*, measurements at ambient pressure) fluorescent yield detection (see below) will be more widely used in the future. Other SEXAFS detection techniques [52, 53] are not generally applicable.

2'6.1. Electron yield detection. For photon energies in the soft-X-ray region and above the photon mean free path is large compared to the escape depth of an electron in a solid. When considering a clean surface, therefore, the surface sensitivity of an electron yield detection will depend on the kinetic energy of the collected electrons. Maximum surface sensitivity is obtained for minimum escape depth which occurs for electron kinetic energies of around 50 eV (cf. fig. 3). But even then only part of the signal originates from the outermost surface layer. In general the effective escape depth is about 50 Å [15], so that electron yield EXAFS on a clean surface then will give bulk information. When atoms are adsorbed on a surface which differ from the substrate atoms, the SEXAFS of the adsorbate atoms can be distinguished from the substrate signal if the substrate atoms do not have an absorption edge in the measured energy range. However, the signal from the adsorbate atoms rides on a large background from the substrate. Both are dependent on the kinetic energy (or the kinetic-energy range) of the detected electrons which determines the signal-to-background ratio and the total signal rate (see below). Structures in the yield spectra due to the monochromator transmission function can be eliminated by dividing the yield spectra of the adsorbate-covered surface and the clean surface. An alternative and dynamic method of normalization uses the signal from a reference monitor [15]. Mainly three different electron yield detection modes are used: Auger yield (AY), partial-yield (PY) and total-yield (TY) detection. In the AY mode Auger electrons corresponding to the core excitation in question are measured with an electron energy analyser using an appropriate energy window. This technique has the largest signal-to-background ratio of all electron yield methods but the smallest signal rates. The AY mode, however, cannot be applied when direct photoemission peaks (*e.g.*, from the substrate atoms) sweep through the Auger peak at certain photon energies, which is normally the case for measurements in the soft-X-ray region. There the PY mode is the technique of choice. In the PY method only electrons with kinetic energies above an adjustable threshold energy are collected with a retarding electron detector (high-pass filter) which in the simplest case can be a channeltron (or channel plate) electron multiplier equipped with a retarding grid. By properly choosing the threshold energy an interference with photoemission peaks can be avoided. As the signal is comprised of elastic and scattered Auger electrons, it is also proportional to the absorption coefficient. The PY mode of detection has a smaller signal-to-background ratio than the AY technique but a larger signal rate. In addition to elastic and scattered Auger electrons the TY also contains elastic and scattered photoelectrons. As the kinetic energy of the photoelectrons changes with photon energy, the TY is no longer proportional to the absorption coefficient [54-56]. However, it still exhibits the same structures and can, there-fore, be used as SEXAFS detector [15]. TY measurements with a channeltron (or channel plate) are generally applicable. They offer the highest signal rates of all electron yield modes but the smallest signal-to-background ratios. In another electron yield mode the low-energy inelastic part of the total yield, the so-called secondary yield (SY), is measured [57]. SY and TY spectra are identical. Because of the signal rate the latter, however, is preferred.

2`6.2. Fluorescence yield detection. With electron yield detection SEXAFS is sensitive to roughly 1/10 monolayers of an adsorbate and restricted to UHV environments. Other shortcomings of the electron yield modes show up in the soft-X-ray region: There are normalization problems because of the pro-nounced structure due to carbon contamination in the monochromator trans-mission function and there are short k-ranges (hence large experimental errors for the determined bond lengths) when studying low-Z molecules. All of these problems can be solved by fluorescence yield detection although for low-Z atoms the nonradiative Auger channel is favoured over the radiative fluorescence channel [58]. This detection mode has already been applied in grazing-incidence geometry for higher photon energies [59] as well as in the soft-X-ray region using a gas-filled proportional counter as detector [14]. In the future windowless Si(Li) detectors will probably be used in the soft-X-ray region which offer a resolution of about 100 eV at photon energies of roughly 700 eV («Quantum» detector of Kevex). Due to the small detector area of about 10 mm^2 and due to a detection efficiency of less than 10% in this photon energy range, higher photon fluxes than presently available are needed for successful experiments.

2.7. Applications.

2.7.1. Ni$\{100\}$$(\sqrt{2} \times \sqrt{2})R\,45°$-S. The Ni$\{100\}$-S system is one of the most thoroughly investigated chemisorption systems in surface science and thus well suited as a test of any new structural probe. It has been characterized by LEED[60-64], low-energy ion scattering (LEIS)[65, 66], photoemission[67], EELS[68], photoelectron diffraction[69-73] and SEXAFS[74, 75]. Cluster slab calculations have also appeared in the literature[76-78].

Sulphur chemisorbs on a clean Ni$\{100\}$ surface in two ordered overlayers, (2×2) and $(\sqrt{2} \times \sqrt{2})R\,45°$ at coverages of 0.25 and 0.5, respectively. Early LEED studies on the $(\sqrt{2} \times \sqrt{2})R\,45°$ phase by DEMUTH et al.[60] showed that the fourfold hollow site is occupied with a distance $Z = (1.30 \pm 0.1)$ Å above the surface corresponding to a S-Ni nearest-neighbour distance of $R = (2.19 \pm 0.06)$ Å. Different Z values obtained in later LEED studies promoted further investigations of this system, which eventually confirmed the value given by DEMUTH et al.[61-63]. A later LEED study by VAN HOVE and TONG[64] showed that the adsorption site in the (2×2) overlayer is exactly the same as in the $(\sqrt{2} \times \sqrt{2})R\,45°$ phase. The LEED results have been quantitatively confirmed in several different photoelectron diffraction studies[69-73] and by a recent LEIS experiment of Fauster et al.[66]. GVB cluster calculations gave S-Ni bond lengths of 2.21 Å for a SNi$_4$ cluster[76] and of 2.15 Å for a SNi$_{20}$ cluster[77]. All these values are also in good agreement with the bond length range of $(2.10 \div 2.33)$ Å determined with X-ray diffraction measurements on Ni-chelate complexes[79]. For comparison, we note that the nearest-neighbour S-Ni bond length in bulk NiS is 2.39 Å [80].

Sulphur K-edge SEXAFS experiments on the Ni$\{100\}$$(\sqrt{2} \times \sqrt{2})R\,45°$-S structure were performed by BRENNAN et al.[74] using Auger yield detection. Spectra were taken for X-ray incidence angles of $\theta = 10°$, $\theta = 45°$ and $\theta = 90°$. For the first time in this field bulklike quality SEXAFS data were obtained in the soft-X-ray region. This is shown in fig. 7 (left), where the 45° spectrum is compared with the total-yield EXAFS spectrum of bulk NiS. The Fourier transforms of these spectra are dominated by a single peak, as shown in fig. 7 (right). Using eq. (2) the backtransforms of the dominant peaks with the dashed lines in fig. 7 (right) as Fourier windows yield the phase shift ψ_i for the bulk sample (with known nearest-neighbour S-Ni distance) and the nearest-neighbour S-Ni distance for the Ni$\{100\}$$(\sqrt{2} \times \sqrt{2})R\,45°$-S system (with the bulk phase shift ψ_i as input). The results for the three SEXAFS spectra were identical to within ± 0.01 Å and indicated a nearest-neighbour S-Ni distance of $R = (2.23 \pm 0.02)$ Å in excellent agreement with the values derived from other techniques. The adsorption site of the sulphur atom was determined by three independent methods: from the second-nearest-neighbour bond length, from relative surface amplitudes and from the absolute surface and bulk amplitudes. A second distance at (4.15 ± 0.10) Å [75] which is clearly observed in the 90° spectrum and which is also visible in the Fourier trasform of the 45° spectrum

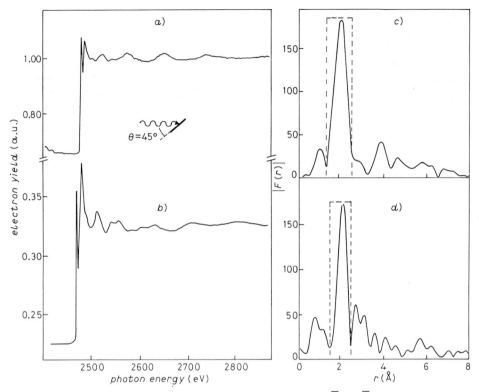

Fig. 7. – Sulphur K-edge SEXAFS spectra for a) Ni {100} $(\sqrt{2} \times \sqrt{2}) R\,45°$-S and b) NiS (left) and their respective Fourier transforms (right) (c) and d)). The spectra were taken in the Auger electron yield and total electron yield modes, respectively. The dashed lines (right) indicate the Fourier windows used for backtransformation (see text). After[74].

(see fig. 7) compares well with the calculated second-nearest-neighbour distance of 4.17 Å for the fourfold hollow site and is in conflict with values of 3.34 Å for a twofold bridge and 3.34 Å for an atop site. The fourfold hollow is also unambiguously determined from SEXAFS amplitude comparisons. According to eq. (8) the SEXAFS amplitude ratio directly yields the corresponding ratio of the effective coordination numbers. This is shown in table I, where the amplitude ratios of the spectra taken at 10°, 45° and 90° X-ray incidence are compared with calculated coordination number ratios for different adsorption sites. The data clearly rule out the twofold bridge and the atop sites and show that the accuracy of relative amplitude determination is better than ± 10%. Finally, the adsorption site was checked by comparing the EXAFS amplitude for bulk NiS with the SEXAFS amplitudes for Ni{100} $(\sqrt{2} \times \sqrt{2}) R\,45°$-S. This is also shown in table I. Due to different exponential damping terms in eq. (5) the accuracy of absolute amplitude comparisons is generally not better than ± 25% but may be sufficient to distinguish between different adsorption sites as in the present case. The data also clearly favour the fourfold hollow adsorption site.

TABLE I. – *Experimental vs. calculated coordination numbers and ratios for* Ni $\{100\}$ $(\sqrt{2} \times \sqrt{2})\,R\,45°$-S.

Incidence angle (degrees)	Experiment	Fourfold hollow	Twofold bridge	Onefold on-top
10/90	1.16 ± 0.10	1.20	4.31	∞
10/45	1.15 ± 0.10	1.09	1.59	1.94
10	4.42 ± 1.04	4.49	4.03	2.91
45	3.77 ± 0.79	4.13	2.53	1.50
90	3.94 ± 0.75	3.75	0.94	0

2˙7.2. Cu $\{110\}$ (2×1)-O. Room temperature adsorption of oxygen on a clean Cu$\{110\}$ surface gives rise consecutively to (2×1) and $c(6 \times 2)$ overlayer structures for coverages of 0.5 and approximately 1.0, respectively. The (2×1) structure has been extensively studied by LEED[81-86], work function measurements[87, 88], AES[89], ellipsometry[88, 89], LEIS[90-93], helium diffraction (HD)[94, 95], EELS[96], UPS[97, 98], high-energy ion scattering (HEIS)[99], SEXAFS[6, 100] and grazing-incidence X-ray scattering (GIXS)[101]. The particular interest in this structure derives from the fact that the adlayer induces a reconstruction of the outermost Cu layer.

It is now commonly accepted that the oxygen atoms reside in the long-bridge sites along the [001] Cu rows. The LEIS results of de Wit *et al.* [90, 91] as well as the HD data of Lapujoulade *et al.* [94] suggested oxygen positions of $Z = (0.6 \pm 0.1)$ Å and $Z = (0.7 \pm 0.1)$ Å, respectively, below the surface corresponding to O-Cu bond lengths of $R = (1.91 \pm 0.03)$ Å and $R = (1.94 \pm 0.04)$ Å, respectively. In the only dynamical LEED study of this system performed by NOONAN and DAVIS[86] unsatisfactory agreement was found between experimental I-V curves and those calculated for models where the oxygen was placed in various positions on a truncated bulk Cu $\{110\}$ substrate. These results pointed towards an oxygen-induced reconstruction of the surface which was subsequently proven with other experimental techniques. The LEIS studies of Bronckers and de Wit[92] and Niehus and Comsa[93] as well as the HD measurements of Lapujoulade *et al.* [95] favoured a missing-row (MR) model for the reconstruction with every second [001] Cu row missing. On the other hand, the HEIS data of Feidenhans'l and Stensgaard[99] and the GIXS studies of Liang *et al.* [101] indicated a buckled-row (BR) model with every second $\langle 100 \rangle$ row displaced vertically. Figure 8 shows schematic drawings of the BR and MR models together with that of the sawtooth (ST) model (rotated by 90° with respect to the original model of Bonzel and Ferrer) which has also been discussed in this context[102].

DÖBLER *et al.* [100] first performed oxygen K-edge SEXFAS measurements on the Cu{110}(2 × 1)-O system using the partial electron yield mode of detection. At the storage ring BESSY on the SX-700 monochromator spectra with bulklike quality were obtained for the first time for photon energies below 1 keV in this work. By orienting the E vector parallel to the [001] and [1$\bar{1}$0] directions, respectively, first- and second-nearest O-Cu distances could be

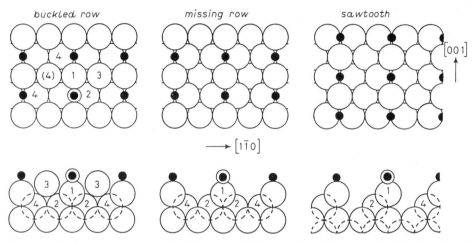

Fig. 8. – Models for the reconstruction of the Cu{110}(2 × 1)-O surface. In the top view of the buckled-row model and in all side views the nearest-neighbour numbering with respect to the absorbing O atom (circled dots) is shown. The label (4) corresponds to the Cu atom in the third layer. After [6].

determined independently: $R_1 = (1.84 \pm 0.02)$ Å and $R_2 = (2.00 \pm 0.05)$ Å. For comparison, the nearest-neighbour O-Cu bond length in Cu$_2$O is 1.85 Å [103]. The SEXAFS data showed that oxygen adsorption occurs in the long-bridge sites at $Z = (0.3 \pm 0.1)$ Å *above* the surface assuming an unrelaxed or only slightly relaxed surface [104]. This is in clear contradiction to the LEIS results [90, 91] which favour an adsorption site *below* the surface. The parameters Z and d_{12} (separation between first and second Cu layers) cannot be determined independently from R_1 and R_2 (at least not with the required accuracy), because R_1 does not depend critically enough on Z. However, an oxygen site in the long bridges 0.6 Å below the surface [90, 91] would require a relaxation of $\Delta d_{12} = +0.9$ Å (expansion) or $\Delta d_{12}/d_{12} = 70\%$ to be consistent with the measured bond lengths. Oxygen adsorption below the surface Cu atoms is also unlikely because of the direction of the work function change [87, 88]. Furthermore, the HD data [95] can equally well be interpreted in terms of oxygen adsorption above the surface [100]. Looking at the conflicting data from LEIS and SEXAFS one should bear in mind that the LEIS experiments [90, 91] were carried out under dynamical equilibrium conditions. Whereas the adsorption site issue now seems to be settled, the question as to the type of reconstruction in the Cu{110}(2 × 1)-O system is still open.

Because of the limited signal-to-noise and signal-to-background ratios no information on surface reconstruction could be obtained in the early SEXAFS study. After improving the experimental conditions the Cu{110}(2 × 1)-O system was reinvestigated by BADER *et al.* [6]. SEXAFS spectra taken at 100 K for X-ray incidence angles of $\theta = 90°$ in the [001] and [1$\bar{1}$0] azimuths are shown in fig. 9 (left), together with their respective Fourier transforms (right). Using the phase shift from bulk Cu_2O [15] as reference and applying the same procedure as in subsect. 2·7.1, the analysis showed that the Fourier peaks correspond to the first $(R_1 = (1.82 \pm 0.02)$ Å$)$, second $(R_2 = (1.99 \pm 0.02)$ Å$)$ and fourth $(R_4 = (4.15 \pm 0.03)$ Å$)$ nearest-neighbour O-Cu distances, respectively, and to the O-O distance of 3.6 Å (multiple-scattering peak) [47].

The measured bond lengths agree with the BR model [99] only if the oxygen atoms sit above the unshifted [001] Cu rows. Third-nearest Cu neighbours should then, however, be observed at a distance of about 3.2 Å with an amplitude larger than those of the second and fourth nearest neighbours in contrast to the

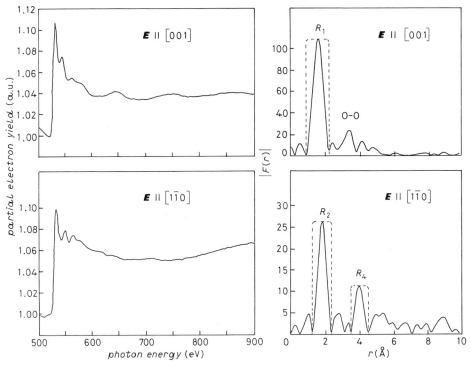

Fig. 9. – Oxygen *K*-edge SEXAFS spectra for Cu{110}(2 × 1)-O recorded in the partial electron yield mode at 100 K with $\boldsymbol{E}\|[001]$ and $\boldsymbol{E}\|[1\bar{1}0]$ (left) and their respective Fourier transforms (right). The numbering of the nearest-neighbour distances is the same as that used in fig. 8. After [6].

experimental result. This indicates that the alternate [001] rows of Cu atoms not containing oxygen atoms are absent ruling out the BR model. The MR and ST models only differ in the number of second- and fourth-nearest neighbours. A SEXAFS amplitude comparison can, therefore, help us to decide between these two models. When comparing amplitudes of different Fourier peaks the exponential damping terms in eq. (5) have to be taken into account. From the k^2-dependence of the measured amplitude ratios A_2/A_1 and A_4/A_2 the contribution of the Debye-Waller terms can be estimated. This leads to the following conclusions: 1) Assuming an isotropic mean free path, λ, the experimental data cannot be satisfactorily fitted to either the MR and ST models, and 2) a fit to the MR and ST models yields «bulk» mean free paths of 5.2 Å and 3.7 Å, respectively, and a «surface» mean free path which is either larger (MR) or smaller (ST) than the corresponding bulk value. As bulk mean free paths of about 5 Å have been derived from bulk EXAFS data [44] and a smaller mean free path on the surface is physically unreasonable, the data tend to rule out the ST model for the reconstruction in agreement with impact collision alkali-ion scattering results of Niehus and Comsa [93].

We conclude this subsection by noting that the SEXAFS technique has contributed significantly to the determination of the structure of the Cu{110}(2 × 1)-O system. As yet there is no equivalent LEED determination for comparison. The oxygen atoms are clearly adsorbed in long-bridge sites 0.3 Å above the surface and a missing-row reconstruction is favoured.

2˙7.3. Ag {111} $(\sqrt{3} \times \sqrt{3}) R 30°$-Cl.   Silver is a unique catalyst for the oxidation of ethylene to ethylene oxide, thus providing a convenient route to ethylene glycol. The selectivity for this particular reaction is increased by adsorbed chlorine. Despite this considerable technical interest in chlorine adsorption on low-index faces of silver, the Ag{111}-Cl system has not been well characterized by surface science techniques to date. Only LEED, AES, TDS and work function change measurements have been performed [105-109]. The SEXAFS experiments [11] were the first to provide detailed structural information. The first adsorbate-induced structure observed on exposing the surface to chlorine is a $(\sqrt{3} \times \sqrt{3}) R 30°$ structure which is somewhat weak and diffuse. BOWKER and WAUGH [109] assigned to it a coverage of about 0.7. At higher chlorine coverages further LEED patterns are observed [105-109]. Prior to the SEXAFS measurements the $(\sqrt{3} \times \sqrt{3}) R 30°$ structure had been explained by a simple overlayer model [105, 107] or by a «reconstruction» model in which chlorine and silver atoms are present in the first layer similar to the {111} face of AgCl [106].

LAMBLE and KING [11] took chlorine K-edge SEXAFS data at 100 K in the total electron yield mode of detection for the weakly ordered phase of chlorine on Ag {111} at $\theta = 0.7$ and for a disordered phase at $\theta = 0.4$. The Fourier transform of the normal-incidence spectrum for 0.7 coverage is shown in fig. 10 (right, solid

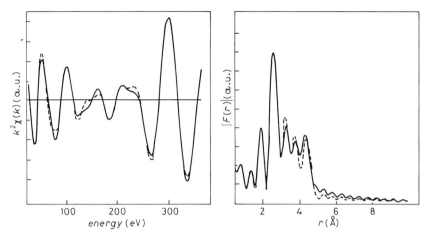

Fig. 10. – Fourier transform of the experimental SEXAFS spectrum (right, solid line) of the Ag {111} ($\sqrt{3} \times \sqrt{3}$) R 30°-Cl system and its backtransform with real-space components above 5.2 Å having been filtered out (left, solid line). The broken lines represent the best fit to the experimental data. After [11].

line). Besides the Ramsauer-Townsend resonance below 2 Å four peaks appear which correspond to different Cl-Ag and Cl-Cl distances. After filtering out real-space components above 5.2 Å by an appropriate Fourier window the backtransform of $|F(r)|$ yields the «experimental» $\chi(k)$ shown in fig. 10 (left, solid line). This $\chi(k)$ is then modelled with a curve-fitting procedure using eqs. (2)-(6) for various configurations of neighbouring substrate and adsorbate atoms. Calculated phase shifts, which are empirically refined by using data from standard compounds, also serve as input. It turned out that for AgCl the EXAFS data had to be taken at 100 K to avoid anharmonicity effects which are present at 300 K and result in Cl-Ag bond lengths being in error by as much as 0.06 Å. The variation of the intershell spacings as well as variation of the coordination number and Debye-Waller factor gives the excellent least-squares fit shown in fig. 10 (left, broken line) with the bond lengths and shell spacings given in table II (fourth column). The latter are in good agreement with those (third column) of the «vacancy» or «honeycomb» structure model depicted in fig. 11, in which the adatoms are adsorbed in threefold hollow sites. The nearest-neighbour Cl-Ag distance is $R = (2.70 \pm 0.01)$ Å with the other spacings being less accurate. No great significance, however, can be attached to variables such as the coordination number and the Debye-Waller factor (see subsect. 2˙1.4 and 2˙3). No other ($\sqrt{3} \times \sqrt{3}$) R 30° structure provided a reasonable fit to the data; a quasi-epitaxial AgCl monolayer was clearly ruled out. The measured Cl-Ag bond length of 2.70 Å is shorter than that (2.77 Å) in bulk AgCl [110]. A similar multishell SEXAFS analysis was also performed for a disordered layer at $\theta = 0.4$ coverage. The nearest-neighbour Cl-Ag distance turned out to be the same as for

TABLE II. – *A comparison of experimental and theoretical real-space components in the* $(\sqrt{3} \times \sqrt{3})R\,30°$ *structure observed when 0.7 monolayers of chlorine are adsorbed on* Ag {111}.

	Shell	N_i	$R_i(\text{Å})$	
			model	experiment
1	Cl-Ag	3	2.70	2.70
2	Cl-Cl	3	2.89	2.93
3	Cl-Ag	3	3.94	3.71
4	Cl-Cl	6	5.00	4.83

the weakly ordered layer at $\theta = 0.7$ coverage. The first Cl-Cl distance located at 2.9 Å for the latter was, however, completely absent and the remaining shells were found at the same spacings for the two coverages. This result suggests that the disordered structure at $\theta = 0.4$ is again an overlayer structure with chlorine in threefold hollow sites but with the nearest-neighbour adatoms in the structure of fig. 11 removed.

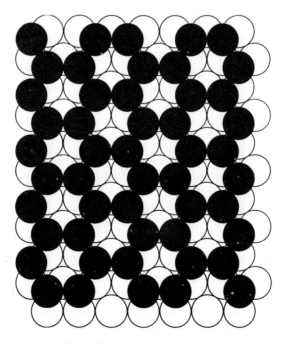

Fig. 11. – Plan view of the $(\sqrt{3} \times \sqrt{3})R\,30°$ «honeycomb» model deduced by SEXAFS for 0.7 monolayers of chlorine adsorbed on Ag{111}. After [11].

3. – NEXAFS.

The near-edge X-ray absorption fine structure (NEXAFS or XANES) is due to dipole-allowed transitions of core electrons into unoccupied electron states both below and above the vacuum level and yields information on both the electronic and the geometric structure [111]. Very close to the edge the fine structure is usually interpreted in terms of the local density of states (cf. eq. (1)), especially when this is large, as in transition metals with unfilled d bands [112] and for adsorbed molecules (see subsect. 3'2) with unfilled antibonding π and σ orbitals. There have been band structure and multiple-scattering approaches to calculate the whole fine structure up to roughly 50 eV above the edge for solids [113, 114] and molecules in the gas phase [115]. As has been shown recently [116], the different approaches are mutually equivalent.

In a scattering picture it is obvious that at low electron kinetic energies the scattering amplitude becomes more isotropic (cf. fig. 4) and the mean free path increases (cf. fig. 3), so that multiple scattering has to be taken into account. As a consequence NEXAFS must be sensitive to the atomic arrangement around the absorbing atom and should in principle be capable to yield information on the absorption site of adsorbate atoms on surfaces. This, indeed, is the case for atomic adsorbates (subsect. 3'1). Multiple-scattering calculations are in close agreement with experimental spectra and adsorption sites can be determined. Attempts to calculate the NEXAFS of molecular adsorbates have been less successful to date. The NEXAFS of molecules and molecular fragments adsorbed on surfaces exhibits characteristic resonances similar to those observed in the corresponding gas phase species [115]. However, by just measuring the intensity of these resonances the molecular orientation with respect to the surface can be obtained from the polarization dependence of the dipole matrix element (subsect. 3'2).

3'1. NEXAFS *of atomic adsorbates.*

3'1.1. Multiple-scattering approach. The multiple-scattering approach has also been formulated in a real-space cluster model [114]. This approach is nothing else than the formulation of the EXAFS problem but with infinitely many scattering events included. Muffin-tin potentials are used whose scattering properties are described by phase shifts which together with the real-space structure serve as input. For the absorbing atom an excited-atom potential can be used and the lifetimes of the core hole and the excited electron are taken into account by a negative imaginary part of a complex (absorptive) potential. The cluster is divided into shells of atoms around a central (absorbing) atom. The multiple-scattering equations are solved within each shell. Finally the multiple scattering between the shells is calculated.

3'1.2. Applications: $Ni\{100\}\,(\sqrt{2} \times \sqrt{2})\,R\,45°$-O, $Ni\{110\}\,(2 \times 1)$-O. The

Ni$\{100\}(\sqrt{2} \times \sqrt{2}) R\,45°$-O system was the first surface system for which multiple-scattering calculations in the real-space cluster formulation were performed[22]. These calculations included about 30 atoms to a distance of > 5.0 Å from the central atom. The O-Ni nearest-neighbour distance was fixed to 1.98 Å [117-119] in all calculations for different adsorption sites. Best agreement between experimental and calculated spectra was obtained for a fourfold hollow adsorption site in accord with LEED[118] and SEXAFS[119] data. The agreement, however, is not perfect. Differences between calculated and measured peak positions might result from improper phase shifts[120]. The agreement is better for normal-incidence data. This seems to be a general result[121] being in part due to the absence of a surface barrier in the

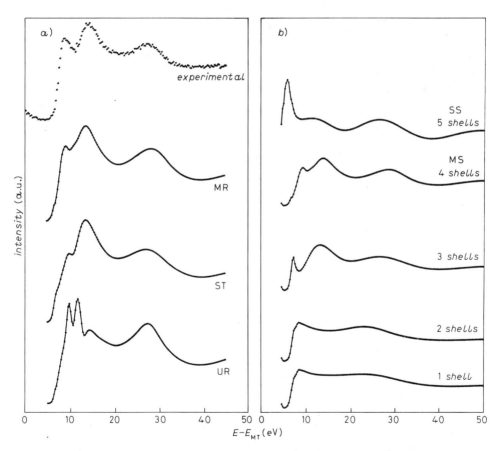

Fig. 12. – a) Comparison between experimental and calculated NEXAFS spectra at normal incidence in the azimuth $\langle 100 \rangle$ for Ni$\{110\}(2 \times 1)$-O. Calculations are presented for the unreconstructed surface (UR) and two reconstruction models, the sawtooth (ST) and missing-row (MR) model. b) Multiple-scattering (MS) calculations for the missing-row reconstruction displaying the effect of including more shells. The topmost curve represents a single-scattering (SS) calculation using 5 shells.

calculations. The experiment demonstrated that adsorption sites can indeed be determined with NEXAFS and stimulated further investigations.

Quite recently the Ni{110}(2 × 1)-O system has been studied[24] for which ion scattering[122] and SEXAFS[5] experiments suggested a missing-row (MR) and a sawtooth (ST) reconstruction, respectively. The calculations were performed for the MR and ST reconstruction models (cf. fig. 8) as well as for an unreconstructed surface (UR). As the result of the multiple-scattering calculations could be influenced by the proper choice of the phase shifts, the latter were taken from a calculation for the «model compound» Ni{100} $(\sqrt{2} \times \sqrt{2}) R 45°$-O. Oxygen was assumed to adsorb in the long-bridge site 0.25 Å above the surface [5] and 5 shells of backscatterers were included in the calculations. A characteristic comparison between an experimental and a calculated spectrum is shown in fig. 12a) for normal X-ray incidence in the azimuth $\langle 100 \rangle$. Obviously the unreconstructed model does not reproduce the experimental spectrum. However, a distinction between the MR and ST is not possible also because nonstructural parameters (phase shifts) change peak shapes and energies[120]. Figure 12b) shows a set of calculations for the MR model in which the number of backscattering shells is successively increased together with a single-scattering approximation. It is interesting to note that the most pronounced change occurs with the inclusion of the third shell which consists of oxygen atoms only and that the calculation with four shells is almost identical to the five-shell result of fig. 12a). The inclusion of the next-nearest oxygen atoms (the fifth shell), therefore, has no big effect. On the other hand, single- and multiple-scattering calculations agree qualitatively from about 10 eV above the absorption threshold onwards, whereas the region closer to the edge cannot be explained by single scattering only. In summary, the above results demonstrate that multiple-scattering NEXAFS calculations for atomic adsorbates are in close agreement with experiment. Compared with SEXAFS, the determination of the adsorption site by use of NEXAFS will in general, however, be not as reliable and accurate.

3˙2. NEXAFS *of molecular adsorbates.*

3˙2.1. π and σ resonances. In a molecular-orbital picture[123, 124] the resonances observed in the near-edge region originate from excitations of core electrons into unoccupied antibonding molecular orbitals of π and σ symmetry. They are, therefore, generally referred to as π and σ resonances which may appear below and above the ionization threshold (cf. fig. 13). In the former case the resonances are so-called bound-state resonances which are generally sharp with linewidths below 1 eV. Resonances showing up above the ionization threshold are so-called continuum resonances and generally are very broad. They become broader the higher they lie in the continuum. In an alternative and equivalent description[115] these features correspond to quasi-bound states in

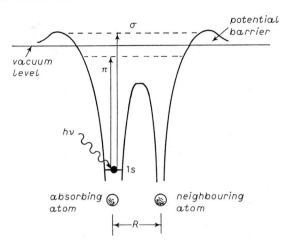

Fig. 13. – π and σ (shape) resonances in a diatomic molecule with an intramolecular bond length R (see text).

the continuum in which certain angular-momentum components of the photoelectron wave are resonantly trapped by a potential barrier due to the interaction of the Coulomb, screening and centrifugal forces. As energy and width of the resonances depend on the shape of the potential, they are frequently called «shape» resonances.

3'2.2. Polarization dependence and symmetry. The orientation determination via the polarization dependence follows directly from the matrix element for an electronic transition in a molecule [125]. According to eq. (1) the absorption intensity is given by

(11)
$$ I \propto |\boldsymbol{E} \cdot \langle f | \boldsymbol{M} | i \rangle|^2 , $$

where $\boldsymbol{M}$ is the electric-dipole vector associated with the transition. For $I \neq 0$ the product of the irreducible representations corresponding to $|f\rangle$, $|i\rangle$ and $\boldsymbol{M}$ must be totally symmetric, or at least contain the totally symmetric representation. For the π resonance of an isolated oriented diatomic (point group $C_{\infty v}$ with $\boldsymbol{z}\|$ molecular axis) at a K-edge

(12)
$$ \langle \pi | M_z | \sigma \rangle = M_z(\pi) = 0 , \qquad \langle \pi | M_{x,y} | \sigma \rangle = M_{x,y}(\pi) \neq 0 . $$

The transition is thus polarized in the (x, y)-plane and the angular dependence of its intensity is given by

(13)
$$ I_\pi \propto |\boldsymbol{E} \cdot \boldsymbol{M}_{x,y}(\pi)|^2 \propto \sin^2 \beta , $$

where β is the angle between the electric vector and the molecular axis. Similarly the transition into a σ final state is polarized along the molecular axis and the angular dependence is given by

$$(14) \qquad\qquad I_\sigma \propto |\boldsymbol{E} \cdot \boldsymbol{M}_z(\sigma)|^2 \propto \cos^2\beta \,.$$

If the molecule is tilted or lies flat, then the point group symmetry is lowered, the degeneracy of the π orbitals is lifted and the simple equations (13) and (14) are no longer valid. In the majority of cases studied so far, however, it appears possible to treat the adsorbed molecule simply as an oriented species and to neglect the interaction with the substrate even for relatively strongly bound adsorbates (see paragraph 3˙2.4.1). It should be pointed out that the polarization dependence of the NEXAFS is always determined by the effective point group of the adsorbed species and by the symmetry of the orbitals involved [20] and that there is no way of deciding *a priori* whether the point group is that of the free molecule or that of the surface species (molecule + substrate). There is no one-to-one correspondence between NEXAFS resonances and the antibonding molecular orbitals because some of the transitions may be symmetry forbidden (see paragraph 3˙4.2.1). Also, the simple approach in which polyatomic molecules are described in terms of diatomic- and ring-like «building blocks» [17] can only be an approximation for those cases where the wave function for an orbital has significant amplitude in certain parts of the molecule only. Generally, however, the orbitals extend over the whole molecule.

Formulae for the polarization dependence of diatomics and rings as a function of the tilt angle (with respect to the surface) and the polarization of the incident radiation as well as of the symmetry of the substrate are given in ref. [17]. They are simply derived from eqs. (13) and (14) and are applicable, however, only in cases where no symmetry lowering on adsorption occurs.

The integrated intensities (peak areas) of the resonances are usually normalized to some kind of edge jump which in NEXAFS is the intensity difference between energies above the σ resonance and before the onset of absorption [126]. As SEXAFS wiggles might interfere in the energy range around the σ resonance energy, there is some degree of arbitrariness with this determination. The same holds for the form of the background under the resonances. When not resolved as separate peaks, scattering features (SEXAFS), Rydberg transitions and multielectron excitations might be hidden in the background. The accuracy of the orientation determination with NEXAFS is, therefore, hardly better than $\pm 10°$.

3˙2.3. σ resonances and bond lengths. For a variety of molecules studied with inner-shell electron energy loss spectroscopy in the gas phase a correlation between σ resonance positions and individual intramolecular bond lengths has been suggested [127]. This correlation was later on also transferred to chemisorbed molecules [18] and generated much interest and doubts [19]. A quantitative relationship would be a very powerful analytical tool. The doubts about this relationship concentrated on the fact that the shape resonance energy depends on the rather complex potential, *i.e.* on the shape, height and width of the potential barrier, which also reflects the electronic configuration, molecular geometry and chemical environment. It is concluded, therefore, in ref. [19] that the resonance energy must also depend on other molecular parameters besides bond length. As in subsect. 3˙2.2 the association of a shape resonance with a particular bond in a complex molecule is regarded as an unproven simplification. The authors of ref. [19] concede, however, that calculations have shown monotonic changes in valence shell shape resonance energies by varying the bond length in a given diatomic molecule and that a resonance energy/bond length relationship exists for series of isostructural molecules. Hopefully, this dispute will stimulate efforts to a better understanding of shape resonances.

3˙2.4. Applications.

3˙2.4.1. Ag {110} (1 × 2)-CO$_3$.   Surface carbonate belongs to a class of molecular adsorbates which includes the nitrate, formate, methoxy and hydroxy species, all having to some extent anionic character. FORCE and BELL [128] appear to have identified it first in infrared measurements of a silver catalyst during the interaction of ethylene with preadsorbed oxygen. Bands due to both mono- and bidentate species were observed. MADIX and co-workers [129, 130] later showed that it can be formed on an Ag {110} surface by the reaction of pre-adsorbed oxygen with carbon dioxide, giving rise to sharp half-order features in LEED indicative of a (1 × 2) surface structure. Their thermal desorption spectra from isotopically labelled species showed that all three oxygen atoms have equal probability of remaining on the surface after decomposition of the carbonate. This could mean that all oxygen atoms are equivalent, which is difficult to conceive because of the twofold symmetry of the surface. Alternatively, the effect could be due to a facile interconversion of mono- and bidentate species. Photoemission experiments [131, 132] indicated that the electronic structure of the surface carbonate strongly resembles that of the anion in the bulk solid. Using an assignment of levels based on this observation PRINCE and PAO-LUCCI [133] applied photoemission selection rules in an attempt to derive structural information. The data indicated a tilted species. Similarly, the EELS studies of Stuve *et al.* [134] provided evidence for a monodentate tilted structure, but then the absence of a loss peak corresponding to the O-C-O scissor mode could not be easily explained.

BADER *et al.* [21] performed NEXAFS measurements on the Ag {110}(1 × 2)-CO$_3$ system. The data were taken above the carbon and oxygen K-edges in the partial electron yield mode. The spectra shown are, in each case, the ratio covered-to-clean. The X-ray incidence angle on the sample could be varied from

70° ($\theta = 20°$) to normal incidence ($\theta = 90°$; E vector parallel to the surface). NEXAFS spectra of the carbonate species adsorbed on Ag{110} are shown in fig. 14 for $\theta = 20°$ and 90° at both the O K- and C K-edges. There was no azimuthal dependence, *i.e.* the spectra were identical in both crystal azimuths $\langle 110 \rangle$ and $\langle 100 \rangle$. The XPS binding energy [131] is referred to the Fermi level.

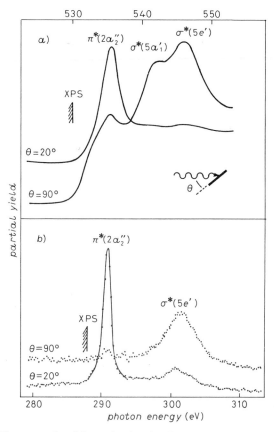

Fig. 14. – NEXAFS spectra for CO_3 on Ag {110} taken at a) the oxygen K-edge and b) the carbon K-edge with the polarization vector of the incident radiation aligned in the $\langle 110 \rangle$ azimuth at polar angles of $\theta = 20°$ and $\theta = 90°$.

The sharp feature at the edge is a core-to-bound resonance. The two broader features at higher photon energies in the oxygen spectrum are clearly continuum resonances; only one is apparent in the carbon spectrum. Since the highest occupied molecular orbital (m.o.) of the surface carbonate species is the $1a_2'$ level [131-133], as in the case of bulk carbonate [135], the first unoccupied m.o. is expected to be the $2a_2''$. This has C $2p_z$ and O $2p_z$ character and is of π type. At higher energies only two further m.o.'s are expected, both of σ type: $5a_1'$ which is strongly C $2s$ and O $2s$ in character and the doubly degenerate $5e'$ which has strong $C\,2p_{x,y}$ and $O\,2p_{x,y}$ character. We, therefore, assign tentatively the observed resonances to transitions into these three levels.

In addition to indicating that azimuthal effects are absent, fig. 14 also shows that the transitions are strongly polarized. The π resonance is excited preferentially by the component of the **E** vector perpendicular to the surface and the σ resonances by the parallel component. Figure 15 shows the results of a

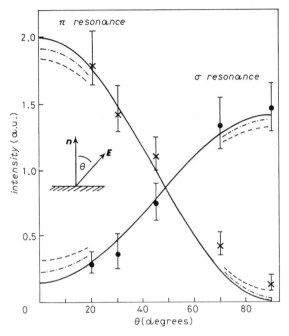

Fig. 15. – π and σ resonance intensities for CO_3 on Ag{110} measured at the oxygen K-edge as a function of the polar angle θ. The calculated curves are given for three different tilt angles α between the surface normal and the C_3 axis of the carbonate: —— $\alpha = 0$, $-\cdot-\cdot-$ $\alpha = 10°$, $---$ $\alpha = 15°$.

quantitative analysis of the data at the O edge. The integrated intensities of the resonances at each angle have been normalized to the height of the edge jump. Since there is some degree of arbitrariness associated with this determination as well as with the form of the resulting background under the resonances, the error bars in this figure indicate the possible extremal values. The experimental data are derived from three separate sets of spectra for the five angles shown. The solid curves represent the expected polarization dependence of the π and σ resonances for an oriented, *isolated* molecule, *i.e.* one in which no interaction takes place with the surface. The unoccupied π orbital is assumed to be oriented with its nodal plane parallel to the surface and the unoccupied σ orbitals randomly oriented in a plane also parallel to the surface. The agrement with experiment is reasonable. However, the lying-down configuration can only be concluded from the experimental results when no symmetry lowering on adsorption occurs. Otherwise the strong polarization dependence of the transitions is no longer necessarily expected. The appropriate point group for a free carbonate anion is D_{3h}. Adsorption in a lying-down configuration would reduce the symmetry to C_{3v}. That the effective point group remains D_{3h} is shown by the spectra at the carbon edge. Only in this case the transition $2a_1' \rightarrow 5a_1'$ is forbidden. At the oxygen edge both σ resonances are allowed. The high effective symmetry is surprising, since the CO_3 species is relatively strongly bound (decomposition temperature 485 K).

The polarization dependence of the NEXAFS data thus tells us that the carbonate species takes up a parallel geometry, $i.e.$ it is adsorbed with its σ_h symmetry plane parallel to the surface. The uncertainty in this determination is $\pm 15°$ and follows from the quality of the fit in fig. 15 as well as from the assumption of a smooth background under the peaks devoid of so-called substrate resonances. A similar polarization dependence was observed at the carbon K-edge, as can be seen in fig. 14b), but, because the height of the edge jump cannot be uniquely ascertained and is, therefore, not available for normalization purposes, it was analysed via the σ/π ratio. The σ/π ratios at the O and C edges are very similar. The fact that the carbonate species lies down on the surface shows that the observed (1×2) surface structure does not derive from the overlayer itself. The Ag atom separation in the $[1\bar{1}0]$ direction is 2.89 Å, whereas the O-O hard-sphere outer limit measures 4 Å [131]. Obviously, an adsorbate-induced reconstruction takes place. The tendency for the clean {110} surfaces of Pt, Au and Ir to reconstruct in this way is well known [136]. More recently, alkali-induced (1×2) reconstructions of Ag{110} and Cu{110} [137] have also been observed. The reasons for this latter behaviour are not well understood. The present data apparently show that an anionic species is as effective as a cationic species in inducing the reconstruction (assuming, of course, that it is of the same type).

3·2.4.2. Cu {110}, Cu {100}-HCOO. The decomposition of formic acid over metal surfaces and the formation of a surface formate species as a reaction intermediate is a well-known model system in studies of heterogeneous catalysis. Formic acid adsorbs in molecular form at low temperature. Just below room temperature (~ 270 K on copper) deprotonation occurs giving rise to the surface formate and the evolution of hydrogen [138, 139]. Structural information about the formate intermediate on both Cu{110} and Cu{100} surfaces has been obtained with EELS [140, 141], infrared reflection-absorption spectroscopy (IRAS) [142] and, most recently and effectively, with NEXAFS and SEXAFS [143-145] as well as with photoelectron diffraction [146]. The EELS study of Sexton [140, 141] on Cu{100} and the IRAS investigation of Hayden et $al.$ [142] on Cu{110} gave clear evidence for bonding of the formate via two oxygen atoms in a bridging (or bidentate) configuration. Whereas the EELS data on Cu{100} were not able to provide any clear information as to the orientation of the formate molecular plane, the IRAS results on Cu{110} indicated that it was aligned perpendicular to the metal surface.

PUSCHMANN *et al.*[143] later performed NEXAFS and SEXAFS measurements on the Cu{110}-HCOO system. Spectra were taken above the oxygen K-edge as a function of the X-ray incidence angle in different azimuths using the Auger electron yield mode for NEXAFS and the partial electron yield mode for SEXAFS. A selection of NEXAFS difference spectra (covered-clean) taken with the polarization vector along the $\langle 110 \rangle$ and $\langle 100 \rangle$ azimuths and at normal $(\theta = 90°)$ and near-grazing $(\theta = 20°)$ X-ray incidence are shown in fig. 16. The

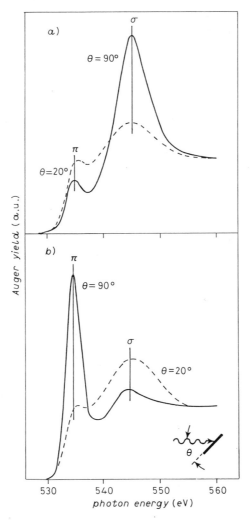

Fig. 16. – Oxygen K-edge NEXAFS spectra for formate on Cu{110} for normal $(\theta = 90°)$ and near-grazing $(\theta = 20°)$ incidence in the *a*) $\langle 110 \rangle$ and *b*) $\langle 100 \rangle$ azimuths. After[143].

polarization dependence of the σ (shape) and π resonance intensities on both the azimuthal and the polar angle of the E vector allowed the following important conclusions to be drawn: Having in mind that the unoccupied π orbital is oriented with its nodal plane parallel to the molecular plane, the strong π resonance seen at normal incidence with the E vector along $\langle 100 \rangle$ and the weak π resonance at both incidence angles with the E vector along $\langle 110 \rangle$ indicate that the molecular plane is perpendicular to the surface and aligned in the $\langle 110 \rangle$ azimuth. More quantitative fitting of the data for the angular dependence of both σ and π resonances confirms this assignment and attributes the nonzero amplitude of the π resonance with radiation incident in the $\langle 110 \rangle$ azimuth to imperfect ($\sim 87\%$) source polarization[126]. Due to background and normalization problems the accuracy of the measured tilt angle of $0°$ was not better than $\pm 10°$. The perpendicular orientation and alignment along $\langle 110 \rangle$ of the formate molecular plane agrees with the IRAS results. In the light of present knowledge, the determination of the O-C-O bond angle in this study is certainly incorrect: the observed σ resonance cannot be considered as deriving from a superposition of two noninteracting C-O bonds. The m.o. diagram for the formate species shows that two σ resonances are expected; these have been resolved recently at the C K-edge[147]. Finally, using the empirical linear relationship between σ resonance position and intramolecular bond length found for gas phase and chemisorbed molecules[18] and interpolating between CO and methoxy chemisorbed on Cu{110} a C-O bond length of 1.28 Å was determined, which is in good agreement with bond lengths of 1.24 Å to 1.30 Å measured in anhydrous cupric formates[148].

SEXAFS data were taken for X-ray incidence angles of $\theta = 20°$ and $\theta = 90°$ in both azimuths $\langle 100 \rangle$ and $\langle 110 \rangle$ (cf. fig. 1). The analysis yielded a single O-Cu bond length of $R = (1.98 \pm 0.07)$ Å using the phase shift from bulk Cu_2O [15] as reference. The large error of ± 0.07 Å results from the limited SEXAFS range for this system. The SEXAFS spetra were truncated at about $700\,eV$ because above this photon energy the C $1s$ photoemission peak intrudes into the partial-yield detector. The measured O-Cu bond length is again in excellent agreement with the values for anhydrous cupric formates[148] (in which four oxygen atoms are spaced at $(1.93 \div 1.99)$ Å from the central copper atom) and with the value of 1.98 Å for a bidentate formate bonded to a binuclear Cu cluster[149]. The adsorption site was established by comparison of the SEXAFS amplitudes with those predicted for various possible geometries. The results clearly favour the on-top configuration (fig. 17, site b) in which the oxygen atoms almost bridge two copper atoms. The bridging configuration (a) is excluded as well as atop and bridging sites involving second-layer copper atoms. The latter would lead to a strong SEXAFS signal for $\theta = 90°$ in the $\langle 100 \rangle$ azimuth, which is not observed. The O-Cu bond length of 1.98 Å, therefore, must be an average value of two O-Cu bond lengths which are within ± 0.10 Å of the measured value.

Almost simultaneously STÖHR et al.[144] performed similar NEXAFS and

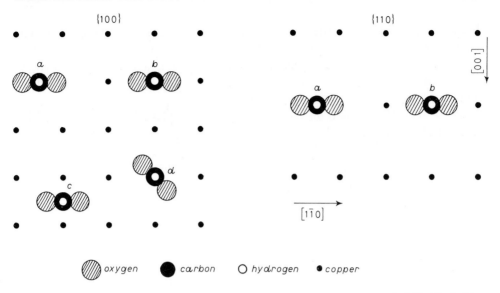

Fig. 17. – Possible adsorption sites for formate on Cu{110} (right) and Cu{100} (left).

SEXAFS measurements for formate adsorbed on Cu{100} using the partial electron yield mode of detection. NEXAFS spectra as a function of the X-ray incidence angle were taken at 90 K and 300 K. The straightforward analysis of the π resonance intensity indicated that the plane of the formate is oriented to within 10° of the surface normal in agreement with the NEXAFS results on Cu{110} discussed above. Because of the higher-symmetry {100} surface the azimuthal orientation cannot be established. Using the empirically established correlation [18] the length of the C-O bond was estimated to be 1.27 Å from the position of the σ resonance in agreement with the value of 1.28 Å found for Cu{110}. At low X-ray incidence angles an additional small peak appeared in the 90 K spectra which is attributed to a minority monodentate formate species. Such a species had already been suggested by the EELS data [140]. The IRAS study [142] on the Cu{110} surface showed that a similar species was clearly not formate, but readsorbed formic acid still present in the chamber after the original dosing of the sample. The analysis of the SEXAFS data for the Cu{100}-HCOO system resulted in an unusually high O-Cu nearest-neighbour bond length of $R = (2.38 \pm 0.03)$ Å and in a cross-bridge adsorption site (fig. 17, site c) completely different from that found on Cu{110}. As this latter site is also present on the {100} surface, it was far from clear why a different adsorption configuration should be adopted on this system. A reanalysis of the Cu{100} SEXAFS spectra by CRAPPER et al. [145] subsequently showed that the data indeed are in good agreement with the diagonal atop adsorption site shown in fig. 17 (site d). Here the formate is atop with the oxygen atoms in off-fourfold hollow sites (referred to as the chelating site in ref. [144]) with a true nearest-neighbour bond length of 1.99 Å. In the original analysis [144] the authors allowed too high a difference in the energy zero for the data taken at $\theta = 20°$ and $\theta = 90°$. The reanalysis thus demonstrates that the adsorption sites and the O-Cu nearest-neighbour distances for formate on the Cu{100} and Cu{110} surfaces are indeed similar.

Quite recently formate on both Cu{100} and Cu{110} has been studied by photoelectron diffraction [146]. Calculations show that the molecule adsorbs on the (short) bridge site (fig. 17, site a) in each case with the oxygen atoms close to atop positions and with a Cu-O nearest-neighbour distance of (1.98 ± 0.04) Å. The bond lengths agree very well with the SEXAFS results but the sites differ.

3'2.4.3. Adsorbed aromatics: benzene and pyridine. – The reactions of unsaturated hydrocarbons over transition metal surfaces are also of particular interest in catalysis. As a result, the adsorption of benzene has been extensively studied with various experimental techniques. In general, the interaction of an unsaturated hydrocarbon with a metal surface may either lead to the formation of a π bond between the C-C π orbitals and the surface or a σ bond involving rehybridization of the carbon $2s$ and $2p$ electrons. For benzene a σ-type dissociative bonding mechanism in which the bond is formed with a ring C atom after the removal of a hydrogen atom has also been proposed. This would necessarily require that the ring plane be inclined with respect to the surface.

The adsorption of benzene on Pt(111) has been studied with LEED [150,151], AES [151], vibrational EELS [152], electronic EELS [153] and NEXAFS [154, 155]. Room temperature adsorption results in two consecutive overlayer structures (given by $\begin{vmatrix} 4 & -2 \\ 0 & 4 \end{vmatrix}$ and $\begin{vmatrix} 4 & -2 \\ 0 & 5 \end{vmatrix}$ in the matrix notation) after doses of approximately 50 L and 300 L, respectively [150, 151]. GLAND and SOMORJAI [150] explained the first structure in terms of a π-bonded species with the ring plane parallel or inclined at some small angle to the surface. Due to the marked work function change at higher coverage the second structure was ascribed to a partially dehydrogenated benzene adsorbed with the ring plane oriented at a large angle or even perpendicular to the surface. From an additional AES and LEED (I-V) analysis STAIR and SOMORJAI [151] concluded that in both structures most of the adsorbed benzene molecules are inclined at an angle to the surface. Vibrational EELS data of Lehwald et al. [152] indicated that no rehybridization at the C atoms takes place upon adsorption on Pt{111} and that π-bonding with the ring plane parallel to the surface dominates. The results further showed that the two phases can be explained by benzene adsorption in on-top and in threefold hollow sites, respectively. Electronic EELS results by NETZER and MATTHEW [153] also suggested benzene adsorption with the ring plane parallel to the surface. In the meantime there is a considerable body of data, particularly photoemission data, for other metal surfaces which indicates that this is in fact always the preferred geometry.

NEXAFS studies on the Pt{111}-C_6H_6 system were originally performed by JOHNSON *et al.*[154] and later reanalysed by HORSLEY *et al.*[155]. The NEXAFS measurements were carried out above the carbon K-edge uşing the partial electron yield mode of detection with the sample at 100 K. Monolayer and multilayer («solid») NEXAFS spectra are shown in fig. 18 together with the carbon K-shell EELS spectrum of the free benzene molecule (top)[156]. The curves of fig. 18 demonstrate the close resemblance of the electron loss spectra of free molecules and the NEXAFS spectra of the same molecules adsorbed on single crystal surfaces. Four main peaks appear labelled *A-D*, the correct assignment of which is still a subject of discussion[155-161]. According to the multiple-scattering calculation of Horsley *et al.*[155] for the free molecule (following the approach of Dehmer and Dill[115]) peaks *A* and *B* are assigned to π resonances, peaks *C* and *D* to σ (shape) resonances. *A* and *B* correspond to

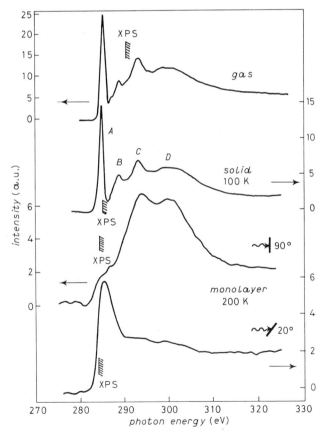

Fig. 18. – Carbon K-shell EELS spectrum of gaseous benzene (top) and carbon K-edge NEXAFS spectra of multilayer («solid») and monolayer coverages of benzene on Pt{111} recorded in the partial electron yield mode. Carbon $1s$ binding energies (referred to the Fermi level) are marked with XPS. After[154].

transitions of the $1s$ core electrons into the two unoccupied π^* orbitals of benzene with e_{2u} and b_{2g} symmetry, respectively; the measured separation of 3.8 eV is in good agreement with electron transmission spectroscopy data [162]. The assignment of the continuum resonances is less clear. The simple NEXAFS concepts which correlate bond length with σ resonance position in diatomics break down for benzene, as might be expected. The calculations show three σ resonances corresponding to transitions into e_{1u}, e_{2g} and a_{2g} unoccupied orbitals of σ-type symmetry. The assignment of peaks C and D as σ resonances is in agreement with calculations of Akimov et al. [160] and seems to be supported by the polarization dependence of the spectra from the chemisorbed monolayer shown in fig. 18: peaks A and B have the opposite polarization dependence to peaks C and D. Remembering that the lobes of the π orbitals in benzene are oriented perpendicular to the ring plane and using the equations of subsect. 3˙2.2, the curves in fig. 18 suggest that the benzene molecules are lying flat on the surface. This orientation was confirmed by a quantitative intensity analysis. Similar results have also been obtained in NEXAFS studies on the Cu{110}-C_6H_6 and Cu{111}-C_6H_6 systems [161], which yield a parallel orientation to within $\pm 10°$. The increased width of the π resonances in the monolayer spectra of fig. 18 compared to the multilayer spectra is explained by the interaction of the benzene molecules with the Pt{111} surface. The broadening is attributed to a direct bonding between benzene π levels and the surface as well as to the reduced lifetime of the final state. As already mentioned above, the correlation between σ resonance position and intramolecular bond length breaks down for benzene where three σ resonances are expected and in fact two are indeed observed experimentally. It should be mentioned, however, that the predicted C-C bond length obtained from the intensity-weighted average energy of peaks C and D is in good agreement with the actual value of 1.40 Å in benzene [163].

Another aromatic species which has provoked considerable interest amongst surface chemists in recent years is pyridine. This has been largely due to the fact that the adsorption of this molecule on silver surfaces is the prototype system of the observation of the surface-enhanced Raman effect [164]. Pyridine might be expected to bond in two quite distinct ways to a metal surface with retention of its molecular integrity: σ-bonding involving the nitrogen «lone pair» could occur as well as π-bonding with the ring plane parallel to the surface. As in the case of benzene, a σ-type dissociative bonding after removal of a hydrogen has also been discussed. From the large work function change on adsorption of pyridine on Pt{111} GLAND and SOMORJAI [150] concluded that nitrogen participates in the adsorbate-substrate bond. Adsorbate-substrate interaction occurring through the lone pair of the nitrogen atom was also established in the UPS measurements of Bandy et al. [165] on the Cu{110}-C_5H_5N system. Consequently tilted or perpendicular pyridine chemisorption states were suggested for all the surfaces studied (see references in [166]). However, on Ag{111} [167], Pt{110} [168], Ni{001} [169] and Pt{111} [166] temperature- and coverage-dependent orien-

tation effects have been found. The Ag {111}-C_5H_5N system has been characterized by vibrational EELS[167], electronic EELS[170, 171], UPS[167, 171, 172],
IPES[171], LEED[172] and NEXAFS[161, 173]. Vibrational EELS studies by
DEMUTH *et al.*[167] indicated a phase transition at about half a monolyer
coverage from a nearly flat-lying π-bonded pyridine species to an inclined N-
bonded phase. The UPS measurements by DUDDE *et al.*[172] revealed a
pyridine adsorption perpendicular to the surface and bonding via the nitrogen
lone-pair orbitals. No sign of a phase transition was observed in agreement with
IPES and electronic EELS results by OTTO *et al.*[171].

BADER *et al.*[173] have performed NEXAFS measurements on the Ag{111}-
C_5H_5N system at 100 K. Spectra were taken above the carbon and nitrogen K-
edges using the partial electron yield mode of detection. Similar to benzene, four
peaks (*A-D*) appear. Two types of experiments were conducted: *a*) continuous
NEXAFS measurements for X-ray incidence angles of $\theta = 20°$ and $\theta = 90°$ during
exposure with pyridine at a rate of ~ 0.1 L/min and *b*) NEXAFS measurements
at various angles after fixed exposures at 100 K. Figure 19 shows that during

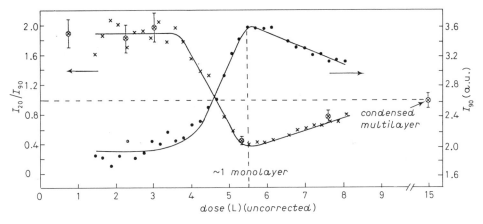

Fig. 19. – Normalized nitrogen K-edge π resonance intensity for normal incidence
($\theta = 90°$), I_{90} (points), and intensity ratio I_{20}/I_{90} (crosses) as a function of dose during the
continuous exposure of pyridine (~ 0.1 L/min) to Ag {111} at 100 K. The I_{20}/I_{90} data points
with error bars were independently measured at fixed pyridine doses. There is a phase
transition near 4.5 L. After[173].

continuous adsorption of pyridine on Ag{111} the normalized π resonance
intensity (peak *A*), I_{90} (points), rises drastically between 4 L and 5.5 L and then
steadily decreases at higher exposures. A similar exposure-dependent
measurement for near-grazing incidence ($\theta = 20°$) also allows the ratio I_{20}/I_{90}
(crosses) to be plotted as a function of dose. I_{20}/I_{90} also changes drastically in the
same exposure range. As the π resonance intensity ratio is a measure of the tilt
angle α between the ring plane and the surface, the results indicate a phase

transition characterized by a change in orientation of the pyridine molecules on the Ag{111} surface as a function of coverage. Around 5.5 L the ratio I_{20}/I_{90} goes through a minimum. Continued exposure results in gradually increasing values, which finally approach $I_{20}/I_{90} = 1$ for a randomly oriented condensed multilayer. A monolayer is, therefore, assigned to an exposure of approximately 5.5 L which is in good agreement with results from IPES and UPS[171]. The phase transition is thus observed for a submonolayer coverage of pyridine in agreement with the EELS data[167]. Fitting the NEXAFS data at fixed doses before (2.3 L) and after (5.3 L), shown in fig. 20, yields tilt angles of $\alpha = (45 \pm 5)°$ and $\alpha = (70 \pm 5)°$, respectively. The measured tilt angles are apparent angles between the molecular plane and the surface. They could result each from a single phase or from a number of coexistent phases with well-defined tilt angles. Also possible is a phase with the pyridine molecules rotated around their C_2 axes which have a well-defined inclination angle with respect to the surface («rotated» model). For the apparent angle of $\alpha = 70°$ all explanations are possible; in the

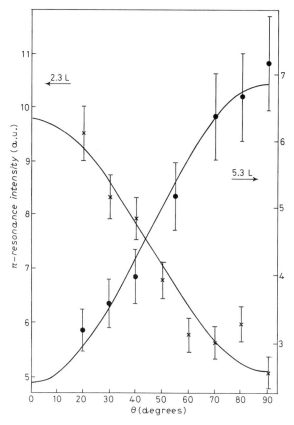

Fig. 20. – Nitrogen K-edge π resonance intensity for submonolayer coverages (2.3 and 5.3 L) of pyridine on Ag{111} at 100 K as a function of the angle of incidence, θ. The full lines are fits with tilt angles of 45° and 70°, respectively.

rotated model the inclination angle would then be 41°. An apparent angle of $\alpha = 45°$ would, however, correspond to an inclination angle of 0° in the rotated model which is physically unreasonable. A comparison between the σ resonance intensity (peak D) with those of benzene on Cu{110} and pyridine on Cu{110} (where pyridine stands exactly upright [162]) indicates that the measured tilt angle of 45° for pyridine on Ag{111} cannot be due to a «mixture» of flat-lying and nearly perpendicular oriented molecules. The measured tilt angles for pyridine on Ag{111} compare well with those found with NEXAFS for pyridine on Pt{111} [166] and on Cu{111} [161]. Moreover, a recent ESDIAD study of pyridine adsorbed on Ir{111} [174] indicated an angle between the ring plane and the surface of 70° in agreement with the NEXAFS results mentioned above. The measured tilt angle for the low-coverage, low-temperature phase of pyridine on Ag{111} of 45° is at variance with that indicated by vibrational EELS, but in this instance we feel that NEXAFS is a much more reliable structural tool.

4. – Comparison LEED/PhD/SEXAFS.

SEXAFS and PhD owe their success in providing surface structural information to the existence of electron storage rings where synchrotron radiation is available. The relative simplicity of interpretation of SEXAFS and PhD, their widespread application and the accuracy they can achieve have often been compared with LEED *I-V* structural analysis, which, of course, can be performed in the laboratory without synchrotron radiation. Because of the use of photons as primary particles, SEXAFS and PhD are essentially nondestructive probes. In LEED the probing particles are electrons from an energy-tunable electron gun. The use of electrons, however, carries with it the inherent problem of damage in molecular adsorbates, *i.e.* causing disorder, desorption and/or dissociation [175]. LEED structural analyses for adsorbed molecules are, therefore, difficult and relatively few have so far been reported. A comparison of LEED, PhD and SEXAFS should, however, begin from a more fundamental viewpoint: in all three methods structure determination is based on electron scattering. It is only the different sources and detectors as well as the different electron kinetic-energy range which give the techniques their particular advantages and disadvantages. A schematic comparison is shown in fig. 21. The following discussion does not attempt, therefore, to single out the «best» method (which probably does not exist) but rather to demonstrate the complementarity of the three methods. In many cases, it is only the application of more than one technique which gives the full structural information required.

In LEED incident electron plane waves from a nonlocal source are scattered from atoms in the surface region and are detected nonlocally outside the sample. In SEXAFS and PhD localized adsorbate atoms simulate an «electron gun» emitting photoelectron spherical waves the amplitude of which decays as the

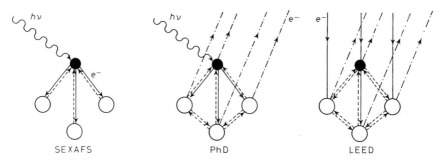

Fig. 21. – Schematic diagram representing the electron scattering processes in SEXAFS, PhD and LEED.

inverse of the distance. The scattering of these photoelectrons is responsible for the observed intensity modulations. In PhD the detector is nonlocal as in LEED, whereas in SEXAFS the absorbing atom is at the same time the «detector». The actual SEXAFS monitor may not even be electron based (*e.g.*, fluorescence). The SEXAFS signal corresponds to measuring the total photoelectron current in 4π solid angle. The photoelectron diffraction effects are entirely averaged out; what remains is the interference between the outgoing spherical wave and the waves backscattered from the surrounding atoms. In PhD a differential (angle-resolved) photoionization cross-section is measured rather than the total (angle-integrated) cross-section monitored by SEXAFS. Therefore, the signal collected in PhD is smaller than that used in SEXAFS. The interference modulations which form the primary data are, however, an order of magnitude larger.

The LEED process relies on the two-dimensional periodicity (long-range order) of the surface. Unless the diffuse «background» of a pattern is analysed [176], the LEED technique cannot be applied to disordered adsorbate systems. This is in contrast to both SEXAFS and PhD, which, mainly because of the inverse distance relation for the emitted photoelectron wave amplitude, are local probes not requiring long-range order. The incident electron plane wave excites every atom in the surface region, so that there is no element specificity in LEED as is the case for SEXAFS and PhD. The scattering problem is thus complex and in the typical LEED region of electron kinetic energies between roughly 30 and 200 eV multiple scattering is very important. The intensities of the diffracted beams contain the detailed information concerning the relative atomic positions at the surface. The structural determination proceeds via a comparison of experimental with calculated *I-V* curves performed for various adsorbate sites and for various adsorbate-substrate distances. This is a trial and error procedure; the necessary multiple-scattering calculations are extensive and require considerable computer time. As the intensities of most of the diffracted beams are only weakly perturbed by the adsorbate, the information content in a single *I-V* LEED curve is relatively low, so that a large number of beams should be included for a satisfactory structural assignment. This is in contrast to SEXAFS and PhD, where due to the photoelectron emission from localized adsorbate atoms the information content of a single spectrum is high, the data base thus small.

SEXAFS is reasonably well described by single-scattering theory. With the increased quality of data now obtainable, multiple-scattering effects, however, cannot be completely ignored. The simple single-scattering picture results mainly from the localization of both the electron «source» and the «detector» at the absorbing atom. The SEXAFS is then purely the results of backscattering: While in the SEXAFS kinetic-energy range between roughly 50 eV and several 100 eV backscattering is actually weak, sideways scattering, however, is even weaker, reducing drastically the importance of multiple-scattering events. As a result of the local electron emission and backscattering, the SEXAFS oscillations decrease as the inverse of the distance squared and are further damped exponentially by inelastic scattering and Debye-Waller-like effects. The latter also tend to increase with distance because of correlated vibrations [177]. Two conclusions can be drawn from this strong distance dependence. Firstly, multiple-scattering contributions are further reduced as they generally involve larger distances than the nearest-neighbour backscattering path length. Secondly, SEXAFS spectra are dominated by a small number (often only one or two) of nearest-neighbour «shells» of atoms. A Fourier transform of the SEXAFS oscillations can often sort out the different nearest-neighbour bond lengths which can then be determined without recourse to complicated theory. Adsorption sites are established by a comparison of measured bond lengths and/or polarization-dependent amplitudes with those calculated for model configurations. The calculations use, however, only simple geometrical relations and are thus also independent of theory. Phase shifts are to some extent a problem. Experience shows that phase shift transferability usually works, but sometimes calculated values have to be used. For adsorbed molecules —particularly those containing more than one atom of the same type—the single-shell analysis may lead to incorrect results. Multishell analysis (curve fitting) as used with success in bulk EXAFS then has to be applied to the sometimes weak SEXAFS oscillations.

Photoelectron diffraction involves the same initial step as SEXAFS also making it a local probe. The single-scattering plane-wave approximation already shows that the PhD oscillations due to a single scattering atom are inversely proportional to its distance from the source atom. This suggests that a larger «cluster» of scatterers is necessary for explaining a PhD spectrum than for describing SEXAFS oscillations. The question as to which of the simplifications of SEXAFS are transferable to the analysis of PhD data is still under discussion. In the general case, the analysis of PhD data has to proceed via a comparison of experimental and calculated spectra as in LEED. The degree of sophistication to which the calculations have to be performed depends on the kinetic-energy range of the PhD data. Early PhD studies in the energy range up to 200 eV required complex LEED-type theories involving extensive calculations. PhD data taken at higher energies, at least in the $(100 \div 400)$ eV range, however, are often well described with single-scattering theory [45, 178] on the spherical-wave or even on the plane-wave level. Inelastic scattering and Debye-Waller-like effects are taken into account and corrections due to correlated vibrations generally improve the agreement between experiment and theory. The single-scattering picture even holds for those geometries in which multiple scattering along rows of atoms in a substrate should be more important. Increased importance of multiple scattering and significant deviations from the single-scattering model occur for geometries in which the polarization is approximately perpendicular to the electron emission angle. In spite of the obvious success of single-scattering calculations for most of the PhD data, the Fourier transform method as a means to directly determine interatomic distances is not generally applicable. Only for those cases where a Fourier peak is related to a single path length difference is the Fourier transform approach reliable. This happens, for example, when detector, adsorbate atom and substrate (backscattering) atom are aligned.

The structural information provided by LEED, PhD and SEXAFS is rather similar. All three techniques are clearly sensitive to the adsorption site. The nature of the site information, however, differs. In PhD the adsorption site configuration is more directly involved in establishing the observed interferences (periodicities) than in SEXAFS where it is obtained via polarization-dependent amplitudes. Whereas LEED (and to a lesser extent PhD) is more sensitive to the vertical distance, SEXAFS determines interatomic separations directly. The accuracy of the vertical distance measurement with LEED and PhD is typically ± 0.1 Å resulting in errors for nearest-neighbour bond lengths of approximately ± 0.05 Å (depending on the system). Errors for bond length determinations with SEXAFS are typically quoted as ± 0.02 Å, although this degree of accuracy has often been doubted. It is interesting to note that, although the intensity variations in PhD are often larger by at least a factor of 10 compared with those in SEXAFS, the accuracy of PhD results appears to be worse even if accurate multiple-scattering calculations are performed. LEED and PhD, on the other hand, are sensitive to surface relaxations and reconstruction, whereas SEXAFS does not generally give information on this phenomenon (cf. subsect. 2˙7.2). Molecular orientation can be established with PhD and NEXAFS, the latter spectrum being obtained automatically together with a SEXAFS spectrum. Measured tilt angles are normally determined by a $\sin^2 \alpha$ or $\cos^2 \alpha$ dependence of the intensity on the angle α between the polarization vector and a symmetry element of the molecule. The accuracy is typically $\pm 5° \div 15°$. In PhD the precision of an orientation determination is controlled by the angular width of the scattering amplitude in the forward direction which, depending on the electron kinetic energy, may be smaller than $\pm 10°$. As already mentioned above, LEED may possess some advantages over SEXAFS for molecular layers. However, the computational effort increases considerably, ordered overlayers are required and the sensitivity of LEED to lateral displacements is lower than for vertical displacements. It is quite possible that PhD will in the future be the most important structural tool for investigating adsorbed molecules.

* * *

I thank my co-workers and colleagues A. PUSCHMANN, M. BADER, B. HILLERT, L. BECKER, C. OCAL, S. D'ADDATO, M. PEDIO, A. M. BRADSHAW, U. DÖBLER, K. BABERSCHKE, M. D. CRAPPER, C. E. RILEY and D. P. WOODRUFF for collaboration and fruitful discussions. The work has been funded by the German Minister for Research and Technology under the contract number 05 390 FX B 2.

REFERENCES

[1] P. H. CITRIN, P. EISENBERGER and R. C. HEWITT: *Phys. Rev. Lett.*, **41**, 309 (1978).

[2] J. STÖHR, D. DENLEY and P. PERFETTI: *Phys. Rev. B*, **18**, 4132 (1978).

[3] P. H. CITRIN: *J. Phys. (Paris)*, C8, 437 (1986); *Surf. Sci.*, **184**, 109 (1987).

[4] L. PAULING: *The Nature of the Chemical Bond*, 3rd edition (Cornell University Press, New York, N.Y., 1960).

[5] K. BABERSCHKE, U. DÖBLER, L. WENZEL, D. ARVANITIS, A. BARATOFF and K. H. RIEDER: *Phys. Rev. B*, **33**, 5910 (1986).

[6] M. BADER, J. HAASE, A. PUSCHMANN and C. OCAL: *Phys. Rev. Lett.*, **57**, 3273 (1986).

[7] D. ARVANITIS, L. WENZEL and K. BABERSCHKE: *Phys. Rev. Lett.*, **59**, 2435 (1987).

[8] P. ROUBIN, D. CHANDESRIS, G. ROSSI, J. LECANTE, M. C. DESJONQUÈRES and G. TRÉGLIA: *Phys. Rev. Lett.*, **56**, 1272 (1986).

[9] F. SETTE, C. T. CHEN, J. E. ROWE and P. H. CITRIN: *Phys. Rev. Lett.*, **59**, 311 (1987).

[10] L. WENZEL, J. STÖHR, D. ARVANITIS and K. BABERSCHKE: *Phys. Rev. Lett.*, **60**, 2327 (1988).

[11] G. LAMBLE and D. A. KING: *Philos. Trans. R. Soc. London, Ser. A*, **318**, 203 (1986).

[12] D. ARVANITIS, K. BABERSCHKE and L. WENZEL: *Phys. Rev. B*, **37**, 7143 (1988).

[13] M. BADER, C. OCAL, B. HILLERT, J. HAASE and A. M. BRADSHAW: *Phys. Rev. B*, **35**, 5900 (1987).

[14] J. STÖHR, E. B. KOLLIN, D. A. FISCHER, J. B. HASTINGS, F. ZAERA and F. SETTE: *Phys. Rev. Lett.*, **55**, 1468 (1985); D. A. FISCHER, U. DÖBLER, D. ARVANITIS, L. WENZEL, K. BABERSCHKE and J. STÖHR: *Surf. Sci.*, **177**, 114 (1986).

[15] J. STÖHR: in *X-Ray Absorption: Principles, Applications, Techniques of EXAFS, SEXAFS, and XANES*, edited by R. PRINS and D. KONINGSBERGER (Wiley, New York, N.Y., 1985), p. 443.

[16] J. STÖHR, K. BABERSCHKE, R. JAEGER, R. TREICHLER and S. BRENNAN: *Phys. Rev. Lett.*, **47**, 381 (1981).

[17] J. STÖHR and D. A. OUTKA: *Phys. Rev. B*, **36**, 7891 (1985).

[18] J. STÖHR, J. L. GLAND, W. EBERHARDT, D. A. OUTKA, R. J. MADIX, F. SETTE, R. J. KOESTNER and U. DÖBLER: *Phys. Rev. Lett.*, **51**, 2414 (1983); J. STÖHR, F. SETTE and A. L. JOHNSON: *Phys. Rev. Lett.*, **53**, 1684 (1984).

[19] M. N. PIANCASTELLI, D. W. LINDLE, T. A. FERRIT and D. A. SHIRLEY: *J. Chem. Phys.*, **86**, 2765 (1987).

[20] A. M. BRADSHAW, J. SOMERS and TH. LINDNER: in *Proceedings of the Solvay Conference on Surface Science*, Austin, Texas, December 1987, edited by F. W. de WETTE, *Springer Series in Surface Science*, Vol. 14, (Springer, Berlin, 1988), p. 469.

[21] M. BADER, B. HILLERT, A. PUSCHMANN, J. HAASE and A. M. BRADSHAW: *Europhys. Lett.*, **5**, 443 (1988).

[22] D. NORMAN, J. STÖHR, R. JAEGER, P. J. DURHAM and J. B. PENDRY: *Phys. Rev. Lett.*, **51**, 2052 (1983).

[23] U. DÖBLER, K. BABERSCHKE, D. D. VVEDENSKY and J. B. PENDRY: *Surf. Sci.*, **178**, 679 (1986).

[24] TH. LINDNER and J. SOMERS: *Phys. Rev. B*, **37**, 10039 (1988).

[25] D. E. SAYERS, E. A. STERN and F. W. LYTLE: *Phys. Rev. Lett.*, **27**, 1204 (1971).

[26] P. A. LEE and J. B. PENDRY: *Phys. Rev. B*, **11**, 2795 (1975).

[27] C. H. ASHLEY and S. DONIACH: *Phys. Rev. B*, **11**, 1279 (1975).

[28] W. SCHAICH: *Phys. Rev. B*, **8**, 4078 (1973).

[29] P. A. LEE, P. H. CITRIN, P. EISENBERGER and B. M. KINCAID: *Rev. Mod. Phys.*, **53**, 769 (1981).

[30] J. E. MÜLLER and W. L. SCHAICH: *Phys. Rev. B*, **27**, 6489 (1983).

[31] S. J. GURMAN, N. BINSTED and I. ROSS: *J. Phys. C*, **17**, 143 (1984).

[32] S. J. GURMAN, N. BINSTED and I. ROSS: *J. Phys. C*, **19**, 1845 (1986).

[33] A. G. MCKALE, B. W. VEAL, A. P. PAULIKAS, S.-K. CHAN and G. S. KNAPP: *J. Am. Chem. Soc.*, **110**, 3763 (1988).

[34] D. E. SAYERS, E. A. STERN and F. W. LYTLE: *Phys. Rev. Lett.*, **27**, 1204 (1971).

[35] B.-K. TEO and P. A. LEE: *J. Am. Chem. Soc.*, **101**, 2815 (1979).

[36] P. H. CITRIN, P. EISENBERGER and B. M. KINCAID: *Phys. Rev. Lett.*, **36**, 1346 (1976).

[37] B. LENGELER: *J. Phys. (Paris)*, **47**, C8, 75 (1986).

[38] G. MARTENS, P. RABE, N. SCHWENTNER and A. WERNER: *Phys. Rev. B*, **17**, 1481 (1978).

[39] N. F. MOTT and H. S. W. MASSEY: *Theory of Atomic Collisions*, 3rd edition (Clarendon, Oxford, 1965).

[40] A. G. MCKALE, B. W. VEAL, A. P. PAULIKAS and S.-K. CHAN: *Phys. Rev. B*, **38**, 10919 (1988).

[41] B. LENGELER and P. EISENBERGER: *Phys. Rev. B*, **21**, 4507 (1988).

[42] P. EISENBERGER and G. S. BROWN: *Solid State Commun.*, **29**, 481 (1979).

[43] G. BENI and P. M. PLATZMAN: *Phys. Rev. B*, **14**, 9514 (1976).

[44] E. A. STERN, B. BUNKER and S. M. HEALD: in *EXAFS Spectroscopy: Techniques and Applications*, edited by B. K. TEO and D. C. JAY (Plenum, New York, N.Y., 1981), p. 59.

[45] C. S. FADLEY: *Phys. Scr.*, **T17**, 39 (1987).

[46] J. J. BOLAND, S. E. CRANE and J. D. BALDESCHWIELER: *J. Chem. Phys.*, **77**, 142 (1982).

[47] A. PUSCHMANN: Thesis, Freie Universität Berlin (1986).

[48] D. J. HOLMES, D. R. BATCHELOR and D. A. KING: *Surf. Sci.*, **199**, 476 (1988).

[49] B. HILLERT, S. D'ADDATO and J. HAASE: BESSY-Report (Berlin, 1987).

[50] A. BIANCONI and R. Z. BACHRACH: *Phys. Rev. Lett.*, **42**, 104 (1979).

[51] F. COMIN, L. INCOCCIA, P. LAGARDE, G. ROSSI and P. H. CITRIN: *Phys. Rev. Lett.*, **54**, 122 (1985).

[52] R. JAEGER, J. FELDHAUS, J. HAASE, J. STÖHR, Z. HUSSAIN, D. MENZEL and D. NORMAN: *Phys. Rev. Lett.*, **45**, 1870 (1980).

[53] R. B. GREEGOR, F. W. LYTLE and D. R. SANDSTROM: SSRL Report 80/01.

[54] P. A. LEE: *Phys. Rev. B*, **13**, 5261 (1976).

[55] G. MARTENS, P. RABE, N. SCHWENTNER and A. WERNER: *J. Phys. C*, **11**, 3125 (1978).

[56] G. MARTENS, P. RABE, G. TOLKIEHN and A. WERNER: *Phys. Status Solidi A*, **55**, 105 (1979).

[57] J. STÖHR: *Jpn. J. Appl. Phys.*, **17**, Suppl. 17-2, 217 (1978).

[58] V. O. KOSTROUN, M. H. CHEN and B. CRASEMANN: *Phys. Rev. A*, **3**, 533 (1971).

[59] S. M. HEALD, E. KELLER and E. A. STERN: *Phys. Lett. A*, **103**, 155 (1984).

[60] J. E. DEMUTH, D. W. JEPSEN and P. M. MARCUS: *Phys. Rev. Lett.*, **31**, 540 (1973); **32**, 1182 (1974).

[61] S. ANDERSSON, B. KASEMO, J. B. PENDRY and M. A. VAN HOVE: *Phys. Rev. Lett.*, **31**, 593 (1973).

[62] C. B. DUKE, N. O. LIPARI, G. E. LARAMORE and J. B. THEETEN: *Solid State Commun.*, **13**, 579 (1973); C. B. DUKE, N. O. LIPARI and G. E. LARAMORE: *Nuovo Cimento B*, **23**, 241 (1974); *J. Vac. Sci. Technol.*, **12**, 222 (1975).

[63] Y. GAUTHIER, D. ABERDAM and R. BAUDOING: *Surf. Sci.*, **78**, 339 (1978).

[64] M. A. VAN HOVE and S. Y. TONG: *J. Vac. Sci. Technol.*, **12**, 230 (1975).

[65] H. H. BRONGERSMA: *J. Vac. Sci. Technol.*, **11**, 231 (1974).

[66] TH. FAUSTER, H. DÜRR and D. HARTWIG: *Surf. Sci.*, **178**, 657 (1986).

[67] E. W. PLUMMER, B. TONNER, N. HOLZWARTH and A. LIEBSCH: *Phys. Rev. B*, **21**, 4306 (1980).

[68] S. ANDERSSON: *Surf. Sci.*, **79**, 385 (1979).

[69] D. H. ROSENBLATT, J. G. TOBIN, M. G. MASON, R. F. DAVIS, S. D. KEVAN, D. A. SHIRLEY, C. H. LI and S. Y. TONG: *Phys. Rev. B*, **23**, 3828 (1981).

[70] J. J. BARTON, C. C. BAHR, Z. HUSSAIN, S. W. ROBEY, J. G. TOBIN, L. E. KLEBANOFF and D. A. SHIRLEY: *Phys. Rev. Lett.*, **51**, 272 (1983).

[71] J. J. BARTON, C. C. BAHR, Z. HUSSAIN, S. W. ROBEY, L. E. KLEBANOFF and D. A. SHIRLEY: *J. Vac. Sci. Technol. A*, **2**, 847 (1984).

[72] J. J. BARTON, S. W. ROBEY, C. C. BAHR and D. A. SHIRLEY: in *The Structure of Surfaces*, edited by M. A. VAN HOVE and S. Y. TONG, Springer Series in Surface Science, Vol. 2 (Springer, Berlin, 1985), p. 191.

[73] P. J. ORDERS, B. SINKOVIĆ, C. S. FADLEY, R. TREHAN, Z. HUSSAIN and J. LECANTE: *Phys. Rev. B*, **30**, 1838 (1984).

[74] S. BRENNAN, J. STÖHR and R. JAEGER: *Phys. Rev. B*, **24**, 4871 (1981).

[75] J. STÖHR, R. JAEGER and S. BRENNAN: *Surf. Sci.*, **117**, 503 (1982).

[76] S. P. WALCH and W. A. GODDARD III: *Solid State Commun.*, **23**, 907 (1977); *Surf. Sci.*, **72**, 645 (1978).

[77] T. H. LOPTON and W. A. GODDARD III: *CRC Crit. Rev. Solid State Mater. Sci.*, **10**, 261 (1981).

[78] PEI-LIN CAO, D. E. ELLIS and A. J. FREEMAN: *Phys. Rev. B*, **25**, 2124 (1982).

[79] R. H. HOHN and M. J. O'CONNOR: in *Progress in Inorganic Chemistry*, Vol. 14, edited by S. J. LIPPARD (Interscience, New York, N. Y., 1971), p. 241.

[80] J. TRAHAN, R. G. GOODRICH and S. F. WATKINS: *Phys. Rev. B*, **2**, 2859 (1970).

[81] G. ERTL: *Surf. Sci.*, **6**, 208 (1967).

[82] G. W. SIMMONS, D. F. MITCHELL and K. R. LAWLESS: *Surf. Sci.*, **8**, 130 (1967).

[83] N. TAKAHASHI, H. TOMITA and S. MATOO: *C.R. Acad. Sci., Ser. B*, **269**, 618 (1969).

[84] G. ERTL and J. KÜPPERS: *Surf. Sci.*, **24**, 104 (1971).

[85] A. OUSTRY, L. LAFOURCARDE and A. ESCAUT: *Surf. Sci*, **40**, 545 (1973).

[86] J. R. NOONAN and H. L. DAVIS: unpublished.

[87] T. A. DELCHAR: *Surf. Sci.*, **27**, 11 (1971).

[88] F. H. P. M. HABRAKEN, G. A. BOOTSMA, P. HOFMANN, S. HACHICHA and A. M. BRADSHAW: *Surf. Sci.*, **88**, 285 (1979).

[89] F. H. P. M. HABRAKEN and G. A. BOOTSMA: *Surf. Sci.*, **87**, 333 (1979).

[90] A. G. J. DE WIT, R. P. N. BRONCKERS and J. M. FLUIT: *Surf. Sci.*, **82**, 177 (1979).

[91] A. G. J. DE WIT, R. P. N. BRONCKERS, TH. M. HUPKENS and J. M. FLUIT: *Surf. Sci.*, **90**, 676 (1979).

[92] R. P. N. BRONCKERS and G. J. DE WIT: *Surf. Sci.*, **112**, 133 (1981).

[93] H. NIEHUS and G. COMSA: *Surf. Sci.*, **140**, 18 (1984).

[94] J. LAPUJOULADE, Y. LE CRUËR, M. LEFORT, Y. LEJAY and E. MAUREL: *Phys. Rev. B*, **22**, 5740 (1980).

[95] J. LAPUJOULADE, Y. LE CRUËR, M. LEFORT, Y. LEJAY and E. MAUREL: *Surf. Sci.*, **118**, 103 (1982).

[96] J. F. WENDELKEN: *Surf. Sci.*, **108**, 605 (1981).

[97] A. SPITZER and H. LÜTH: *Surf. Sci.*, **118**, 121 (1982).

[98] R. A. DIDIO, D. M. ZEHNER and E. W. PLUMMER: *J. Vac. Sci. Technol. A*, **2**, 852 (1984).

[99] R. FEIDENHANS'L and I. STENSGAARD: *Surf. Sci.*, **133**, 453 (1983).

[100] U. DÖBLER, K. BABERSCHKE, J. HAASE and A. PUSCHMANN: *Phys. Rev. Lett.*, **52**, 1437 (1984).

[101] K. S. LIANG, P. H. FUOSS, G. J. HUGHES and P. EISENBERGER: in *The Structure of Surfaces*, edited by M. A. VAN HOVE and S. Y.TONG, Springer Series in Surface Science, Vol. 2 (Springer, Berlin, 1985), p. 246.

[102] H. P. BONZEL and S. FERRER: *Surf. Sci.*, **118**, L263 (1982).

[103] G. TUNELL, E. POSNJAK and C. J. KSANDA: *Z. Kristallogr.*, **90**, 120 (1930).

[104] U. DÖBLER: PhD thesis, Freie Universität, Berlin (1986).

[105] G. ROVIDA, F. PRATESI, M. MAGLIETTA and E. FERRONI: *Jpn. J. Appl. Phys.*, Suppl. 2, Pt. 2, 117 (1974).

[106] G. ROVIDA and F. PRATESI: *Surf. Sci.*, **51**, 270 (1975).

[107] P. J. GODDARD and R. M. LAMBERT: *Surf. Sci.*, **67**, 180 (1977).

[108] YUNG-YI TU and J. M. BLAKELY: *J. Vac. Sci. Technol.*, **15**, 563 (1978).

[109] M. BOWKER and K. C. WAUGH: *Surf. Sci.*, **134**, 639 (1983).

[110] R. W. G. WYCKOFF: *Crystal Structure*, Vol. 1 (Interscience, New York, N. Y., London, Sydney, 1963).

[111] EXAFS *and Near Edge Structure IV*, *J. Phys.* (Paris), **47**, C8 (1986).

[112] B. CORDTS, D. M. PEASE and L. V. AZAROFF: *Phys. Rev. B*, **22**, 4692 (1980).

[113] J. E. MÜLLER and J. W. WILKINS: *Phys. Rev. B*, **29**, 4331 (1984).

[114] P. J. DURHAM, J. P. PENDRY and C. H. HODGES: *Comp. Phys. Commun.*, **25**, 193 (1982); D. D. VVEDENSKY, D. K. SALDIN and J. B. PENDRY: *Comp. Phys. Commun.*, **40**, 421 (1986).

[115] J. L. DEHMER and D. DILL: *J. Chem. Phys.*, **65**, 5327 (1976).

[116] C. R. NATOLI and M. BENFATTO: *J. Phys.* (Paris), **47**, C8, 11 (1986).

[117] J. E. DEMUTH, D. W. JEPSEN and P. M. MARCUS: *Phys. Rev. Lett.*, **31**, 540 (1973).

[118] M. VAN HOVE and S. Y. TONG: *J. Vac. Sci. Technol.*, **12**, 230 (1975).

[119] J. STÖHR, R. JAEGER and T. KENDELEWICZ: *Phys. Rev. Lett.*, **49**, 142 (1982).

[120] M. BENFATTO, C. R. NATOLI, A. BIANCONI, J. GARCIA, A. MARCELLI, M. FANFANI and I. DEVOLI: *Phys. Rev. B*, **34**, 5774 (1986).

[121] D. D. VVEDENSKY, J. B. PENDRY, U. DÖBLER and K. BABERSCHKE: *Phys. Rev. B*, **35**, 7756 (1987).

[122] H. NIEHUS and G. COMSA: *Surf. Sci.*, **151**, L171 (1985).

[123] P. W. LANGHOFF, S. R. LANGHOFF, T. N. RESCIGNO, J. SCHIRMER, L. S. CEDERBAUM, W. DOMCKE and W. VAN NIESSEN: *Chem. Phys.*, **58**, 71 (1981).

[124] F. A. GIANTURCO, C. GIDALLI and U. LAMANNA: *J. Chem. Phys.*, **57**, 840 (1972).

[125] J. S. SOMERS, TH. LINDNER, M. SURMAN, A. M. BRADSHAW, G. P. WILLIAMS, C. F. MCCONVILLE and D. P. WOODRUFF: *Surf. Sci.*, **183**, 576 (1987).

[126] J. STÖHR and R. JAEGER: *Phys. Rev. B*, **26**, 4111 (1982).

[127] F. SETTE, J. STÖHR and A. P. HITCHCOCK: *J. Chem. Phys.*, **81**, 4906 (1984).

[128] E. L. FORCE and A. T. BELL: *J. Catalysis*, **38**, 440 (1975); **40**, 356 (1975).

[129] M. BOWKER, M. A. BARTEAU and R. J. MADIX: *Surf. Sci.*, **92**, 548 (1980).

[130] M. A. BARTEAU and R. J. MADIX: *J. Chem. Phys.*, **74**, 4144 (1981).

[131] M. A. BARTEAU and R. J. MADIX: *J. Electron Spectrosc. Relat. Phenom.*, **31**, 101 (1983).

[132] K. C. PRINCE and A. M. BRADSHAW: *Surf. Sci.*, **126**, 49 (1983).

[133] K. C. PRINCE and G. PAOLUCCI: *J. Electron Spectrosc. Relat. Phenom.*, **37**, 181 (1985).

[134] E. M. STUVE, R. J. MADIX and B. A. SEXTON: *Chem. Phys. Lett.*, **89**, 48 (1983).

[135] J. A. CONNOR, M. CONSIDINE and I. H. HILLER: *J. Chem. Soc. Faraday Trans.*, 2, 1285 (1978).

[136] W. MORITZ and B. WOLF: *Surf. Sci.*, **88**, L29 (1979).

[137] B. E. HAYDEN, K. C. PRINCE, P. J. DAVIE, G. PAOLUCCI and A. M. BRADSHAW: *Solid State Commun.*, **48**, 325 (1983).

[138] D. H. S. YING and R. J. MADIX: *J. Catalysis*, **61**, 48 (1980).

[139] M. A. BOWKER and R. J. MADIX: *Surf. Sci.*, **102**, 542 (1981).

[140] B. A. SEXTON: *Surf. Sci.*, **88**, 319 (1979).

[141] B. A. SEXTON: *J. Vac. Sci. Technol.*, **17**, 141 (1980).

[142] B. E. HAYDEN, K. C. PRINCE, D. P. WOODRUFF and A. M. BRADSHAW: *Phys. Rev. Lett.*, **51**, 475 (1983); *Surf. Sci.*, **133**, 589 (1983).

[143] A. PUSCHMANN, J. HAASE, M. D. CRAPPER, C. E. RILEY and D. P. WOODRUFF: *Phys. Rev. Lett.*, **54**, 2250 (1985); M. D. CRAPPER, C. E. RILEY, D. P. WOODRUFF, A. PUSCHMANN and J. HAASE: *Surf. Sci.*, **171**, 1 (1986).

[144] J. STÖHR, D. A. OUTKA, R. J. MADIX and U. DÖBLER: *Phys. Rev. Lett.*, **54**, 1256 (1985); D. A. OUTKA, R. J. MADIX and J. STÖHR: *Surf. Sci.*, **164**, 235 (1985).

[145] M. D. CRAPPER, C. E. RILEY and D. P. WOODRUFF: *Phys. Rev. Lett.*, **57**, 2598 (1986); *Surf. Sci.*, **184**, 121 (1987).

[146] D. P. WOODRUFF, C. F. MCCONVILLE, A. L. D. KILCOYNE, TH. LINDNER, J. SOMERS, M. SURMAN, G. PAOLUCCI and A. M. BRADSHAW: *Surf. Sci.*, **201**, 228 (1988).

[147] J. SOMERS, A. ROBINSON, TH. LINDNER and A. M. BRADSHAW: *Phys. Rev. B*, **40**, 2053 (1989).

[148] G. A. BARCLAY and C. H. L. KENNARD: *J. Chem. Soc.*, 3289 (1961).

[149] D. M. L. GOODGAME, N. J. HILL, D. F. MARSHAM, A. C. SKAPSKI, M. L. SMART and P. G. H. TROUGHTON: *Chem. Commun.*, 629 (1969).

[150] J. L. GLAND and G. A. SOMORJAI: *Surf. Sci.*, **38**, 157 (1973).

[151] P. C. STAIR and G. A. SOMORJAI: *J. Chem. Phys.*, **67**, 4361 (1977).

[152] S. LEHWALD, H. IBACH and J. E. DEMUTH: *Surf. Sci.*, **78**, 577 (1978).

[153] F. P. NETZER and J. A. D. MATTHEW: *Solid State Commun.*, **29**, 209 (1979).

[154] A. L. JOHNSON, E. L. MUETTERTIES and J. STÖHR: *J. Am. Chem. Soc.*, **105**, 7183 (1983).

[155] J. A. HORSLEY, J. STÖHR, A. P. HITCHCOCK, D. C. NEWBURY, A. L. JOHNSON and F. SETTE: *J. Chem. Phys.*, **83**, 6099 (1985).

[156] A. P. HITCHCOCK and C. E. BRION: *J. Electron Spectrosc. Relat. Phenom.*, **10**, 317 (1977).

[157] W. BUTSCHER, W. H. E. SCHWARZ and K. H. THUNEMANN: in *Inner-Shell and X-Ray Physics of Atoms and Solids*, edited by D. J. FABIAN, H. KLEINPOPPEN and L. WATSON (Plenum, New York, N.Y., 1981), p. 841.

[158] F. SETTE, J. STÖHR and A. P. HITCHCOCK: *J. Chem. Phys.*, **81**, 4906 (1984).

[159] J. C. GIORDON, J. H. MOORE and J. A. TOSSEL: *J. Am. Chem. Soc.*, **107**, 5600 (1985).

[160] V. N. AKIMOV, A. S. VINOGRADOV, A. A. PAVLYCHEV and V. N. SIRKOV: *Opt. Spectrosc. (USSR)*, **59**, 206 (1985).

[161] M. BADER, J. HAASE, K.-H. FRANK, C. OCAL and A. PUSCHMANN: *J. Phys. (Paris)*, C8, 491 (1986).

[162] I. NENNER and G. J. SCHULZ: *J. Chem. Phys.*, **62**, 1747 (1975).

[163] E. G. COX: *Rev. Mod. Phys.*, **30**, 159 (1958).

[164] R. K. CHANG and T. E. FURTAK, Editors: *Surface Enhanced Raman Scattering* (Plenum Press, New York, N.Y., 1983).

[165] B. J. BANDY, D. R. LLOYD and N. V. RICHARDSON: *Surf. Sci.*, **89**, 344 (1979).

[166] A. L. JOHNSON, E. L. MUETTERTIES, J. STÖHR and F. SETTE: *J. Phys. Chem.*, **89**, 4071 (1985).

[167] J. E. DEMUTH, K. CHRISTMANN and P. N. SANDA: *Chem. Phys. Lett.*, **76**, 201 (1980).

[168] N. V. RICHARDSON: *Vacuum*, **33**, 787 (1983).

[169] N. J. DiNARDO, PH. AVOURIS and J. E. DEMUTH: *J. Chem. Phys.*, **81**, 2169 (1984).

[170] J. E. DEMUTH and P. N. SANDA: *Phys. Rev. Lett.*, **47**, 57 (1981); PH. AVOURIS and J. E. DEMUTH: *J. Chem. Phys.*, **75**, 4783 (1981).

[171] A. OTTO, K.-H. FRANK and B. REIHL: *Surf. Sci.*, **163**, 140 (1985).

[172] R. DUDDE: DESY Report No. F 41 (1984), unpublished; R. DUDDE, E. E. KOCH, N. UENO and R. ENGELHARDT: *Surf. Sci.*, **178**, 646 (1986).

[173] M. BADER, J. HAASE, K.-H. FRANK, A. PUSCHMANN and A. OTTO: *Phys. Rev. Lett.*, **56**, 1921 (1986).

[174] J. U. MACK, E. BERTEL and F. P. NETZER: *Surf. Sci.*, **159**, 265 (1985).

[175] M. A. VAN HOVE, W. H. WEINBERG and C.-M. CHAN: *Low-Energy Electron Diffraction*, Springer Series in Surface Sciences, Vol. 6 (Springer-Verlag, Berlin, 1986).

[176] K. HEINZ, D. K. SALDIN and J. B. PENDRY: *Phys. Rev. Lett.*, **55**, 2312 (1985); K. HEINZ: *Appl. Phys. A*, **41**, 3 (1986).

[177] E. A. STERN, D. E. SAYERS and F. W. LYTLE: *Phys. Rev. B*, **11**, 4836 (1975).

[178] D. P. WOODRUFF: *Surf. Sci.*, **166**, 377 (1986).

Photoemission Spectroscopy of Interfaces.

G. Margaritondo

Department of Physics and Synchrotron Radiation Center
University of Wisconsin-Madison - Madison, WI

1. – Introduction.

Synchrotron radiation is an almost ideal instrument for the study of surfaces and interfaces [1, 2]. This is primarily due to the short mean free path of electrons excited by soft-X-ray synchrotron radiation inside a solid—rather than to the properties of the photons themselves. A beam of ultraviolet or soft-X-ray photons can travel over a relatively large distance in the solid before being completely absorbed. The magnitude of this distance is determined by the reciprocal of the optical-absorption coefficient, and is typically of the order of several hundred ångström. Thus the photon beam can probe a region whose thickness is large compared to the distances between adjacent atomic planes, a few angström.

On the other hand, the electrons that are excited by absorption of soft-X-ray photons in a solid can travel only over a few ångström or tens of ångström before losing part of their energy. This means, for example, that electrons which overcome the surface barrier and reach the vacuum, becoming photoelectrons, originate from a very thin layer close to the surface. Thus synchrotron radiation experiments based on the analysis of photoelectrons are highly surface sensitive. As such, they are very powerful in the study of surface chemical and physical phenomena, and in particular in the study of the formation of solid interfaces.

Synchrotron radiation studies of interfaces have been, since the 1970's, one of the most active areas of solid-state science [1-4]. Although these studies involve a wide variety of different experiments, their basic procedure is always the same. First, one or more surface-sensitive experimental techniques are used to characterize the properties of an ultraclean surface. Next, the same techniques are used to characterize the surface after it has been covered by a thin layer of another solid. The experiment is repeated for different thicknesses of the overlayer, until the signal from the interface region is attenuated too much by the overlayer and can no longer be detected.

Specific aspects of this field have been described by many recent reviews [1-4].

The purpose of this lecture is not to present a complete overview of results—a rather impossible task in the limited time allowed for it. I will try, instead, to provide up-to-date information on two specific, important problems. In the first case, I will discuss how the very extensive experimental information on semiconductor interfaces can be finally organized in a unified picture. In the second case, I will discuss a recent example of synchrotron radiation research in which surface sensitivity is perhaps an impediment rather than an advantage. This is the photoemission study of high-temperature superconductors.

Before discussing the two problems, I will briefly review the synchrotron radiation techniques that are most widely used in solid interface research. The general principles of synchrotron radiation experiments are covered by other distinguished lecturers in this school, and need no further treatment. For a more detailed discussion of the technique outlined here, the reader may want to use ref. [1].

The final part of my presentation is dedicated to future prospectives. In particular, I will try to answer the question: *How should we use for interface studies the new, ultrabright synchrotron radiation sources under development?* Specifically, I will discuss novel spectromicroscopies, that can potentially revolutionize interface research.

2. – Experimental techniques.

As we have seen, the photoelectric effect provides the basic instrument for most synchrotron radiation studies of interfaces. The initial state for this well-known phenomenon is the ground state of the electrons in the solid, $|i\rangle$. The phenomenon involves the annihilation of a photon of energy $\hbar\omega$ and momentum p_{ph}, and ends with a final state that is the combination of a free-electron state $|k, s\rangle$ for the ejected photoelectron and of a collective state of the solid minus one electron. The free-electron state is entirely determined, of course, by its k-vector, k, and by its spin polarization, s.

Experimentally, one collects and analyzes the ejected photoelectron, determining its state. Then, with the help of theory, one tries to retrieve the properties of the state $|i\rangle$ from those of the state $|k, s\rangle$. The analysis of the state $|k, s\rangle$ can be limited to measuring its (kinetic) energy, *i.e.* to determining the magnitude of k. This is performed with angle-integrating electron energy analyzers that collect and analyze photoelectrons emitted in many different directions from the solid. In angle-resolved photoemission, instead, the analysis includes the determination of both the direction and the magnitude of k. In spin-polarized photoemission, the analysis also includes measurements of the spin polarization of the photoelectrons. This last technique still suffers from low signal levels, although the situation is somewhat improving with the high brightness of the new undulator sources of synchrotron radiation. As a consequence, its applications in interface physics, although extremely interesting, are still limited.

The complete identification of k requires, of course, three scalar variables. These can be, for example, the kinetic energy, E_k, and the angles θ and ϕ shown in fig. 1. Similarly, the photon momentum p_{ph} is identified with three scalar variables, such as the photon energy and two angles that identify the direction.

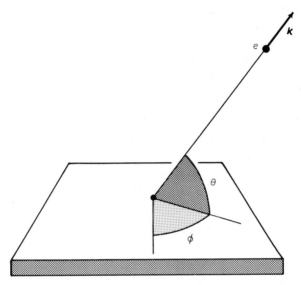

Fig. 1. – Definition of the angles θ and ϕ used to define the direction of emission of a photoelectron of k-vector, k.

Including the spin polarization and the photon polarization, the photoelectric process is characterized by eight scalar variables. In a given photoemission experiment, one or more of these eight variables are scanned and analyzed, while other variables are kept constant or averaged. This choice of variables gives the experimentalists considerable flexibility in performing the experiments, and in retrieving information on the electronic structure of the system under investigation. Each mode of photoemission, corresponding to a specific selection of the variables that are scanned, kept constant or averaged, is particularly effective in the study of a particular aspect of the electronic structure [1].

We will now summarize the features of photoemission modes that are most commonly used in interface research:

1) EDC mode. The acronym EDC is derived from the name of the spectra produced in this mode, «energy distribution curves». The mode consists of measuring the number of collected photoelectrons per unit time while scanning

E_k and keeping constant $\hbar\omega$. Depending on how the remaining variables are handled, one has the following submodes:

A) Angle-integrated EDC mode. The spectra are taken collecting electrons over a large spherical angle. In first approximation [5], the curves produced by this mode correspond to the density of states of the electrons in the ground state of the system under investigation. This correspondence is limited because of the limitations in the angular integration—rigorously, one should integrate over 4π steradians, while the maximum possible is, of course, 2π steradians, and the practical analyzers cover spherical angles smaller than this. Furthermore, it is affected by the photoelectronic cross-section [5].

B) ESCA. This acronym means electron spectroscopy for chemical analysis [6]. This is an angle-integrated EDC mode of photoemission based on the analysis of electrons excited from core levels. As is known, the energy of a core level depends in first approximation on the corresponding atomic level of the atom to which it belongs. Thus the detection of a given core level can be used for qualitative and quantitative analysis of the chemical composition of the surface or interface. The energy is also affected by the chemical status of the element, and its study delivers information on such a status.

C) Cross-section techniques. These are specialized techniques that enhance the spectral contributions of certain electronic states, by selecting the photon energy so that the corresponding cross-section is maximized [5]. We will see in sect. 4 that one of the cross-section techniques, the so-called *photoemission resonance*, is very important in the study of superconductors.

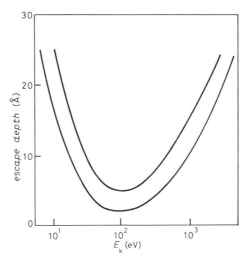

Fig. 2. – The photoelectron escape depth, l, as a function of the kinetic energy [1, 5]. Different materials correspond to different escape depths, but most results fall within the area outlined in this figure.

D) EDC mode with variable surface sensitivity. As shown in fig. 2, the «escape depth» for photoelectrons, determined by their mean free path, depends on the kinetic energy, E_k. In turn, because of energy conservation, E_k depends on $\hbar\omega$. This can be exploited to enhance the surface sensitivity of the probe, by selecting $\hbar\omega$ so that E_k corresponds to the minimum escape depth. Furthermore, it can be used to vary the surface sensitivity by varying $\hbar\omega$, thereby obtaining a layer-by-layer picture of the electronic states [7]. Note that, for valence electrons and shallow core levels, the maximum surface sensitivity corresponds to $\hbar\omega$ values that can only be obtained from synchrotron radiation sources [5].

E) Band structure mapping. This technique determines the electronic bands of crystalline systems [5]. The direction of k is measured besides E_k. Using well-established techniques, it is possible to derive from k the crystal momentum k_0 of the electron in the ground state. The corresponding initial-state energy is given by $E_i = E_k - \hbar\omega$. The curves $E_i(k_0)$ define the band structure of the system. In the case of interface studies, this technique is often applied to the study of the two-dimensional band structure of ordered overlayers.

2) CFS mode. This name stands for constant final state (photoemission) [5, 8]. The mode is implemented by scanning $\hbar\omega$ while keeping constant E_k, thereby scanning E_i. CFS experiments are typically performed without angular resolution. The CFS curves, like in the EDCs, correspond to the ground electronic density of states of the system under investigation. They are, however, partially immune from the convolution between initial and final states that is intrinsic to the EDCs [5].

3) CIS mode. In this constant-initial-state mode [5, 8], E_i is kept constant by scanning both $\hbar\omega$ and E_k under the condition $E_k - \hbar\omega = \text{const}$. CIS experiments are again typically performed without angular resolution. The CIS spectra primarily reflect the density of final states inside the solid, and the photoexcitation cross-section. In certain cases, the CIS mode of photoemission has been used to detect the extended X-ray absorption fine structure (EXAFS) with high surface sensitivity. This last technique is known as «photoemission EXAFS» [9].

4) Multibody techniques. In general, the photoemission process involves more than just the photon and the photoelectron [5]. It involves the collective states of all electrons in the solid, as well as quasi-particles such as phonons. Many-body effects are present in all of the previously illustrated techniques. For example, multiparticle excitations produce satellites of one-electron photoemission peaks, and these satellites can be a very important source of information. *Secondary* electrons, that have lost part of their energy before becoming photoelectrons, produce a characteristic tail and a low-E_k peak, present in the EDCs and in other photoemission spectra. The secondary electrons are exploited in a technique known as partial-yield spectroscopy [5, 10],

which measures the secondary-electron yield as a function of $\hbar\omega$. Such curves correspond to measurements of the optical-absorption coefficient as a function of $\hbar\omega$, performed with high surface sensitivity.

5) Spin polarization techniques. Many of the modes discussed above can be implemented with simultaneous analysis of the spin polarization. In practice, most spin-polarized photoemission experiments are performed in the angle-integrated EDC mode [5, 11].

Some photoemission techniques have been specifically developed for the study of interfaces. Many are highly specialized, but the following are now widely applied:

1) Schottky barrier measurements [5, 12]. For a semiconductor, the leading high energy of the angle-integrated EDCs corresponds to $E_v + \hbar\omega$, where E_v is the valence band edge energy. Thus the position of E_v can be derived from the spectral edge. When referred to the Fermi energy, E_F, the edge position measured in this way generally deviates from the bulk value (which is determined by doping). This is due to the fact that photoemission is highly surface sensitive, and the band structure of the semiconductor is bent as it approaches the surface, due to the extra charge in localized surface states. When a metal is deposited on the semiconductor forming a metal-semiconductor interface, the band bending is modified by the metal adatoms. The new distance in energy between E_v and E_F is, by definition, the Schottky barrier height in the

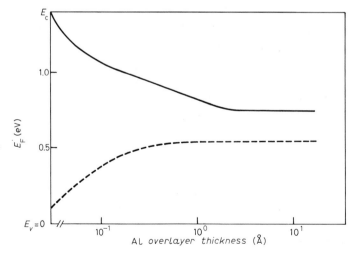

Fig. 3. – The leading edge of angle-integrated photoemission EDCs can be used to derive the surface/interface position of the Fermi level. In this case [13], the shift of E_F with respect to the valence and conduction band edges, E_v and E_c, is monitored during the formation of a metal-semiconductor interface: —— n-type, --- p-type. The final distance between E_F and E_v is, by definition, the Schottky barrier height ϕ for the p-type GaAs-Al interface. For the p-type case, the barrier is E_g minus the distance between E_c and E_F.

case of a p-type semiconductor (and E_g minus the Schottky barrier height for n-type materials, where E_g is the forbidden gap). Thus photoemission spectroscopy can be used to measure the Schottky barrier height in a direct way, bypassing all the difficulties of space-averaging techniques such as the transport measurements. Furthermore, this approach can be used to follow the establishment of the Schottky barrier step by step, for different thicknesses of the metal overlayer. Figure 3 shows an example of the applications of this technique.

2) Band lineup measurements [3, 4]. Assume now that the experiment involves the deposition on a semiconductor substrate of a *semiconductor* overlayer, rather than of a metal overlayer. Such an experiment studies the formation of a heterojunction instead of a metal-semiconductor junction. If the overlayer is sufficiently thin and photoelectrons emitted from the substrate can still be observed, the angle-integrated EDC has a double leading edge, due to the presence of two E_v's. From the double edge, it is possible to measure the distance in energy of the two E_v's, thereby deriving the interface valence band discontinuity, ΔE_v, and the lineup of the band structures of the two semiconductors that determines such a discontinuity. The discontinuity so measured is a feature of extreme importance for the technological use of heterojunctions. In many cases, unfortunately, the double edge is not resolved. However, ΔE_v can still be derived with photoemission methods [3, 4], essentially by measuring the valence band edge of the substrate, then following its changes due to band-bending modifications during the interface formation, and finally relating the final position to that of the overlayer valence band edge. This method is ordinarily implemented by deriving band-bending changes from the shifts in energy of core level photoemission peaks. Figure 4 shows an example of double-edge spectrum for a heterojunction interface.

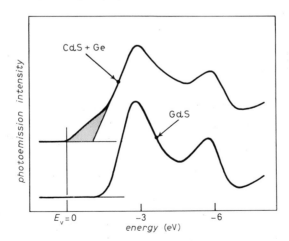

Fig. 4. – The leading double edge of EDCs taken on Ge-covered CdS [14] reveals the positions of the two valence band edges, and can be used to derive the corresponding discontinuity ΔE_v.

3) Morphology studies from core level peak intensities. Assume that an interface is formed by depositing an overlayer on a substrate. Consider the intensity of EDC core level peaks from the substrate and from the overlayer. The intensity depends, of course, on the thickness of the overlayer. The specific form of such a dependence contains information on the morphology of the interface[15]. Consider the simplest case of a smooth, nondiffusive overlayer of thickness τ. It is easy to demonstrate that the intensity for a substrate core level peak is proportional to $\exp[-\tau/l]$, where l is the photoelectron escape depth. Similarly, the intensity of an overlayer core level peak increases like $1 - \exp[-\tau/l]$. Deviations from such functional dependences indicate that the overlayer is either nonsmooth or diffusive or both. Detailed studies of the intensity as a function of τ, complemented by theoretical models, can be used to identify morphological features such as the formation of islands. This method is enhanced by the study of different components of the same core level, corresponding to different chemical configurations of the corresponding element. In this way, one can obtain at the same time chemical and morphological information. Morphological information is also obtained from the study of core level intensities as a function of the angle θ[16].

In summary, modern photoemission spectroscopy provides a powerful array of techniques for the study of interface formation processes. In principle, by using many or all of the above techniques one could obtain very extensive information. The information can be further extended by using nonphotoemission techniques such as surface vibrational spectroscopy and electron microscopy. One should not forget, however, that interfaces are unstable systems, easily contaminated even at ultrahigh-vacuum pressures. In practical experiments, only a limited number of techniques can be applied to characterize an interface, before it becomes contaminated.

3. – A unified picture of semiconductor interfaces.

Photoemission experiments on interface formation processes have been, for fifteen years, one of the most active areas of solid-state research[1-4, 15]. This is due, on the one hand, to the powerfulness and flexibility of the techniques outlined in the previous section. On the other hand, solid interfaces are systems of very high interest, fundamental as well as technological. They attract the attention of an increasingly large number of scientists, from theorists, to experimentalists, to technologists.

These fifteen years of activity have produced an impressive amount of data. Dealing with such a mass of information is difficult and may generate some discouragement. Solid interfaces are, after all, very complicated systems, whose properties are determined by the interplay of many chemical, physical and morphological factors. The proliferating mass of data may suggest that no generalized theoretical scheme can describe the interface properties—and that interface science is on the way to becoming a mere collection of uncorrelated data and recipes.

We would like to show that this need not be true. Using semiconductor interfaces as an example, we would like to argue that a unified description is still possible even with the current, large mass of data. The unified description might not be able to account for all phenomena of all interfaces, but it does explain the general trends of diverse classes of semiconductor interfaces, and accounts for many data with a small number of physical mechanisms.

The basic problems in semiconductor interface research are 1) the identification of the mechanisms that determine the Schottky barrier height at metal-semiconductor interfaces, 2) the identification of the mechanisms that determine the band lineups at heterojunction interfaces. The first problem has been treated by theorists since the 1930's, and the second since the late 1950's. In general, theories which have been developed to treat the first problem can be generalized to deal with heterojunctions as well. Most such theories belong to three main classes: 1) Schottky-like models, 2) midgap energy models, 3) defect models.

3´1. *Basic models of semiconductor interfaces.* – The basic philosophy of the Schottky-like models is to simulate the properties of a metal-semiconductor interface by using simple combinations of the properties of the two interfaces between each component and vacuum[17]. The most important parameters of the interfaces with vacuum are the *work function* for the metal, ψ, and the *electron affinity* for the semiconductor, χ. As shown in fig. 5, if one neglects the

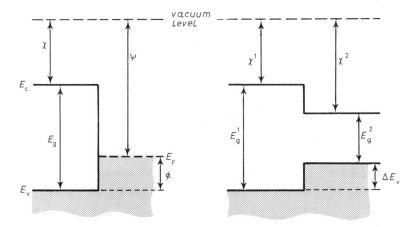

Fig. 5. – Elementary explanation of the Schottky model[17] for metal-semiconductor interfaces (left) and of the corresponding Anderson model[18] for semiconductor-semiconductor interfaces. For simplicity, the region shown is very close to the interface, and the band bending is neglected.

microscopic properties of the «real» interface, the Schottky model simply gives

$$\phi = \chi - \psi + E_g,\tag{1}$$

where ϕ is the Schottky barrier for the p-type semiconductor-metal interface. Note that a similar expression is valid for n-type semiconductor-metal interface. For simplicity, we will assume p-type semiconductor-metal interfaces throughout our discussion.

Similarly, fig. 5 shows that a Schottky-like approach gives, for the valence band discontinuity caused by the lineup at a heterojunction interface,

$$\Delta E_v = (\chi^1 - \chi^2) + (E_g^1 - E_g^2),\tag{2}$$

where the χ^i's are the electron affinities of the two semiconductors, and the E_g^i's are their gaps. Equation (2) is known as the *electron affinity rule*, and was introduced in 1962 by a classic work of Anderson[18].

There are at least two fundamental problems in Schottky-like models. First, eqs. (1) and (2) are derived neglecting screening effects. Second, they neglect the real processes that take place during the formation of a real interface. It is not surprising, therefore, that their predictions are typically not in agreement with the experimental findings. For example, the linear relation between Schottky barrier height and metal work function which is predicted by eq. (1) is not observed for the most important semiconductor materials, such as silicon. Specifically, the slope of the $\phi(\psi)$ curve is much smaller, in most cases, than the value of unity predicted by the equation—and often very close to zero, *i.e.* ϕ is completely independent of ψ.

The «real» interface will include, at least in some cases, localized interface phases that are created by chemical reactions between the interface components. Such phases can be more or less extended in the direction perpendicular to the interface, depending, for example, on the chemical reactivity of the components. In an attempt to provide a more realistic description of these effects, FREEOUF and WOODALL[19] modified the basic Schottky and Anderson models, developing their *effective work function model*. This model replaces the work function of the metal with an effective value, determined by reacted interface phases. For binary semiconductors, FREEOUF and WOODALL argued that the anion species dominate the interface phases and the effective work function.

In general, the «real»-interface formation process strongly modifies the local electronic structure of the components, and can introduce localized electronic states. In the late 1940's, BARDEEN[20] used the concept of Fermi-level pinning by localized states to explain the failure of the Schottky model. Assume that the density of interface states is of the order of one per interface atom, and that such states are localized within a few atomic planes of the interface. The equivalent bulk density is many orders of magnitude larger than the density of doping impurities. Hence, the local position of the Fermi level is determined by the interface states. This is still true even if the interface states are much less than one per interface atom.

Several factors can determine the nature of the interface electronic states. Even in the case of a clean surface, localized states can be created by the interruption of the bulk periodic potential. In the late 1970's, SPICER and co-workers [21] proposed that defects are responsible for the localized states that determine Schottky barrier heights. Perhaps the most important fact leading to this conclusion was the rapid formation of the Schottky barrier when a metal overlayer is deposited on a III-V substrate. Using photoemission methods, the Fermi level was seen to move from its clean-surface position in the III-V gap to a position consistent with the final Schottky barrier height—and to reach such a position for overlayer thicknesses of the order of 1/10 monolayer or even less. This suggested that the states responsible for the Fermi level pinning are related to defects, *e.g.*, vacancies or antisites, induced by the adsorption process on the III-V surface.

The relevant defect states in the defect model, therefore, are determined by the substrate and do not depend much on the nature of the overlayer. Photoemission measurements of many Schottky barrier heights involving different metals and the same semiconductors seemed to confirm this prediction. For example, many of the measured *p*-type Schottky barriers could be expressed as

$$(3) \qquad\qquad \phi = E_D,$$

where E_D is the interface Fermi-level pinning position for the semiconductor, interpreted by the defect model as due to defects. In eq. (3), E_D is measured from the valence band edge.

The unified defect model could be easily extended to heterojunction interfaces [22]. Assume that surface defects determine the position of the Fermi level in the gap of each of the two semiconductors. Since E_F is the same for both sides of the interface, the band lineup is automatically determined. As shown in fig. 6, after requiring continuity of E_F and, therefore, the coincidence of E_D^1 and E_D^2, the valence band discontinuity is

$$(4) \qquad\qquad \Delta E_v = E_D^1 - E_D^2,$$

where the E_D^i's are the defect-induced Fermi level pinning positions of the two semiconductors.

The defect model cannot explain some relevant experimental results. For example, not all the Schottky barriers between a given semiconductor and different metals can be explained by eq. (3). Furthermore, the Fermi level is often found to reach a given position for ultralow overlayer thicknesses, but then

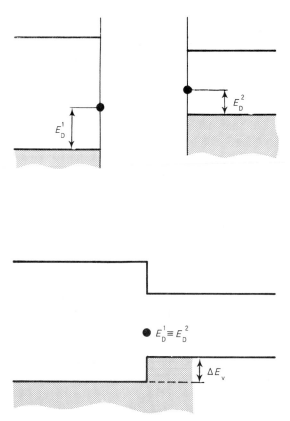

Fig. 6. – The defect model [21] applied to heterojunction band lineups.

to move from this to another position for larger thicknesses, as shown in fig. 7. In the case of heterojunctions, it was found [3, 4, 23, 24] that controlled interface contamination can modify the value of ΔE_v. This is in disagreement with the prediction of eq. (4), since in the defect model each pinning position should be independent of the specific nature of the other side of the interface. These facts could be explained by more complicated versions of the defect model, based on the interplay of several different kinds of defects. This, however, jeopardizes the simplicity that is one of the most appealing characteristics of this approach.

From a more fundamental point of view, the difficulty encountered by the defect model in the case of metal-semiconductor interfaces is that it neglects the effects of the metal overlayer on the localized electronic states. For example, the overlayer can modify the local screening. Furthermore, the bulk metallic wave function can produce effects that are very similar to those of localized states—and influence E_F in competition with the defect states. This concept was introduced by HEINE [25], and then elaborated by TERSOFF [26, 27] and by FLORES and his co-workers [28]. Consider, for example, a bulk metal wave

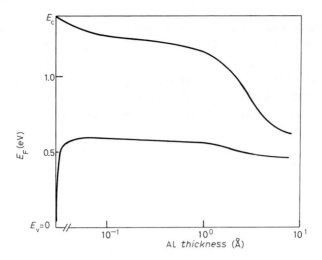

Fig. 7. – In many cases, the Fermi level at a metal-semiconductor interface reaches a given position in the gap for a low thickness of the metal overlayer—but then it moves to a new position for higher coverages. The data were taken[13] by depositing Al on low-temperature GaAs substrates. The upper curve is for an *n*-type substrate, and the other is for a *p*-type substrate.

function at a metal-semiconductor interface, and assume that its energy corresponds to a forbidden energy in the gap of the semiconductor. The wave function does not correspond, therefore, to a propagating state in the semiconductor. It can, however, penetrate to some extent the semiconductor side by tunneling, producing what TERSOFF calls «metal-induced gap states» or MIGS[26, 27]. Such states influence and can even determine the interface position of the Fermi level.

Specifically, TERSOFF argued that the MIGS pin E_F at a *midgap energy* level E_M, characteristic of the semiconductor and independent of the metal. Thus the effects of MIGS are somewhat similar to those of defect states, and are described by equations similar to eqs. (3) and (4):

$$(5) \qquad\qquad \phi = E_M,$$

$$(6) \qquad\qquad \Delta E_v = E_M^1 - E_M^2,$$

where the energies E_M, E_M^1 and E_M^2 are again measured from the corresponding valence band edge. A recent article by CARDONA and CHRISTENSEN introduces interesting and illuminating ideas on the nature of the midgap energy[29].

Midgap energy theories explain a number of experimental facts. For example, in cases like that illustrated by fig. 7, it was found that the Fermi level reaches its final interface position at overlayer thicknesses for which the interface begins to exhibit metallic character. This was demonstrated, for example, by high-resolution electron energy loss experiments in the case of Al overlayers on Si [30]. On the other hand, the fact that a given semiconductor can produce different Schottky barriers is in disagreement with eq. (5) as well as with eq. (3). Similarly, the contamination-induced modifications in the band lineup at the interface between two given semiconductors are in disagreement with eq. (6) as well as with eq. (4).

3˙2. *Some relevant experimental facts.* – We can summarize the previous subsection by saying that each one of the three fundamental classes of semiconductor interface theories has strong points and is supported by some experimental evidence—but also weak points and negative experimental evidence. Thus none of these theories can lead by itself to a unified description of semiconductor interfaces. It should be noted that, by themselves, measurements of Schottky barrier heights and heterojunction band discontinuities do not provide clear-cut evidence in favor of one model. In many cases, the predicted E_F pinning positions are similar for different models, and the agreement with measured values is not a strongly discriminating test.

The way to a unified description is opened by a number of recent experiments. This subsection describes some of them. I must clearly state that this is not an exhaustive description. Many additional experiments lead to the same conclusions. I am merely presenting results of experiments in which I was personally involved, because I am most familiar with their details.

The first series of experiments treated here explores the dependence of the Schottky barrier formation process on the temperature of the substrate [13]. These experiments produced a large mass of results, and I will limit the description to the most relevant messages for our present objective. It was found that, for a given metal-semiconductor pair, 1) the dependence of the Fermi level position on the metal overlayer thickness is different for different substrate temperatures, and 2), as we have seen, the final E_F pinning position is often different from the submonolayer value, and is reached when metallic character appears at the interface.

The first fact can be explained assuming that defect states influence the Fermi level position, and that the formation of such states depends on the temperature. This last assumption is quite reasonable. The formation of defects requires much energy, and the condensation of metal clusters has been proposed as a possible source of the required energy [31]. In turn, the formation of clusters depends on the surface diffusion of the adatoms, which depends on the substrate temperature.

The second fact, on the contrary, is explained by midgap energy models. Therefore, the explanation of these experiments requires *both* MIGS and defect states. The substrate temperature influences the interplay of MIGS and defect states, by affecting, for example, the rate of defect formation. Depending on their density, the defect states can either dominate the Fermi level pinning position or have negligible effects with respect to the MIGS.

The second series of experiments confirms the interplay between MIGS and defect states. The experiments investigated the deposition of semiconductor overlayers on metal substrates [32]. Paradoxically, almost all previous photo-emission studies of metal-semiconductor interfaces investigated metal over-layers on semiconductor substrates. There is no fundamental reason for this complete asymmetry—the practical motivation is that semiconductor substrates are somewhat easier to clean and be kept clean than metal substrates.

We will consider, specifically, the case of Si deposited on single-crystal Al(111) [32]. Previous experiments on Al overlayers on Si(111) had shown that the final Fermi level pinning is related to the appearance of interface metallic character [30], in agreement with the MIGS model. On the other hand, when Si is deposited on Al(111), photoemission measurements give a Schottky barrier height much different from that obtained by depositing Al on Si. Si overlayers on Al(111) are likely to have a large defect density, and in fact the measured E_F pinning position is consistent with an acceptor defect level caused by vacancies. Once again, the Schottky barrier height is determined by the interplay of MIGS and defect states whose relative weights are determined, in this case, by the deposition sequence.

In first approximation, therefore, we must describe metal-semiconductor and heterojunction interfaces using either eqs. (1), (2) or eqs. (3), (4), depending on which factor—MIGS or defects—prevails. Note that both sets of equations have

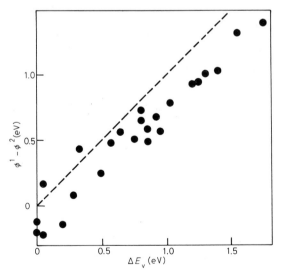

Fig. 8. – General correlation between the measured ΔE_v for each heterojunction and the difference of the p-type Schottky barrier heights for the corresponding metal-semicon-ductor interfaces involving the two components of the heterojunction [3, 4]. In this case, the metal was Au. The dashed line corresponds to the predictions of eq. (7).

one important consequence in common. They predict that Schottky barrier heights and the heterojunction band lineups are related to each other. In fact, both sets give

(7)
$$\Delta E_{\mathrm{v}} = \phi^1 - \phi^2 ,$$

where ΔE_{v} refers to the interface between two semiconductors, and the ϕ^i's are the p-type Schottky barrier heights between each semiconductor and a given metal. Note that, in principle, such a relation should be valid even without using the same metal.

The prediction of eq. (7) was tested using ΔE_{v}'s measured with synchrotron radiation photoemission and Au-based Schottky barrier heights obtained from transport properties [3, 4]. The results of the test are shown in fig. 8. We see that there is a clear correlation between ΔE_{v} and $\phi^1 - \phi^2$. We also see that the correlation deviates from the simple proportionality predicted by eq. (7). This discrepancy is the key for a complete generalization of the description of semiconductor interfaces.

3˙3. *The unified picture.* – In order to understand the discrepancy between eq. (7) and the experimental facts, a series of tests were performed to directly relate the properties of Schottky barriers and those of heterojunctions [14]. The experiments began with the study of the interface between two semiconductors, A and B. They continued with the study of interfaces A-M-B, involving A and B plus an ultrathin metal intralayer M between them. In the final stage, the metal intralayer thickness was increased to the point where the system was equivalent to two back-to-back, noninterfering Schottky barriers. The tests were performed for the systems ZnSe-Al-Ge, CdS-Al-Ge and GaP-Al-Si.

The basic results of these are the following. First, as we already mentioned [23, 24], the intralayer modifies the valence band discontinuity with respect to the intralayer-free interface. This fact *per se* confirms that the relation between Schottky barriers and heterojunctions deviates from the simple predictions of eq. (7). We note, *inter alia*, that these intralayer-induced changes of ΔE_{v} have great potential interest in technology. Second, the final value of ΔE_{v} is reached for very low values of the intralayer thickness, as shown in fig. 9. The value of ΔE_{v} measured for intralayer thicknesses of the order of 1/2 monolayer is already indistinguishable from the value for very large thicknesses. In turn, the latter value coincides with $\phi^1 - \phi^2$. Thus the back-to-back Schottky barrier limit, *i.e.* agreement with eq. (7), is reached for submonolayer intralayer thicknesses. However, eq. (7) does not explain the initial ΔE_{v}, measured for the intralayer-free interface.

The failures of eq. (7) can be understood by re-introducing the Schottky-like philosophy into the unified picture of semiconductor interfaces. In essence, we cannot neglect the role of the metal work functions and of the semiconductor

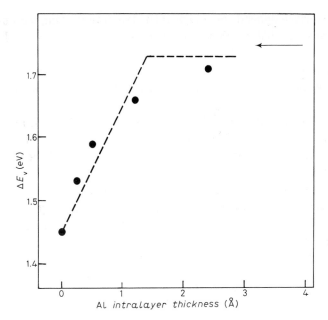

Fig. 9. – Changes in ΔE_v induced by a thin intralayer at the ZnSe-Ge heterojunction[14]. The dashed line emphasizes that the changes saturate at approximately 1/2 monolayer thickness of the intralayer. The horizontal arrow shows the back-to-back Schottky barrier, $i.e.$ $\Delta E_v = \phi^1 - \phi^2$.

electron affinities. Such a role, however, is not as dramatic as it was predicted by eqs. (1) and (2). This is due to the effects of screening[33], that can be described in first approximation by multiplying the Schottky-like terms by a so-called «pinning strength parameter», S. Thus, for example, eq. (5) becomes (using Cardona's and Christensen's theoretical framework[14, 29])

$$(8) \qquad\qquad \phi = E_M + S(\chi - \psi + E_g - E_M).$$

Note that, for $S \to 0$, the Schottky-like term is negligible and eq. (8) becomes equivalent to eq. (5). On the other hand, for $S = 1$, eq. (8) becomes equivalent to the Schottky limit, eq. (1). In the case of heterojunctions, eq. (6) must be replaced by

$$(9) \qquad \Delta E_v = E_M^1 - E_M^2 + S\left[(\chi^1 - \chi^2) + (E_g^1 - E_g^2) - (E_M^1 - E_M^2)\right],$$

that again converges to the Schottky-Anderson limit (eq. (2)) for $S = 1$, and to the MIGS limit (eq. (6)) for $S = 0$.

Similar equations are easily derived when the Schottky-like terms are introduced in the defect model, and they coincide with eqs. (8) and (9) after replacing the E_D^i's with the E_M^i's. The specific set of equations that must be used for a given interface depends on which of the two interplaying factors, defects or MIGS, prevails for that particular system.

Consider now the difference between the heights of two back-to-back Schottky barriers, $\phi^1 - \phi^2$. Assuming that S^1 and S^2 are the S-parameter of the corresponding metal-semiconductor interfaces, from eq. (8) we have

(10) $$\phi^1 - \phi^2 = E_M^1 - E_M^2 + S^1(\chi^1 - \psi + E_g^1 - E_M^1) - S^2(\chi^2 - \psi + E_g^2 - E_M^2)$$

(note that ψ is the same for both interfaces, since the metal is the same). In order to compare this result to the ΔE_v for the intralayer-free heterojunction A-B, we will assume that the S-parameter for the heterojunction is the average of those of the two Schottky barriers: $S = (S^1 + S^2)/2$. After introducing this expression in eq. (9) and comparing the result with eq. (10), with a few trivial steps we find

(11) $$\phi^1 - \phi^2 = \Delta E_v + \xi(S^1 - S^2),$$

with $\xi = (\chi^1 + E_g^1 - E_M^1 + \chi^2 + E_g^2 - E_M^2 + 2\psi)/2$.

The S^i's depend on screening and increase with the reciprocal of the dielectric constant of the corresponding semiconductor [34]. In turn, the dielectric constant increases with the reciprocal of the gap, and, therefore, with the reciprocal of the midgap energy, since E_M increases with E_g. Assuming $S^i = aE_M^i + b$, with a and b constants, we have $S^1 - S^2 = a(E_M^1 - E_M^2)$. If the Schottky-like correction is small, this last expression is close to ΔE_v, thus eq. (11) becomes

(12) $$\phi^1 - \phi^2 \approx \Delta E_v(1 + K),$$

with $K = \xi a$, i.e.

(13) $$K = \frac{a}{2}(\chi^1 + E_g^1 - E_M^1 + \chi^2 + E_g^2 - E_M^2 + 2\psi).$$

Equation (12) replaces eq. (7), and in conjunction with eq. (13) explains its failures. In the case of fig. 8, it explains why the slope is different from unity, by predicting a slope $1 + K$. Similarly, it explains intralayer-induced changes like those of fig. 9. Note that the sign of K is primarily determined by the metal work function, ψ. Numerical estimates of K [14] are in qualitative agreement with the slope of fig. 8, which implies $K < 0$, and with the fact that Al intralayers increase the valence band discontinuity.

The quantitative predictions of eqs. (12) and (13) have limited accuracy, as should be expected. The main limitation of this approach is the treatment of the screening effects. The above treatment was based on the dielectric constant, and it is questionable, since the interface is a highly localized system. On the other hand, the Schottky-like correction is predicted by eqs. (12) and (13) with the correct sign and with the correct order of magnitude.

In summary, eqs. (8) and (9) and their counterparts derived from the defect model provide a unified description of many phenomena observed for different kinds of semiconductor interfaces. The basic philosophy of this description is that the properties of each specific interface are primarily determined by either MIGS or defects states, depending on the density of the defects, with an additional Schottky-like term whose magnitude depends on screening. As already mentioned, the same conclusion is supported by the results of other authors. Particularly noteworthy is the work of Mönch [35], who provided strong evidence for this unified picture by analyzing data obtained on a large number of silicon-based interfaces.

4. – High-temperature superconductivity.

Surfaces and interfaces based on metals and/or semiconductors play the leading role in the applications of photoemission spectroscopy to solid-state science. Since early 1987, however, a new kind of application has been growing steadily—the study of the electronic structure of high-temperature superconductors. This field was activated almost overnight by Paul Chu's announcement of superconductivity above liquid-nitrogen temperature [36], and has recently produced some very exciting results [37].

High-temperature superconductors are a tough challenge for photoelectron spectroscopists. One of the problems is the complex chemical composition of the samples, that include at least four different elements. This makes it difficult to distinguish from each other the spectral contributions of different elements. Synchrotron radiation provides powerful help in solving this problems. In fact, by taking EDCs at different photon energies, one can exploit the photon energy dependence of the cross-section of each spectral contribution to identify the corresponding element.

The most spectacular application of this technique is based on photoemission resonances [5]. In general, a resonance occurs, for a valence band spectral feature due to a given element, when the photon energy is close to one of the core level optical-absorption thresholds of the element. Such photons, in fact, can produce the same photoelectron state $|k, s\rangle$ of energy E_k in two different ways. The first process is the direct excitation of the photoelectron from the initial state $|i\rangle$. The second is the optical absorption of the photon by an electron in the core level, followed by decay to the same level and transfer of the energy to another unexcited electron, that is thus excited to the state $|i\rangle$. The quantum interference between the two processes gives rise to rapid variations in the photoionization cross-section as a function of $\hbar\omega$. Similar resonance phenomena occur for spectral features caused by multibody processes.

Consider the $YBa_2Cu_3O_{6.8}$ spectra of fig. 10, that were taken at three different photon energies. One of such energies is close to a core level threshold of copper.

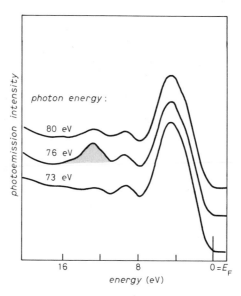

Fig. 10. – EDCs taken on the high-temperature superconductor $YBa_2Cu_3O_{6.8}$ reveal the resonant behavior of a Cu-related peak for photon energies close to a Cu core level absorption threshold [38].

We can see the spectral contribution from copper resonating. This immediately identifies the position in energy of the contribution, that is much weaker in the other spectra. The feature is a well-known satellite of the main Cu peak, and from its lineshape it has been possible to derive information on the valence state of copper [38]. This information is extremely important for the interpretation of the mechanism of high-temperature superconductivity.

Another difficult problem encountered in the photoemission studies of high-temperature superconductors is the quality of the materials. Most of the early samples were sintered pellets. As we have seen, photoemission spectroscopy is highly sensitive to the surface properties, and it must be performed on ultraclean surfaces. Methods such as fracturing or scraping under vacuum have been applied with varying degrees of success to the production of clean surfaces from superconductor pellets. In general, however, the use of these methods leaves doubts about the quality of the surfaces and of the corresponding results.

Late in 1987, significant progress was made when single crystals of the high-temperature superconductor $YBa_2Cu_3O_{7-x}$ were successfully cleaved under ultrahigh vacuum [39]. Figure 11 shows one of the first low-energy electron diffraction patterns obtained from cleaved $YBa_2Cu_3O_{7-x}$. The cleaved single-crystal surfaces enable the experimentalists to perform photoemission studies in the angle-resolved mode. For example, they searched extensively for evidence of angular dispersion in the photoemission spectra, that is a signature of the band structure.

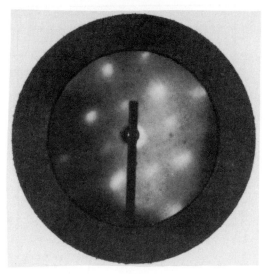

Fig. 11. – The first low-energy electron diffraction (LEED) patterns obtained [39] from cleaved single-crystal $YBa_2Cu_3O_{7-x}$.

As shown in fig. 12, hardly any dispersion is seen in the angle-resolved spectra of $YBa_2Cu_3O_{7-x}$. The only significant exception is a small upward shift in energy of the leading edge when the collection angle is moved away from the direction normal to the sample surface. This shift is qualitatively consistent with the dispersion of the upper valence bands when one moves away from the Γ-point in k-space [39]. Quantitatively, however, the shift is much less than theoretically predicted.

The lack of dispersion in fig. 12 can be explained, in part, by the large number of electrons in the unit cell. This corresponds to a large number of bands, and makes it difficult to observe the dispersion of a single band. Even so, the almost total lack of dispersion is somewhat puzzling, and it raises serious questions about the kind of information provided by photoemission for these materials. Can, for example, a single-particle picture be used at all for these systems? Other experimental features fueled these doubts. In particular, one sees from the $YBa_2Cu_3O_{7-x}$ spectra no evidence of signal at the Fermi edge. This could be explained by a strong role of correlation, that would make the conventional one-electron description of photoemission spectra totally inadequate for high-temperature superconductors.

Some of these crucial questions have been attenuated by the recent advent of a new class of high-temperature superconductors, $i.e.$ the Bi-Ca-Sr-Cu-O (BCSCO) family. The photoemission spectra of the low-temperature phase of BCSCO [40], contrary to those of $YBa_2Cu_3O_{7-x}$, exhibit a small but clearly visible Fermi edge, as shown in fig. 13. It is not true, therefore, that the Fermi edge cannot be detected for high-temperature superconductors.

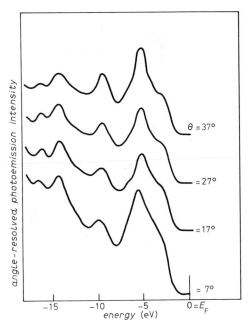

Fig. 12. – Angle-resolved EDCs taken on cleaved single crystals of $YBa_2Cu_3O_{7-x}$, at different values of the angle θ along the Γ-M direction of the two-dimensional Brillouin zone, photon energy 30 eV [39].

The observation of the Fermi edge for BCSCO recently led to a very exciting result [41]. For the first time, the changes in the electronic structure near E_F that accompany a superconducting transition were directly observed in photoemission spectroscopy. These changes included the opening of the superconduc-

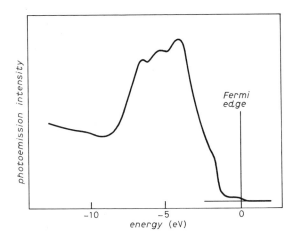

Fig. 13. – Contrary to $YBa_2Cu_3O_{7-x}$, the EDCs of the new superconductor BCSCO exhibit a clearly visible Fermi edge [40].

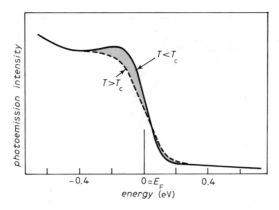

Fig. 14. – For the first time, the changes in the density of states near the Fermi edge that accompany the onset of superconductivity are observed with photoemission spectroscopy. The EDCs were taken on single-crystal BCSCO [41].

tivity gap and the growth of a peak in the one-electron density of states immediately below the gap. Figure 14 shows the changes in the Fermi edge spectrum of BCSCO when the temperature is lowered below the critical value. Figure 15 shows a comparison between the well-known BCS density of states, the same curve convoluted with the instrumental response and an experimental curve.

The preliminary data suggest a superconductivity gap of the order of a few ten meV, that is quite reasonable when compared to the results of tunneling experiments. Note, however, that the BCS function is *not* a good description of our data. First, the one-electron density of states does not account for the photoemission process. A more complete quasi-particle description [41] indicates that the photoemission spectrum for the superconducting state is proportional to the BCS density of states plus a step function. Second, both the BCS density of states and the quasi-particle picture are developed assuming weak coupling, while the ratio between Δ and the critical temperature (85 K) is large, and indicates strong coupling. In fact, the data of fig. 15 show that the area is not conserved during the superconducting transition, consistent with strong coupling.

The results of fig. 15 open a new chapter in photoemission spectroscopy. For many years, photoemission studies of superconductivity were severely limited by the huge difference between the magnitude of the superconductivity gap and that of the instrumental resolution of electron analyzers. The improvements in the signal level, primarily due to better synchrotron radiation sources, have pushed up the resolution limits. The new high-temperature superconductors produced a gigantic increase in the gap width. Figure 15 shows that the two improvements are now converging. This opens up excellent new possibilities for the study of the mechanism of superconductivity.

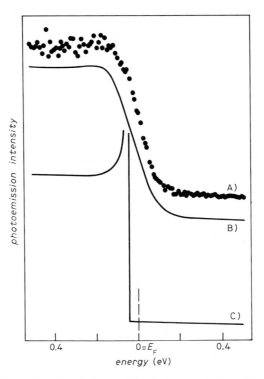

Fig. 15. – (Bottom) BCS density of states, for a half-gap of 40 meV. (Middle) The same function, after convolution with the instrumental broadening function. (Top) Experimental EDC taken on BCSCO below the critical temperature [41]. A quasi-particle picture of the photoemission process shows that the BCS function correctly describes the photoemission lineshape for weak coupling [41]. This corroborates the conclusion that the experiment is consistent with the opening of a superconductivity gap a few ten meV wide.

5. – Spectromicroscopy.

The recent breakthroughs in the photoemission studies of superconductivity occurred on the 100th anniversary of Hertz's discovery of the photoelectric effect [42, 43]. Photoemission spectroscopy is a rather remarkable technique, that after one century of prime contributions to science is again rejuvenating rather than becoming obsolete. Superconductor studies are only one of the many aspects of this rejuvenation.

Other exciting developments are produced by the advent of new, high-brightness sources of synchrotron radiation in the soft-X-ray spectral region. The two most significant examples are the Trieste storage ring and the Advanced Light Source (ALS) at Berkeley. In this section, we briefly review spectromicroscopy, that is one of the most advanced projects based on ultrabright synchrotron radiation.

Virtually all the present spectroscopy experiments performed with synchrotron radiation have had limited lateral resolution. This is primarily due to the low signal level. Roughly speaking, a spectrum taken with energy resolution of $(200 \div 300)$ meV required $(5 \div 10)$ min with the storage rings operating in the 1970's and early 1980's. The area of sample probed by such experiments had a lateral dimension of the order of a few hundred micrometers. With the improvements in the instrumentation and with the advent of intermediate-brightness sources such as the UV ring at Brookhaven, BESSY and Aladdin at Wisconsin, the time per spectrum was pushed down to seconds or less. Even so, the lateral resolution achieved by using a pinhole to limit the photon beam size cannot easily exceed 10 μm without increasing too much the time per experiment. Consider, in fact, that in the scanning mode of microscopy the time to obtain a complete micrograph is at least a factor 10^4 larger than that required for a single pixel. Even taking less data points per spectrum, the time would be too long for practical applications.

This situation can be improved in different ways. First, rather than using a pinhole, the lateral resolution can be achieved with focussing devices. One can use either photon focussing or electron focussing. In the first case, important breakthroughs were recently obtained. The use of multilayer coatings [44] increased the near-normal reflectivity in the soft-X-ray range well above 50%. This makes it possible to develop new and powerful optical components without requiring grazing incidence.

The second improvement is the brightness of the source. In the existing storage rings, the brightness can be increased by several orders of magnitude

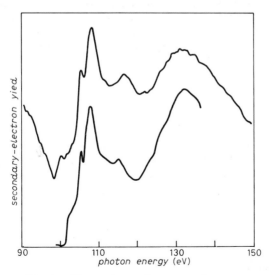

Fig. 16. – (Top) XANES (*i.e.* X-ray absorption near-edge structure) curve taken on a silicon oxide surface with a photoemission microscope at Wisconsin. The lateral resolution was of the order of 10 μm. The bottom curve shows, for comparison, a conventional XANES curve [45].

using undulators. Conservative estimates show that it is now feasible to reach lateral resolutions better than one micrometer while keeping good energy resolution. For example, the photoemission microscope MAXIMUM[44], under development at the Wisconsin Synchrotron Radiation Center with the collaboration of Minnesota, Berkeley and Xerox, will reach a lateral resolution of the order of 1000 ås and an energy resolution of 200 meV at 100 eV. Figure 16 shows an example of near-edge optical-absorption structure taken with good lateral resolution, also at Wisconsin, by the group of Tonner[45]. The spectrum was taken in the partial-yield mode of photoemission, with a focussing electron optical system.

The performances of these spectromicroscopes will further improve with the undulator radiation of the new, ultrabright sources of synchrotron radiation, reaching physical limitations not related to the signal levels. For example, an instrument like MAXIMUM will reach the diffraction limit of a few hundred ångström.

The applications of the novel spectromicroscopies made possible by these instruments are wide and very exciting. The term «spectromicroscopy» distinguishes these techniques from other microscopies, emphasizing that they couple lateral resolution and energy resolution. Thus they can provide information that neither pure spectroscopy nor pure microscopy can provide.

Consider, for example, photoemission microscopy. This technique will provide micromaps of the chemical composition of the specimen, plus micromaps of the chemical status of each element. The alternate technique, Auger microprobe, has limited perfomances and applications. Photoemission microscopy, on the contrary, will be extensively applied to diverse areas including medicine and biology.

In the case of interface research, photoemission spectromicroscopy will break the barriers that make it impossible to study interface properties with lateral resolution. Consider, for example, Schottky barriers and heterojunction band lineups. What is the lateral dependence of these properties? The answer to such a question could shed new light into the problem of semiconductor interfaces, broadening the description outlined in sect. 3. This, however, requires considerable lateral resolution. The relevant length reference in this case is the Debye length. In cases of relatively low doping, the Debye length can become of the order of thousands of ångström. Thus the new spectromicroscopes will be able to explore this fascinating and, until now, unexplored aspect of interface science.

* * *

The author's research described in this lecture was supported by the National Science Foundation, by the Office of Naval Research, by the Wisconsin Alumni Research Foundation and by Bellcore. Much of the detailed development of the unified descripton outlined in sect. 3 must be credited to D. W. NILES, and it greatly profited from interactions with J. TERSOFF, W. MÖNCH and M.

Cardona. I am grateful to B. TONNER for releasing the results of fig. 16 prior to publication. The University of Wisconsin Synchrotron Radiation Center is a national facility supported by the National Science Foundation.

REFERENCES

[1] G. MARGARITONDO: *Introduction to Synchrotron Radiation* (Oxford, New York, N.Y., 1988).

[2] E.-E. KOCH: *Handbook of Synchrotron Radiation* (North-Holland, Amsterdam, 1983).

[3] G. MARGARITONDO: *Electronic Structure of Heterojunction Interfaces* (Jaca-Reidel, Milan, 1989), and the references therein.

[4] F. CAPASSO and G. MARGARITONDO: *Heterojunction Band Discontinuities: Physics and Device Applications* (North-Holland, Amsterdam, 1987), and the references therein.

[5] For a detailed discussion of this point, see Chapt. IV in ref. [1].

[6] K. SIEGBAHN, C. NORDLING, A. FAHLMAN, R. NORDBERG, K. HAMRIN, J. HEDMAN, G. JOHANSSON, T. BERGMARK, S.-E. KARLSSON, I. LINDGREN and B. LINDBERG: *ESCA: Atomic, Molecular and Solid State Structure Studied by Means of Electron Spectroscopy* (Almqvist & Wiksells, Uppsala, 1967); K. SIEGBAHN, C. NORDLING, G. JOHANSSON, J. HEDMAN, P. F. HEDEN, K. HAMRIN, U. GELIUS, T. BERGMARK, L. O. WERME, R. MANNE and Y. BAER: *ESCA Applied to Free Molecules* (North Holland, Amsterdam, 1969).

[7] L. J. BRILLSON, A. D. KATNANI, M. K. KELLY and G. MARGARITONDO: *J. Vac. Sci. Technol. A*, **2**, 551 (1984).

[8] G. J. LAPEYRE, A. D. BAER, J. HERMANSON, J. ANDERSON, J. A. KNAPP and P. L. GOBBY: *Solid State Commun.*, **15**, 1601 (1974); G. L. LAPEYRE, J. ANDERSON, P. L. GOBBY and J. A. KNAPP: *Phys. Rev. Lett.*, **33**, 1290 (1977).

[9] K. M. CHOUDHARY, P. S. MANGAT, A. E. MILLER, D. G. KILDAY, A. FILIPPONI and G. MARGARITONDO: *Phys. Rev. B*, **38**, 1566 (1988).

[10] W. GUDAT and C. KUNZ: *Phys. Rev. Lett.*, **29**, 169 (1972).

[11] M. CAMPAGNA, D. T. PIERCE, F. MEIER, K. SATTLE and H. SIEGMANN: in *Advances in Electronics and Electron Physics*, Vol. 41, edited by L. MARTON (Academic Press, New York N.Y., 1976), p. 113; E. KISKER, K. SCRÖDER, R. CLAUBERG, W. GUDAT and M. CAMPAGNA: *Phys. Rev. Lett.*, **52**, 2285 (1984); *Phys. Rev. B*, **61**, 453 (1985), and the references therein.

[12] G. MARGARITONDO: *Solid-State Electron.*, **26**, 499 (1983).

[13] M. K. KELLY, A. KAHN, N. TACHE, E. COLAVITA and G. MARGARITONDO: *Solid State Commun.*, **58**, 429 (1986); *J. Vac. Sci. Technol. A*, **4**, 882 (1986); K. STILES, A. KAHN, D. G. KILDAY and G. MARGARITONDO: *J. Vac. Sci. Technol. A*, **5**, 1527 (1987); *J. Vac. Sci. Technol. B*, **5**, 987 (1987).

[14] D. W. NILES, M. TANG, J. McKINLEY, R. ZANONI and G. MARGARITONDO: *Phys. Rev. B*, **38**, 10949 (1988).

[15] G. MARGARITONDO: *Vuoto*, **16**, 3 (1986), and the references therein.

[16] L. J. BRILLSON: *Surf. Sci. Rep.*, **2**, 123 (1982).

[17] W. SCHOTTKY: *Z. Phys.*, **113**, 367 (1939); **118**, 539 (1942).

[18] R. L. ANDERSON: *Solid-State Electron.*, **5**, 341 (1962).

[19] J. L. FREEOUF and J. M. WOODALL: *Appl. Phys. Lett.*, **39**, 727 (1981); *Surf. Sci.*, **168**, 518 (1986).

[20] J. BARDEEN: *Phys. Rev.*, **71**, 717 (1947).
[21] W. E. SPICER, P. W. CHYE, P. SKEATH, C. Y. SU and I. LINDAU: *J. Vac. Sci. Technol.*, **16**, 1422 (1979); W. E. SPICER, I. LINDAU, P. SKEATH and C. Y. SU: *J. Vac. Sci. Technol.*, **17**, 1019 (1980), and the references therein.
[22] A. D. KATNANI and G. MARGARITONDO: *Phys. Rev. B*, **28**, 1944 (1983).
[23] D. W. NILES, G. MARGARITONDO, P. PERFETTI, C. QUARESIMA and M. CAPOZI: *Appl. Phys. Lett.*, **47**, 1092 (1985).
[24] P. PERFETTI, C. QUARESIMA, C. COLUZZA, C. FORTUNATO and G. MARGARITONDO: *Phys. Rev. Lett.*, **57**, 2065 (1986).
[25] V. HEINE: *Phys. Rev. A*, **138**, 1689 (1965).
[26] J. TERSOFF: *Phys. Rev. Lett.*, **52**, 465 (1984).
[27] J. TERSOFF: *Phys. Rev. B*, **30**, 4874 (1984).
[28] F. FLORES and C. TEJEDOR: *J. Phys. C*, **12**, 731 (1979).
[29] M. CARDONA and N. E. CHRISTENSEN: *Phys. Rev. B*, **35**, 6182 (1987).
[30] M. K. KELLY, G. MARGARITONDO, L. PAPAGNO and G. J. LAPEYRE: *Phys. Rev. B*, **34**, 6011 (1986).
[31] R. R. DANIELS, A. D. KATNANI, TE-XIU ZHAO, G. MARGARITONDO and A. ZUNGER: *Phys. Rev. Lett.*, **49**, 895 (1982).
[32] Y. CHANG, E. COLAVITA, N. TACHE and G. MARGARITONDO: *J. Vac. Sci. Technol. A*, **6**, 1971 (1988), and the references therein.
[33] W. A. HARRISON and J. TERSOFF: *J. Vac. Sci. Technol. B*, **4**, 1068 (1986), and the references therein.
[34] Most of this derivation must be credited to J. TERSOFF (unpublished).
[35] W. MÖNCH: *Phys. Rev. Lett.*, **58**, 1260 (1987).
[36] C. W. CHU, P. H. HOR, R. L. MENG, L. GAO, Z. J. HUANG and Y. K. WANG: *Phys. Rev. Lett.*, **58**, 405 (1987).
[37] For an extensive collection of photoemission results in this field see G. MARGARITONDO, R. JOYNT and M. ONELLION, Editors: *High-T$_c$ Superconducting Thin Films, Devices and Characterization, AIP Conf. Proc.* (New York, N.Y., 1989); J. A. HARPER, R. J. COLTON and L. C. FELDMAN, Editors: *Thin Film Processing and Characterization of High-Temperature Superconductors, AIP Conf. Proc.*, **165** (1988).
[38] N. G. STOFFEL, J. M. TARASCON, Y. CHANG, M. ONELLION, D. W. NILES and G. MARGARITONDO: *Phys. Rev. B*, **36**, 3986 (1987).
[39] N. G. STOFFEL, Y. CHANG, M. K. KELLY, L. DOTTL, M. ONELLION, P. A. MORRIS, W. A. BONNER and G. MARGARITONDO: *Phys. Rev. B*, **37**, 7952 (1988).
[40] M. ONELLION, M. TANG, Y. CHANG, G. MARGARITONDO, J. M. TARASCON, P. A. MORRIS, W. A. BONNER and N. G. STOFFEL: *Phys. Rev. B*, **38**, 881 (1988).
[41] Y. CHANG, M. TANG, R. ZANONI, M. ONELLION, G. MARGARITONDO, P. A. MORRIS, W. A. BONNER, J. M. TARASCON and N. G. STOFFEL: *Phys. Rev. B*, **39**, 4740 (1988).
[42] H. HERTZ: *Sitzungsber. Berl. Akad. Wiss.*, 9 June 1887; *Ann. Phys. (Leipzig)*, **31**, 983 (1887); *Electric Waves* (McMillan, London, 1900).
[43] G. MARGARITONDO: *Phys. Today*, **41**, No. 4, 66 (1988).
[44] F. CERRINA, G. MARGARITONDO, J. H. UNDERWOOD, M. HETTRICK, M. A. GREEN, L. J. BRILLSON, A. FRANCIOSI, H. HÖCHST, P. M. DELUCA jr. and M. N. GOULD: *Nucl. Instrum. Methods A*, **266**, 303 (1988), and the references therein.
[45] B. P. TONNER and G. R. HARP: *Rev. Sci. Instrum.*, **59**, 853 (1988), and unpublished; the conventional XANES curve of fig. 16 is derived from A. BIANCONI and R. S. BAUER: *Surf. Sci.*, **99**, 76 (1980).

Photoemission Experiments on Small Particles in Gas Suspension.

H. C. Siegmann and H. Burtscher

Swiss Federal Institute of Technology - CH-8093 Zürich, Switzerland

Small particles of liquid or solid matter are usually investigated on a substrate. However, small particles in gas suspension are also familiar to the physicist, and have played an important role in the development of modern physics. We recall the discovery of the Brownian motion, the study of light scattering which explains the spectacular phenomena when the Sun is rising or setting, and finally the famous determination of the elementary charge by Millikan. However, this lecture will deal with much smaller particles in the submicrometre range with diameters between 0.1 and 0.001 μm. These particles are practically invisible as the scattering cross-section for light becomes very small. Yet, particles of this size range are omnipresent in the air in which we live. Normally, their concentration varies within $(10^4 \div 10^6)$ particles/cm^3. Depending on their chemical composition, they can be toxic and lead to a series of undesired chemical reactions. However, without these particles, there would also be no cloud formation and hence no rain. It is incredible how little is known today about these particles, specifically their surface properties. The reason for this gap in our knowledge is the lack of experimental methods to investigate particularly the surface properties. This has to be done *in situ*, as the particles, when precipitated in a filter, will react with the substrate and the other particles in the filter and thus change their surface chemical composition, so that not much insight can be gained from the chemical analysis of precipitates regarding the initial state of the surface when the particle is still suspended in its original environment. Therefore, it is essential to develop experimental methods with which these submicrometre particles can be studied in their natural gaseous environment and with which the change of the surface chemistry can be observed when spurious amounts of other gases are admitted or when light is incident inducing photochemical reactions. One of the important methods to achieve this objective is photoelectric charging, as we shall show below. This technique has been developed at the Swiss Federal Institute of Technology by A. Schmidt-Ott, H. Burtscher, U. Müller, L. Scherrer, P. Cohn, the late

B. FEDERER and a number of students and simultaneously also at the University of Dortmund by Prof. R. NIESSNER and his collaborators.

1. – Putting electrical charges on particles.

Particles, in the submicrometre range, cannot be observed by scattering of light. However, they can be counted and size selected by measuring the electric mobility as soon as they carry an electric charge. So far, electrical charging of the particles has been achieved by generating ions in the gas in which the particles are suspended and letting the ions diffuse onto the particles. The ions are generated by a radioactive source, for instance in fire warning systems, or by an electrical discharge, as in the electrical filters that are widely used to clean exhaust gases from industrial processes. This charging of the particles by diffusion depends on the size of the particles, but not on their chemical nature or their surface properties. The size dependence of diffusion charging can be easily derived by considering a spherical particle of radius R, and a large sphere of radius r concentric to the particle. The current of charged gas molecules through the sphere with radius r is given by

$$(1) \qquad 4\pi r^2 \left(D \frac{dn}{dr} + b \cdot n(r) \cdot \frac{dU}{dr} \right) = \text{const} ,$$

where n is the number of charged gas molecules per cm^3, D the diffusion constant and b the electrical mobility. U is the potential energy between the diffusing gas molecule and the particle. For initially uncharged particles, $U = 0$, and (1) can easily be integrated yielding $n(r) = -\text{const}/4\pi rD + n(\infty)$ and $i = 4\pi r^2 \cdot dn/dr = 4\pi RD(n(\infty) - n(R))$, where $n(\infty)$ is the concentration of charged gas molecules at infinite distance from the particle and $n(R)$ the concentration of charged gas molecules at the particle surface. This can, of course, be applied also to uncharged gas molecules diffusing to the particle surface; $n(R)$ depends then on the sticking probability of these gas molecules at the particle surface. In the case of charged gas molecules, the sticking probability is unity and $n(R) = 0$. The rate of diffusion charging is then given by $i = 4\pi RDn(\infty)$. It becomes smaller as the particle radius R decreases. This explains then why very small particles cannot be efficiently removed from industrial exhaust in electrofilters. These particles are then found in great densities in the atmosphere and often have lifetimes of months or even years particularly if they are hydrophobic, that is if they cannot act as condensation nuclei for water. It should be noted that a correction to the above formula becomes necessary if the particle radius R is much smaller than the mean free path of the molecules. There is then a ballistic fall of the gas molecule close to the particle just as in vacuum and the charging rate decreases even faster with R for $R < 100$ nm, as was shown by HOPPEL[1].

This can now be contrasted with photoelectric charging of the particles, where the gas containing the particles is irradiated with ultraviolet light. The energy $h\nu$ of the photons has to be below the ionization threshold of the gas molecules, but above the photoelectric work function of the particle. In this way, the gas molecules remain electrically neutral, but the particle can absorb a photon and emit a photoelectron. This process obviously depends on the chemical composition of the particle and particularly on the state of its surface, as the photoelectric work function is a very sensitive function of the nature and amount of adsorbates at the surface. However, whether or not the particle will remain with a positive charge depends on whether or not the photoelectron diffuses back to it. Hence photoelectric charging depends also on diffusion charging, but in an opposite sense. It is large when the backdiffusion probability of the photo-electron is small. This shows that precisely those particles that cannot be charged by diffusion charging will be charged with great efficiency by photoelectric charging and *vice versa*. It turns out that in air at ambient pressure and temperature the limiting size range for photoelectric charging is just around a radius of $R \approx 1\,\mu\mathrm{m}$. This arises because $R/\lambda_e \approx 1$ for $R = 1\,\mu\mathrm{m}$, where λ_e is the mean free path of electrons. It should also be noted that the photoelectron attaches itself to an oxygen molecule after being thermalized by collision with the air molecules and forms a negative oxygen ion in $\sim 10^{-5}\,\mathrm{s}$ with much smaller electrical mobility and diffusion constant. The fact that the natural limit of photoelectric charging is around a particle size of $1\,\mu\mathrm{m}$ means that exactly the particle size range that can penetrate through the human filtering system in the nose far into the deepest part of the lung may be measured selectively. Hence photoelectric charging should become important for surveying the air quality just for this reason alone. There are, however, more important reasons why it has very great potential for improving and surveying the quality of the air, which has deteriorated to a deplorable extent in recent years due to various factors including excessive use of the automobile.

2. – General characteristics of photoelectric charging.

It is then expected that photoelectric charging of particles in gas suspension has the following 3 important characteristics:

1) It is effective only for particles below a certain radius R_0. R_0 depends on the diffusion of the electrons; it is different in carrier gases where negative ions are formed such as air as opposed to gases in which the photoelectrons remain free such as He or clean N_2.

2) It depends sensitively on the energy $h\nu$ of the light and on the photo-electric work function of the particles. Of particular interest are electropositive adsorbates, that is adsorbates that donate part of their electrons to the particle

when adsorbed at the surface. In this case an electric-dipole layer is formed at the surface with the positive pole at the particle-gas interface. This means that the photoelectric work function Φ is lowered when the molecule is adsorbed. If now $h\nu$ is chosen such that it is below the work function Φ_0 of the «naked» particle surface, but above the work function Φ of the surface with adsorbates, only those particles can be photoelectrically charged that have adsorbates on air surface. This leads to an incredibly sensitive yet very simple detection scheme for electropositive adsorbates. For instance, in the interesting case of polycyclic aromatic hydrocarbons adsorbed on carbon particles, detection sensitivities of $10^{-10}\,\mathrm{g/m^3}$ have been realized with a small commercial low-pressure mercury lamp as light source.

3) For weak light intensities, photoelectric charging should be proportional to the photoelectric yield Y of the particles, to the density of the photon current $j_{h\nu}$ and to the part F of the total surface of the particles that has a work function $\Phi \leqslant h\nu$. The number of photoelectrically charged particles can simply be measured by letting the gas carrying the particles flow through a filter in which all particles, charged and uncharged, are caught, and by measuring the current i that flows to ground potential. We have then for weak light intensity

(2) $i = Y j_{h\nu} F \,.$

If, on the other hand, the intensity of the light is strong, the particles acquire a limiting charge $p_{\max}(h\nu)$ that is reached when a subsequent further photoelectron can no longer escape from the Coulomb potential of the previously charged particle. The photoelectric work function of the particle is given by $\Phi(R) = \Phi(\infty) + (p+1)\,e^2/4\pi\varepsilon_0 R - 5e^2/32\pi\varepsilon_0 R$, where $\Phi(\infty)$ is the work function of a flat surface and p the number of positive charges which are already on the particle. This was shown by WOOD[2] and others; the last term is the image potential. From the condition $h\nu \leqslant \Phi(R)$ for the emission of a photoelectron from a particle, we obtain for the limiting charge obtainable in photoelectric charging

(3) $p_{\max}(h\nu) = (h\nu - \Phi(\infty))\,4\pi\varepsilon_0 R/e^2 + 5/8 \,.$

We see that much larger charges are expected in photoelectric charging as compared to diffusion charging, where the limiting charge is given by the thermal energy of the diffusing ion. With sufficiently large photon energies $h\nu$ as available in modern synchrotron radiation sources, the charge on the particle could be large enough to let it explode by the Coulomb repulsion of the charges.

3. – Some tools of aerosol science.

If one looks at particles on a substrate, one usually is stuck with particles of various sizes. However, in the case of particles in gas suspension, one has the

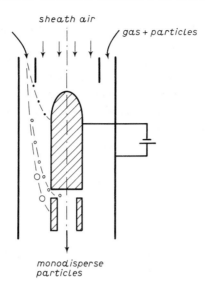

sheath air

gas + particles

monodisperse
particles

Fig. 1. – Differential mobility analyser (DMA) to select a narrow size range of charged particles.

possibility to select a quite narrow size range of the particles by means of a differential mobility analyser (DMA) shown in fig. 1. An electric field is generated between the inner rod and the outer tube by applying an electrical voltage. Gas which is free of particles («sheath air») flows through the cylindrical condenser in a laminar flow. The carrier gas with electrically charged particles enters through an outer ring, and the particles acquire a drift velocity $v = b \cdot E$ in the electric field E according to their electrical mobility b. As b depends on the friction in the gas, it is related to the aerodynamic radius R of the particles. The drift velocity v is perpendicular to the carrier gas flow and only particles with a specific velocity v will enter through the slit in the inner electrode and can thus be separated.

Another remarkable instrument of aerosol technology is the diffusion battery. Due to the Brownian motion, the penetration P of a screen-type diffusion battery [3] is given by

$$(4) \qquad P = \exp[-cD^{2/3}],$$

where D is the diffusion constant of the particles. The constant c depends on the flow rate and the geometry of the battery. For $R \leqslant 100 \, \mathrm{nm}$, the free-molecule relation $D \propto R^{-2}$ is a good approximation. If the penetration rises from P to P', the radius increases by $\Delta R = R_0[(\ln P'/\ln P)^{-3/4} - 1]$. If the initial penetration is small, say $P = 10^{-2}$, even a very small increase in the size of the particles by $\Delta R = 0.1 \, \mathrm{nm}$ will lead to a measurable change of P. This means that the growth of a particle of, *e.g.*, $R = 10$ nm by adsorption of 1 monolayer of molecules can still be detected.

A further very useful instrument is the condensation nucleus counter with which as little as 1 particle/cm³ can readily be detected. The particle is brought into a gas with a supersaturated vapour for which it acts as a condensation nucleus. After sufficient condensation of vapour, the particle grows to an optical size and can be detected by light scattering.

4. – The photoelectric yield of metal particles suspended in He.

We will now examine to what extent the theoretical expectations on photoelectric charging of submicrometre particles can be experimentally verified. It turns out that the general agreement is good, but there are few observations that are at present not yet explained. For the basic investigations, experiments with Ag and Au particles suspended in very clean He gas seem

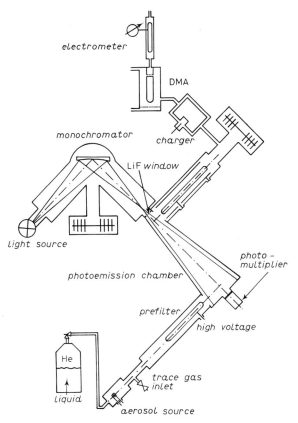

Fig. 2. – Apparatus to measure the photoelectric yield of metal particles suspended in He gas as a function of the radius of the particles.

most appropriate. Figure 2 shows the apparatus used to measure the absolute photoelectric yield Y from the metal particles *vs.* particle diameter [4]. Before each measurement, the apparatus is evacuated to ultra high vacuum (10^{-9} mbar) and baked. Then very clean He gas as obtained by evaporating liquid He is admitted. The He gas is at atmospheric pressure and temperature, and a flow of 5 l/min through the apparatus is maintained. A silver wire of 99.999% purity is heated in the flow of He. The Ag atoms evaporating from the hot Ag wire are cooled by collisions with the He gas and condense by homogeneous nucleation into particles ranging within $(2 \div 10)$ nm. Electron microscopy on precipitated particles reveals that the particles obtained by this method have the form of spheres. The smaller particles from $(2 \div 5)$ nm are single crystals, but in the larger ones more than one crystalline orientation may exist.

There is then a possibility to admit traces of a gas like O_2 into the He. The gas has about 10 s to adsorb on the particle surface. Simultaneously, the particles charged in the process of production are removed from the flow in the electric field of the prefilter. The neutral particles enter the photoemission chamber which is penetrated by monochromatic ultraviolet light up to a photon energy of 11 eV. If a particle emits a photoelectron, it will quickly diffuse to the walls of the chamber and a positively charged Ag particle is left behind. Some of the photoelectrons also attach themselves to neutral Ag particles thereby forming negatively charged Ag particles. Only 0.1% of the particles are photoelectrically charged. Therefore, eq. (2) applies and doubly charged particles can be neglected. One can also neglect recombination of photoelctrons with positively charged particles as well as neutralization by coagulation of negative and positive metal particles under these conditions.

The photoelectrically charged positive particles are size selected in a DMA as shown in fig. 1 and then electrically precipitated, which gives rise to an electric current which yields the density of those Ag particles that have emitted a

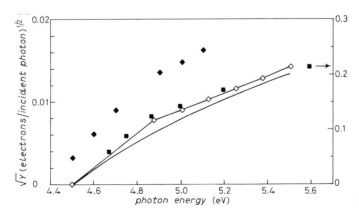

Fig. 3. – Yield of a plane Ag surface compared to the yield from Ag particles with a radius of 3.8 nm: $\diamond$ Ag(111), $\blacklozenge$ Ag(001), —— theory, plane, $\blacksquare$ 3.8 nm particle.

photoelectron. To obtain the absolute magnitude of the photoelectric yield Y, the density of the uncharged particles of the same size range must also be known. This is achieved by diffusion charging of the particles in the charger. The charger contains a radioactive foil of ^{63}Ni. The β-rays generate He ions in the carrier gas that charge the metal particles according to their concentration and size. The percentage of this diffusion charging is determined in an independent calibration. In this way, the total concentration of metal particles is obtained for each range of sizes and can be compared to the concentration of photoelectrically charged metal particles.

Figure 3 shows the yield Y of photoelectrons per unit surface area from a plane Ag surface *vs.* photon energy and compares it to the yield from Ag particles with $R = 3.8$ nm [5]. It is remarkable that the yield of the particles shows the same spectral dependence as the one of the plane surface, hence, apart from a kink at $\hbar\omega = 4.9$ eV, both are well described by the law of Fowler-Nordheim. Yet the yield from the Ag particles is substantially larger by a factor of ~ 200, compared to the plane surface. The enhancement turned out to be independent of R in the range 2.7 nm $\leqslant R \leqslant 5.4$ nm.

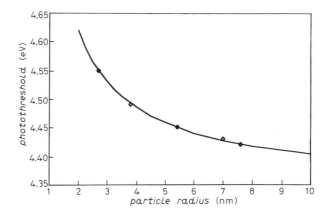

Fig. 4. – Photoelectric threshold Φ vs. radius of Ag particles. The full curve is obtained from simple electrostatics according to WOOD [2], ◇ experiment.

Figure 4 shows that the photoelectric threshold $\Phi(R)$ which is obtained from extrapolating $Y \to 0$ is in perfect agreement with electrostatics [2].

Figure 5 finally shows the photoelectric yield of Ag particles with $R = 5$ nm up to a photon energy of 11 eV. It is remarkable that photoelectric yields close to unity can be reached at photon energies of 11 eV. The full curve is derived on a simple application of the familiar 3-step model of photoemission, yet accounting for the spherical shape of the particles [6]. It explains the anomalous shape of the yield curve compared to a plane Ag surface. However, these geometrical arguments cannot explain the enhancement of the yield in the small particles.

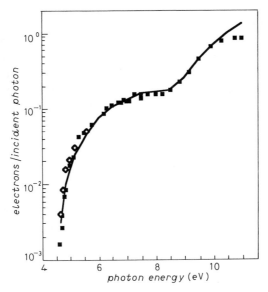

Fig. 5. – Photoelectric quantum yield of Ag particles with $R = 5\,\text{nm}$. The full curve is obtained from a three-step model assuming an escape function adapted to the spherical shape of the particles and a large mean free path of the photoexcited electrons [6]. The theoretical curve is further multiplied by a constant factor to achieve best agreement with the data. The experimental data are taken from Burtscher, Schmidt-Ott and Siegmann [7] (squares) and Müller [4] (diamonds).

The absolute value of the yield obtained from the simple theory is ~ 40 times lower than the observed yield over the entire range of photon energies. When the escape probability is set to unity, which is equivalent to an infinite mean free path of the photoexcited electrons, the yield increases only by a factor of ≈ 7. At least this residual enhancement factor has to be explained by different arguments. An increase of the Mie absorption cross-section by an order of magnitude, constant over a wide energy range, also seems unlikely.

5. – Adsorption of gas molecules at the surface of metal particles.

With the apparatus shown in fig. 2, the adsorption of oxygen molecules to the surface of free ultrafine Ag particles could also be observed [8]. The coverage of a surface by an adsorbate is defined as $\Theta = N/N_0$, where N_0 and N are the total and the occupied absorption site densities, respectively. In the case of O_2 adsorption on Ag, it is known that the work function increase $\Delta\Phi$ is proportional to Θ, hence one can determine Θ by measuring $\Delta\Phi/\Delta\Phi_{\text{max}}$, where $\Delta\Phi_{\text{max}}$ is the increase of the work function when all adsorption sites are occupied. From Θ, one can in turn evaluate the sticking parameter α, as the partial pressure of O_2 in the He gas is determined with a mass spectrometer. In contrast to the findings on macroscopic surfaces, it was found that $\alpha = \alpha_0 (1 - \Theta)^2$. This indicates simple dissociative adsorption with no physisorbed precursor state. Additionally, α_0 is smaller by 2 orders of magnitude for small particles compared to the flat Ag surface. It is clear that such measurements are crucial to the understanding of catalytic phenomena on finely dispersed catalysts, and it is definitely necessary to do more measurements of this type on different materials.

6. – Adsorption of perylene on ultrafine C particles.

An even more refined experiment concerning the adsorption of molecules onto fine particles was performed by BURTSCHER and SCHMIDT-OTT[3] and by NIESSNER[9].

The experimental set-up is shown in fig. 6. Fine C particles are generated by electric sparks between C electrodes in Ar gas (impurities <1 p.p.m.). The particles are of spherical shape with a radius $R \approx 50$ nm. The particles are carried by the flow of the Ar carrier gas into a neutralization chamber, in which a radioactive source produces enough ionized Ar^+ and electrons to establish a well-defined charge distribution on the particles; with the temperature and particle sizes present, doubly charged particles do not occur. Hence the following DMA

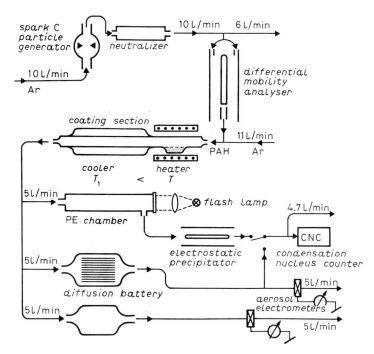

Fig. 6. – Experimental set-up for observing the adsorption and condensation of perylene on C particles, from ref. [3].

(compare fig. 1) selects one size of negatively charged C particles. The stream of the Ar gas containing these particles now passes through a section in which an elevated partial pressure of perylene is maintained by appropriately heating a perylene container. In the subsequent cooler section, the molecules condense onto the C particles. The stream containing the particles can now enter a photoionization chamber in which the negatively charged particles may be neutralized by emission of a photoelectron. After removal of all particles that have not emitted a photoelectron, the number of neutral particles, that is the number of the photoelectrically active particles, is determined in a condensation nucleus counter. The stream of gas molecules can also be directed through a diffusion battery, and the transmission of this battery can simply be measured by registering the current of negatively charged particles. If the particles have grown in size by adsorption of perylene, the transmission of the diffusion battery changes and the physical growth can be determined. The penetration of the diffusion battery is directly related to the growth ΔR of the aerodynamic radius of the particles in the coating section. Finally, the flow of the charged and coated particles can also be directly measured to ensure and check the constancy of the concentration of the C particles before they enter the diffusion battery. Figure 7 shows recent results [10] obtained with this set-up in the range of the monolayer adsorption of perylene on C particles with $R = 6.7$ nm. Plotted is the photo-electric activity ε with a pulsed mercury high-pressure arc *vs.* the increase of the

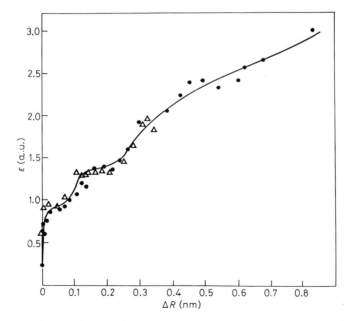

Fig. 7. – Probability ε of photoelectric removal of a previously acquired negative charge of a C particle of $R = 6.7$ nm *vs.* the increase ΔR of the aerodynamic radius of the particle by adsorption of perylene, from ref. [10].

aerodynamic radius ΔR. First, it is noted that adsorption of perylene obviously increases ε. Second, one observes 2 flat sections in ε vs. ΔR, which indicates 2 different adsorption sites. Considering the size of the perylene molecule and the absolute values of ΔR, and assuming that perylene lies flat on the surface, one can calculate that the adsorption exhibits plateaux before the surface is fully covered. This indicates that the perylene settles only on specific adsorption sites such as steps or other deviations from the ideal surface.

7. – Multiple charging by photoemission.

According to the general characteristics of photoelectric charging, it is expected that more than one electrical charge can be put on fine particles in gas suspension if only the photon energy $h\nu > \Phi(R)$, where $\Phi(R)$ is the work function of the particle and R its radius. JUNG, BURTSCHER and SCHMIDT-OTT [11] have shown that this is indeed the case. Figure 8 shows the experimental set-up used in this experiment. The upper part indicates how monodisperse neutral Ag particles were generated, charged by photoemission and subsequently analysed

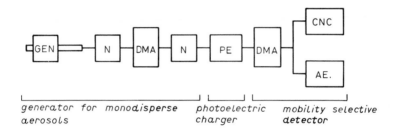

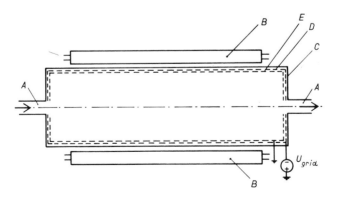

Fig. 8. – Schematic arrangement for observing multiple charging in photoemission (upper half) (GEN = heated-wire Ag particle generator; N = neutralizer; PE: photoelectric charger; AE: aerosol electrometer, measures current produced by charged particles) and construction of the photoemission chamber (lower part), from ref. [11].

for their spectrum of electrical mobilities. The mobility spectrum of a monodisperse aerosol must reveal the relative abundance of multiply charged particles.

To achieve multiple charging, the chamber in which photoionization occurs must be designed in a special way (lower part of fig. 8). It is particularly important that photoelectrons created at the wall of the chamber cannot discharge the highly charged particles to which they are attracted. Furthermore, there must be an absolute absence of electrical fields in the interior of the chamber. Even spurious small fields at edges or oxidized surfaces would attract the highly mobile multiply charged particles and remove them from the carrier gas. These objectives are achieved by 2 grids E and D covering the wall of the chamber C. A voltage is applied to grid D with respect to E which is held at ground potential. The outer tube C consists of quartz glass and is, therefore, transparent to the light of photon energy 4.9 eV from 2 low-pressure 15 W mercury lamps. Photoelectrons generated at the grids D and E by the UV light are removed by the electric field between the grids. Electric fields existing at the surface of the quartz tube are effectively screened by the grids. In this way, the inner part of the photoionization chamber is free of electric fields, and the photoelectrons emitted by the particles are quickly removed by diffusion to the grids. Figure 9 shows that multiple charging of the particles is indeed observed. The maximum number of charges on the particles is found to agree with the predictions of eq. (3).

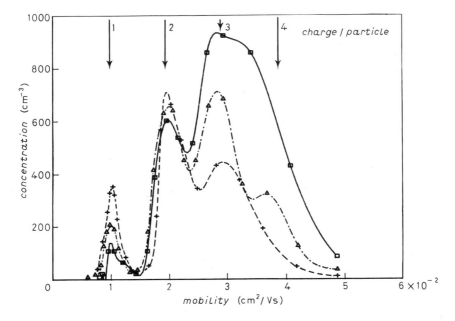

Fig. 9. – Aerosol concentration *vs.* electrical mobility b for monodisperse Ag particles of radius 7.3 nm. The voltage applied to grid D is $+$ 10 V, $\triangle$ 50 V, $\square$ 150 V.

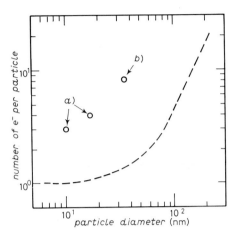

Fig. 10. – Number p of elementary charges on a particle *vs.* particle diameter as obtained in field charging after DENNIS[12], and as obtained by *a)* photoelectric charging of Ag particles with a low-pressure mercury discharge (4.9 eV)[11] and *b)* photoelectric charging of C particles with an excimer laser of 6.4 eV photon energy, from ref.[10].

Figure 10 compares the number of elementary charges p that can be put on a particle of radius R by field charging, which is the most efficient of the commonly used charging mechanisms[12], to the one obtainable in photoelectric charging. It is obvious that much higher charging is achieved by photoemission, and that the limiting charge is given by the photon energy $h\nu$ and the particle radius R, as predicted by eq. (3).

The experiments then clearly show that photoemission of electrons can easily be observed on particles in gas suspension. The general characteristics of photoelectric charging, namely its dependence on the particle radius R and on the presence of adsorbates that affect the photoelectric work function, agree with the expectations. The spectral dependence of the photoelectric yield of small particles is also understood in detail. However, the large enhancement of the photoelectric yield with Ag particles is not understood at present.

8. – Applications of photoemission of electrons from particles in gas suspension.

Several interesting applications of photoemission on gas-suspended particles have been identified. They include the observation of submicrometre particle properties in the plume of volcanoes[13] and the investigation of particle formation in combustion processes[14]. These applications make use of the fact that photoelectric charging can be achieved without electrodes, hence can be done in aggressive environments, and is sensitive to the condensation or chemical reactivity of matter on the particle surface as the hot gases containing the particles mix with air and/or become cooler. Other applications such as monitoring the air quality make use of the fact that photoelectric charging of particles is selective for respirable particle size ranges and also responds specifically to particles coated with polycyclic aromatic hydrocarbons that contain highly carcinogenic species such as benzo(a)-pyrene [15, 16].

As an example it will be shown how photoelectric charging of particles can be used to control the combustion of oil in a stove and may serve to achieve maximum efficiency of the combustion at a minimal rate of pollution of the environment with soot particles [14]. Efficient control of furnace and boiler combustion processes becomes increasingly important as improved furnace operation can provide significant cost savings through higher combustion efficiency and reduced flue gas emissions. Usually this is accomplished by continuous monitoring of excess oxygen, CO concentrations, or both. These quantities are then used to adjust the amount of air admitted to the combustion [17].

Measuring, for instance, the excess oxygen with a zirconium oxide probe and keeping this quantity constant by a feedback control means that the air/fuel ratio is kept constant. No information on the emissions is obtained in this approach. Yet changes in the combustion due to different oil quality, contamination of the furnace, etc. may require a different air/fuel ratio for optimum combustion. This problem can be solved by using the oxygen measurement in combination with a CO measurement. CO probes are, however, very expensive and/or hardly reliable. Therefore, this conventional technique can only be applied cost effectively to very large furnaces.

In the new method of photoelectric charging, the exhaust gases are irradiated with ultraviolet light from a low-pressure mercury discharge. If the work function Φ of the particles generated in the combustion is lower than the energy $h\nu = 4.9$ eV of the light, photoelectrons are emitted from the particles and a positive charge of the aerosol is detected. The experiments with adsorption of perylene on C particles [3] described above lead to the hypothesis that polycyclic aromatic hydrocarbons (PAH) with three or more rings are present at the surface of exhaust particles if the combustion process proceeds with lack of oxygen, and causes the lowering of Φ below 4.9 eV. Hence the presence of PAH in the exhaust gas can readily be detected by photoelectric charging and indicates that the combustion was incomplete due to lack of oxygen. A feedback mechanism can then regulate the air inlet in such a way that the PAH signal just disappears. It should be noted that excess of air is also unwanted as it simply cools the furnace.

However, other substances may also condense on the exhaust particles and quench photoemission. For example, films of water or acids have been found to be very effective in quenching photoelectric charging of particles. Hence the photoemission chamber has to be maintained at an elevated temperature to prevent the condensation of acids and water. A detailed description of an

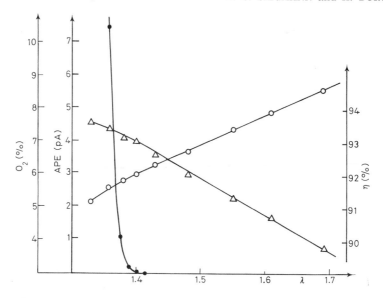

Fig. 11. – The positive current from photoelectrically charged exhaust particles (APE)
(●—●—●), the efficiency η of the combustion (△—△—△) and the oxygen concentration
(○—○—○) in the exhaust gases of an oil furnace are plotted against the air/fuel ratio λ.

appropriate device to achieve photoelectric charging of particles generated in
combustion is given in ref. [14].

Figure 11 shows results obtained with a small domestic oil stove. The
photoelectric signal (aerosol photoemission, APE) shows a steep increase close
to the maximum value of the efficiency η of the burner. Comparing the APE
signal with the oxygen signal reveals that the changes in the concentration of
oxygen are much smaller. Therefore, the measurement of the oxygen concentration for combustion regulation must be very precise, in fact to $\approx 0.1\%$, to
achieve reliable combustion regulation. However, even a very rough measurement of photoelectric charging provides an accurate criterium to find the
optimum value of the air/fuel ratio λ.

The sharp increase of the photoelectric signal occurs when soot and CO are
produced in the combustion. The soot production is commonly measured by the
Bacharach number, which is a logarithmic optical scale for the amount of soot.
The relation between log (photoelectric signal), Bacharach number and log (CO
concentration) was found to be linear over more than 3 decades. It should be
noted that relatively small changes in the air supply already cause a large
increase in the amount of soot and CO, and also the photoelectric charge of the
aerosol.

The PAH concentration on filter samples of the exhaust has also been
determined by gas chromatography [18]. A linear relation between the photoemission signal and the chemically determined total PAH concentration has been

found. This is in agreement with the hypothesis that the lowering of the work function is actually due to condensation of PAHs or due to formation of PAHs on the surface of the exhaust particles, which are in fact submicroscopic soot particles. The linear relation between PAH concentration and photoelectric signal is predicted by eq. (2). With the current from a photoelectric sensor, the air supply to the combustion must be regulated to keep the photoelectric signal just below the steep rise where soot and CO production starts. The effect of this regulation is operation of the burner close to the point of maximum efficiency by minimizing excess air, while contamination of the environment is kept to a minimum.

Particles with photoelectric work functions below 4.9 eV occur in large quantities in the ambient air in cities, near traffic areas and in cigarette smoke. It can easily be shown with a photoelectric aerosol sensor that practically all of these particles are produced in incomplete combustion, particularly by Diesel engines. It is very likely that the lowering of the work function is produced by the presence of PAHs at the surface of the particles. As the PAHs contain extremely carcinogenic species such as benzo(a)-pyrene and as the photoelectric detected particles are small enough to penetrate into the human lung, these particles may present a hazard to human health besides being responsible for unwanted photochemical reactions in the atmosphere. The photoelectric aerosol sensor is a simple tool to detect the particles and can help to reduce their concentrations by properly adjusting the various combustion processes in which they are generated.

In conclusion, we note that photoemission experiments on particles suspended in gases are only at the beginning. There are further very promising possibilities in combination with modern synchrotron radiation sources [19]. The study of photoemission properties of submicroscopic particles is at present the only way to obtain information on surface chemistry in the natural gas environment of the particles.

REFERENCES

[1] W. A. HOPPEL: in *Electrical Processes in Atmospheres*, edited by H. DOLEZALEK and R. REITER (Steinkopf, Darmstadt, 1977), p. 60.
[2] D. M. WOOD: *Phys. Rev. Lett.*, **46**, 749 (1981).
[3] H. BURTSCHER and A. SCHMIDT-OTT: *J. Aerosol Sci.*, **17**, 699 (1986).
[4] U. MÜLLER: Diss. No. 8544 ETH Zürich (1988).
[5] U. MÜLLER, A. SCHMIDT-OTT and H. BURTSCHER: *Z. Phys. B*, **73**, 103 (1988).
[6] U. MÜLLER, H. BURTSCHER and A. SCHMIDT-OTT: *Phys. Rev. B*, **38**, 7814 (1988)
     G. FARACI, A. R. PENNISI, V. PRIVITERA, H. BURTSCHER and A. SCHMIDT-OTT: *Phys. Rev. B*, **37**, 10542 (1988).
[7] H. BURTSCHER, A. SCHMIDT-OTT and H. C. SIEGMANN: *Z. Phys. B*, **56**, 197 (1984).

[8]   U. MÜLLER, A. SCHMIDT-OTT and H. BURTSCHER: *Phys. Rev. Lett.*, **58**, 1684 (1987).

[9]   H. NIESSNER: *J. Aerosol Sci.*, **17**, 705 (1986).

[10] A. SCHMIDT-OTT: *Experimente an kleinen Teilchen in Gassuspension*, Habizitationsschrift ETH-Zürich.

[11] TH. JUNG, H. BURTSCHER and A. SCHMIDT-OTT: *J. Aerosol. Sci.* **19**, 485 (1988).

[12] R. DENNIS: *Handbook on Aerosols* (National Technical Information Service, US Department of Commerce, 1976).

[13] H. BURTSCHER, P. COHN, L. SCHERRER, H. C. SIEGMANN, G. FARACI, A. R. PENNISI, V. PRIVITERA, R. CRISTOFOLINI and V. SCRIBANO: *J. Volcanol. Geotherm. Res.*, **33**, 349 (1987).

[14] H. BURTSCHER, A. SCHMIDT-OTT and H. C. SIEGMANN: *Aerosol Sci. Technol.*, **8**, 125 (1988).

[15] H. C. SIEGMANN: *Universitas*, **40**, 1057 (1985).

[16] J. SANTODONATO, P. HOWARD and D. BASU: *J. Environ. Pathol. Toxicol.*, **5**, 1 (1981).

[17] C. PRESSNER and M. G. SEMERJIAN: *Evaluation of Industrial Combustion Control Systems* (National Bureau of Standards, Gaithersburg, Md., January 1984).

[18] R. McDOW, W. GIGER, H. BURTSCHER and A. SCHMIDT-OTT: *3rd International Conference on C-particles in the atmosphere, Berkeley, Cal., October 5-8, 1987*, Lawrence Berkeley Laboratory, Report LBL-23997.

[19] U. MÜLLER, A. SCHMIDT-OTT, H. C. SIEGMANN, S. KRUMMACHER and W. NIEMANN: *Proposal for an experiment to measure photoemission from small particles with synchrotron radiation* (unpublished).

Clusters—Structure, Electronic Structure and Spectroscopy.

R. O. JONES

Institut für Festkörperforschung der Kernforschungsanlage Jülich
D-5170 Jülich, B.R.D.

1. – Introduction.

Clusters of atoms have been a recurrent theme in condensed-matter physics for many years. Often viewed as a bridge between the familiar areas of atomic and molecular physics, on the one hand, and extended systems with translational invariance, on the other, they are more difficult to characterize than either. However, the last few years have seen a spectacular increase in interest in this area, with important progress being made, particularly in experimental techniques. This interest is reflected by the choice of «clusters» as the theme of two recent Enrico Fermi Schools here in Varenna[1, 2], including the one which has just finished[2]. Last week Aix-en-Provence was the venue of the Fourth International Meeting on Small Particles and Inorganic Clusters. It would be pointless to try to survey all the topics covered at these meetings. In the time available to me, however, I shall discuss some of the developments in both experimental and theoretical techniques in the field. It is by no means obvious that the difficulties facing the theorists are appreciated fully by many experimentalists who perform cluster measurements, and I hope to point to areas where the cooperation between theory and experiment is likely to be most fruitful.

The term «cluster» means different things to different people, and even different things to the same person at different times. JORTNER[3] provided a useful classification into 1) microclusters, with $N \leq 13$, 2) small clusters up to $N = 100$, 3) large clusters with $100 \leq N \leq 1000$ and 4) small particles. In the first category it should be possible to use the experience of molecular physics, and solid-state properties are progressively better developed as the systems grow larger. In the present lecture I focus on aggregates of up to one to two hundred atoms, which are rather modest by some standards. Others prefer to exclude very small molecules or insist on including much larger systems. Most of the systems I consider comprise only one or two different elements.

In the first part of this lecture (sect. 2) I examine some of the experimental techniques that are currently used to study atomic clusters, and give examples of results obtained with different techniques. My selection is by no means exhaustive. A detailed picture of the present situation can be found in the contributions to the 1986 NEC Symposium [4] or in other reviews [5-7]. The sort of problems that the experimentalists have presented to the theorists is indicated in sect. 3, where we see that the task of understanding the properties of clusters at a microscopic level is a daunting one. Nevertheless, theorists have also been active in developing techniques for the calculation of electronic and geometrical structures. I shall emphasize one of them, the combined molecular dynamics/density functional technique of Car and Parrinello [8] and present some results for clusters of selenium [9] and sulphur [10] in sect. 4. My concluding remarks follow in sect. 5.

2. – Generation and spectroscopy of atomic clusters.

The experimental problem can be viewed as having two separate steps. In the first, a gas or vapour is generated with an adequate concentration of the species of interest. In the second, the clusters are identified and separated, if necessary, and characterized using a suitable spectroscopy technique. While there is no ideal way to classify the different methods of cluster preparation, it is natural to focus on one of these features. The scheme I adopt here is based on the former, the method of cluster production. It would also have been possible to focus on the method of analysis, and some overlap in the classifications is unavoidable.

2`1. *Thermal vaporization.* – If a sample is heated to a sufficiently high temperature (generally in a Knudsen cell), the resulting vapour contains small clusters as well as atoms. If the vapour is now ionized by an electron beam, it can be analysed in a mass spectrometer to find the ratio of the cluster to the atomic species. The dependence of this ratio on the temperature enables one to determine thermodynamic information about the cluster, including its binding energy.

This method has been used for a large number of small clusters, and a review of the technique and the results obtained has been given by GINGERICH [11]. Analysis of the data obtained requires assumptions to be made about the bond in the cluster, such as the equilibrium separation of the atoms, the vibration frequency and the nature of the ground state, particularly its spin degeneracy. If this information is available from calculations and/or other measurements, reliable estimates can be made of the binding energy.

There have been substantial improvements in the sensitivity of this technique during the past few years, and a recent example is the work by HILPERT and RUTHARDT [12] on the Cr_2 molecule. This molecule provides a stringent test of methods for calculating molecular electronic structures and the comparison with experiment was made difficult by the large error bars on earlier measurements using Knudsen cell techniques. The recent measurements show that the dissociation energy of Cr_2 is (1.44 ± 0.03) eV, and the improved accuracy has pointed to a lack of reliability in some calculational schemes [13].

2'2. *Evaporation onto a thin film.* – Vaporization of a solid followed by condensation onto an inert substrate often results in the formation of clusters with a range of sizes. These can be investigated by electron microscopy or by other means. A recent example is the work of Ijima [14], who deposited gold particles on spherical particles of single-crystal Si. He was able to observe structural instabilities taking place, presumably due to the influence of the electron beam. The structural changes take place in clusters down to $N \sim 460$, where the cuboctahedra structures are favoured.

An interesting application of this technique is the study of the melting temperature of small particles as a function of particle size. BUFFAT and BOREL [15] evaporated gold particles and condensed them on a thin amorphous carbon film. The loss of crystalline order in the particles was monitored by measuring the intensity of the (220) electron diffraction ring, which is the most intense ring for which no liquid-associated rings are superimposed. The dependence of the melting temperature on particle diameter is shown in fig. 1.

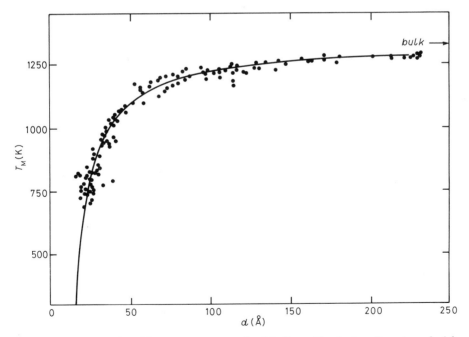

Fig. 1. – Variation of melting temperature of gold, T_M, with cluster diameter, d. After ref. [15].

The behaviour is reproduced rather well by a phenomenological model (full curve). The lowering in the melting point with decreasing particle size is quite dramatic, and can be related to the larger fraction of surface atoms in small particles.

2'3. *Gas aggregation.* – If a sample is vaporized in an inert-gas atmosphere (usually He, Ar or Xe), the latter can be used as both temperature control and carrier. Expansion through a nozzle results in cluster formation, and variants of this technique have become popular in recent years.

KNIGHT *et al.* [16] coupled this technique with time-of-flight mass spectroscopy and found conspicuous peaks for Na clusters with $N = 8, 20, 40, 58$ and 92. This «quantum shell structure» could be understood in terms of a nearly-free-electron picture where the valence electrons of Na are confined in a spherically symmetrical potential (see subsect. 3'1). The shell structure is also found in ionization potential measurements of both Na and K, and the results for potassium are shown in fig. 2[17]. The ionization potential falls from the atomic value (4.33 eV) to the work function of the bulk (2.22 eV) with pronounced breaks at the cluster sizes shown.

Photoionization of mercury clusters has been carried out by RADEMANN *et al.* [18], who co-expanded Hg vapour with argon or xenon through a nozzle into a vacuum chamber. The cluster distributions can be varied by changing the

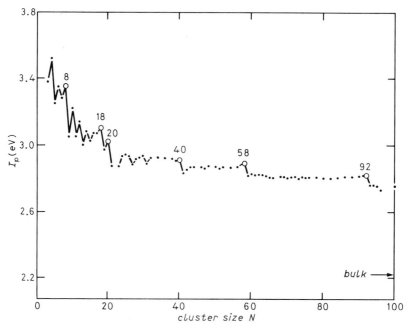

Fig. 2. – Ionization potential, I_p, of potassium as a function of cluster size, N (after ref. [17]). Major discontinuities are shown. I_p for the atom is 4.339 eV, and the bulk work function is 2.24 eV.

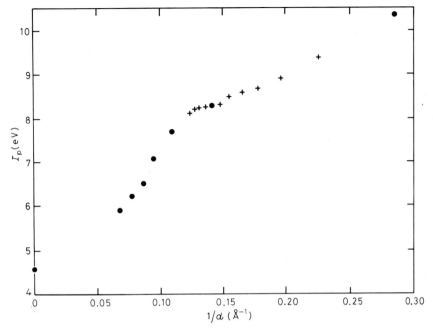

Fig. 3. – Experimental ionization potential data for Hg plotted against d^{-1}, where d is the diameter of a sphere with the same volume as the cluster. After ref. [18]. The bulk work function is 4.50 eV.

expansion conditions, and any desired mass spectrum between $N = 1$ and 100 can easily be obtained. In fig. 3 we show the ionization potential as a function of the inverse of the diameter of a sphere with the same volume as the cluster. The curve shows a pronounced change in the size dependence of I_p in the range $20 < N < 70$. The authors have interpreted this as a transition between «Van der Waals» and «metallic» bonding.

A variation of this technique was used by MARTIN [19] to study reactions between Cs vapour and other gases or vapours (Cl_2, O_2, H_2O, S_8) generated in an adjacent oven. He used He gas at a pressure of 1 mbar as carrier, and the clusters and microcrystals condensed out of the quenched vapour were analysed in a mass spectrometer. This method generated a vast range of mixed clusters of the form Cs_mX_n. Certain clusters showed up strongly in the mass spectra and appear to be particularly stable. In the case of the mass spectrum of pure sulphur quenched in He, clusters containing from 2 to 56 atoms could easily be identified, and the peaks corresponding to clusters of 8, 16, 24, 32 and 40 sulphur atoms were very pronounced. MARTIN inferred that clusters tend to form out of S_8 building blocks, presumably ring molecules.

2˙4. *Isolation in an inert matrix*. – Co-condensation of metal and rare-gas atoms on a cold substrate results in metal clusters in an inert matrix. The cluster size depends on the deposition conditions, and these can be chosen to give a narrow size distribution. An example of this technique is provided by the work of Montano *et al.* [20], who generated Cu clusters of diameter $(25 \div 150)$ Å ± 5 Å in an argon matrix. Analysis of the clusters was performed using synchrotron radiation and extended X-ray absorption fine structure (EXAFS). The authors found evidence for the existence of a cluster of 13 atoms with f.c.c. structure, and the nearest-neighbour distances for different cluster sizes are shown in fig. 4. The bond length in Cu_2 is much shorter than in the larger clusters, where the deviations from the bulk nearest-neighbour separation are small.

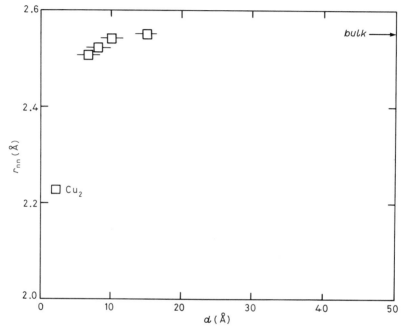

Fig. 4. – Variation of nearest-neighbour separation r_{nn} for Cu with cluster diameter, d. After ref. [20].

A further example of matrix isolation is provided by the work on small Ag clusters by BECHTHOLD *et al.* [21, 22]. The technique provides a higher particle density than available in other cluster techniques, and opens up the possibility of resonance Raman spectroscopy. In this case the exciting laser is tuned to an absorption band of the particle under study, and some vibrational modes are strongly enhanced. Under favourable conditions it is possible to obtain information about the resonant electronic state and its coupling to the vibrations in the system. Matrix-induced perturbations, in particular, can be studied in some detail. From the anharmonicities it is also possible to deduce information about the dissociation energies of the clusters.

BECHTOLD *et al.* studied small Ag particles (mostly Ag_2 and Ag_3) in Xe, Kr and Ar matrices, and found that dimers can occupy one, two, or three trapping sites, respectively. In fig. 5a) we show the resonant Raman spectrum of Ag_2 in Kr matrices at 20 K with 406.74 nm excitation of a Kr ion laser. The Raman lines at 194 cm^{-1} and 203 cm^{-1} correspond to the internal stretching vibrations of Ag_2 at the two trapping sites. If the matrix is annealed at 50 K for a short time and then cooled to 20 K, one observes (fig. 5b)) relative increases in the intensity at

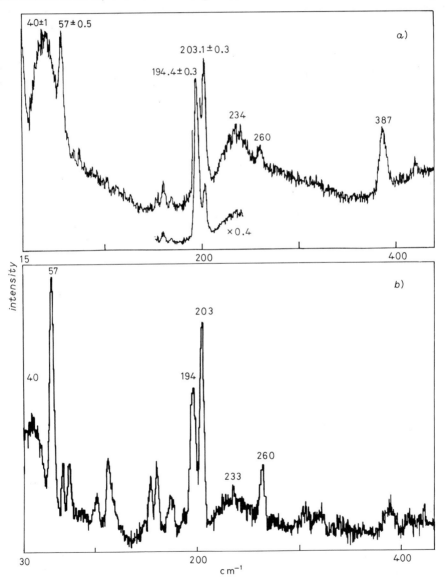

Fig. 5. – Resonance Raman spectra of Ag_2/Kr at 20 K: a) freshly prepared matrix, b) after a short annealing at 50 K (ref. [21]).

57, 203 and 260 cm^{-1}, as well as additional modes due to larger clusters. The low-energy excitations (40 and 57 cm^{-1}) provide a probe of the dimer-lattice interaction and of the local environment of the cluster. Computer simulations have been able to reproduce the experimental data very well, so that a detailed picture of the geometries of the trapping sites has resulted.

2˙5. *Supersonic expansion.* – The adiabatic expansion of a gas from a region of high pressure into vacuum results in a dramatic lowering of the temperature. If the thermal energy is sufficiently low, dimer formation can result, followed by a rapid aggregation. This is a simple technique, but is restricted to those gases that have a high vapour pressure (several hundred Torr) at the temperature of generation. As an example, the expansion of argon at 10 bar through a 0.05 mm nozzle results in clusters with a mean size of $N = 150$ [23].

ECHT et al. [24] have performed adiabatic-expansion measurements on Xe, with a nozzle diameter of 0.2 mm. The time-of-flight mass spectra they obtained show marked intensity drops after $N = 13, 19, 25, 55, 71, 87$ and 147. The authors interpret these results («magic numbers») as demonstrating the existence of energetically stable clusters with regular packings. Complete shell icosahedra correspond to $N = 13, 55$ and 147, and a double icosahedron to $N = 19$. Maxima for all these values are clearly apparent in the mass spectra. The observation of magic numbers implies that the geometrical structure is well defined for each N, *i.e.* that the clusters are solid.

2˙6. *Laser vaporization.* – This technique was introduced by SMALLEY and co-workers [25]. It can be used for all solids and has become very popular in recent years. In the work of Bondybey and English [26, 27], a pulsed laser beam is used to evaporate the sample (usually a metal), and the hot plasma produced is diluted by an excess of cold He gas and the supercooled vapour expanded through an orifice. The cluster beam is analysed by a second pulsed laser, with a delay to allow for the time of flight of the sampled molecules. While the amount of material vaporized is very small, it corresponds to more than 10^{13} atoms per pulse. In the case of materials with low vapour pressures or high melting points, such as Be (1551 K), the fraction of clusters in the beam ($> 10\%$) is much greater than can be obtained, for example, in a Knudsen cell.

BONDYBEY and ENGLISH [27] used this technique to advantage in their study of the beryllium dimer. While this molecule had previously been detected in the vapour phase, no spectroscopic information about it had been found using Knudsen cell methods. If the temperature of the gas flow is ~ 77 K (liquid N$_2$), two strong band systems appear. The higher-energy transition (fig. 6) is characterized by a vibrational frequency of ~ 510 cm^{-1} and a very short lifetime, and is consistent with the $B\,^1\Sigma_u^+ \rightarrow X\,^1\Sigma_g^+$ transition in Be$_2$. A second progression with $\omega_e = 668$ cm^{-1} is due to the $A\,^1\Pi_u \rightarrow X\,^1\Sigma_g^+$ transition. The spectra could be analysed to obtain the first detailed spectroscopic data about this molecule.

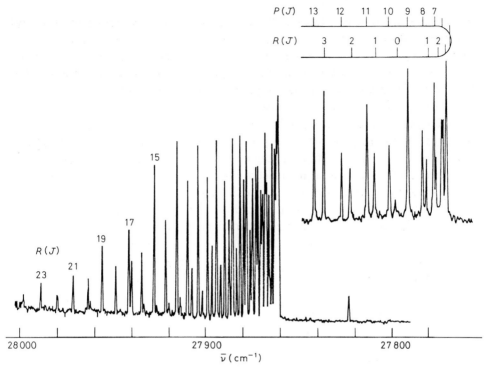

Fig. 6. – Fluorescence excitation scan of the 0-0 band in Be$_2$, $B\,^1\Sigma_u^+ \leftrightarrow X\,^1\Sigma_g^+$ transition. The J numbering is shown. The inset shows the band origin at higher resolution (ref. [27]).

Another striking success of this technique is the case of Cr$_2$, where the half-filled shells in the Cr atom lead to a great variety of possible low-lying states. Experimental work of several groups [26, 28, 29] using laser vaporization has

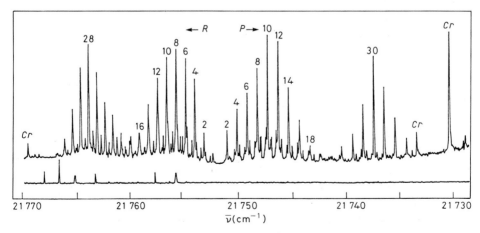

Fig. 7. – High-resolution scan of the Cr$_2$ $A'\,\Sigma_u^+ \leftrightarrow X\,^1\Sigma_g^+$ 0-0 band. Rotational numbering is shown for ^{52}Cr$_2$ (ref. [26]).

resulted in a consistent picture of the ground state of this molecule. As an example, we show in fig. 7 the laser-induced fluorescence spectrum of Cr_2[26]. The analysis of such measurements has shown that the bond length $((1.68 \pm 0.01)$ Å) is much shorter than the nearest-neighbour separation in the bulk (2.5 Å). The Cr_2 molecule is a critical test of methods of molecular-structure calculation, and such data have been very useful.

A further example of this technique is provided by the work of Rohlfing *et al.*[30] on small iron clusters. They evaporated iron into an inert carrier gas (*e.g.*, He), and expansion into vacuum results in the formation of small clusters. They measured photoionization spectra for cluster sizes between 2 and 25, and the results are shown in fig. 8. As in the other examples we have seen, the ionization energy falls from the atomic value to the value in the bulk (the work function). I_p shows an irregular size dependence for which there appears to be no simple explanation.

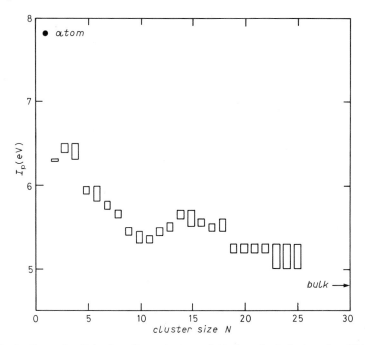

Fig. 8. – Ionization potentials, I_p, of iron clusters plotted against cluster size, N (ref. [30]).

2˙7. *Solid-state clusters.* – There are some cases where solid-state phases are composed of well-separated clusters, so that the properties of the cluster units are important. In contrast to laser vaporization, this technique is far from universal. In cases where the condensed phase is a single crystal, however, the structure can be determined by X-ray diffraction, so they give the most precise structural information currently available for any clusters.

Examples of this category are the regular crystalline forms, or «allotropes», of sulphur and selenium and some of their compounds. Sulphur has by far the largest number of allotropes of any element, and many of these consist of well-separated cyclic molecules [31, 32]. At the present time, sulphur clusters with $N = 6, 7, 8, 10, 11, 12, 13, 18$ and 20 have been prepared in single-crystal form, and S_9 has been found as microcrystals. S_{18} is the only one which exists in two distinct forms. The preparation of these clusters is part of the art of inorganic chemistry, and details are given in the review by STEUDEL [32]. S_7, for example, can be prepared by the reaction of titanocene pentasulphide with dichloro-disulphane in organic solvents at $0°$, yielding (24%)

$$(1) \qquad (C_5H_5)_2TiS_5 + S_2Cl_2 \rightleftharpoons S_7 + (C_5H_5)_2TiCl_2.$$

Ring structures of selenium (Se_6, Se_8) are also known [31], as well as several mixed crystals of the form Se_nS_m [33] and a range of sulphur oxides (S_nO, $n = 5 \div 10$; S_7O_2, $S_{12}O_2$) and ions. The decisive advantage of this method of cluster preparation is that the *structure* is known very accurately in most cases, although it is appropriate to note at this point that the packing of the molecular units results in structures that may differ from those of the molecules in free space. This effect is more pronounced in some molecules (*e.g.*, S_7) than in others (*e.g.*, S_6, S_8). Sulphur and selenium clusters appear currently to provide the best tests of the predictive capability of theoretical methods, and we discuss these systems in some detail below. The central role played by the structure in understanding the microscopic properties should be evident from the title of this lecture.

3. – The task of the theoretician.

3`1. *Qualitative considerations*. – The results presented above are only a sample from the extensive and expanding literature in the field. However, they raise several questions which require a response from the theoreticians. A qualitative understanding of some of the problems is not difficult, but there are important areas where little is known and the prospects are not at all good. Before we get too discouraged, let us focus on some of the simpler problems.

The fraction of surface atoms in a cluster falls dramatically with increasing size. If the characteristic radius of the atom is r and that of the cluster is R and geometrical effects are ignored, then it is easy to show that the fraction of surface atoms $N_s/N \sim r/R$. For cluster radii of around 10 Å (with $r = 2 \text{ Å} \sim 100$ atoms), about 80% of the atoms are on the surface, and the percentage is still around 8% for a cluster of radius 100 Å. The dramatic change in melting temperature noted above for Au particles can be correlated with the parallel increase in the fraction of surface atoms. Surface atoms have fewer neighbours than atoms in the bulk, and their motion is correspondingly less constrained. In line with the Lindemann criterion for melting, the larger r.m.s. thermal vibration amplitudes which result favour surface melting over bulk melting.

We have seen two examples above of a common (while not universal) behaviour of bond lengths in clusters. In both Cu_2 and Cr_2 the equilibrium internuclear separation is substantially shorter than in the bulk, and in Cu clusters we have seen that the nearest-neighbour separation rapidly approaches the bulk value with increasing cluster size. This can also be seen as a consequence of the removal of constraints in the small clusters. In the dimer, the only requirement on the wave function is that it be cylindrically symmetrical about the axis of the molecule. The bond can usually be strengthened if charge is transferred to the region between the nuclei, and this is perfectly consistent with the symmetry of a diatomic molecule. In a larger cluster or bulk system, there may be several nearest neighbours with similar claims, and it is quite plausible that the result is usually a longer bond.

The results shown previously indicate that the ionization potential, I_p—the energy required to remove an electron from the cluster (see fig. 9)—, generally decreases from the atomic value to the bulk work function. In the case of metallic clusters, this can be made plausible by considering the energy required to remove an electron from a conducting sphere of radius R. The distortion of the electric field by the sphere can be determined by the method of images, and we obtain $\Delta I_p = -5e^2/8R$. Ionization of a free cluster produces an electron and, therefore, an additional contribution to the electrostatic energy, e^2/R. In this

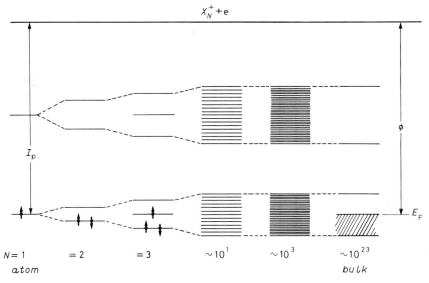

Fig. 9. – Schematic behaviour of the energy spectrum of clusters with increasing size, N. I_p is the ionization potential of the atom, and ϕ is the bulk work function.

simple model, therefore,

$$(2) \qquad I_{\mathrm{p}}(R) = \phi + \frac{3e^2}{8R} \approx \phi + \frac{5.40\,\mathrm{eV}}{R\,[\text{Å}]}.$$

A similar relationship is found for the electron affinity, the change in energy upon *adding* an electron to the system. These simple dependences on cluster radius, R, hold surprisingly well for many metals. It is obvious that the behaviour of *nonmetallic* clusters will be quite different. Indeed, the departure in $I_{\mathrm{p}}(R)$ in Hg from the simple behaviour for a conducting sphere was interpreted as evidence for an absence of «metallic» bonding [18].

The above arguments show that atomic and molecular properties will change into those characteristic of extended systems as the cluster size increases, but the way this transition occurs depends on the property in question. More interesting, and certainly more difficult to explain, are the *irregularities* which have been found. As noted above, the irregularities found in the mass spectra of Na and K by KNIGHT *et al.* [16,17] can be understood in terms of a nearly-free-electron picture. These authors assumed a spherically symmetrical potential well of the form

$$(3) \qquad U(r) = -\frac{U_0}{\exp\left[(r-r_0)/\varepsilon\right]+1},$$

where U_0 is the sum of the Fermi energy, E_{F}, and the work function ϕ of the bulk; r_0 is the effective radius of the cluster and assumed to be $r_{\mathrm{s}}N^{1/3}$, where r_{s} is the radius of the sphere containing one electron in the bulk. The parameter ε determines the variation of the potential at the surface of the sphere, and was estimated from calculations for a jellium surface. The Schrödinger equation is solved numerically for this potential, leading to discrete electronic energy levels which shift down slowly as N increases. If the electronic energy is associated with the sum of the eigenvalues, clusters of relative stability should result when an energy level is just filled at a certain N and the next orbital starts to be occupied in the cluster with $N+1$ atoms. The prominent peaks found in the mass spectra can then be correlated with the electronic properties, without invoking structural packing patterns.

While this is an interesting observation, and the correlation with the experimental spectra is clearly apparent, these arguments can only be used in simple metals. Furthermore, there is a wealth of experimental evidence, particularly for transition metal clusters, which points to structural variations with changing cluster size. It seems to be widely accepted in the cluster community that better and more detailed structural information is required before the solutions of many of the problems can be found. This comes as no surprise to the solid-state or molecular theorist, for whom the *structure* of the system is an essential input for the calculation of any electronic property.

3˙2. *Structure determination.* – The problem of determining the ground-state structure of any system of electrons and ions can be simply stated. One needs to calculate the contributions to the total energy of the system for all geometries, and locate the minimum value. Since the number of geometrical configurations grows rapidly with increasing cluster size (there are $3N-6$ independent *coordinates* and a mesh of points would need to be constructed for *each*), it might be simpler to locate all the local energy minima, and then to order these minima to find the lowest. Even this problem is, however, one of breathtaking complexity.

HOARE and MCINNES[34] have performed a study of clusters up to $N = 13$ interacting with pairwise forces of the Lennard-Jones and Morse forms. They located *all* the structures corresponding to local energy minima, and the number of these is given in table I. The differences between the two potential forms and the resulting structural types are interesting, but it also significant for the present discussion that the number of minima grows approximately exponentially with increasing N. If this were true, the task of finding the ground state by the above method would be of no value. There is worse to come. WILLE and VENNIK[35] have shown that, for clusters of identical particles interacting with a pairwise potential, there is no known algorithm for which the exact ground-state energy (and hence the structure) can be found with a polynomial time dependence. The exponential explosion indicated by table I would appear to be guaranteed. In the language of the specialists, the problem belongs to the class NP[36] and may be viewed as «intractable». It is rather sobering when mathematicians tell you that there is no solution to the problem which most of

TABLE I. – *Comparison of the multiplicities of Morse and Lennard-Jones (LJ) clusters* (ref. [34]).

N	Morse	LJ
6	1	2
7	3	4
8	5	8
9	8	18
10	16	57
11	24	145
12	22	366
13	36	988

the physicists and chemists consider essential for a microscopic explanation of the experiments.

This is not all. Much of the work on cluster simulations has been based on simple force laws such as those mentioned above. It is well known, however, that *no* pairwise force can stabilize the diamond structure, and there are many other examples where three- and four-body forces are essential. Parametrized force laws have further disadvantages in that they are semi-empirical by nature, and it is difficult to know how much the final results are affected by the choice of parameters.

In principle, the last problem raised would disappear if we could solve the Schrödinger equation for the interacting system of electrons and ions. From the many-particle wave function which results we would then be able to determine the total energy and most other properties of interest. This is the basic aim of much work in theoretical chemistry, and I do not have time to discuss the method in detail (for a review of some aspects, see ref. [13]). The situation has been summarized by MARTIN *et al.* [37], who noted that «an *ab initio* calculation of the total energy of a ten-atom metal cluster, including the effects of correlation, is a state-of-the-art calculation even if it is performed for only one point on the energy surface». Nevertheless, there have been some interesting applications of the traditional methods of quantum chemistry to small clusters. As an example, RAGHAVACHARI [38] has calculated the structure of small Si clusters (Si_3 to Si_7 and Si_{10}), using the Hartree-Fock method with correlation corrections estimated by the method of Møller and Plesset. For $N > 6$ the search for the structure with lowest energy was restricted to particular symmetries. Silicon is a material of great technological importance, and it is unfortunate that almost no experimental information is available about the structure of small silicon clusters.

The above discussion shows that the theorist interested in calculating the structure of a cluster by means of a total-energy calculation is faced with two problems: 1) The number of structures corresponding to a local energy minimum increases dramatically with increasing N, even for simplified force laws, 2) the determination of the *exact* energy from a solution of the many-particle wave function for a *single* geometry is itself a problem which is intractable for all but the smallest clusters.

4. – The molecular dynamics/density functional approach.

The discussion of sect. 3 has shown that there are two distinct problems to be solved, and we now turn to ways of overcoming some aspects of both. We discuss the density functional formalism as a scheme for calculating the *total energy* of the system of ions and electrons. In this scheme there is no attempt to calculate the exact wave function of the interacting system of electrons and ions.

Furthermore, the emphasis on the applications of this formalism is on the calculation of energy *differences*. While the exact calculation of the total energy E is presently out of the question for most clusters of interest, this is a quantity which is usually inaccessible to experiment. Most quantities of interest (cohesive, binding or dissociation energies, ionization and other excitation energies, etc.) are determined by energy differences. Secondly we introduce «simulated annealing» as a strategy for determining low-lying energy minima. Calculations on sulphur clusters (subsect. 4'3.3) indicate that the equilibrium structures found are, in fact, close to the structures determined experimentally.

4'1. *Density functional formalism.* – The two basic theorems of the density functional formalism were derived by HOHENBERG and KOHN[39]. They showed that: 1) Ground state (GS) properties of a system of electrons and ions in an external field, V_{ext}, can be expressed as functionals of the electron density, $n(r)$, *i.e.* they are determined by a knowledge of the density *alone*. The total energy is such a functional, *i.e.* $E = E[n]$. 2) $E[n]$ satisfies the variational principle, $E[n] \geqslant E_{GS}$, and the density for which the equality holds is the ground-state density, n_{GS}. Simple and general proofs of these assertions have been provided by LEVY[40].

The *applicability* of this scheme results from the observation by KOHN and SHAM[41] that the minimization of the functional $E[n]$ can be simplified if the following separation is adopted:

$$(4) \qquad E[n] = T_0[n] + \int dr\, n(r)\left(V_{ext}(r) + \frac{1}{2}\Phi(r)\right) + E_{xc}[n],$$

where T_0 is the kinetic energy which a system with density n would have in the absence of electron-electron interactions, $\Phi(r)$ is the Coulomb potential, and E_{xc} defines the exchange-correlation energy.

The variational principle now yields

$$(5) \qquad \frac{\delta E[n]}{\delta n(r)} = \frac{\delta T_0}{\delta n(r)} + V_{ext}(r) + \Phi(r) + \frac{\delta E_{xc}[n]}{\delta n(r)} = \mu,$$

where μ is the Lagrange multiplier associated with the requirement of constant particle number. Comparing this with the corresponding equation for a system with an effective potential $V(r)$ but *without* electron-electron interactions,

$$(6) \qquad \frac{\delta E[n]}{\delta n(r)} = \frac{\delta T_0}{\delta n(r)} + V(r) = \mu,$$

we see that the mathematical problems are identical, provided that

$$(7) \qquad V(r) = V_{ext} + \Phi(r) + \frac{\delta E_{xc}[n]}{\delta n(r)}.$$

The solution of (7) can be found in a straightforward fashion by solving the Schrödinger equation for noninteracting particles,

$$(8) \qquad \left(-\frac{1}{2}\nabla^2 + V(\boldsymbol{r})\right)\psi_i(\boldsymbol{r}) = \varepsilon_i \psi_i(\boldsymbol{r}),$$

yielding

$$(9) \qquad n(\boldsymbol{r}) = \sum_{i=1}^{N} |\psi_i(\boldsymbol{r})|^2.$$

It is necessary to satisfy condition (7), and this can be achieved in a self-consistent procedure.

The problem has, therefore, been reduced *exactly* to the solution of a single-particle equation of Hartree form. The solution of the equations leads to the energy and density of the lowest state, and all quantities derivable from them. It shares the main advantage of the HF method, namely the single-particle interpretation of the results. However, in contrast to the Hartree-Fock potential

$$(10) \qquad V_{\mathrm{HF}}\psi(\boldsymbol{r}) = \int d\boldsymbol{r}' \, V_{\mathrm{HF}}(\boldsymbol{r}, \boldsymbol{r}')\psi(\boldsymbol{r}'),$$

the effective potential, $V(\boldsymbol{r})$, has a *local* dependence on the density.

The only term in (4) for which an approximation is necessary is E_{xc}, so that approximations for the exchange-correlation energy play a central role in applications of the method. If E_{xc} has a local dependence on the electron density, the equations present no more numerical complications than the solution of Hartree's equations. The most widely used approximation for E_{xc} is the local spin density (LSD) approximation

$$(11) \qquad E_{\mathrm{xc}}^{\mathrm{LSD}} = \int d\boldsymbol{r}\, n(\boldsymbol{r})\, \varepsilon_{\mathrm{xc}}[n_\uparrow(\boldsymbol{r}), n_\downarrow(\boldsymbol{r})].$$

Here $\varepsilon_{\mathrm{xc}}[n_\uparrow, n_\downarrow]$ is the exchange and correlation energy per particle of a homogeneous, spin-polarized electron gas with spin-up and spin-down densities $n_\uparrow$ and $n_\downarrow$, respectively. The LSD approximation is free of adjustable parameters, but its application to atoms, molecules and solids cannot be justified by small departures from homogeneity. For a detailed discussion, see the review article by JONES and GUNNARSSON [13]. The parametrization of Vosko *et al.* [42] is used in the calculations on S_n clusters described below.

There have been many applications of DF calculations with the LSD approximation [13]. In the context of molecules containing group VI A elements, the DF approach has been applied with success to triatomic molecules [43] and to helical chains of sulphur and selenium [44]. Calculated geometries agree well with experimental values where available, and the systematic overestimate of bond strengths (by $\sim 1\,\text{eV}$ per bond) means that bonding trends are given reliably. These are general features of DF/LSD calculations for sp-bonded systems.

An interesting application to larger clusters was the work of Martins *et al.* [45] to Na and Na$^+$ clusters with $N \leqslant 8$ and $N = 13$. They started from randomly generated geometries and let them relax under the action of the forces on the atoms. From the structures corresponding to local minima in the energy surfaces they could then determine the ground state. They found that clusters of up to five atoms are planar, Si$_6$ is quasi-planar, and that real three-dimensional structures occur only for larger clusters. The calculated ionization energies reproduce the experimentally observed zig-zag trend.

Density functional calculations of Si clusters up to $N = 14$ have been performed by TOMÁNEK and SCHLÜTER [46]. These authors used a parametrized tight-binding Hamiltonian to locate the minimum energy of each cluster, and then determined the DF energies for several stable and metastable structures of Si$_2$ to Si$_{10}$. No open structures and no diamond lattice fragments were found, and particularly stable structures were found for Si$_6$ and Si$_{10}$, «magic numbers» for small Si clusters.

4˙2. *MD/DF method.* – In systems where the ground state is unknown or there are many local minima, it is necessary to develop alternative methods for finding solutions which are *near* to optimal. KIRKPATRICK *et al.* [47] noted the connection between statistical physics and the minimization of a function of many variables, and suggested «simulated annealing» based on a Monte Carlo sampling as a generally applicable way of finding such solutions. It has been shown recently [48] that this approach can lead to *nearly* optimal solutions of special *NP*-complete problems in polynomial average time.

Molecular dynamics (MD) provides an alternative to the Monte Carlo approach, and CAR and PARRINELLO [8] showed that it could be combined with the density functional (DF) scheme for calculating total energies. The energy minimization can be performed using MD techniques, since E can be viewed as a function of two interdependent sets of degrees of freedom, $\{\psi_i\}$ and $\{R_I\}$,

$$
(12) \quad E[\{\psi_i\}, \{R_I\}] = \sum_i \langle \psi_i(r) | -\frac{\nabla^2}{2} | \psi_i(r) \rangle + \int dr\, n(r) \left(V_{\text{ext}}(r) + \frac{1}{2} \Phi(r) \right) +
$$
$$
+ E_{\text{xc}}[n(r)] + \frac{1}{2} \sum_{I \neq J} \frac{Z_I Z_J}{|R_I - R_J|}.
$$

Dynamical simulated-annealing techniques can then be used to follow the trajectories of $\{\psi_i\}$ and $\{R_I\}$ given by the Lagrangian

$$
(13) \quad \mathscr{L} = \sum_i \mu_i \int_\Omega dr\, |\dot{\psi}_i^* \psi_i| + \sum_I \frac{1}{2} M_I \dot{R}_I^2 - E[\{\psi_i\}, \{R_I\}] + \sum_{ij} \Lambda_{ij} \left(\int_\Omega dr\, \psi_i \psi_j^* - \delta_{ij} \right)
$$

and the corresponding equations of motion

(14)
$$\begin{cases} \mu \ddot{\psi}_i(\mathbf{r}, t) = -\dfrac{\delta E}{\delta \psi^*(\mathbf{r}, t)} + \sum_k \Lambda_{ik}\, \psi_k(\mathbf{r}, t)\,, \\[2ex] M_I \ddot{\mathbf{R}}_I = -\nabla_{R_I} E\,. \end{cases}$$

Here M_I and $\mathbf{R}_I$ denote the ionic masses and coordinates, μ_i are fictitious «masses» associated with the electronic degrees of freedom, dots denote time derivatives, and the Lagrangian multipliers Λ_{ij} are introduced to satisfy the orthonormality constraints on the $\psi_i(\mathbf{r}, t)$. From these orbitals and the resultant density $n(\mathbf{r}, t) = \sum_i |\psi_i(\mathbf{r}, t)|^2$ we can use eq. (12) to evaluate the total energy E, which acts as the classical potential energy in the Lagrangian (13). The artificial second-order Newton's dynamics for the electronic degrees of freedom, together with the assumption $\mu_i \ll M_I$, effectively prevent transfer of energy from the classical to the quantum degrees of freedom over long periods of simulation.

The MD procedure is initiated by displacing the atoms randomly from an arbitrary geometry, with velocities $\dot{\psi}_i$ and $\dot{\mathbf{R}}$ set equal to zero. For this geometry we then determine those ψ_i that minimize E via an efficient self-consistent iterative diagonalization technique. With the electrons initially in their ground state, the dynamics (14) generate Born-Oppenheimer (BO) trajectories over several thousand time steps without the need for additional diagonalization/self-consistency cycles for the electrons. If we define a «temperature» T by the mean classical kinetic energy of the atoms, the atomic configurations observed during the annealing process then represent physical molecules at elevated temperatures.

The combination of MD and DF techniques results in a parameter-free method for calculating electronic properties that makes no assumptions about ground-state geometries. There are two main advantages of this approach. Firstly, it extends the range of application of density functional techniques beyond calculations at $T = 0$, allowing simulations of liquid and amorphous phases and even of transitions between them. In the case of cluster calculations, where the structure and nature of the ground state is not generally known, finite-temperature MD techniques allow an efficient sampling of the potential-energy surface. It becomes possible to discriminate between the large number of local energy minima that usually arise. Secondly, the DF method[13], with the local spin density (LSD) approximation for the exchange-correlation energy, does not require the parametrization of the interatomic forces common in MD schemes[49]. This combination sounds very nice in principle, but the usefulness of any technique depends on how it performs in practice.

4`3. Applications.

4`3.1. Silicon clusters. The MD/DF approach has been applied to amorphous and liquid Si with impressive results[50], with the mean coordination number and radial distribution function being in good agreement with experiment. The first application to small Si clusters was described by CAR *et al.*[51], who found a number of structures with comparable energy. While the structures at $T = 0$ tend to be symmetrical, those at higher energies appear to be disordered. This is an interesting result, since calculations using other techniques[46] had generally considered only symmetric structures as alternatives to the ground state. We have noted above that almost nothing is known experimentally about the structure of small Si clusters.

4`3.2. Sulphur clusters. Experimental data on the structure of S and Se clusters are very extensive. Application of the MD/DF approach to selenium clusters Se_n ($n = 3 \div 8$)[9] yielded excellent agreement with X-ray structural analyses for Se_6 and Se_8, and the predictions for other structures should be reliable. I now turn to S_n molecules ($n = 2 \div 13$), which have been discussed in detail by HOHL *et al.*[10] and for which much experimental structural information is available.

An example of the results are low-lying structures for S_6, S_7 and S_8, shown in fig. 10. Crystalline rhombohedral S_6 contains six-membered rings with D_{3d} symmetry, and X-ray data show that these molecular units are well separated. The annealing process was initiated from slightly displaced planar hexagon and linear chains, *i.e* the initial conditions did *not* exploit available experimental information for S_6. Nevertheless, the $T \to 0$ ground-state geometry matches the experimental configuration very well[10]. A «boat» isomer was not observed during the annealing process, but was checked via steepest-descent quenches and low-temperature annealing. The minimum energy of this isomer was ~ 0.5 eV above the ground state.

Homocyclic S_7 is a constituent of sulphur vapour, liquid sulphur and quenched liquid sulphur, of irradiated sulphur solutions and of some chemically synthesized sulphur modifications[52]. There are four crystalline allotropes[52], and two have been characterized structurally. The molecular units found have almost identical geometries and approximately C_s symmetry. The shortest intermolecular distances (6.43 a.u.) are slightly less than in S_6, and the crystal packing effects are correspondingly greater. We have started from a perturbed planar heptagon, and found very good agreement with experiment. Other pronounced minima were not observed. If the cluster in its ground state is coupled to a heat bath at $T = 1500$ K, simultaneous motion about several of the bonds could be observed, corresponding to a migration of the bond with torsion angle $\gamma = 0$ around the molecule. Such a «pseudorotation» has been inferred by STEUDEL *et al.*[53] from room temperature Raman and ^{77}Se NMR spectra in S_7

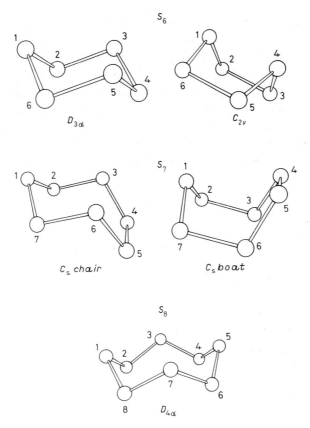

Fig. 10. – The structures of S_6, S_7 and S_8.

and Se_2S_5. Our calculations require a substantially higher temperature to overcome the short time scale in our simulations ($\sim 10^{15}$ shorter than reality).

Another candidate for the ground state of S_7 is the boat isomer shown in fig. 10. This structure has been checked by steepest-descent quenches and low-temperature annealing and found to be the lowest excited state, only $0.1\,\mathrm{eV}$ above the ground state. The boat form of S_7 may, therefore, be observable experimentally.

Cyclooctasulphur S_8 is the most common and best characterized allotrope of elemental sulphur and a large amount of experimental data is available for the three crystalline forms [32]. The minimum ratio of intra- to intermolecular distances in the crystal is 1.64 and the small crystal field distorts the ring marginally from perfect D_{4d} symmetry. The annealing process was initiated from a planar octagon with small random perturbations, and we found the familiar «crown» structure (fig. 10) as the $T \to 0$ ground state. Apart from a small overestimate of the bond lengths, the agreement with experimental data is very satisfactory. In order to determine the vibration frequencies, the converged ground-state geometry was randomly displaced with sufficient amplitude to excite all vibrational modes, and the subsequent constant-energy MD evolution followed for 35000 time steps. The peak positions in the vibrational DOS agree well with the experimental Raman and IR spectra of S_8[54].

The S_9 molecule is particularly interesting. While it can be prepared as microcrystals[32, 55], the absence of single crystals has ruled out an X-ray structure determination. However, the Raman spectra are very different from those of all other S molecules studied to date, and STEUDEL et al.[55] concluded that the constituent molecules have nearly identical structures with S-S bonds that are neither unusually long nor short. From the predicted values (between 3.84 a.u. and 3.95 a.u.) and the empirical relationship between d_{SS} and the corresponding torsion angle (see ref.[10]), they concluded that $70° \leqslant \gamma \leqslant 130°$, i.e. geometries with four consecutive atoms nearly in a plane can be excluded. This rules out C_s symmetry, and allows only C_1 and C_2. Threefold axes can be excluded.

The calculations were started from a randomly displaced planar ring, and led to a ground-state structure (fig. 11) that fulfils all of the above criteria. It has C_2 symmetry, approximately equal bond lengths and torsion angles between 66° and 114°. Three low-lying structures result by adding S_2 units to S_7 (C_s «boat», C_s «cage») and replacing a single atom in S_8 (C_s «chair») by a dimer. The fully relaxed structures are also shown in fig. 11. None of these conforms to the above spectroscopic findings, and they lie $\geqslant 0.2$ eV above the ground states according to our calculations.

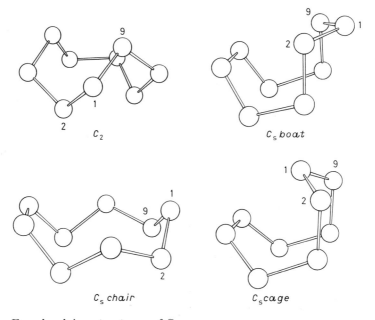

Fig. 11. – Four low-lying structures of S_9.

4˙3.3. S_7O. The MD/DF method can locate energy minima for geometries far from the starting structure. In S_7O [56], we started with a random perturbation of the calculated ground-state structure of $S_8 [D_{4d}]$ and located the nearest stable minimum, which is a ring structure with symmetry D_{2d}. If the mean kinetic energy of the nuclei is raised to 2000 K, we observe the dramatic structural structural changes shown in fig. 12a)-i) at intervals of 150 time steps ($\Delta t = 1.7 \cdot 10^{-16}$ s). The S_1-O and S_7-O bonds become longer, the S_1-O-S_7 angle is reduced sharply, and the O atom is pushed out of the ring. The structure at this point (fig. 12i)) is an oxygen atom outside the S_7 ring structure. However, the topology is different from that of the measured ground state.

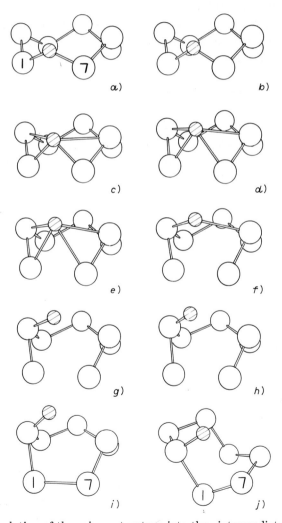

Fig. 12. – a)-i) Evolution of the «ring» structure into the «intermediate» structure. The interval between each frame is 150 time steps. j) The ground-state structure from the same perspective.

We have continued the high-temperature simulation from the «intermediate» geometry shown in fig. 12i), reducing the temperature in stages to $T < 100$ K. The structure changes continuously into the ground-state form, shown in fig. 12j) from the same perspective as in fig. 12a)-i). The lowest energy is a stable minimum 0.13 eV below that of the ring structure, and the calculated geometry is in very good agreement with that determined from X-ray diffraction [57]. Experimental bond and torsion angles are generally reproduced to within $(2 \div 3)°$, lengths of shorter bonds are overestimated by $(1 \div 4)\%$, while the particularly long bonds d_{12} and d_{23} are overestimated by slightly more. The height of the barrier between ring and «intermediate» structures is ~ 5 eV, substantially more than the thermal energy available initially. The energy fed into the system to maintain the average kinetic energy of the nuclei means that the system is far from equilibrium during most of the simulation.

The above results provide a remarkable demonstration of the value of MD methods in probing complicated energy surfaces in systems where the ground-state structure is not known. An essential feature is, of course, the introduction of kinetic-energy («temperature») terms in eq. (13). It is unrealistic to expect that conventional methods, either using correlated wave function or density functional techniques, will be able to find a path in configuration space between low-lying minima in such cases. This will be true even if gradient techniques are used, as the analytical properties of the energy as a function of the coordinates are by no means simple. In principle, the MD method could be applied to Hartree-Fock or configuration interaction techniques, but the CPU time required with current implementations of these methods would be much greater than the 5 s required with the present approach.

5. – Concluding remarks.

There have been exciting developments in experimental techniques in the production, detection and analysis of small clusters in the last decade. Spectroscopic data have resulted on a great range of clusters, and the prospects for even more improvements are excellent. There is, however, relatively little detailed information about the *structure* of most clusters, and this fact makes a microscopic analysis of the properties very difficult.

The cluster theorist is then faced with problems even more difficult than those faced by his colleagues in solid-state physics and, to some extent, in surface physics. The difficulties resemble those in surface physics about 20 years ago, before the development of reliable methods for surface structure determination (*e.g.*, low-energy electron diffraction (LEED), EXAFS, medium-energy ion scattering (MEIS), scanning tunnelling microscopy (STM)). It would be an intriguing development if the structure could be determined from theoretical considerations alone. We have seen, however, that the traditional methods of calculating total energies become intractable as the cluster size increases. Moreover, the number of structures corresponding to local energy minima becomes so great that it is impossible to determine the ground state by locating and sorting them all.

Nevertheless, the discussion in sect. 4 shows that the prospects for progress are far from hopeless. Most properties of interest, including the structures of low-lying states, are determined by energy *differences*, and we have seen examples where these are given satisfactorily by density functional calculations with a simple, local density approximation for the exchange-correlation energy. These calculations are relatively easy to perform and can be extended to larger systems without difficulty. The second problem, locating the *exact* minimum of the total energy, will probably remain «intractable». Our experience with simulated-annealing techniques shows, however, that the combination of molecular dynamics with density functional techniques provides a useful method of determining minima which should be close to the exact value. With appropriately constructed basis functions, the technique should also be applicable to situations, such as transition metals, where the pseudopotential approach we have used is difficult to apply.

Whatever developments take place, it would benefit everyone if theory *and* experiment would take note of each other's strengths and weaknesses. It is not obvious how much is gained, for example, by performing measurements on systems which are far beyond the current capabilities of theory, while ignoring others where predictive calculations are already possible. On the other hand, oversimplified «explanations» of complicated data can sometimes be more confusing than useful.

* * *

The calculations on sulphur clusters and S_7O discussed here were performed in collaboration with D. HOHL, R. CAR and M. PARRINELLO. I thank P. S. BECHTHOLD for useful discussions on the experimental situation.

REFERENCES

[1]   J. JORTNER, D. SCHARF and U. LANDMAN: in *Excited-State Spectroscopy in Solids*, *Proc. SIF*, Course XCVI, edited by U. M. GRASSANO and N. TERZI (North-Holland, Amsterdam, 1987), p. 438.

[2]   G. SCOLES, Editor: *The Chemical Physics of Atomic and Molecular Clusters*, *Proc. SIF*, Course CVII (North-Holland, Amsterdam, 1990).

[3]   J. JORTNER: *Ber. Bunsenges. Phys. Chem.*, **88**, 1 (1984).

[4]   S. SUGANO, Y. NISHINA and S. OHNISHI, Editors: *Microclusters, Proceedings of the First NEC Symposium, Tokyo, 1986* (Springer-Verlag, Berlin, 1987).

[5]   P. JENA, B. K. RAO and S. N. KHANNA, Editors: *The Physics and Chemistry of Small Clusters*, NATO ASI Series B, Vol. **158** (Plenum, New York, N.Y., 1987).

[6] H. HABERLAND: in *Proceedings of the NATO Advanced Study Institute on Fundamental Processes of Atomic Dynamics*, edited by J. S. BRIGGS, H. KLEINPOPPEN and H. O. LUTZ (Plenum, New York, N. Y., to be published).

[7] G. BENEDEK and T. P. MARTIN, Editors: *Elemental and Molecular Clusters, Proceedings of the International School of Materials Science and Technology, Erice* (Springer, Berlin, to be published).

[8] R. CAR and M. PARRINELLO: *Phys. Rev. Lett.*, **55**, 2471 (1985).

[9] D. HOHL, R. O. JONES, R. CAR and M. PARRINELLO: *Chem. Phys. Lett.*, **139**, 540 (1987).

[10] D. HOHL, R. O. JONES, R. CAR and M. PARRINELLO: *J. Chem. Phys.*, **89**, 6823 (1988).

[11] K. A. GINGERICH: in *Current Topics in Materials Science*, Vol. **6**, edited by E. KALDIS (Elsevier, Amsterdam, 1980), p. 345.

[12] K. HILPERT and K. RUTHARDT: *Ber. Bunsengs. Phys. Chem.*, **91**, 724 (1987).

[13] R. O. JONES and O. GUNNARSSON: *Rev. Mod. Phys.*, **61**, 689 (1989).

[14] S. IJIMA: in *Microclusters, Proceedings of the First NEC Symposium*, Tokyo, 1986, edited by S. SUGANO, Y. NISHINA and S. ONISHI (Springer-Verlag, Berlin, 1987), p. 186.

[15] P. BUFFAT and J.-P. BOREL: *Phys. Rev. A*, **13**, 2287 (1976).

[16] W. D. KNIGHT, K. CLEMENGER, W. A. DE HEER, W. A. SAUNDERS, M. Y. CHOU and
M. L. COHEN: *Phys. Rev. Lett.*, **52**, 2141 (1984).

[17] W. D. KNIGHT, W. A. DE HEER, W. A. SAUNDERS, K. CLEMENGER and M. L. COHEN: *Chem. Phys. Lett.*, **134**, 1 (1987).

[18] K. RADEMANN, B. KAISER, U. EVEN and F. HENSEL: *Phys. Rev. Lett.*, **59**, 2319 (1987).

[19] T. P. MARTIN: *J. Chem. Phys.*, **81**, 4427 (1984).

[20] P. A. MONTANO, G. K. SHENOY, E. E. ALP, W. SCHULZE and J. URBAN: *Phys. Rev. Lett.*, **56**, 2076 (1986).

[21] P. S. BECHTHOLD, U. KETTLER, H. R. SCHOBER and W. KRASSER: *Z. Phys. D*, **3**, 263 (1986).

[22] P. S. BECHTHOLD and H. R. SCHOBER: in *The Physics and Chemistry of Small Clusters*, NATO ASI Series B, Vol. **158**, edited by P. JENA, B. K. RAO and N. KHANNA (Plenum, New York, N.Y., 1987), p. 583.

[23] H. P. BIRKHOFER, H. HABERLAND, M. WINTERER and D. R. WORSNOP: *Ber. Bunsenges. Phys. Chem.*, **88**, 207 (1984).

[24] O. ECHT, K. SATTLER and E. RECKNAGEL: *Phys. Rev. Lett.*, **47**, 1121 (1981).

[25] T. G. DIETZ, M. A. DUNCAN, D. E. POWERS and R. E. SMALLEY: *J. Chem. Phys.*, **74**, 6511 (1981).

[26] V. E. BONDYBEY and J. H. ENGLISH: *Chem. Phys. Lett.*, **94**, 443 (1983).

[27] V. E. BONDYBEY and J. H. ENGLISH: *J. Chem. Phys.*, **80**, 568 (1984).

[28] D. L. MICHALOPOULOS, M. E. GEUSIC, S. G. HANSEN, D. E. POWERS and R. E. SMALLEY: *J. Phys. Chem.*, **86**, 3914 (1982).

[29] S. J. RILEY, E. K. PARKS, L. G. POBO and S. WEXLER: *J. Chem. Phys.*, **79**, 2577 (1983).

[30] E. A. ROHLFING, D. M. COX, A. KALDOR and K. H. JOHNSON: *J. Chem. Phys.*, **81**, 3846 (1984).

[31] J. DONOHUE: *The Structures of the Elements* (Wiley, New York, N.Y., 1974), Chapt. 9.

[32] R. STEUDEL: in *Studies in Inorganic Chemistry*, Vol. **5**, edited by A. MÜLLER and B. KREBS (Elsevier, Amsterdam, 1984), p. 3.

[33] R. STEUDEL and E. M. STRAUSS: in *The Chemistry of Inorganic Homo- and Heterocycles*, Vol. 2 (Academic Press, London, 1987), p. 769.

[34] M. R. HOARE and J. A. MCINNES: *Adv. Phys.*, **32**, 791 (1983).

[35] L. T. WILLE and J. VENNIK: *J. Phys. A*, **18**, L419, L1113 (1985).

[36] M. R. GAREY and D. S. JOHNSON: *Computers and Intractability: A Guide to the Theory of NP-Completeness* (Freeman, San Francisco, Cal., 1979).

[37] T. P. MARTIN, T. BERGMANN and B. WASSERMANN: in *Microclusters, Proceedings of the First NEC Symposium, Tokyo, 1986*, edited by S. SUGANO, Y. NISHINA and S. OHNISHI (Springer-Verlag, Berlin, 1987), p. 152.

[38] K. RAGHAVACHARI: *J. Chem. Phys.*, **84**, 5672 (1986).

[39] P. HOHENBERG and W. KOHN: *Phys. Rev.*, **136**, B864 (1964).

[40] M. LEVY: *Proc. Nat. Acad. Sci. USA*, **76**, 6062 (1979).

[41] W. KOHN and L. J. SHAM: *Phys. Rev.*, **140**, A1133 (1965).

[42] S. VOSKO, L. WILK and M. NUSAIR: *Can. J. Phys.*, **58**, 1200 (1980).

[43] For a discussion of O_3, SO_2, SOS and S_3, see R. O. JONES: *Adv. Chem. Phys.*, **67**, 413 (1987), and references therein.

[44] M. SPRINGBORG and R. O. JONES: *Phys. Rev. Lett.*, **57**, 1145 (1986); *J. Chem. Phys.*, **88**, 2652 (1988).

[45] J. L. MARTINS, J. BUTTET and R. CAR: *Phys. Rev. B*, **31**, 1804 (1985).

[46] M. TOMÁNEK and M. SCHLÜTER: *Phys. Rev. Lett.*, **56**, 1055 (1986).

[47] S. KIRKPATRICK, C. D. GELATT jr. and M. P. VECCHI: *Science*, **220**, 671 (1983).

[48] G. H. SASAKI and B. HAJEK: *J. Assoc. Comput. Mach.*, **35**, 387 (1988).

[49] F. STILLINGER, T. A. WEBER and R. A. LAVIOLETTE: *J. Chem. Phys.*, **85**, 6460 (1986).

[50] R. CAR and M. PARRINELLO: *Phys. Rev. Lett.*, **60**, 204 (1988).

[51] R. CAR, M. PARRINELLO and W. ANDREONI: in *Microclusters, Proceedings of the First NEC Symposium, Tokyo, 1986*, edited by S. SUGANO, Y. NISHINA and S. OHNISHI (Springer-Verlag, Berlin, 1987), p. 134.

[52] R. STEUDEL, J. STEIDEL, J. PICKARDT, F. SCHUSTER and R. REINHARDT: *Z. Naturforsch., Teil B*, **35**, 1378 (1980).

[53] R. STEUDEL, M. PAPAVASSILIOU and D. JENSEN: *Z. Naturforsch., Teil B*, **43**, 245 (1988).

[54] For references to the experimental data, see ref. [10].

[55] R. STEUDEL, T. SANDOW and J. STEIDEL: *Z. Naturforsch., Teil B*, **40**, 594 (1985).

[56] D. HOHL, R. O. JONES, R. CAR and M. PARRINELLO: *J. Am. Chem. Soc.*, **111**, 825 (1989).

[57] H. BITTERER, Editor: *Schwefel: Gmelin Handbuch der Anorganischen Chemie*, 8. Auflage, Ergänzungsband 3 (Springer-Verlag, Berlin, 1980), p. 8.

Surface Magnetism and Spin-Resolved Photoemission.

E. KISKER

Institut für Angewandte Physik, Universität Düsseldorf - D-4000 Düsseldorf 1

C. CARBONE

Institut für Festkörperforschung, Forschungszentrum Jülich - b-5170 Jülich, B.R.D.

1. – Introduction.

Substantial amount of research activity has been concentrated in recent years on the magnetic properties of surfaces, interfaces and thin films. Major breakthroughs have occurred because of improvement and developments of new experimental methods such as spin-resolved photoemission[1] and inverse photoemission[2] which allow us to determine the electronic structure of ferromagnets. Success has also been achieved in the ability to prepare well-characterized thin films and surfaces. Theoretical models and computational methods have been developed which allow a better understanding of the basic phenomena.

Electron spectroscopy is best suited to investigate surface and bulk phenomena due to the dependence of probing depth on energy (see fig. 1)[3]. The minimum probing depth of only a few Å, or $1 \div 2$ atomic layers, is obtained in the 30 to 100 eV energy range. The small probing depth of the photoemitted electrons also allows one to study the interplay between electronic and magnetic character of ultrathin films in the monolayer regime and of interface systems.

Fundamental questions regarding ferromagnetism can be studied by testing the *valence band* electronic structure, which has been calculated by modern theories for the elemental ferromagnets, for their surfaces and for ultrathin films deposited on various substrates. A field which just becomes experimentally accessible is the spin-dependent electronic structure of *core levels*. In contrast to the valence bands, the core electrons are localized at the atomic site, and, therefore, provide information on the local electronic structure.

This lecture is intended to give an introduction to the experimental principles behind spin- and angle-resolved photoemission from magnetic surfaces. On the basis of selected results, the capabilities of this new technique for investigating the bulk and surface electronic properties of ferromagnets are demonstrated. For a more comprehensive review on this already substantially wide field, see [1].

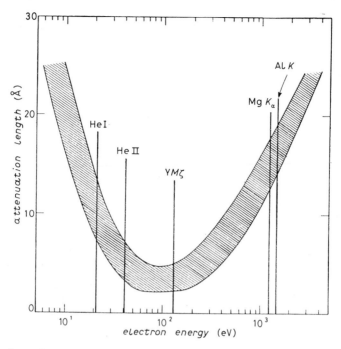

Fig. 1. – The dependence of probing depth on energy in electron spectroscopy from metals (after [1]).

2. – Bulk and surface magnetism.

Magnetism in the $3d$ transition metals Fe, Co and Ni is caused by the unbalance of *itinerant* (collective) d-electrons whose magnetic moments are pointing either parallel or antiparallel to the magnetization direction. The numbers of these two electron groups are referred to as $n^\uparrow$ and $n^\downarrow$, and the magnetization is $M \simeq (n^\uparrow - n^\downarrow)\mu_B$. Within the itinerant-electron (band) model the nonintegral number of Bohr magnetons in Fe, Co and Ni (2.2, 1.72 and $0.6\,\mu_B$) is easily accounted for. The magnetism of the rare earths, however, is due to the spin and orbital moment of the *localized* $4f$ electrons. Most investigations have in the past concentrated on the $3d$ transition metals, because of controversies regarding their electronic structure [4]. However, there is no reason to expect that the electronic structure of the rare earths is much simpler [5, 6].

The electronic and magnetic properties of the surface may significantly differ from those of the bulk because of the reduced number of neighbouring atoms at the surface. This is easily seen in the mean-field approximation: Considering a f.c.c. lattice, the number of next neighbours of an atom located in the bulk is 12, but only 8 for a surface atom in a (001) plane. In the mean-field approximation, the magnetization at temperature T is given by

(1) $$M = N \mu_B B\{\mu\lambda M/kT\} \,,$$

where B is the usual Brillouin function, N the number of atoms, and λ is the mean-field constant, assumed to be independent of temperature. Introducing $T_C = N\mu^2\lambda/k$, $t = T/T_C$, $m = M/N\mu$, eq. (1) reads

(2) $$m = B(m/t).$$

The magnetization at the layer with index i is accordingly

(3) $$m_i = B\{\textstyle\sum m_k/zt\} \,,$$

where z is the number of next neighbours and the sum extends over the next neighbours. It is also assumed that the exchange constant is the same at the surface as in the bulk. This yields a set of n equations for each layer with index i:

(4) $$\begin{cases} m_1 = B\{(4m_1 + 4m_2)/zt\} \,, \\[2mm] m_i = B\{4(m_{i-1} + m_i + m_{i+1})\}/zt\} \,. \end{cases}$$

Equations (4) have to be solved by iteration until self-consistency is achieved, starting from an arbitrary m_i distribution. The result of such a calculation is shown in fig. 2. It is seen that, at $T = 0$, the magnetization is uniform

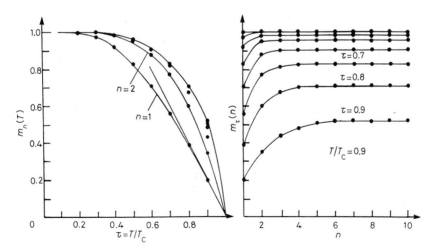

Fig. 2. – Result of a mean-field calculation of the layer- and temperature-dependent magnetization in a (100) oriented f.c.c. lattice.

everywhere within the semi-infinite solid. But at elevated temperatures, the surface magnetization is less than the magnetization in the bulk. As a function of temperature, a more or less linear decrease in magnetization is observed for the first layer. The temperature dependence of the 3rd layer is already close to that of the bulk.

Due to the break in symmetry at the surface, self-consistent electronic-structure calculations to determine the surface magnetization are a difficult task [7]. Starting from a simpler tight-binding approach, it can be seen that the band width at the surface will be more narrow than in the bulk. For a metal where the Fermi energy falls near the middle of the bands (as in Fe), this results in an increase of the surface magnetic moment. This is also borne out in the self-consistent calculations [8]. Similar arguments as for the surface of a semi-infinite crystal apply to ultrathin films of a ferromagnet on a nonmagnetic substrate, *i.e.* the reduced coordination number also is expected to cause an enhanced magnetic moment for films in the monolayer (ML) thickness regime.

For Ni, the fact that the Fermi level falls near the top of the *d*-bands and is above the top of the majority spin bands complicates the theoretical treatment considerably [7]. However, recent self-consistent calculations agree in predicting an enhanced surface magnetization [9].

When the surface magnetization differs from the bulk magnetization (at $T = 0$), this would correspond in the mean-field approximation to an enhanced coupling of the magnetic moments in the surface layer. This can be introduced also in the mean-field calculations (eq. (4)) [10]. The coupled equations (4) can then be solved in the same way, and the temperature dependence of the magnetization of each layer can be calculated and compared with experimental data.

Recently, it has become possible to calculate surface electronic properties and the electronic structure of ultrathin films [11]. The reduced symmetry at the surface and in ultrathin films is expected to affect also the surface magnetic properties, such as the local magnetic moment, the surface magnetization direction and the Curie temperature (T_C) [12, 13].

3. – Principles of spin- and angle-resolved photoemission.

3'1. *Angle-resolved photoemission.* – In spin- and angle-resolved photo-emission, the intensity I^σ the photoelectron beam, measured at the angles Θ and Φ from the surface normal, depends on several experimental parameters:

$$(5) \qquad\qquad I^\sigma = I^\sigma(E_k, \Theta, \Phi, h\nu, \vartheta, \varphi, \mathscr{A}),$$

where E_k is the kinetic energy of the electron and σ the electron spin. The direction of the incident light is defined by ϑ and φ, $h\nu$ is the photon energy and $\mathcal{A}$ the light vector potential, describing the light polarization direction with respect to the crystal lattice. In a spin- and angle-resolved photoemission experiment employing polarized light, all the above quantities are experimentally determined. The energy E_k is related to the binding energy E_B and the Fermi energy E_F by the relation $E_B = h\nu - E_k - (E_V - E_F)$. The emission angles Θ and Φ are determined by the internal wave vector (k_{int}) of the photoelectron. Because of diffraction at the surface, it is, however, not always easy to determine k_{int} from these externally observed angles. The most simple experimental situation arises when Θ and Φ are zero, *i.e.* for normal emission. Most of the spin-resolved photoemission experiments have been performed for normal emission.

3˙2. *Spin detection.* – It is not possible to measure the electron spin by classical methods [14]. Therefore, either the (relativistic) spin-orbit interaction or the quantum-mechanical exchange interaction have been used for spin analysis. It is the effect of spin-orbit interaction which is used most commonly for spin analysis, referred to as Mott scattering.

Spin analysis works because the scattering cross-section for a fixed scattering angle is spin-dependent due to the interaction of the electron magnetic moment with the magnetic field which results from the relative motion of the electron with respect to a target atom. By symmetry, the increase in cross-section for electrons with spin σ, scattered by angle $+\Theta$, is the same as the decrease in cross-section for the electrons being scattered by the angle $-\Theta$ (as compared to the case without considering the spin). Because of this symmetry, it is convenient to use a pair of electron detectors in the Mott spin analyser, one detector being set up at scattering angle $+\Theta$, the other one at $-\Theta$, and to determine the spin polarization from the intensity asymmetry obtained at the scattering angles $\pm\Theta$.

The spin polarization P for an electron beam is defined as

(6) $$P = (N^\uparrow - N^\downarrow)/(N^\uparrow + N^\downarrow),$$

where $N^\uparrow$ ($N^\downarrow$) are the numbers of electrons with their magnetic moments pointing parallel (antiparallel) to the magnetization direction, respectively. For a beam of given spin polarization P, the measured scattering asymmetry $A'(P)$ is defined as

(7) $$A'(P) = (N_1 - N_r)/(N_1 + N_r),$$

where $N_{1,r}$ are the numbers of electrons detected by the pair of detectors at scattering angles $\pm\Theta$. $A'(P)$ generally depends on energy and on the scattering angle [14]. From A', the spin polarization is obtained as

$$(8) \qquad\qquad p = 1/S_{\text{eff}} A' ,$$

where S_{eff} is the so-called effective Sherman factor. Obviously, $1/S_{\text{eff}}$ determines the spin sensitivity of the spin detector.

Two regimes of Mott scattering can be defined: Scattering at energies $10 \text{ keV} < E < 150 \text{ keV}$ (high-energy Mott scattering) and scattering at low energies ($10 \text{ eV} < E < 200 \text{ eV}$), referred to as low-energy Mott scattering. In both regimes, spin-orbit interaction is the reason for the intensity asymmetry with respect to the spin.

In high-energy Mott scattering, A' is largest for scattering at nuclei with large atomic number, as, *e.g.*, Au, and for about 100 keV electron energy, at $\theta = \pm 120°$ scattering angles. A thin Au film is generally used as scattering target. The spin sensitivity depends on the thickness of the scattering foil because of multiple scattering [14]. Typical values of S_{eff} are between 0.2 and 0.3 for gold foils of 1000 Å thickness.

Only a fraction of $10^{-3} \div 10^{-4}$ of the incoming electron beam intensity is backscattered into the collection angle of the spin-sensitive detectors. Therefore, *spin-* and angle-resolved photoemission is much more difficult to perform than ordinary angle-resolved photoemission. However, with well-designed electron spectrometers and spin detectors, count rates ranging from 50 to several 1000 counts per second are achieved in the $3d$ metal valence bands.

A spin-resolved, angle-dependent energy distribution curve is measured at a fixed photon energy for a given emission angle by recording simultaneously the signal from the two electron detectors at $\pm 120°$ scattering angle. The stronger spin-integrated signal from the electrons transmitted through the gold foil is measured by a third detector placed directly behind the scattering foil. A voltage ramp is applied to the sample or to the analyser electrodes to scan the photoelectron initial-state energy.

A high-energy (100 keV) Mott detector requires some minimum air gap between the high-voltage parts and the surrounding. Its minimum overall dimensions, therefore, are about $1 \times 1.2\varPhi \text{ m}^2$ as in a new apparatus which has been set up recently in our group at BESSY. Smaller spin detectors would be desirable in many circumstances (for instance, when the angle of photoelectron collection has to be varied). Effort has been devoted in miniaturizing the Mott detector by using only a moderate acceleration voltage ($(10 \div 30) \text{ kV}$) [15, 16]. In this energy range, surface barrier electron detectors are not suitable, since the signal cannot be separated well enough from the detector and the amplifier noise. Channeltrons, with a retarding field in front for energy discrimination can be used for electron detection. Values for $S_{\text{eff}} \approx 0.1$ and $S_{\text{eff}}^2 I \approx 3 \cdot 10^{-5}$ have been obtained [15].

Recently, also low-energy Mott scattering is used as a spin detector in spin-

and angle-resolved photoemission. Diffuse scattering on a polycrystalline Au film is used [17], or low-energy diffraction from a W single crystal [18]. The electron beam energy in this case is of the order of 100 eV.

When the photoelectron spin is analysed, one spin-resolved energy distribution curve (SREDC) is obtained for each spin component $(I^\uparrow(E_B), I^\downarrow(E_B))$.

It is common to define the spin polarization curve as

(9) $$P(E_B) = (I^\uparrow(E_B) - I^\downarrow(E_B))/(I^\uparrow(E_B) + I^\downarrow(E_B))$$

and the spin-averaged intensity curves $I_0(E_B) = I^\uparrow(E_B) + I^\downarrow(E_B)$. Obviously,

(10) $$I^{\uparrow,\downarrow}(E_B) = 0.5\, I_0(E_B)\,(1 \pm P(E_B)).$$

3'3. *Experimental systems.* – Various kinds of light sources have been employed: high-pressure arc lamps in the photothreshold region, rare-gas resonant lamps and synchrotron radiation. Synchrotron radiation is now more widely accessible and has the advantage of producing intense, bright, con-

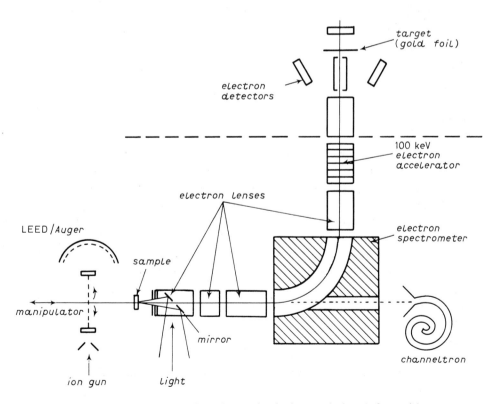

Fig. 3. – Apparatus for spin- and angle-resolved photoemission (schematic).

tinuously tunable and polarized light. In fig. 3, a sketch of the experimental set-up used in Berlin at the storage ring BESSY is shown[19]. Monochromatic synchrotron radiation light is reflected by a plane mirror on the sample at normal incidence for obtaining the so-called s-polarized light. The electrons emitted from the sample surface at a given angle are then analysed as a function of their kinetic energy by a multilens system with a 90° spherical condensor which acts as a dispersive element. At the exit slits of the analyser the electron beam is accelerated to 100 keV into the Mott detector for performing the spin analysis.

3`4. *Sample configurations.* – A specific problem in spin- and angle-resolved photoemission concerns the magnetization of the samples. Samples have to be magnetically saturated within the sampled surface area to avoid (trivial) spin mixing due to electron emission from several macroscopic domains with different magnetization direction. Indeed, the size of the light spot and of the analyser acceptance area are typically much larger than a single Weiss domain of a demagnetized sample. Therefore, magnetically saturated samples have to be used. For obtaining good angle resolution, magnetic stray fields have to be avoided. Samples of suitable shapes with a minimal stray field are thin single-crystal plates, picture-frame-shaped crystals or epitaxial films which can be remanently magnetized in plane, *i.e.* they remain magnetically saturated over a large enough surface area after the removal of the external magnetizing field.

4. – Selected results.

4`1. *Fe(001).* – According to eqs. (9), (10), the results from a spin-resolved photoemission experiment can be presented in two equivalent ways, as the spin-integrated intensity and the spin polarization curves, or as the two SREDCs. In fig. 4a)[19], a spin-integrated spectrum from the Fe(100) surface at 60.eV photon energy is shown, as recorded by the transmitted-electron detector. That is just what would be ordinarily measured in conventional photoemission. Figure 4b) shows the spin polarization as a function of the electron binding energy, for the data points of the EDC. The spin-resolved energy distribution curves are shown in fig. 4c). These curves, for samples fully magnetized and at low temperature, often allow a direct identification of the spin character of each feature which appears in the spin-integrated EDC. The results shown in fig. 4c) allow direct determination of the exchange splitting at the $\Gamma_{25'}$ critical point, as will be discussed further in the next section.

Figure 5 is a collection of spin-resolved EDCs for normal emission from a Fe(001) surface, measured at various photon energies with s-polarized light. In this geometry, the initial-state wave vector is varied along the high-symmetry

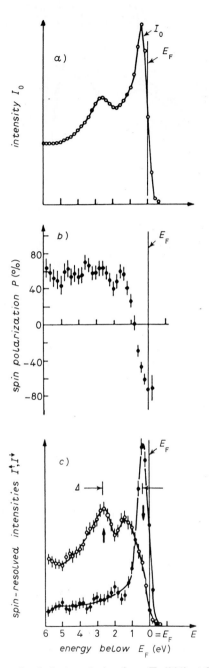

Fig. 4. – Spin- and angle-resolved photoemission from Fe(001) at 60 eV photon energy, for normal emission and normal incident light: a) spin-averaged, angle-resolved intensity distribution; b) the simultaneously measured spin polarization; c) spin- and angle-resolved energy distribution curves.

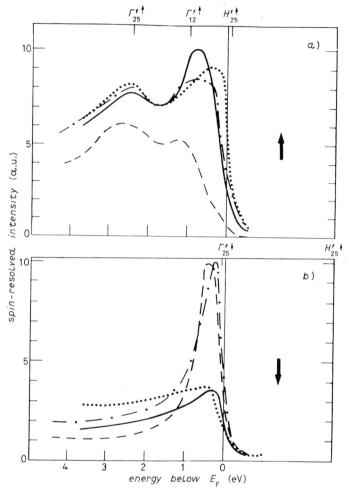

Fig. 5. – Spin- and angle-resolved energy distribution curves for normal emission and normal incident light from Fe(001) at photon energies between 20 and 60 eV: $\cdots$ $h\nu = 20\,\text{eV}$, —— $h\nu = 31\,\text{eV}$, $-\cdot-\cdot-$ $h\nu = 35\,\text{eV}$, $---$ $h\nu = 60\,\text{eV}$.

direction Γ-H in the Brillouin zone. From the final-state band dispersion, using a free-electron-like approximation or calculated bands[20, 21] (see fig. 6), one would expect emission from initial states near Γ, at $h\nu = 60$ eV, and from the vicinity of the H point, at about 20 eV photon energy. Dipole selection rules[22] indicate that only transitions from the Δ_5-symmetry bands are allowed with s-polarized light and for normal emission.

Figure 7 shows SREDCs from Fe(001) at 60 eV for normal emission and s-polarized light, similar to the data in fig. 2. The top panel shows the Fe band structure along Γ-H[23]. The binding energy of the peaks in the SREDCs compares well with the calculated critical-point energies at Γ. The features of the

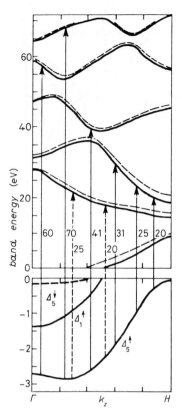

Fig. 6. – Initial and final states in normal photoemission from Fe(001) with s-polarized light[21].

spectra at 60 eV can be well interpreted within this frame. The majority spin peak at 2.6 eV binding energy and the minority spin peak at 0.3 eV below E_F are the pair of exchange-split states at $\Gamma_{25'}^{\uparrow}$ and $\Gamma_{25'}^{\downarrow}$. The other peak in the majority spin EDC, at 1.3 eV below E_F, corresponds to the emission from $\Gamma_{12}^{\uparrow}$, which, although forbidden under ideal conditions, still contributes to the spectra. The reason for its occurrence may be attributed to experimental considerations: incomplete light polarization and limited angular resolution. From the band structure one would expect that, when the initial-state k-vector is scanned from Γ to H, emission from $\Delta_5^{\downarrow}$ to move up, approaching the Fermi level, crossing it and finally to vanish. This is actually observed when the photon energy is lowered, from 45 eV to 31 eV, as shown in fig. 5.

The minority spin peak near E_F at 60 eV photon energy (cf. fig. 7) not only is very narrow in energy, but also in its angular distribution[24] which is shown in fig. 8. Figure 8 shows the spin-averaged EDCs as a function of emission angle, for s-polarized light. The half-width of the peak near E_F is less than 5°. Since the emission cone is that narrow, photoemission from Fe(001) at 60 eV photon

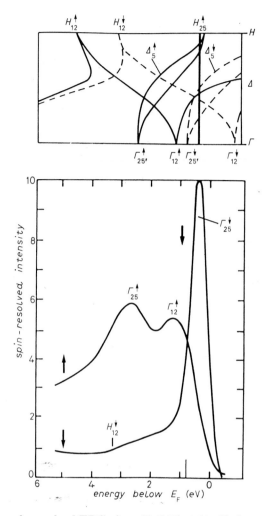

Fig. 7. – Spin- and angle-resolved EDCs from Fe(001) at 60 eV photon energy for normal emission and normal incident light (spline fits to the data points shown), compared with the band structure along Γ-H of Fe; $\tau = 0.3$.

energy is used routinely to align the spin-resolving photoelectron spectrometer at BESSY.

Between 31 and 21 eV photon energy, the minority spin spectra are dominated by indirect transitions from the point where the $\Delta_5^{\downarrow}$ band crosses the Fermi level, whereas the majority spin spectra are dominated by direct transitions from initial states at about 0.7 times the Γ-H separation. All the essential features of the experimental data have been reproduced by quantitative theoretical calculations using a nonrelativistic Green's function formalism on the basis of a one-step model (see fig. 9)[24].

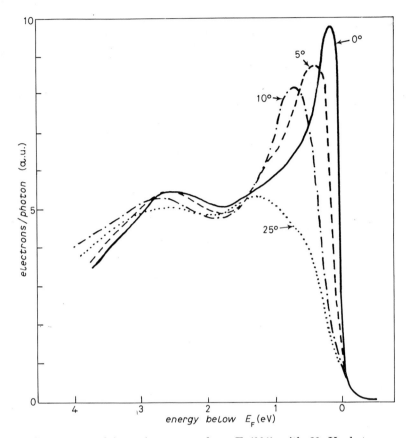

Fig. 8. – Spin-averaged intensity curves from Fe(001) with 60 eV photon energy for various emission angles. The angle has been varied by rotating the sample around the direction of the light electric vector to maintain s-polarized light.

Although the data on Fe(001) at 60 eV photon energy can be explained well in terms of the bulk band structure, there are small quantitative problems. As compared to the band structure calculations, the experimental binding energy of $\Gamma_{25'}^{\uparrow}$ is too large by about 0.3 eV. FEDER et al. [21] attributed this shift to self-energy effects.

On the other hand, the larger experimental separation of $\Gamma_{25'}^{\uparrow}$ and $\Gamma_{25'}^{\downarrow}$ (2.3 eV) than expected from the bulk band structure might be seen as a larger experimental exchange splitting than predicted by the electronic-structure calculations for the bulk (2.0 eV [20, 23]). In view of the surface sensitivity at 60 eV photon energy (which has been inferred directly from the measurement of the spin polarization vs. temperature, see subsect. 4·3), it could be expected that the data exhibit the predicted enhanced exchange splitting of surface resonances at $\bar{\Gamma}$, the centre of the 2D Brillouin zone [8]. Actually, the positions of the exchange-split states agree quantitatively with calculated electronic state

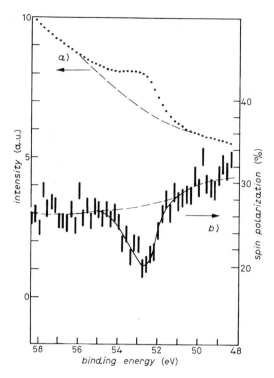

Fig. 9. – Spin-averaged intensity distribution (a)) and spin polarization (b)) in the vicinity of the Fe(001) 3p emission (normal emission), $h\nu = 92\,\text{eV}$. Dashed lines indicate the assumed backgrounds [25].

energies at Γ which are concentrated to 70% in the first two layers [8]. It had also been suggested by DURHAM et al. [26] that the large splitting might be caused by an enhanced surface exchange splitting.

4'2. *Spin-resolved photoemission from the* Fe 3p *core level.* – A theory from first principles of core level photoemission from magnetic 3d transition metals has to deal with the difficulty of describing the physical situation properly, taking into account the itinerant character of the 3d electrons.

An electron energy distribution curve taken at $h\nu = 92\,\text{eV}$ over an energy region which includes the Fe 3p emission is shown in fig. 9a) [25]. Because of the low cross-section for Fe 3p photoionization [27] the Fe 3p signal is superimposed on a large background of inelastically scattered electrons.

The photoelectron polarization in the kinetic-energy region including the Fe 3p subshell emission is shown in fig. 9b). The magnitude of the dip in the polarization implies that the net polarization of the 3p emission at the peak position is negative and that it amounts to about -20%. After background subtraction, the spin-resolved Fe 3p SREDCs are shown in fig. 10. The minority

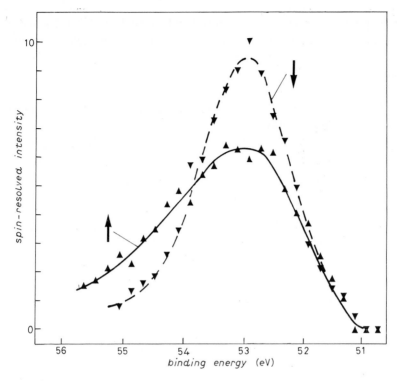

Fig. 10. – Spin-resolved Fe $3p$ core level EDCs after subtracting the backgrounds [25].

and majority spin EDCs have a full width at half maximum of (1.7 ± 0.3) eV and (2.7 ± 0.4) eV, respectively. The majority spin EDC is more asymmetric than the minority spin curve. The difference in binding energies between the peaks of the two Fe $3p$ SREDCs (Δ) is much smaller than their intrinsic width ($\Delta \simeq 0.5$ eV).

Calculations within the local spin density functional formalism predict an initial-state exchange splitting of 2.5 eV for the Fe $3p$ states [28], similar in magnitude to the average valence band exchange splitting. Including the relaxation shift in an impuritylike calculation, using the Slater transition state, the splitting between the two spin components is reduced to 1.5 eV [28]. This is still significantly larger than the measured value.

The physical situation is very complicated with the core level exchange splitting and photoemission lineshape depending critically on the ratio of U/W, where U is the intratomic Coulomb interaction and W the $3d$ band width [29]. However, by a simple argument it is expected that the width of the Fe $3p$ core level line should be spin dependent. Because of spin conservation a majority spin core hole will be refilled by a majority spin valence band electron and a minority spin core hole by a minority spin valence electron. The ratio of the numbers of majority spin and minority spin valence band electrons is 1.7 in Fe and a similar ratio is, therefore, expected for the spin-dependent core hole relaxation rates. This value agrees well with the observed ratio (1.6) of the spin-resolved linewidths.

4'3. *Finite-temperature effects*. – It is instructive to estimate how the spin-resolved energy distribution curves are expected to change at elevated temperatures, up to T_C. At $T = 0$, the magnetization is the same in magnititude and direction at any lattice site. Due to the influence of thermal disorder, the uniform spin alignment might break up, yielding fluctuating local magnetization directions (see fig. 11 [30]). If the electronic structure (the exchange splitting) in the

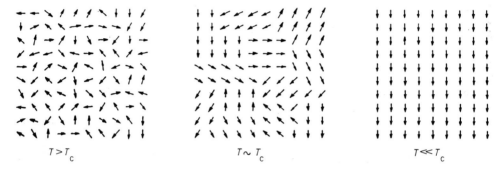

Fig. 11. – Model for the spin disorder occurring in Fe at elevated temperatures [30].

small regions would not change, the effect of an increase in temperature would be simply a uniform depolarization, which could be observed in the spin polarization curve as a function of binding energy. In the spin-resolved EDCs, this would result in a spin mixing. New («extraordinary») peaks would emerge in each SREDC at the binding energies where the other one has a peak at low temperature (see fig. 12) [31].

By comparing with the measured change of the SREDCs from Fe(001) as a function of temperature (cf. fig. 13), some correspondence is actually observed to the simple model calculation.

A proper theoretical interpretation of the experimental data on Fe(001) has been performed by HAINES, CLAUBERG and FEDER [32]. They performed a tight-binding calculation for a cluster of 2000 atoms, allowing for spin fluctuations with a parametrized spin correlation length. The spin-resolved photo-electron spectra are calculated for a set of correlation length parameters and for different values of the long-range magnetization, corresponding to the magnetization at elevated temperatures. The «extraordinary peaks» are observed in the calculated spectra for a spin correlation length exceeding 4 Å.

In agreement with the surface sensitivity of photoemission, the spin

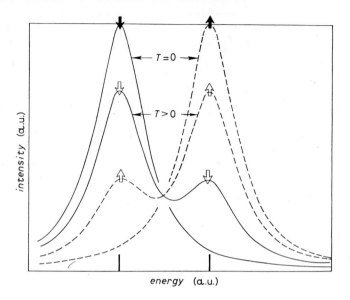

Fig. 12. – Model spin-resolved energy distribution curves at $T = 0$ and at an elevated temperature $T < T_C$, showing the effect of transverse fluctuations of magnetic moments.

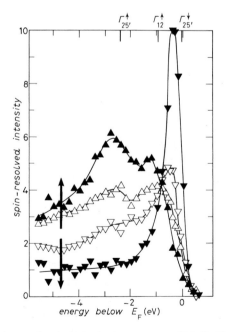

Fig. 13. – Spin- and angle-resolved photoelectron energy distribution curves from Fe(001) at 60 eV photon energy, at two temperatures: ▼, ▲ $\tau = T/T_C = 0.3$; ▽, △ $\tau = T/T_C = 0.85$.

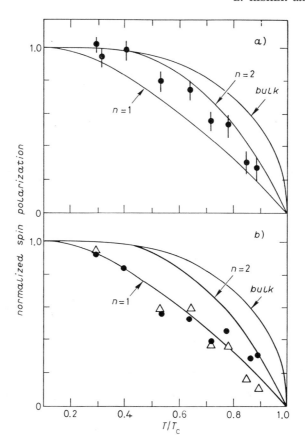

Fig. 14. – Dependence of spin polarization on temperature in photoemission from Fe(001): a) $h\nu = 60\,\text{eV}$, $E_B = 2.6\,\text{eV}$; b) $h\nu = 30\,\text{eV}$, $E_B = 0.8\,\text{eV}$. For comparison, the mean-field layer dependence of magnetization is also shown.

polarization, measured at 60 eV and at 30 eV photon energies, at constant binding energy, decreases more linearly than does the spontaneous bulk magnetization (see fig. 14)[19]. The data points follow the magnetization as calculated in the mean-field approximation for the first and the second layer.

4˙4. Ni. – The exchange splitting of Ni has been subject to controversies[4]. Its value, predicted from self-consistent calculations, is about 0.6 eV. From angle-resolved photoemission and spin-resolved threshold photoemission data, a much smaller value, $\approx 0.3\,\text{eV}$, has been inferred[33, 34]. Explicitly, the exchange splitting has been resolved by spin- and angle-resolved photoemission by RAUE et al. [35]. Figure 15 shows data obtained at temperatures between room temperature and the Curie temperature, T_C. The splitting is, indeed, very small ($\approx 0.2\,\text{eV}$ at room temperature). The reason for the much smaller experimental

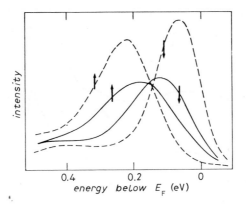

Fig. 15. – Spin- and angle-resolved photoelectron energy distribution curves from Ni(110) at 21.2 eV (He I) photon energy, at two temperatures: $T/T_C = 0.6$ (broken curves) and 0.8 (full curves), respectively [35].

exchange splitting than that predicted by modern theories is thought to be in screening effects in the photoemission process [36, 37].

At elevated temperatures, the peak separation in the SREDCs (fig. 15) decreases. This might suggest a Stoner-like decrease of the exchange splitting. However, a similarity to the the Fe(001) data and to the simple model based on transverse fluctuations of magnetic moments is obvious. The data have been interpreted in terms of the fluctuating band picture by KORENMANN and PRANGE [38]. From electron energy loss data, a more or less constant magnetic moment through T_C has been inferred [39].

4‘5. *Thin-film magnetism.* – Fundamental questions regarding 2D ferro-magnetism theory have been raised long time ago [40]. Theoretical results (see, *e.g.*, ref. [11]) have been obtained on ideal systems, *i.e.* assuming a perfectly flat surface and no interdiffusion. Real systems are generally unlikely to meet these assumptions. For instance, strong interdiffusion has been observed in Fe on GaAs(110) [41]. Also, in Fe on $Cu_3Au(001)$ [42], diffusion of the substrate constituents into the Fe film has been observed, with a strong temperature dependence.

A dispute is on the possibility of Cu diffusion into Fe films on Cu(001) [43]. It is beyond the scope of this lecture to discuss the experimental information on this system which is very important in relation to the electronic properties of γ-(f.c.c.) Fe.

4‘6. *Fe/Ag(001).* – As a model for a weakly interacting system the growth of epitaxial Fe films on Ag(100) has been studied. Recently, the spin-split electronic band structure of one and two epitaxial monolayers of Fe on the Ag(001) surface has been calculated [11]. Both groups predict that the reduced

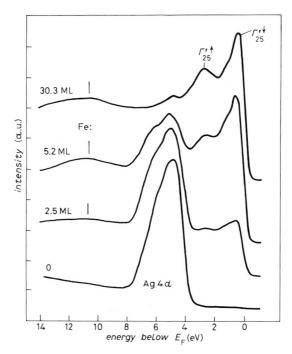

Fig. 16. – Energy distribution curves for different coverages of Fe on Ag(001) taken at 60 eV photon energy for normal emission and normal incident light. Coverages are given in equivalent Fe(001) monolayers.

atomic coordination and the weak interaction with the substrate cause a substantial enhancement of the Fe magnetic moment. However, these calculations are unable to specify the easy magnetization direction, since the underlying local spin density approximation contains no inherent directional information.

The Fe/Ag(001) films grow pseudomorphically, with a rotation of the surface nets by 45° [44]. The growth mode is layer by layer up to 3 ML [45]. The epitaxial growth proceeds via island formation at higher coverage. The epitaxial system of Fe on Ag is spectroscopically well suited for a detailed study of the electronic structure of very thin films, since Ag has only a very weak (sp) emission between the Fermi energy and 3.5 eV binding energy, the energy range over which emission from Fe $3d$ bands occurs. Figure 16 shows a series of energy distribution curves for various Fe coverages deposited on the Ag(100) surface [46]. In fig. 17 difference EDCs evidence the Fe $3d$ emission and its development at very low coverage. These data compare favourably with the overlayer calculated electronic band structure in the range from 0.95 to 2.5 ML [11]. The most striking result of this study is, however, that a net in-plane polarization is not observed up to 2.5 ML coverage (fig. 18). Although an exchange-split electronic structure may be inferred from the comparison with

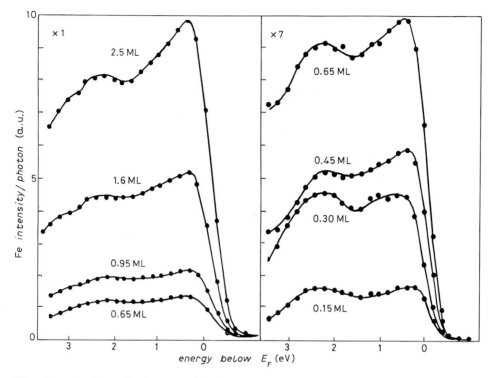

Fig. 17. – Angle-resolved energy distribution curves for low coverages of Fe on Ag(001). The emission from the Ag has been subtracted [46].

the calculations, the absence of spin polarization indicates that these ultrathin films cannot be magnetized along the in-plane (100) direction. The polarization sets in at the next coverage that has been studied, corresponding to 5.2 Fe ML.

In the experiment described above, it has not been possible to determine the onset curve of the spin polarization as a function of the film thickness. In the similar system Fe on $Cu_3Au(001)$, this has been done recently [42]. It was found that the spin polarization sets in steplike within less than 1 Å change of film thickness at 4 ML, at $T = 8\,°C$, and then increases gradually further by about 20% with increasing film thickness.

A reason for the sudden onset of polarization might be in a dependence of Curie temperature on film thickness [46]. But also, a change of the easy magnetization axis from in-plane to normal to the surface might occur at a certain film thickness [47].

The Fe/Ag(001) system has been studied also by spin-resolved threshold photoemission [48]. In this kind of spectroscopy, the spin polarization of the *total* photoyield is measured, with the photon energy slightly exceeding the work function of the sample. A static magnetic field between 0 and several kOe can be applied to the sample during the spin polarization measurements for providing

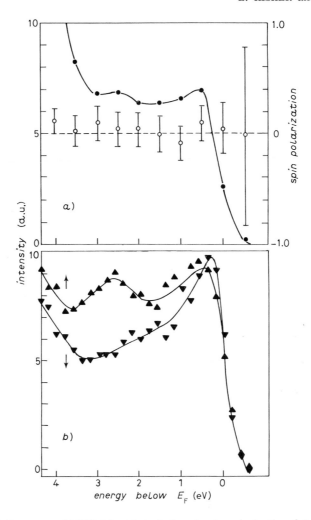

Fig. 18. – *a*) Spin-averaged EDC (closed circles) and spin polarization data (open circles) for a 2.5 ML Fe film on Ag(001); *b*) spin-resolved EDCs for a 5.3 ML Fe film on Ag(001).

magnetic saturation in cases where no remanence parallel to the surface normal exists. The dependence of spin polarization on the magnetic-field strength for ultrathin Fe/Ag(001) films is shown in fig. 19. The sample temperature was 30 K. It is seen that for 3.5 and 5 ML thick films, the samples are easily magnetized along the surface normal. 3.5 ML films even have a large magnetic remanence. In contrast, the 10 ML thick film has no remanence along the surface normal. In agreement with the experiment by JONKER *et al.* [46], it can be assumed that for thicker films the easy magnetization direction lies in the film plane.

From ferromagnetic resonance measurements it has been concluded that the magnetic moment in Fe/Ag(001) films of $(1 \div 2)$ ML thickness is oriented along

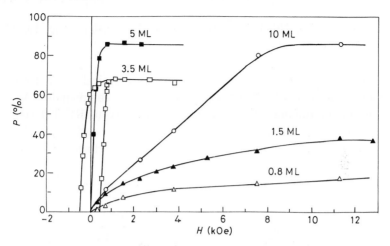

Fig. 19. – Magnetization curves taken at $T = 30\,\mathrm{K}$ for epitaxial films of b.c.c. Fe on Ag(001) of various thicknesses. The field H is applied perpendicular to the surface[48].

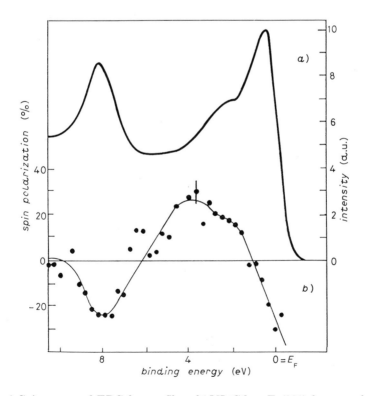

Fig. 20. – a) Spin-averaged EDC from a film of 1 ML Gd on Fe(001) for normal emission and normal incident 70 eV light at $T = 175\,\mathrm{K}$. b) Spin polarization distribution as measured simultaneously with a)[49].

the surface normal[50]. This is in contrast to the observation in fig. 19, where no remanence is observed.

4˙7. Fe/Gd. – Materials with a large magneto-optical Kerr rotation are ferrimagnetic rare-earth(RE)-transition-metal (TM) compounds. As a step towards the understanding of the electronic structure of RE-TM compounds, spin- and angle-resolved photoemission studies on evaporated Gd and Tb films on Fe(100) for a wide range of Gd coverages have been performed[49]. It has has

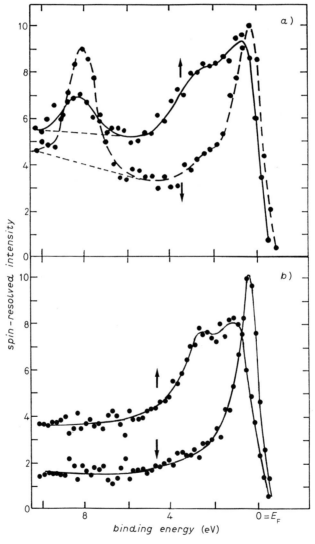

Fig. 21. – a) Spin-resolved EDCs corresponding to fig. 20. b) Spin- and angle-resolved EDCs from clean Fe(001) before Gd evaporation, for the same conditions otherwise.

been shown previously by spin-resolved Auger electron spectroscopy [51] that the Gd magnetic moment is aligned opposite to the Fe moment, the interface forming a ferrimagnetic system. The Gd $4f$ level and the Fe-derived valence bands originating from the interfacial region can be observed together, as shown in fig. 20a) for 1 ML Gd on Fe(001), at 70 eV photon energy and $T = 175$ K. The clean Fe(100) SREDCs taken prior to Gd evaporation are shown in fig. 20b).

The clean Fe(100) valence band SREDCs shown in fig. 21b) are understood according to ref. [19]. The inelastic background at low binding energy is small and smooth in both spin channels. Gd has the electronic configuration $4f^7(5d\,6s)$. The peak at about 8 eV binding energy (E_B) in fig. 20a) is due to the Gd $4f$ emission. The Gd $5d\,6s$ valence bands are overlapping the Fe valence bands, but they give rise to a negligible contribution ($\approx 1\%$) at 70 eV photon energy because of the very low cross-section of the Gd $5d6s$ emission as compared to that of the Fe $3d$ emission. Since the dominating Gd $4f$ signal is superimposed onto the Fe minority spin EDC, it is concluded that the Gd $4f$ moment is aligned opposite to that of the Fe. After subtracting a background of inelastically scattered electrons as indicated in fig. 21a), a spin polarization $P \approx 60\%$ is obtained for the Gd $4f$ peak.

By comparing fig. 21a), b), we see that also the Fe-derived valence band emission is affected by the Gd adsorbate. Especially, in the minority spin EDC, a shoulder grows around $E_B = 3$ eV. It can be shown (fig. 22) that the different appearance of the Fe-valence-band-derived SREDCs in fig. 21a) as compared to the clean Fe(100) SREDCs (cf. fig. 21b)) is essentially due to depolarization [49].

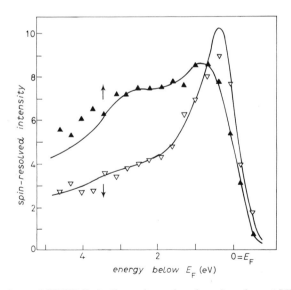

Fig. 22. – Comparison of SREDCs in the valence band regime from 1 ML Gd on Fe(001) (data from fig. 21a)) with the results of a model calculation based on the SREDCs from the clean Fe(001) surface, by considering a uniform depolarization from 1 to 0.6.

The Gd 4f SREDCs in fig. 21a) (with the backgrounds subtracted) are subject to the same depolarization, assuming initially fully polarized Gd 4f emission. This suggests that the reductions of the Fe and the Gd polarizations are strongly correlated.

In principle, two different mechanisms could produce the changes in the Fe-derived valence bands which have been discussed above: i) a reduced interface magnetization or ii) spin-flip scattering. Regardless of which of the above-mentioned depolarizing mechanisms is at work, the conclusion is that the «local» Fe valence band electronic structure in the interfacial region is only weakly affected by the Gd adsorption.

5. – Outlook.

The investigations of bulk and surface electronic properties of ferromagnets are just in the beginning. New experiments are being set up at many places, and systematic studies will help to clarify many of the problems which have just been touched. New experimental developments are heading towards better energy resolution in valence band spectroscopy to be able to study in more detail the states in the vicinity of the Fermi level, their dependence on temperature and adsorbate-induced changes. Other developments aim towards higher photon energies to accomplish spin-resolved core level photoemission spectroscopy. An exciting problem is to uncover the reason for the Fe 3s level splitting, which has been the origin of many speculations.

REFERENCES

[1] For a recent review, see E. KISKER: *3d-metallic magnetism and spin-resolved photoemission*, in *Metallic Magnetism, Topics in Current Physics*, edited by H. CAPELLMANN (Springer-Verlag, Berlin, 1987), p. 513

[2] For a recent review see V. DOSE and M. GLÖBL: *Spin dependent inverse photoemission from ferromagnets*, in *Polarized Electrons in Surface Physics* (World Scientific, Singapore, 1985).

[3] M. SEAH and W. A. DENCH: *Surf. Interf. Anal.*, **1**, 2 (1979).

[4] For a review see J. CALLAWAY: *Inst. Phys. Conf. Ser.*, **55**, 1 (1980).

[5] B. ACKERMANN, R. FEDER and E. TAMURA: *J. Phys. F*, **14**, L173 (1984); M. POMAREV: *J. Magn. Magn. Mater.*, **61**, 129 (1986).

[6] W. NOLTING: *Phys. Status. Solidi B*, **96**, 11 (1979).

[7] For a recent review see J. MATHON: *Rep. Prog. Phys.*, **51**, 1 (1988).

[8] S. OHNISHI, A. J. FREEMAN and M. WEINERT: *Phys. Rev. B*, **28**, 6741 (1983).

[9] E. WIMMER, A. J. FREEMAN and H. KRAKAUER: *Phys. Rev. B*, **30**, 3113 (1984); ZHU XUE-YUAN, J. HERMANSON, F. J. ARLINGHAUS, J. G. GAY, R. RICHTER and J. R. SMITH: *Phys. Rev. B*, **29**, 4426 (1984); O. JEPSEN, J. MADSEN and O. K. ANDERSON: *Phys. Rev. B*, **26**, 2790 (1982).

[10] R. FEDER, S. F. ALVARADO, E. TAMURA and E. KISKER: *Surf. Sci*, **127**, 83 (1983).

[11] R. RICHTER, J. G. GAY and J. R. SMITH: *Phys. Rev. Lett.*, **54**, 2704 (1985); C. L. FU, A. J. FREEMAN and T. OGUCHI: *Phys. Rev. Lett.*, **54**, 2700 (1985).

[12] U. GRADMANN, J. KORECKI and G. WALLER: *Appl. Phys. A*, **39**, 101 (1986).

[13] J. GAY and R. RICHTER: *Phys. Rev. Lett.*, **56**, 2728 (1986).

[14] J. KESSLER: *Polarized Electrons*, 2nd edition, *Springer Series on Atoms and Plasmas*, Vol. **1** (Springer-Verlag, Berlin, 1976).

[15] F. B. DUNNING, L. G. GRAY, J. M. RATCLIFF, F.-C. TANG, X. ZHANG and G. K. WALTERS: *Rev. Sci. Instrum.*, **58**, 1706 (1987).

[16] D. M. CAMPBELL, C. HERMANN, G. LAMPEL and R. OWEN: *J. Phys. E*, **18**, 664 (1984).

[17] J. UNGURIS, D. T. PIERCE and R. J. CELOTTA: *Rev. Sci. Instrum.*, **57**, 1314 (1986).

[18] H. P. OEPEN, K. HÜNLICH, J. KIRSCHNER, A. EYERS, F. SCHÄFERS, G. SCHÖNHENSE and U. HEINZMANN: *Phys. Rev. B*, **31**, 6846 (1985).

[19] E. KISKER, K. SCHRÖDER, W. GUDAT and M. CAMPAGNA: *Phys. Rev. B*, **31**, 329 (1985).

[20] K. B. HATHAWAY, H. J. F. JANSEN and A. J. FREEMAN: *Phys. Rev. B*, **31**, 7603 (1985).

[21] R. FEDER, A. RODRIGUEZ, E. BAIER and E. KISKER: *Solid State Commun.*, **52**, 57 (1984).

[22] J. HERMANSON: *Solid State Commun.*, **22**, 9 (1977).

[23] V. L. MORUZZI, J. F. JANAK and A. R. WILLIAMS: *Calculated Electronic Properties of Metals* (Pergamon, New York, N.Y., 1978).

[24] E. KISKER and K.-H. WALKER: unpublished.

[25] C. CARBONE and E. KISKER: *Solid State Commun.*, **65**, 1107 (1988).

[26] P. J. DURHAM, J. STAUNTON and B. L. GYORFFY: *J. Magn. Magn. Mater.*, **45**, 38 (1984).

[27] J. J. YEH and I. LINDAU: *At. Data Nucl. Data Tables*, **32**, 1 (1985).

[28] B. DRITTLER, M. WEINERT and P. H. DEDERICHS: private communication.

[29] Y. KAKEHASHI, K. BECKER and P. FULDE: *Phys. Rev. B*, **29**, 16 (1984).

[30] H. A. GERSCH, C. G. SHULL and M. K. WILKINSON: *Phys. Rev.*, **103**, 525 (1956).

[31] E. KISKER: *J. Magn. Magn. Mater.*, **45**, 23 (1984).

[32] R. CLAUBERG, E. HAINES and R. FEDER: *Z. Phys. B*, **62**, 31 (1985).

[33] I. D. MOORE and J. B. PENDRY: *J. Phys. C*, **11**, 4615 (1978).

[34] D. E. EASTMAN, F. J. HIMPSEL and J. A. KNAPP: *Phys. Rev. Lett.*, **44**, 95 (1980).

[35] H. HOPSTER, R. RAUE, E. KISKER, M. CAMPAGNA and G. GÜNTHERODT: *Phys. Rev. Lett.*, **50**, 70 (1983); R. RAUE, H. HOPSTER and R. CLAUBERG: *Phys. Rev. Lett.*, **50**, 1623 (1983).

[36] A. LIEBSCH: *Phys. Rev. Lett.*, **43**, 1431 (1979).

[37] R. H. VICTORA and L. M. FALICOV: *Phys. Rev. Lett.*, **55**, 1140 (1985).

[38] V. KORENMANN and R. E. PRANGE: *Phys. Rev. Lett.*, **53**, 186 (1984).

[39] J. KIRSCHNER and E. LANGENBACH: *Solid State Commun.*, **66**, 761 (1988).

[40] See, *e.g.*, M. PRZYBILSKI and U. GRADMANN: *Phys. Rev. Lett.*, **59**, 1152 (1987).

[41] C. CARBONE, B. T. JONKER, K.-H. WALKER, G. A. PRINZ and E. KISKER: *Solid State Commun.*, **61**, 297 (1987).

[42] D. TILLMANN, TH. DODT, R. ROCHOW and E. KISKER: *Europhys. Lett.*, **6**, 375 (1988).

[43] See, *e.g.*, D. A. STEIGERWALD and W. F. EGELHOFF jr.: *Phys. Rev. Lett.*, **60**, 2558 (1988).

[44] D. C. HOTHERSALL: *Philos. Mag.*, **15**, 1023 (1967).

[45] G. C. SMITH, G. A. PADMORE and C. NORRIS: *Surf. Sci*, **119**, L287 (1982).

[46] B. T. JONKER, K.-H. WALKER, E. KISKER, G. A. PRINZ and C. CARBONE: *Phys. Rev. Lett.*, **57**, 142 (1986).

[47] U. GRADMANN: *Appl. Phys.*, **3**, 161 (1974).

[48] M. STAMPANONI, A. VATERLAUS, M. AESCHLIMAN and F. MEIER: *Phys. Rev. Lett.*, **59**, 2483 (1987).

[49] C. CARBONE and E. KISKER: *Phys. Rev. B*, **36**, 1280 (1987).

[50] B. HEINRICH, K. B. URQUHARDT, S. T. ARROT, J. F. COCHRAN, K. MYRTLE and S. T. PURCELL: *Phys. Rev. Lett.*, **59**, 1756 (1987).

[51] R. ALLENSPACH, M. TABORELLI, M. LANDOLT and H. C. SIEGMANN: *Phys. Rev. Lett.*, **56**, 953 (1986).

Photoelectron Spectroscopy.

J. J. BARTON and F. J. HIMPSEL

IBM Research Division, T. J. Watson Research Center
Box 218, Yorktown Heights, NY 10598

How can one characterize a solid as completely as possible? We can study the electronic structure, how the atoms interact, and we can study the geometrical structure, where the atoms are. Here we will discuss how photoelectron spectroscopy contributes to both studies.

For electronic structure, we probe the electronic distribution, using photons to remove valence electrons. In this case we concentrate on initial-state properties, trying to avoid final-state effects. For geometrical structure, we probe deep into the cores of atoms in the solid, using photons to remove electrons from near the nuclear centers. Then we concentrate on final-state properties, accelerating the electrons to high velocities so they interact with the cores of neighboring atoms and are little perturbed by valence electronic structure. Using these two extremes, photoelectron spectroscopy can provide a rather complete understanding of solids and surfaces. We will start with the electronic structure problem.

I. – Electronic Structure of Solids and Surfaces.

1. – Introduction.

The ground state of a solid is fully characterized by the total energy (*e.g.*, the cohesive energy) and the total charge density of all electrons[1]. The charge density may be measured by elastic scattering of particles (X-rays, electrons) without exciting the solid. Such integral quantities, however, give only a crude description of a solid because in most real-life situations one has a probe particle interacting inelastically with the solid. The interaction creates an excited state. Therefore, we have to ask ourselves how to characterize the simplest and most fundamental excited states of a solid. These can be described by quasi-particles. Some of them are mainly associated with the atoms (*e.g.*, phonons), others with the electrons, *e.g.* an extra electron in the solid, a hole, an electron-hole pair (exciton), etc. These quasi-particles are characterized by quantum numbers, which reflect symmetries of the solid. The quasi-particle energy E is related to the invariance with respect to translation in time, the reduced momentum $\hbar k$ is related to the invariance with respect to translation by a lattice vector in a crystal. In addition, there are angular symmetries, given by the point groups, and spin.

spectroscopies
for electronic states

		outgoing	particle	
		photon	electron positron	ion atom
particle (incoming)	photon	optical spectroscopy luminescence Raman spectroscopy	photoemission Auger spectroscopy	photon stimulated desorption
	electron positron	inverse photoemission appearance potential spectroscopy positron annihilation	electron energy loss spectroscopy Auger spectroscopy appearance potential spectroscopy tunneling spectroscopy	electron stimulated desorption
	ion atom	chemoluminescence	ion neutralization spectroscopy Penning ionization spectroscopy	secondary-ion spectroscopy

Fig. 1. – Grid of various spectroscopies. Photoemission and inverse photoemission are distinguished by their capability of determining all quantum numbers of an electron in a solid.

For measuring the quantum numbers of a quasi-particle, one may excite a solid with a photon, electron, atom, or ion and look for outcoming particles. The corresponding grid (fig. 1) contains many well-known spectroscopies. However, only few of them are useful for determining quantum numbers of a given quasi-particle. Concentrating on electron (hole) quasi-particles, we can leave out atoms/ions as probe particles since they complicate the analysis by introducing extra atomic degrees of freedom. Among the remaining techniques there are two which make it possible to determine all quantum numbers of an electron. They are photoemission [2] and inverse photoemission [3]. The energy is determined by energy conservation from the kinetic energy of the electron E_{kin}. To measure the momentum $\hbar\boldsymbol{k}$ one needs three extra parameters, which will turn out to be two angles ϑ, φ characterizing the electron and the photon energy $h\nu$. As shown below, the photon energy is related to the momentum component perpendicular to the surface $\hbar k_{\perp}$. Therefore, one needs a tunable light source (such as synchrotron radiation) to map three-dimensional electron states. The spin quantum number in a ferromagnetic solid can be determined by spin-polarized photoemission, the point group symmetry can be inferred from the orientation of the vector potential $\boldsymbol{A}$ of the photon via dipole selection rules. The polarized nature of synchrotron radiation is a great help in this case. The experimental and theoretical information about electronic states is usually summarized by energy band dispersions $E(\boldsymbol{k})$, where energy is plotted *vs.* momentum for different energy bands, which are distinguished by their spin and point group labels.

2. – Instrumentation.

In order to detect energy and momentum of photoelectrons one may use standard electron spectrometers by restricting their angular acceptance with apertures. A very popular type of spectrometer is based on the hemispherical deflection analyzer [4]. With traditional designs, one needs a smaller angular aperture as well as a smaller source size in order to achieve better angular resolution. Liouville's theorem, however, suggests that one should be able to increase the source size when decreasing the angular acceptance since the product of cross-section, angular divergence and kinetic energy remains constant over an electron beam trajectory. Advanced designs (fig. 2) take advantage of these implications of Liouville's theorem by demagnifying the source with a zoom lens. This type of spectrometer has the capability of high energy and momentum resolution, but it is slow since each angle has to be measured separately.

Simultaneous detection of several energies greatly improves the speed of spectrometers. A well-known method in X-ray spectrometers is to insert a position-sensitive detector into the image plane of a deflection-type spectrometer (such as that shown in fig. 2). This detector is usually a microchannel plate

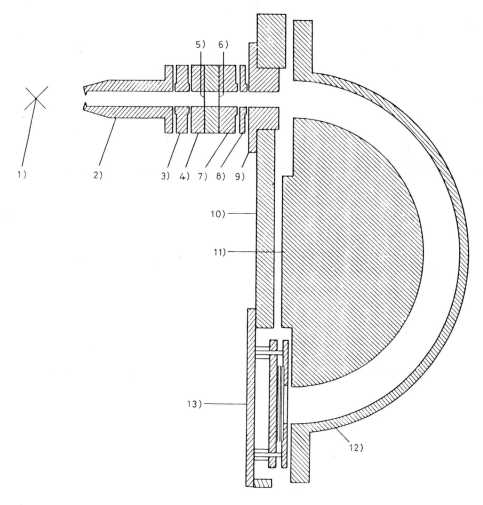

Fig. 2. – Angle-resolving hemispherical deflection analyzer (11, 12) with two lenses at the input (2-9) and a position-sensitive detector (13) (from KEVAN, see ref. [4]).

electron multiplier coupled with one of several types of read-out: an optical read-out using a TV camera, a resistive anode strip with charge-sensing devices at each end, or an array of anodes each coupled to separate counting channels. The first method is fast but somewhat plagued with the nonideal response of a vidicon tube (dark current, nonlinearity); the second method is simple but limited to about $10^5 \, \mathrm{s}^{-1}$ overall count rate; the third method has probably the best performance but the highest complexity.

A fundamentally different way of parallel energy detection is the time-of-flight method which is made possible by the time structure of synchrotron radiation. Light bursts of less than 1 ns duration are separated by up to 1 μs. The off-time can be used for energy analysis. Such a spectrometer[5] basically consists of a drift tube for the photoelectrons which is long enough to match the spread in drift time arising from the spread in kinetic energies to the interval between synchrotron radiation pulses, and to the speed of the detector electronics. It is useful to add a two-grid retarding state in front of the drift tube in order to higher improve the energy resolution for high-kinetic-energy electrons.

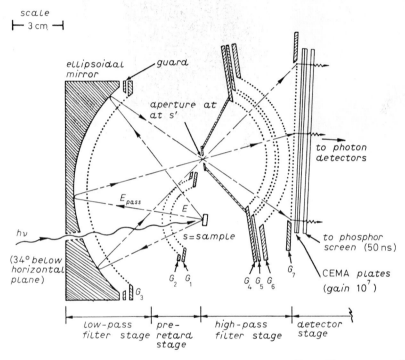

Fig. 3. – Scale drawing of an ellipsoidal mirror spectrometer (from EASTMAN *et al.*, see ref. [6]).

Instead of parallel detection of various energies, one can also detect several escape directions simultaneously. This can be achieved by a display-type spectrometer[6] as shown in fig. 3. The energy is selected by a retarding grid high pass combined with a repelling mirror low pass. Figure 4 shows pictures of the angular distribution of photoelectrons obtained with such a spectrometer. The ellipsoidal mirror spectrometer may also be operated in the time-of-flight mode[7], thus detecting the three parameters (E, θ, ϕ) simultaneously. There is at present no pressing need for more efficient systems since an oversupply of information can be as much a hindrance as a lack thereof. Nevertheless the number of different spectrometer designs is increasing rapidly as many innovative ideas are realized. For example, toroidal spectrometers have been

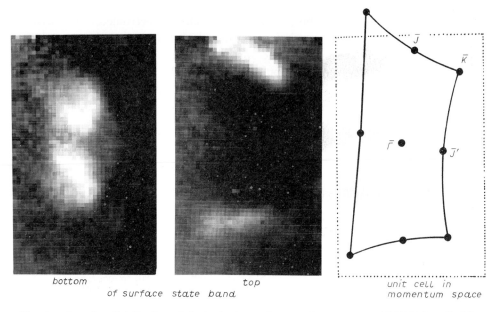

bottom   top   unit cell in
            of surface   state   band   momentum space

Fig. 4. – Angular distribution of photoelectrons from surface states on Si(111)-(2 × 1). The rectangular surface Brillouin zone (right) is distorted by the imaging properties of the spectrometer (after ref. [8]).

designed which are capable of parallel detection of energy and one escape angle [9].

3. – Theoretical basis.

A consensus has been established on the formulation of a first-principles theory of photoemission in the so-called single-step model [10, 11]. The differential cross-section for the primary photoelectron emission is given by a *golden rule* expression

$$\sigma = \frac{1}{N_i} \sum_{i,f} |\langle \psi_f | H_{int} | \psi_i \rangle|^2 , \tag{1}$$

where the matrix element of the interaction Hamiltonian H_f is averaged over all initial states i and summed over all final states f. The nonrelativistic matrix element takes the form

$$(2) \quad \begin{cases} \langle \psi_f | H_{int} | \psi_i \rangle \sim e \int \boldsymbol{j}_{fi}(r) \cdot \boldsymbol{A}(r) \, \mathrm{d}^3 r , \\[2mm] \boldsymbol{j}_{fi} \sim \dfrac{1}{m} \left[(i\boldsymbol{\nabla}\psi_f)^* \psi_i + \psi_f^* (i\boldsymbol{\nabla}\psi_i) \right]. \end{cases}$$

It may be transformed into the form

$$\langle \psi_f | H_{\text{int}} | \psi_i \rangle \sim i \, \frac{e}{m} \int \psi_f^* \cdot (\mathbf{A} \cdot \mathbf{\nabla} + \mathbf{\nabla} \cdot \mathbf{A}) \, \psi_i \, \mathrm{d}^3 r \tag{3}$$

by shifting the $\mathbf{\nabla}\psi_f$ derivative and dropping surface integrals. Other forms are obtained using the commutation relations with the Hamiltonian (see ref. [11]). In this overview, we will not be concerned with the actual calculation of photoelectron spectra from the wave functions (which has been perfomed successfully [12]) but with the reverse, *i.e.* the determination of energy bands (and possibly wave functions) from angle-resolved photoelectron spectra.

4. – Bulk energy bands.

The main idea in determining energy band dispersions $E(\mathbf{k})$ from photoelectron spectra is to use the conservation laws for the energy E and for the reduced momentum parallel to the surface $\hbar k_\parallel$:

$$E_i = E_f - h\nu = E - h\nu, \qquad k_{i\parallel} = k_{f\parallel} = k_\parallel, \tag{4}$$

$(E, k_\parallel)$ are the measured (energy, momentum) of the photoelectron. Indices i and f refer to the initial and final state. A useful relation is

$$|k_\parallel|/\text{Å}^{-1} = 0.51 \sin \vartheta \, \sqrt{E_{\text{kin}}/\text{eV}} \,, \tag{5}$$

where ϑ is the electron escape angle from the sample normal, and E_{kin} the kinetic

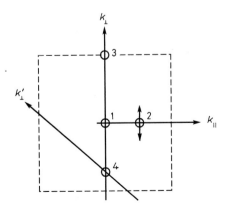

Fig. 5. – Various methods for determining the momentum $(\hbar k_\parallel, \hbar k_\perp)$ of bulk electronic states. The dashed line depicts the Brillouin zone boundary. $k_\parallel$ is given by momentum conservation (eqs. (4), (5)), $k_\perp$ is obtained by varying the photon energy (see text).

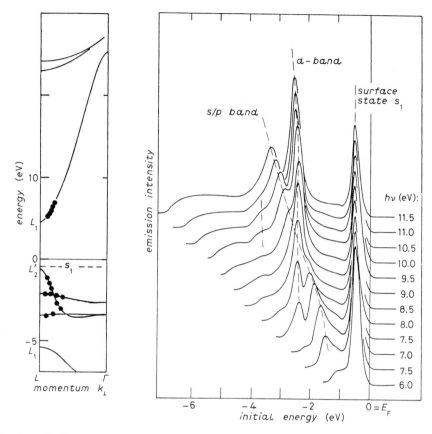

Fig. 6. – Surface and bulk states for Cu(111). The structure S_1 at $E_F - 0.4$ eV in the normal-emission photoelectron spectra is a surface state; the other peaks are bulk states. Two characteristics of a surface state can be seen: 1) S_1 is located in the $L_2' - L_1$ gap of bulk states, and 2) the binding energy of the surface state is independent of the photon energy $h\nu$ in contrast to the bulk states which move due to dispersion along the $k_\perp$ axis as shown on the left-hand side (from ref. [13]). An extended photon energy range is shown in fig. 9 (ref. [14]).

energy of the photoelectron. The momentum perpendicular to the surface $\hbar k_\perp$ is not conserved due to the lack of periodicity in this direction. However, there is a well-defined momentum transfer perpendicular to the surface, which can be determined experimentally. There are several methods which may be summarized by considering the location of interband transitions in the $(k_\parallel, k_\perp)$ plane as shown in fig. 5. By fixing $k_\parallel$ to zero one varies k along the $k_\perp$ axis when changing the photon energy (compare fig. 6). When a symmetry line is crossed (e.g., at $k_\perp = 0$) the sign of the band dispersion reverses. Thus one has nailed down one point in k-space (point 1 in fig. 5). The same procedure can be repeated for a different $k_\parallel$ (point 2 in fig. 5). Thus the band dispersion along the symmetry line is obtained. A second method is to keep moving with $k_\perp$ until one encounters a second symmetry line (point 3 in fig. 5). Between the symmetry points 1 and 3 one then interpolates the final-state band by using a nearly-free-electron approximation (see fig. 6). The initial-state band is obtained by subtracting the photon energy from the data points on the final-state band (see fig. 6). A third method is triangulation from two different crystal surfaces. One tries to find the same transition on both surfaces, identified by identical photon and electron energies. This is accomplished by varying $k_\perp$ and $k'_\perp$ for the two crystal surfaces, respectively. The corresponding lines intersect in a well-defined point (point 4 in fig. 5). For general-purpose band mapping, the approximate interpolation method is the most convenient. The absolute methods are more tedious since a whole series of spectra is required for each $E(k)$ point. They are useful for refining a band structure that is already known qualitatively.

A few examples [15] of experimentally obtained band dispersions are given in fig. 7. Band mapping with angle-resolved photoemission has reached a state of maturity such that a comprehensive set of data is being collected for a new issue of Landolt-Börnstein [16], a major collection of physical data.

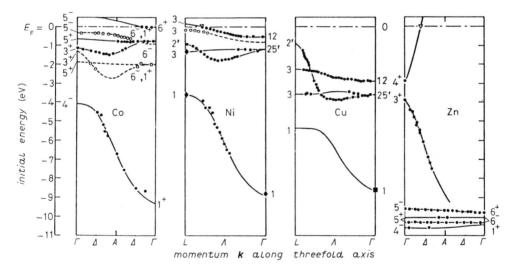

Fig. 7. – Overview of three-dimensional exerimental energy band dispersions along the first row of transition and noble metals (from ref. [15]). The flat $3d$ bands become filled with increasing atomic number and move down relative to the strongly dispersing s, p bands.

5. – Surfaces.

Surfaces, interfaces and adsorbates exhibit two-dimensional electronic states, which are distinguished from three-dimensional bulk states by the absence of periodicity in the direction perpendicular to the surface. Consequently, $k_\perp$ is not a good quantum number for these states. How can surface states be distinguished from bulk states experimentally? There are three criteria in common use. The first is the «crud test». By modifying a surface (*e.g.*, putting an adsorbate onto a clean surface) the intensity of surface states is changed, that of bulk states not. This test does not always work, since certain surface states are insensitive to adsorbates and since the transmission coefficient of bulk states through the surface can be affected by an adsorbate. The second criterion is almost trivial: Two-dimensional states have to be located in regions of $(E, k_\parallel)$ space where bulk states of the same symmetry are absent. If one has mapped out the bulk bands, one can project them along $k_\perp$ into a projected $(E, k_\parallel)$ bulk band structure (see dashed lines in fig. 8) and look for two-dimensional states in the gaps of bulk states. The third criterion is the absence of an $E(k_\perp)$ band dispersion for surface states. This is demonstrated in fig. 6. The $k_\perp$ component is scanned by changing the photon energy $h\nu$ with $k_\parallel$ fixed. Bulk states change their energy with $h\nu$ being varied, the surface state near E_F does not. Here again, the tunability of synchrotron radiation is an essential tool in identifying two-dimensional states.

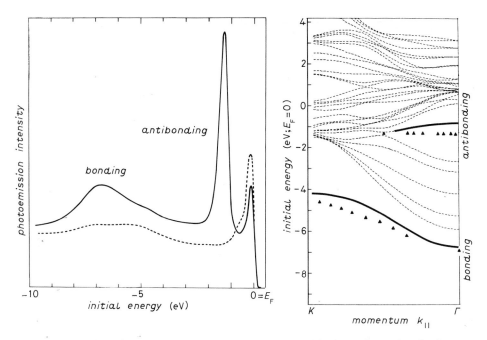

Fig. 8. – Interplay of two-dimensional adsorbate states with three-dimensional substrate states for H on Ti(0001) (after ref. [17]); normal emission, $h\nu = 22$ eV: —— Ti(0001)-H(1×1), – – – Ti(0001). The full (dashed) lines in the right panel are adsorbate (bulk) states, respectively, which were calculated from first principles, the triangles are H-induced states observed in angle-resolved photoelectron spectra (left panel).

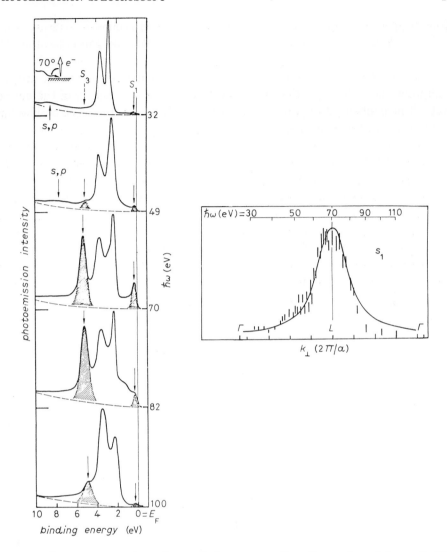

Fig. 9. – Photon energy dependence of the cross-section of a surface state on Cu(111) (from ref. [14]; compare fig. 6 for the low-photon-energy regime). Intensity maxima are seen when the $k_\perp$ of the bulk final-state band matches the $k_\perp$ Fourier components of the surface state; —— theory, | experiment.

By tuning $h\nu$ over a wide range, as in fig. 9, a new phenomenon shows up. The intensity of the surface states varies dramatically with $h\nu$. There are photon energies where it is easy to miss the surface state due to its low intensity. This effect is quite common, and can be explained (see ref. [14]) by considering the Fourier spectrum $k_\perp$ that is obtained from the z-dependence of the wave function. Since this surface state penetrates substantially into the bulk, it has a

Fourier spectrum of $k_\perp$ values which is fairly peaked around one value, which corresponds to the L point in $\boldsymbol{k}$ space where the bulk band gap is located. The width of this Fourier spectrum is roughly given by the inverse of the decay length. Therefore, we have a large overlap in the matrix element (eq. (2)), $i.e.$ high intensity if the $k_\perp$ of the final state matches the average $k_\perp$ of the surface state. This happens at $h\nu = 70$ eV and at $h\nu \lesssim 6$ eV where free-electron-like final states exist at the L point.

The electronic structure of adsorbates can also be characterized by two-dimensional states, although a conceptual problem has to be considered: In many cases an adsorbate orbital does overlap bulk states of the same energy and $\boldsymbol{k}_\parallel$. Therefore, it cannot be considered as a truly two-dimensional state, although it shows up as a well-defined peak in the spectra. Such «surface resonances» also occur at clean semiconductor surfaces with dangling-bond-like surface states. One may argue that a surface and a bulk state form two linear combinations, one of them with predominantly surface character, the other with bulk character—provided the mixing is weak.

A relatively simple adsorbate system is hydrogen interacting with a titanium surface (fig. 8 and ref. [17]). To describe the occupied states, one can start with two orbitals, the H1s and the Ti3d. The noninteracting Ti3d level can be seen in the spectrum for the clean surface near E_F. At $\boldsymbol{k}_\parallel = 0$ the H1s and Ti3d states form bonding and antibonding combinations, which fall into gaps of the projected bulk band structure and, therefore, form well-defined two-dimensional states. The lower state has predominantly H1s character, since it lies close to the atomic H1s orbital, the upper has mainly Ti3d character. For increasing $\boldsymbol{k}_\parallel$ the upper state vanishes since the gap in the projected bulk band structure closes. The lower state can be followed up to the Brillouin zone boundary. Its dispersion tells us about the lateral H-H interaction. At the zone center one has a bonding H-H interaction, whereas at the zone boundary one has an antibonding H-H interaction, since the sign of the wave function alternates between adjacent H atoms. The observed difference of about 2 eV between bonding and antibonding H-H states is typical for adsorbates. Thus one can obtain the interaction energies, both substrate-adsorbate and adsorbate-adsorbate, from the two-dimensional band dispersions. The first-principles calculations shown in fig. 8 demonstrate that such prototype adsorbate systems can be understood rather well.

6. – Future directions.

The methods of mapping energy band dispersions have become routine since the first attempts about a decade ago. Band structure measurements have proceeded to very complicated materials (such as A15 superconductors [18]) and to very simple materials (such as Na metal [19]), which can be very difficult to prepare. One may ask whether there is anything new to be done, apart from establishing complete collections of band structures. A principal limitation of present band mapping work is energy and momentum resolution. With current light sources one obtains typically a resolution of 0.1 eV and 0.1 Å^{-1}, respectively. Better resolution can be achieved by available monochromators and spectrometers, but the count rates become too low to be usable. With new undulator-based storage rings this restriction will be removed. A glimpse of things to come can be obtained by looking at electron energy loss spectroscopy (EELS) data (fig. 10, ref. [20, 21]) where an order of magnitude better energy and momentum resolution is already achieved routinely. Here, the differences in

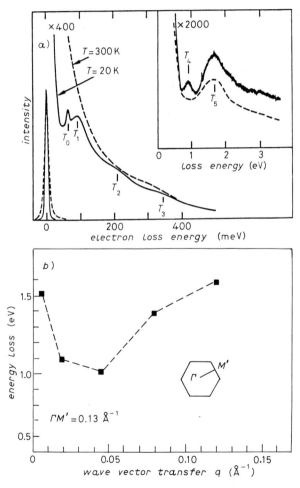

Fig. 10. – Prospects for spectroscopy with high energy and momentum resolution. *a*) An electron energy loss spectrum of the Si(111)7 × 7 surface exhibiting fine structure at the 50 meV scale (from ref. [20]); *b*) $\Delta E(\Delta k_{\parallel})$ dispersion of an energy loss feature on Si(111)7 × 7 with structure within the tiny 7 × 7 surface Brillouin zone (from ref. [21]).

energy and momentum between two energy bands are measured. It is clear from fig. 10 that there is fine structure that escapes present photoelectron spectroscopy. The surface Brillouin zone of surfaces with large unit cells such as Si(111)7 × 7 can be very small, and band dispersion is detected at the corresponding momentum scale (fig. 10b)). A case for high energy resolution can be made in all systems which exhibit electronically driven phase transitions, e.g. charge density waves, metal-insulator transitions, superconductors, heavy fermions, Kondo systems. The relevant energy scale is given by the transition temperature kT_c, which can be on the order of meV.

All high-resolution experiments are limited by the intrinsic width of electronic states. A predominant broadening effect is lifetime broadening due to Auger decay, although electron-phonon interaction also plays a role, particularly in superconductors[18] and ionic crystals. In addition, there is broadening due to scattering at defects[22]. The lifetime broadening induced by such interactions is described theoretically by the imaginary part of the self-energy of the quasi-particle[1]. Experimentally, the lifetime broadening has been determined[23] from the width of the peaks in angle-resolved photoelectron spectra (fig. 11, 12).

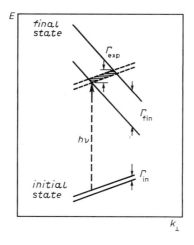

Fig. 11. – Influence of initial- and final-state lifetime broadening ($\Gamma_{in}, \Gamma_{fin}$) on the experimental width of interband transitions (Γ_{exp}).

The contributions from the electron and the hole lifetime are generally convoluted as shown in fig. 11. However, the electron lifetime broadening can be measured separately[24] by starting from states just below the Fermi level (or the valence band maximum in semiconductors), which have negligible hole lifetime. The lifetime broadening goes to zero as one approaches the Fermi level (or the band edges in semiconductors). Therefore, one will always find sharper states by going closer to the Fermi level.

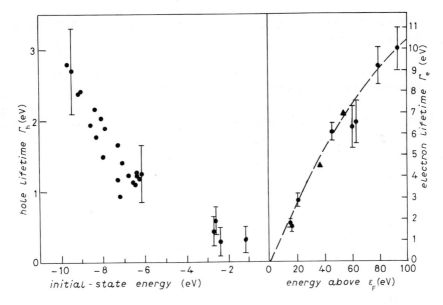

Fig. 12. – Lifetime broadening of electron and hole states in ● Al and ▲ Zn (from ref. [23]). Note the change in scales between electrons and holes.

II. – Photoelectron Diffraction and Holography.

1. – Introduction.

Photoelectron diffraction [25] and photoelectron holography are structure determination methods based on photoelectron interference. Core level photoelectron wavelets are temporally and spatially coherent. When they scatter from the ion core potential of atoms surrounding the photoemitter, these wavelets have multiple paths to the same final destination at the detector. Thus we expect interference to influence photoemission intensity. This part discusses photoelectron interference and methods of using the interference oscillations to determine surface geometrical structure. The generic term for these methods is «photoelectron diffraction». We will discuss several approaches to measuring and analyzing the diffraction patterns, including the new two-angle photoelectron holography methods which yield fully three-dimensional images of surface structure.

We start with the basic physics of photoelectron interference to isolate the ingredients needed to detect and use the effect. With this background, we describe the various established photoelectron diffraction techniques and the newest of these methods, photoelectron holography. Finally, we compare the photoelectron methods to other electron-based surface structure methods, SEXAFS and LEED.

2. – Photoelectron interference physics.

Figure 13 illustrates the fundamental photoelectron interference effect. Core level photoemission creates a photoelectron wave centered on the emitter. The expanding spherical wave propagates to the electron detector and to atoms in

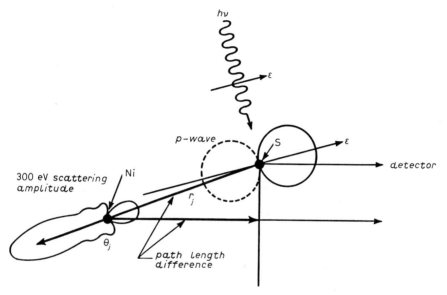

Fig. 13. – Basic principles of photoelectron diffraction illustrated for S 1s photoemission from $c(2 \times 2)$S/Ni(100). The photoelectron wave emanating from the sulfur finds multiple paths to the detector. The path along the line straight from the emitter to the detector is the primary or reference wave. Other paths involve elastic scattering of the primary wave from nearby ion cores to create object or secondary waves. The bond length r_j and the scattering angle θ_j for the j-th atoms are also indicated in the figure.

the vicinity of the emitter. The wave scatters from the surrounding atoms, leading to secondary waves which can propagate to the detector and interfere with the primary wave. Thus the measured photoemission intensity will change when we change the relative phase of the primary and secondary wavelets. This is the photoelectron diffraction effect.

The secondary, scattered wave interferes constructively with the primary wave whenever the extra path length d of the secondary wave and the electron wave number k satisfy

$$2\pi n = kd, \qquad\qquad n = 0, \pm 1, \pm 2, \dots .$$

The extra path length d has three components, the distance r_j from the photoemitter to the j-th scattering atom, an angle-dependent part which can be written in terms of the j-th scattering angle θ_j, and a small scattering potential part ϕ_j which we will discuss later. Thus

$$d_j = r_j - r_j \cos \theta_j + \phi_j.$$

While this is the conventional expression for the path length difference, for photoelectron holography we use the three-dimensional electron wave vector $\boldsymbol{k}$ and the three-dimensional relative position vector $\boldsymbol{r}_j$ to write the constructive-interference condition as

$$2\pi n = |\boldsymbol{k}||\boldsymbol{r}_j| - \boldsymbol{k} \cdot \boldsymbol{r}_j + \phi_j, \qquad n = 0, \pm 1, \pm 2, \ldots .$$

All the photoelectron diffraction techniques determine structure by modulating the electron wave vector $\boldsymbol{k}$, observing the diffraction oscillations in the photoemission intensity, and working back through the interference condition to deduce the relative position vector $\boldsymbol{r}_j$. Various techniques use various methods to modulate the electron wave vector and to deduce the relative position vectors from the oscillations. We will return to examine the differences between techniques after we discuss their common features.

2˙1. *Photoemission.* – The experimental measurement in all photoelectron diffraction techniques is core level angle-resolved photoemission intensity. Absorption of a photon with energy $h\nu$ in a level with binding energy E_b below the vacuum level creates a photoelectron with momentum

$$hk = \sqrt{2m_e(h\nu - E_b)}.$$

In contrast to the previous part, now we are interested in deep core levels which are highly localized in space and whose shape is largely atomic, little perturbed by the formation of the solid. Furthermore, we use high electron kinetic energies to obtain short electron wavelengths. For a fixed photon energy, the photoemission peak at kinetic energy $h\nu - E_b$ is characteristic of an oxidation state of a single element in the sample. Thus the underlying photoelectron spectroscopy guarantees that the photoelectron diffraction pattern selectively contains information on a single species. This selectivity dramatically simplifies the connection between the observed diffraction pattern and the microscopic geometry.

2˙2. *Space coherence.* – Soft X-rays excite core level electrons with probabilities given by the dipole selection rule. The excitation is atomic in character when the initial orbital is a tightly bound core level orbital. A spherical wave centered on the emitting atom results with an angular-momentum distribution dominately one quantum number higher than the initial orbital. For example, an initial $1s$ core level is photoexcited to a p_z spherical wave with its axis parallel to the photon's electric-dipole vector. Thus each photoelectron wave is centered on a single atom in the sample: the primary wave is effectively emitted from an atomic-scale point source.

Note that each photoelectron is emitted from a different atom. Each photoelectron adds incoherently to build the final diffraction pattern. If the selected photoemitting species exists on the sample in more than one environment, each environment will produce a different pattern and only the sum of the patterns can be observed.

2˙3. *Time coherence.* – The photoabsorption event creates a photoelectron wave pulse. The width of the pulse is related to the lifetime of the photoion final state through the time-energy uncertainty relation

$$\Delta t \, \Delta E \sim h \, .$$

For linewidths typically 0.1 eV to 1.0 eV the electron pulse width will be between 4 and 40 fs, during which time the electron can travel many times the mean free path in any material. Therefore, despite the time delay of propagation and scattering, the secondary, scattered wavelets will always be coherent with the primary wave.

2˙4. *Scattering.* – While the exact calculation of photoemission intensity in a solid is a complex dynamic quantum-mechanical problem, adequate results can be obtained with greatly simplified models[26]. Photoabsorption occurs in fractions of a femtosecond: we can assume that the photoelectron-photoion pair is created instantaneously. The photoelectron pulse width is long compared to the transit times to all nearby atoms but short compared to the motion of atoms in the solid. Hence we use a time-independent model with frozen nuclear positions and average over an ensemble of nuclear positions.

Since the primary wave dominates the sum of all scattered waves, perturbation theory applies. We take atomic photoabsorption as the zeroth-order solution. The first-order corrections are called «single scattering», second order becomes double scattering and so on. Within each perturbation order we break up the solid into nonoverlapping spheres and compute the elastic scattering within each sphere (atom) separately. While this description implies that we sum first over all atoms, and then over scattering order, we can just as well re-arrange the sums to follow a scattering «event» through one or more scattering atoms and sum over scattering events. In doing so we often discuss a scattering event as if the electron traveled from the photoemitter to the first scattering atom, then on to the second and so forth, but this only a convenient pictorial description of terms in a perturbation theory. Note each scattering event has a path length difference, so that this approach is the most natural one for connecting diffraction oscillations to structure.

We treat the electron scattering within each atom as electron-potential scattering and furthermore we use spherically symmetric potentials. These approximations are well justified by the high kinetic energy of the photoelectron: there is little deviation from a simple density-dependent exchange and correlation interaction and the scattering occurs primarily in the deeper core regions of the atoms. Furthermore we approximate the inelastic scattering throughout the solid by an isotropic mean-free-path damping.

By this series of approximations we reduce our original problem to a sum of scattering events each of which is not much more complex than simple plane-wave scattering from a spherical potential. The only significant complication is the spherical character of the incident waves upon each atom. Fortunately good approximations exist which allow the spherical character to be built into the scattering on each atom [27].

The final orbital is a sum of the primary wave ψ_0 and the secondary wave ψ_s which in turn is a sum over scattering events. Each event j has an amplitude, F_j, a scattering phase, ϕ_j, and a path length difference (propagation delay), $kr_j - \mathbf{k} \cdot \mathbf{r}_j$:

$$\psi_s = \sum_j \psi_j = \sum_j F_j \exp[ikr_j - \mathbf{k} \cdot \mathbf{r}_j + \phi_j].$$

2`5. *Interference*. – We detect wave intensity, I, the square of the total wave amplitude:

$$I = (\psi_0 + \psi_s)^* \times (\psi_0 + \psi_s) = \psi_0^* \psi_0 + 2\,\mathrm{Re}\,(\psi_0^* \psi_s) + \psi_s^* \psi_s.$$

The first term, also called I_0, is just the cross-section for isolated atom photoemission; the last term is a double sum of small terms which we ignore. The interference term in the center leads to photoelectron diffraction. We isolate this term by writing

$$\chi(\mathbf{k}) = \frac{I - I_0}{I_0} = 2 \sum_j \frac{|F_j|}{I_0} \cos(kr_j - \mathbf{k} \cdot \mathbf{r}_j + \phi_j).$$

This cosine interference structure is the photoelectron diffraction. In the next section we will describe the various ways of measuring the diffraction oscillations and using the resulting data to infer geometries.

3. – Measurement techniques.

The photoelectron diffraction pattern has three degrees of freedom. These can be chosen to be two emission angles and photoelectron energy, or directly the k vector values. The experimental techniques vary one or two of the degrees of freedom while holding the remainder constant.

3`1. *Energy-dependent methods.* – The energy-dependent photoelectron diffraction methods fix the electron emission angles, vary the electron kinetic energy, and measure the photopeak intensity. Much of the early work was limited to lower kinetic energies and to normal emission, and came to be called normal photoelectron diffraction[28]. As the diffraction phenomenon becomes better understood, off-normal angles and much higher and wider energy ranges were used. This form goes by the term ARPEFS: Angle-Resolved Photoemission Extended Fine Structure[29], to acknowledge some similarity to EXAFS.

The photoelectron kinetic energy is swept in these methods by sweeping the photon energy with a monochromator at a synchrotron light source. The underlying quantity being varied is the electron's wave number k. The conjugate quantity available from the analysis of the diffraction is then the path length difference $d_j = r_j - r_j \theta_j$. In fact, in the ARPEFS form of the experiment, it has been shown[29] that the Fourier transform of suitably normalized experimental data directly yields a curve of scattering intensity against geometrical path length difference. Verification of adsorption sites deduced from these path length difference analyses and precise geometrical parameters are obtained by trial-and-error fits of scattering simulations using model geometries[30].

The direct method of analysis is a key advantage of the energy-dependent approach. Another important advantage is accuracy and precision with which photon energy—and hence electron energy—can be varied. Furthermore, this technique requires exactly the same apparatus as might already be available for valence band angle-resolved photoemission measurement described in the previous section.

The key disadvantage of the energy-dependent diffraction techniques is the indirect connection between the derived path length difference and the actual geometrical quantities. Even with exact values of the path length differences for several emission angle directions, deduction of an unknown structure might be difficult. And the experimental curves rarely give clearly resolved peaks for individual path length differences. Furthermore, to obtain good-quality experimental data for Fourier transformation requires special monochromators and considerable time to scan over wide photon energy ranges. The method has limited applicability for buried photoemitters whose short path length differences lead to low-frequency oscillations which cannot be separated from the background intensity.

An example of normal photoelectron diffraction data and analysis from the

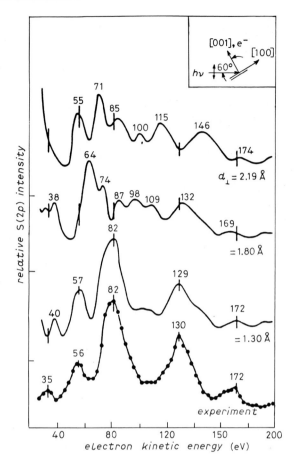

Fig. 14. – An example of normal photoelectron diffraction data and analysis from the work of Rosenblatt *et al.* (ref. [28]): $c(2 \times 2)$Si-Ni(001), $V_0 = 14$ eV, $E_B^v = 170$ eV.

work of Rosenblatt *et al.*, ref. [28], is shown in fig. 14. The bottom curve is the experimental S$2p$ core level intensity into a narrow range of emission angles near normal emission for electron kinetic energies from 30 to 200 eV. The remaining curves are simulations of the data using scattering theory and various trial geometries, parameterized by the S-Ni interlayer spacing. Visual or numerical comparisons of the theory with experiment yield geometries with typical precisions around 0.1 Å.

An example of ARPEFS data from ref. [29] is shown in fig. 15. S$1s$ emission intensity was measured along the [011] direction for both $c(2 \times 2)$S/Ni(001) and $p(2 \times 2)$S/Cu(001); the SEXAFS data of Brennan, Stohr and Jaeger[31] for $c(2 \times 2)$S/Ni(001) were shown for comparison. Note the much wider energy range used in ARPEFS compared to the NPD data of fig. 14; also note that the peaks in fig. 14 lie near valleys in this figure due to the dominance of d-wave

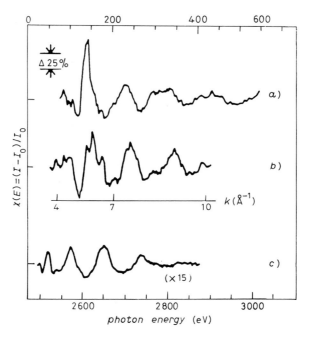

Fig. 15. – An example of ARPEFS data from ref. [29]: *a*) S/Cu ARPEFS, *b*) S/Ni ARPEFS, *c*) S/Ni SEXAFS.

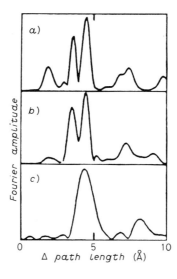

Fig. 16. – Fourier analysis of the ARPEFS data of figure 15, also taken from ref. [29]. The SEXAFS data transform is from ref. [31]: *a*) S/Cu ARPEFS, *b*) S/Ni ARPEFS, *c*) S/Ni SEXAFS.

photoemission in the $2p$ core level data compared to p-wave emission for the $1s$ case. These data are discussed in detail in ref. [30].

Fourier analysis of the ARPEFS data of fig. 15 is shown in fig. 16, also taken from ref. [29]. The SEXAFS data transform from ref. [31] was included in this figure. Each of the peaks below 5 Å in the ARPEFS transforms can be assigned to nearest-neighbor Ni atoms. While it is evident from this analysis that S/Cu is structurally similar to S/Ni, it is not clear whether the structure of S/Cu could have been deduced from the Fourier transform data alone. However, with several transforms, auxiliary information and scattering simulations completely unknown structures can be solved [32].

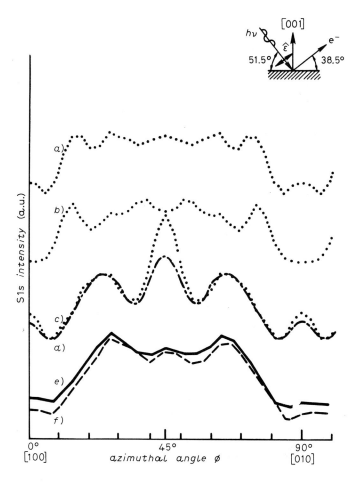

Fig. 17. – An example of azimuthal photoelectron diffraction data from ref. [33]: $c(2 \times 2)$S on Ni(001), $\theta_{e^-} = 38.5°$, $\theta_{h\nu} = 51.5°$, $E_{kin} = 282$ eV: a) theory: bridge, $z = 1.80$ Å, $\Delta I/I_{max} = 22.8\%$; b) atop, $z = 2.19$ Å, $\Delta I/I_{max} = 23.3\%$; c) fourfold, $z = 1.35$ Å, $\Delta I/I_{max} = 36.1\%$; d) correlated vibration, experiment; e) averaged; f) first quadrant, $\Delta I/I_{max} = 31.7\%$.

3˙2. *Angle-dependent methods*. – The angle-dependent photoelectron diffraction techniques fix the electron energy (or equivalently electron wave number) and one of the emission angles while varying the remaining angle. This variable angle can be a grazing-emission azimuth (Azimuthal Photoelectron Diffraction, APD[33]) or a polar angle. Most if not all of the angle-dependent diffraction measurements have been made by rotating the sample. Accurate angular motion in ultra-high vacuum can be obtained but requires special sample manipulators.

Qualitative analysis of the angle-dependent oscillations has not been concerned with the phase of the oscillations, but with the amplitude of the diffraction structure. The largest dependence in the scattering factors F_j is on the scattering angle θ_j. The largest amplitude occurs in the forward ($\theta_j = 0$) direction, and in this direction $\mathbf{k} \cdot \mathbf{r}_j$ is always zero, so the cosine oscillation is also

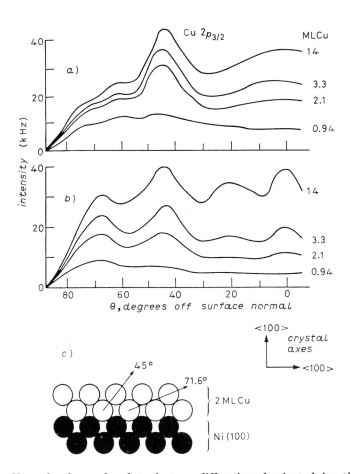

Fig. 18. – Use of polar-angle photoelectron diffraction dominated by the forward scattering peaks to study epitaxial growth of Cu on Ni(001) taken from ref.[34]: *a*) epitaxial Cu on Ni(100), $E_k = 317$ eV; *b*) CuCVV Auger, $E_k = 917$ eV; *c*) angles for enhanced forward scattering.

maximum. Thus the angular distributions will have peaks corresponding to atoms intervening between the photoemitter and the detector $(\theta_j = 0)$. In favorable cases, simple analysis of the experimental data can yield structure information by this means. As in the energy-dependent approach, accurate work requires scattering calculations compared to experiment.

The primary advantage of the angle-dependent methods is the use of a fixed photon energy, avoiding the need for synchrotron radiation and wide-energy-range monochromators. The potential for qualitative analysis makes angle-dependent methods particularly attractive for buried photoemitters, although the method's sensitivity for forward scatterers can be misleading.

The major disadvantage of the angle-dependent methods is the present inability to directly analyze the data for structure information. The azimuthal methods also suffer from a lack of sensitivity to vertical displacement.

An example of azimuthal photoelectron diffraction data from ref. [33] is shown in fig. 17. The experimental data are commonly measured in a full rotation and then averaged over symmetry equivalent sections. Several scattering simulation results are also shown.

Use of polar-angle photoelectron diffraction dominated by the forward-scattering peaks to study epitaxial growth of Cu on Ni(001) is illustrated in fig. 18; this is taken from ref. [34]. Most of the peaks in the data can be assigned to atoms aligned by the formation of the Cu overlayer.

3`3. *Photoelectron holography.* – The latest method for exploiting the photoelectron diffraction phenomenon to determine structure is photoelectron holography. Two angles are varied in this method, while the electron wave number (kinetic energy) is fixed. But what distinguishes the holography method from a simple elaboration of the angle-dependent methods is the analysis of the results. By exploiting the similarity between holography and photoelectron diffraction, the hologram can be Fourier analyzed to give an image of the crystal structure directly.

Holography captures three-dimensional image information by recording the interference between a reference wave and object waves. To be successful the reference and object waves must be spatially and temporally coherent. Most familiar are laser holograms in which a beam splitter divides highly coherent laser light into reference and illuminating beams. Light scattered from the illuminating beam by the objects to be imaged interferes with the reference beam on a film detector. When the film is developed and re-illuminated with the reference beam, full three-dimensional images of the original objects appear. In photoelectron holography [35], the reference beam is the primary, atomiclike photoemission. The objects are nearby atoms, illuminated by the photoemission wave. The scattered waves interfere with the primary wave in just the same fashion as in holography.

Once this connection is recognized, analysis of the hologram to give a three-

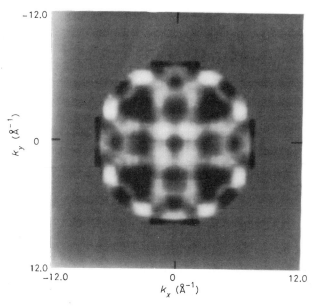

Fig. 19. – Simulated photoelectron hologram for S $1s$ photoemission from $c(2 \times 2)$S/Ni(001). The modulation in S $1s$ intensity is shown as a function of $\hat{k}_x$ and $\hat{k}_y$, where the k values are zero in the center of the frame, and reach a value of 12.0 Å^{-1} at the edge. The photon electric vector is pointing to the right, 10° from the center. The simulation includes curved-wave corrections and multiple scattering. From ref. [36].

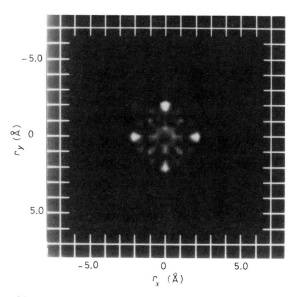

Fig. 20. – Holographic reconstruction of fig. 19 (also from ref. [36]). Each figure is a map of image intensity in a plane 1.29 Å below the photoemitter and parallel to the hologram (and hence parallel to the surface plane) and spans 16.7 Å. The lines on the edges of the photographs are spaced by 1.0 Å.

dimensional image can follow the methods developed for optical holograms. As the orginator of holography, D. GABOR, demonstrated, obtaining the required image magnification with an optical reconstruction of electron holograms is quite difficult [37]. Today, synthetic reconstruction and display by computer allows arbitrary magnification though the final image must be rendered onto two dimensions [36].

In addition to the considerable advantage of a direct image of the surface structure, measurement of the hologram does not require scanning the photon energy.

The main disadvantage of the photoelectron holography method is the large number of data points which must be measured. Extension of the conventional single-angle measurement to two angles is not likely to be feasible. Fortunately, fully two-dimensional analyzers do exist (ref. [6]) which allow all hologram points to be measured at once.

While experimental measurements have not been tried as yet, a hologram simulating the available experimental conditions has been computed using a standard photoelectron diffraction theory. The simulated hologram for S 1s

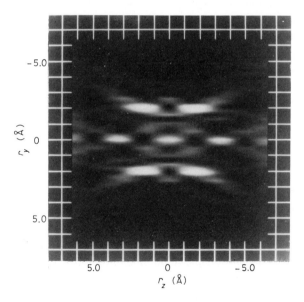

Fig. 21. – A view in a plane perpendicular to the surface of the reconstructed image from the hologram in fig. 19. The plane is a $\langle 100 \rangle$ plane containing the S photoemitter. The surface runs from top to bottom in the center of the picture. The image has inversion symmetry with all intensity to the left of center representing the conjugate image and that to the right of center being the real crystal image. The intensity maximum in the center of the image coresponds to the S atom visible through multiple scattering. Moving to the right, the two largest intensity features are the two Ni nearest neighbors in the $\langle 100 \rangle$ plane. Further to the right and vertically at the same height as the S atom appears the second nearest neighbor Ni directly below S and in the second Ni layer.

photoemission from $c(2 \times 2)$S/Ni(100) is shown in fig. 19. Reconstruction of the image displayed as a slice through the image intensity parallel to the surface and 1.3 Å below the photoemitter is shown in fig. 20. The four nearest-neighbor Ni atoms are clearly visible. Figure 21 shows a cross-sectional slice through the image, containing both the real and virtual reconstructed images (which are mirror images). In addition to two nearest neighbors, a third Ni atom directly below the S in the second Ni plane is visible, as well the S atom.

4. – Comparison to SEXAFS and LEED.

There are by now numerous surface structure determination methods, each with its own advantages and problems. Rather than making a comprehensive comparison here, we shall concentrate upon two electron scattering techniques, surface extended X-ray absorption fine structure (SEXAFS) and low-energy electron diffraction (LEED). Our primary aim is to emphasize that, particular practical problems aside, the fundamental nature of the information measured by the electron scattering techniques determines the completeness of the resulting structure determination and the ease with which the information can be extracted.

SEXAFS is an angle-integrated ARPEFS: if the energy-dependent photo-electron diffraction would be measured without any angular resolution the result would be a SEXAFS curve [38]. In this angle integration process, the scattering events with angle dependence cancel, leaving only certain multiple-scattering events which return to the photoemitting atom. To first order the oscillations which remain have a path length difference of $2r_j$:

$$\chi(k) = \frac{I - I_0}{I_0} = 2\sum_j \frac{|F_j|}{I_0} \sin(2kr_j + \phi_j).$$

In this view, there is only one SEXAFS curve, and the only information which can be derived is the distance between the photoemitter and nearby scattering atoms. This information is directly available via the Fourier transform because the distance is directly conjugate to the independent variable, k.

To first order, the oscillations have no information about the orientation of the atoms around the photoemitter. However, there is an angle dependence in the intensity of the oscillations determined by the angle between the photon's electric vector and the position of the the the scattering atom. By clever use of this dependence orientation information can be obtained from SEXAFS [39], but it is not always unambiguous. The photoelectron diffraction curves contain a great deal more orientation information and this information has been used to clarify structures that SEXAFS has not been able to solve [40]. However, this added information comes at a considerable cost in experimental and analytic complexity.

LEED is a further step up in complexity. Both SEXAFS and photoelectron diffraction give information about the environment near the photoemitting point source. Because incoming plane waves are used in LEED to probe the structure, all the uniquely situated atoms contribute comparable information to the data. While the photoemission techniques can only infer the relative position of two nonemitting atoms by subtraction of the coordinates from the photoemitter, all interatomic distances are roughly equivalent in LEED. Recently, it has been shown that the LEED signal consists of Bragg-like oscillations, ARPEFS-like oscillations and oscillations with both characteristics [41]. Thus potential information content in LEED is enormous, but direct analysis seems unlikely. Nevertheless, indirect analysis methods, spurred by the relative experimental ease of the LEED measurements, have advanced sufficiently for LEED to verify comprehensive structure models with high reliability.

This discussion has centered on the phase information available in these techniques and it neglects both the symmetry information and the amplitude information. The symmetry information is vital indeed, and in this regard LEED provides rapid, direct information on the two-dimensional lattice symmetry, while the photoemission techniques provide information on the photoemitter local site symmetry. Amplitude information is also important but generally requires detailed computations to analyze. There are important cases where amplitude information provides considerable aid in structure determination and we have mentioned some of them in the preceding sections.

Photoelectron diffraction lies somewhere between SEXAFS and LEED in its information content. To some degree, SEXAFS does not provide enough information to pin down completely unknown structures, while LEED provides too much information for direct analysis. Until recently, photoelectron diffraction was thought to lie in the middle of this spectrum, providing information about one photoemitter in such a jumbled way that direct analysis was still not feasible. Coupled with the more sophisticated apparatus needed for photoemission, this meant that photoelectron diffraction has been reserved for special systems which had not yielded to other methods. The advent of photoelectron holography may well alter this view. If the experimental measurements can be made practical, the holograms can be directly analyzed with the ease of SEXAFS, while providing structure information nearly as complete as LEED.

REFERENCES

[1] See A. LIEBSCH: this volume, p. 1.
[2] For reviews on (angle-resolved) photoemission see E. W. PLUMMER and W. EBERHARDT: *Adv. Chem. Phys.*, **49**, 533 (1982); F. J. HIMPSEL: *Adv. Phys.*, **32**, 1 (1983).

[3]   For reviews on inverse photoemission see V. DOSE: *Surf. Sci. Rep.*, **5**, 337 (1985); N. V. SMITH: *Rep. Prog. Phys.*, **51**, 1227 (1988); F. J. HIMPSEL: *Comments Condens. Matter Phys.*, **12**, 199 (1986); *J. Phys. Chem. Solids*, **49**, 3 (1988).

[4]   C. E. KUYATT and J. A. SIMPSON: *Rev. Sci. Instrum.*, **38**, 103 (1967); S. D. KEVAN: *Rev. Sci. Instrum.*, **54**, 1441 (1983).

[5]   M. G. WHITE, R. A. ROSENBERG, G. GABOR, E. D. POLIAKOFF, G. THORNTON, S. H. SOUTHWORTH and D. A. SHIRLEY: *Rev. Sci. Instrum.*, **50**, 53 (1979); R. Z. BACHRACH, F. C. BROWN and S. B. M. HAGSTROM: *J. Vac. Sci. Technol.*, **12**, 309 (1975).

[6]   D. E. EASTMAN, J. J. DONELON, N. C. HIEN and F. J. HIMPSEL: *Nucl. Instrum. Methods*, **172**, 327 (1980); D. RIEGER, V. SAILE, R.-D. SCHNELL and W. STEINMANN: *Nucl. Instrum. Methods*, **208**, 777 (1983).

[7]   J. F. VAN DER VEEN, F. J. HIMPSEL, D. E. EASTMAN and P. HEIMANN: *Solid State Commun.*, **36**, 99 (1980); R. STOCKBAUER *et al.*: to be published.

[8]   F. J. HIMPSEL, P. HEIMANN and D. E. EASTMAN: *Phys. Rev. B*, **24**, 2004 (1981).

[9]   R. G. SMEENK, R. M. TROMP, H. H. KESTERN, A. J. H. BOERBOOM and F. W. SARIS: *Nucl. Instrum. Methods*, **195**, 581 (1982); H. A. ENGELHARDT, W. BÄCK and D. MENZEL: *Rev. Sci. Instrum.*, **52**, 835 (1981).

[10] I. ADAWI: *Phys. Rev. A*, **134**, 788 (1964).

[11] See G. D. MAHAN: this volume, p. 25.

[12] D. J. SPANJAARD, D. W. JEPSEN and P. M. MARCUS: *Phys. Rev. B*, **15**, 1728 (1977); J. F. L. HOPKINSON, J. B. PENDRY and D. J. TITTERINGTON: *Comput. Phys. Commun.*, **19**, 69 (1980).

[13] J. A. KNAPP, F. J. HIMPSEL and D. E. EASTMAN: *Phys. Rev. B*, **19**, 4952 (1979).

[14] S. G. LOUIE, P. THIRY, R. PINCHAUX, Y. PETROFF, D. CHANDESRIS and J. LECANTE: *Phys. Rev. Lett.*, **44**, 549 (1980).

[15] D. E. EASTMAN and F. J. HIMPSEL: *Inst. Phys. Conf. Ser.*, **55**, 115 (1980).

[16] LANDOLT-BÖRNSTEIN, *Numerical Data and Functional Relationships, New Series III/23a, Electronic Structure of Solids: Photoemission Spectra and Related Data*, edited by A. GOLDMANN and E. E. KOCH (Springer, Berlin, 1989).

[17] P. J. FEIBELMAN, D. R. HAMANN and F. J. HIMPSEL: *Phys. Rev. B*, **22**, 1734 (1980).

[18] M. AONO, F. J. HIMPSEL and D. E. EASTMAN: *Solid State Commun.*, **39**, 225 (1981).

[19] I.-W. LYO and E. W. PLUMMER: *Phys. Rev. Lett.*, **60**, 1558 (1988).

[20] J. E. DEMUTH, B. N. J. PERSSON and A. J. SCHELL-SOROKIN: *Phys. Rev. Lett.*, **51**, 2214 (1983).

[21] J. M. LAYET, J. Y. HOARAU, H. LÜTH and J. DERRIEN: *Phys. Rev. B*, **30**, 7355 (1984).

[22] S. D. KEVAN: *Phys. Rev. Lett.*, **50**, 526 (1983).

[23] H. J. LEVINSON, F. GREUTER and E. W. PLUMMER: *Phys. Rev. B*, **27**, 727 (1983).

[24] D. E. EASTMAN, J. A. KNAPP and F. J. HIMPSEL: *Phys. Rev. Lett.*, **41**, 825 (1978).

[25] For recent reviews of photoectron diffraction see C. S. FADLEY: *Prog. Surf. Sci.*, **16** 275 (1984); Y. MARGONINSKI: *Contemp. Phys.*, **27**, 203 (1986).

[26] J. J. BARTON, S. W. ROBEY and D. A. SHIRLEY: *Phys. Rev. B*, **34**, 778 (1986).

[27] J. J. BARTON and D. A. SHIRLEY: *Phys. Rev. B*, **32**, 1906 (1985).

[28] D. H. ROSENBLATT, J. G. TOBIN, M. G. MASON, R. F. DAVIS, S. D. KEVAN, D. A. SHIRLEY, C. H. LI and S. Y. TONG: *Phys. Rev.*, **23**, 3828 (1981).

[29] J. J. BARTON, C. C. BAHR, Z. HUSSAIN, S. W. ROBEY, J. G. TOBIN, L. E. KLEBANOFF and D. A. SHIRLEY: *Phys. Rev. Lett.*, **51**, 272 (1983).

[30] J. J. BARTON, C. C. BAHR, S. W. ROBEY, Z. HUSSAIN, E. UMBACH and D. A.

SHIRLEY: *Phys. Rev. B*, **34**, 3807 (1986); C. C. BAHR, J. J. BARTON, Z. HUSSAIN, S. W. ROBEY, J. G. TOBIN and D. A. SHIRLEY: *Phys. Rev. B*, **35**, 3773 (1987).

[31] S. BRENNAN, J. STOHR and R. JAEGER: *Phys. Rev. B*, **24**, 4871 (1981).

[32] S. W. ROBEY, C. C. BAHR, Z. HUSSAIN, J. J. BARTON, K. T. LEUNG, J.-R. LOU, A. E. SCHACH VON WITTENAU and D. A. SHIRLEY: *Phys. Rev. B*, **35**, 5657 (1987).

[33] C. S. FADLEY: *Phys. Scr.*, T**17**, 39 (1987).

[34] W. F. EGELHOFF: *Phys. Rev. B*, **30**, 1052 (1984).

[35] A. SZOKE: in *Short Wavelength Coherent Radiation: Generations and Applications*, *Am. Inst. Phys. Conf. Proc.*, No. 147 (New York, N.Y., 1986), p. 361.

[36] J. J. BARTON: *Phys. Rev. Lett.*, **61**, 1356 (1988).

[37] D. GABOR: *Nature (London)*, **161**, 777 (1948).

[38] P. A. LEE: *Phys. Rev. B*, **13**, 5261 (1976).

[39] J. STOHR, D. A. OUTKA, R. J. MADIX and U. DOBLER: *Phys. Rev. Lett.*, **54**, 1256 (1985).

[40] D. WOODRUFF, C. F. MCCONVILLE, A. L. D. KILCOYNE, TH. LINDNER, J. SOMERS, M. SURMAN, G. PAOLUCCI and A. M. BRADSHAW: *Surf. Sci.*, **201**, 228 (1988).

[41] J. J. BARTON, M.-L. XU and M. A. VAN HOVE: *Phys. Rev. B*, **37**, 10475 (1988).

Laser Photoemission.

F. Meier, M. Aeschlimann, M. Stampanoni and A. Vaterlaus

Laboratorium für Festkörperphysik, ETH Hönggerberg - CH-8093 Zurich, Switzerland

1. – Introduction.

The marriage of new techniques and instrumentation can give rise to promising offspring. First experiences give confidence that spin-polarized photoemission and laser technology provide a good example. In this lecture the prospects of using high-intensity pulsed lasers as light sources in spin-polarized photoyield measurements are discussed and a few examples are presented: 1) laser-induced photoemission from iron and 2) laser-induced thermomagnetic writing combined with photoemission reading using magnetic recording materials, including epitaxial iron films which are about 6 monolayers thick. Above all, it is shown that the capability of producing incredibly high intensities of radiation—more than $10\,\mathrm{GW/cm^2}$ in the laser focus—can be exploited in measurements of the spin polarization of the photoyield. Other types of photoemission suffer from the adverse effects of space charge.

2. – Laser-induced spin-polarized photoemission.

Since its discovery [1], spin-polarized photoemission from solids has been an ever-growing branch of electron spectroscopy. Up to the present day it played an important role in the investigation of magnetism [2]. The realization of spin-polarized photoemission from nonmagnetically ordered solids resulted in the development of highly efficient sources of polarized electrons [3] together with new developments in exploring symmetry-related properties of electronic states, like band hybridization [4]. Recently, a laser-induced photoemission experiment has been set up using a pulsed KrF excimer laser as light source. The photon energy is $h\nu = 5\,\mathrm{eV}$ at an output energy of $250\,\mathrm{mJ/pulse}$ corresponding to about 10^{17} photons. The pulse duration is $16\,\mathrm{ns}$. Focused on a spot of $100\ \mu\mathrm{m}$ diameter the average intensity (or fluence) is $200\,\mathrm{GW/cm^2}$, sufficient to disintegrate any material instantaneously.

The polarization P of the photoelectric yield is defined as $P = (N\uparrow - N\downarrow)/(N\uparrow + N\downarrow)$, where $N\uparrow$ $(N\downarrow)$ is the total number of electrons emitted with spin parallel (antiparallel) to the surface normal. An external field perpendicular to the surface is applied to align the magnetic domains.

Compared to conventional spin-polarized photoemission[2] some special adaptations are necessary for the laser experiment. The reason is that in laser photoemission all electrons photoemitted by a given pulse are monitored collectively in the detectors of the spin analyser. The solid-state Schottky-barrier detectors have been proved to respond linearly up to 10^5 electrons—and probably more—incident simultaneously, *i.e.* within a few nanoseconds. In order to protect the preamplifiers against overloading, the conversion gain is made adjustable according to the number of electrons collected. Following the modified preamplifier, the signal enters the main amplifier. From there it is fed into a pulse height analyser where its amplitude—corresponding to the number of electrons per pulse—is measured. After analog-digital conversion, the signal from each detector is transmitted via fiber optics into the data acquisition system of a desk computer for further processing.

In a conventional polarization measurement single electrons are counted, each giving a single pulse. This is a digital measurement. In laser photoemission all electrons per pulse are measured at once and their number is derived from the measured pulse height. Evidently, this is an analog measurement.

Laser-induced spin-polarized photoemission was first applied to polycrystalline iron[5]. The work function of the sample was (4.2 ± 0.1) eV, about 0.8 eV less than the photon energy of the laser. All photoemitted electrons were collected and spin-analysed together.

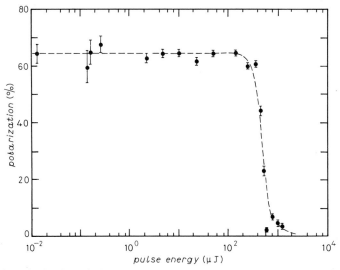

Fig. 1. – Spin polarization of photoelectrons emitted from polycrystalline iron *vs.* pulse energy. Around 1 mJ the sample is heated up to the Curie temperature causing the drop of the polarization. In measurements of the photoelectric yield, space charge effects become noticeable above 1 μJ. This figure shows that the space charge does not affect the polarization.

Figure 1 shows the polarization as a function of pulse energy. Over many decades—up to 1/10 mJ—the polarization stays constant. It is equal to the one obtained with a conventional light source where single electrons are detected. However, approaching a laser pulse energy of 1 mJ a rapid drop of the polarization occurs. A simple estimate [6] shows that the average temperature of the irradiated sample area gets close to the Curie point of Fe, which is about 1000 K.

The average temperature increase produced by a laser pulse is

(1) $$\Delta T \sim (I - R)It_{\mathrm{p}}/C_{\mathrm{v}}\rho L_0$$

with

R = reflectivity,

I = intensity,

t_{p} = pulse duration,

ρ = density,

C_{v} = specific heat,

L_0 = thermal diffusion length during laser pulse.

The sample is considered to be irradiated by a pulse of constant intensity I during the time t_{p}. Then, the energy deposited per unit area of the surface is $(I - R)It_{\mathrm{p}}$. The denominator $C_{\mathrm{v}}\rho L_0$ is the specific heat of that volume of Fe which is effectively warmed up by the propagation of heat from the surface.

For Fe the appropriate data are: $R = 0.35$, focus area $5.3 \cdot 10^{-4}\,\mathrm{cm}^2$, pulse energy $10^{-1}\,\mathrm{J}$, pulse duration $16\,\mathrm{ns}$, $\rho = 7.86\,\mathrm{g/cm}^3$, $C_{\mathrm{v}} = 0.45\,\mathrm{J/K\,cm}^3$ and $L_0 = \sqrt{2Dt_{\mathrm{p}}} = 8.9 \cdot 10^{-5}\,\mathrm{cm}$. D is the thermal diffusivity of Fe: $0.24\,\mathrm{cm}^2/\mathrm{s}$. Then, a pulse of 1/10 mJ causes an average temperature rise of 350 K. Consequently, above 1/10 mJ pulse energy, the temperature at the sample surface becomes high enough to reduce the magnetization significantly. In accordance, also the measured polarization of the photoelectrons decreases. Note that ΔT is proportional to I, i.e. the energy of the pulse (with all other parameters constant). By putting the same energy into a pulse of shorter duration ΔT increases as $(\sqrt{t_{\mathrm{p}}})^{-1}$.

Figure 2 shows the region of the polarization drop on a linear scale. The full line indicates the polarization of the photoelectrons using instead of eq. (1) an exact temporal and spatial temperature profile across the laser focus [7]. The static magnetization $M(T)$ of iron has been used to calculate P for each time—and space—element. Obviously, the fit of the measured data is excellent.

When measuring $P(I)$, a curious effect occurs: At low intensities the number of electrons counted is small and the statistics correspondingly poor. Upon increasing the pulse energy, the number of photoelectrons first increases proportionally. However, at still higher energies, the number of electrons

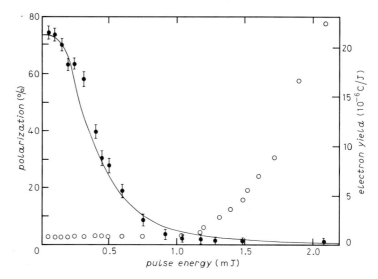

Fig. 2. – Pulse energy dependence of the spin polarization of the photoelectrons emitted from polycrystalline iron shown on a linear scale. The solid curve has been calculated from the temporal and spatial temperature profile of the sample area hit by the laser pulse under the assumption of thermal equilibrium with the magnetization of Fe. The open circles show the photoelectric yield which rises steeply above 1 mJ due to positive-ion emission.

emitted per pulse levels off or even decreases! In fig. 3, this effect occurs above 1 µJ laser pulse energy.

The reason is that at sufficiently high pulse energies the large number of photoelectrons gives rise to a noticeable space charge in front of the sample surface. This repulsive-charge distribution prevents excited electrons from leaving the sample. Ultimately, the potential energy of the electrons in the space charge cloud is transformed into kinetic energy of the photoelectrons: thereby the photoelectrons lose all memory about the distribution of their individual energies and momenta just after excitation.

For laser-induced photoemission space charge effects have serious consequences [8, 9]. Angle- and energy-resolved photoemission is restricted to low intensities, at least if the measured energies and momenta parallel to the surface should bear any meaning to related crystal properties as studied in conventional photoemission. Why, then, do not anomalies of $P(I)$ occur in the spin-polarized photoyield experiment? From fig. 1 no indication can be inferred that anything like space charge actually exists.

Spin-polarized photoyield measurements are space charge insensitive because they profit from conservation of the total spin angular momentum of the photoelectrons. In the space charge cloud all possible interactions take place: spins of individual electrons may change their orientation wildly. However, there is the stringent restriction that the total spin polarization of the charge

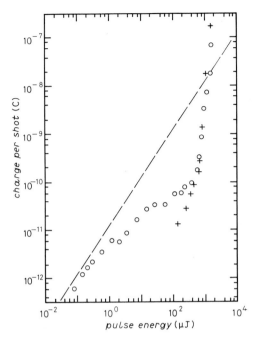

Fig. 3. – Energy dependence of the negative, electronic charge (open circles) and positive, ionic charge (crosses) emitted by a laser pulse. At low energies, the electronic charge is equal to the one emitted by a conventional light source (dashed line) with the same number of photons. Above 1 μJ, the number of electrons emitted by a laser pulse is markedly reduced with respect to the extrapolation from low energies: this is the effect of space charge. Above 1 mJ, the surface melts and positive ions are emitted. They partially neutralize the space charge, thus giving rise to a steep increase of the number of emitted electrons.

cloud remains constant. Within the charge cloud only internal forces act, and internal forces must conserve the total angular momentum. Evidently, this applies to the total energy and momentum as well—however, no application has been found where these quantities are of interest.

Negative space charge reduces the photoelectric yield. Interestingly, above a certain intensity threshold positive charges are ejected from the surface, too. Then, due to space charge neutralization, the photoelectric yield again increases enormously, as shown in fig. 2. Positive-ion emission is a sign that the focus area has become at least partially molten[10]. Note that the increase of the yield in fig. 2 occurs at pulse energies where P has already dropped to zero. This is in accordance with the fact that the melting point of iron, 1809 K, is markedly higher than the Curie point $T_C = 1044$ K.

Recently, thermomagnetic perpendicular recording materials have been investigated using laser photoemission. The magnetization in these materials is directed perpendicular to the surface. As a consequence, very high bit densities can be achieved. In longitudinal recording the adverse effects of the demagnetizing field between adjacent domains put severe restrictions on the minimum size of stable domains [11].

For a typical recording material, amorphous TbFe [12], an upper limit for the switching time of the magnetization could be established. For this purpose the uniformly magnetized film was put on the unstable branch of its hysteresis loop, *i.e.* a field smaller than the coercive field was applied opposite to the direction of the magnetization. Then a laser pulse of appropriate energy was directed onto the sample. The energy of the pulse was chosen such that the reduction of the coercive field due to heating was sufficient to reverse the magnetization in the bias field. The polarization was measured of those electrons which were emitted by the very same pulse causing the magnetization reversal. For suitable pulse energies the polarization turned out to be exactly the same as when the sample was in the reversed state from the beginning. The equality of the two polarizations shows that the reversal of the magnetization takes place in a time much shorter than the pulse duration, *i.e.* 16 μs. Incidentally, this also means that thermomagnetic writing can be done faster than the reading of the stored information by means of the magneto-optic Kerr effect. Although the reading time has the tendency of becoming shorter with modern optical technologies [13], it is fundamentally limited by the small value of the Kerr rotation angle, which is typically of the order of 1′.

Reading of the magnetically stored information by photoemission is possible with a single laser pulse. Figure 4 shows the distribution of P-values from a remanently magnetized TbFe film, once the magnetization direction being «up», once «down». The average polarization is $\pm 31\%$. Obviously, this P-value corresponds to a large signal, opposite to what is found in magneto-optic Kerr reading. Whereas, however, in magneto-optic reading highly powerful optical

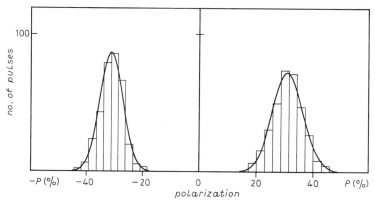

Fig. 4. – Distribution of polarization values per single pulse. The a-TbFe sample was in remanence, either «up» or «down», giving an average polarization of $\pm 31\%$. The average number of electrons reaching each detector of the spin analyser was about 4000 per shot. The distribution of P-values is somewhat larger than expected from the statistics of electron scattering in the spin analyser alone.

analysers can be used—with efficiency 1—the efficiency of the polarization analyser for electrons is deplorably low, of the order $10^{-4} \div 10^{-5}$. Although this small number does not arise from a fundamental law, nothing better has yet been invented. For the reading by photoemission, as shown in fig. 4, it means that only about 4000 of the 10^8 electrons transmitted to the Mott polarization analyser per pulse are impinging on the detectors. The overwhelming majority of the electrons is just wasted. Therefore, the uncertainty of the polarization is rather large. It turns out that it is somewhat larger than expected from statistical scattering by the polarization detector alone. Many other random events like fluctuations of the laser intensity, space charge and small variations of the electron trajectories towards the polarization analyser may be responsible for that.

Thermomagnetic writing and photoemission reading have been combined using a new highly interesting magnetic material: epitaxial iron films grown on Cu(001). Details on the magnetic properties of these films which have f.c.c. structure may be found in ref. [14].

The focused laser beam was scanned by a step-motor-controlled deflection prism on the surface of a 6 ML thick epitaxial iron film. The film was held at 30 K, where the coercive field of the perpendicular magnetization is several kG. Applying a small bias field—opposite to the magnetization—a bit can be written into the film by local heating with a laser pulse. In this area of increased temperature, the coercitivity becomes smaller and the bias field is able to reverse the magnetization locally. Then, at reduced intensity, the laser is again scanned over the surface of the sample and the polarization of the emitted electrons is determined. For each location, the bit magnetization is displayed on a screen either black or white, depending on the sign of the polarization. In this way the written information can be retrieved using exclusively the instrumentation of laser-induced spin-polarized photoemission.

3. – Conclusions.

Spin-polarized photoemission with a high-intensity, pulsed laser as light source opens new prospects in electron spectroscopy: the time-resolved investigation of matter under the extreme conditions imposed by high-fluence electromagnetic radiation. Although still in its infancy, the feasibility of the experiment is proved and first encouraging results are obtained.

* * *

We thank G. L. BONA for his decisive contributions in setting up the experiment and H. C. SIEGMANN for his continuous support. We profited greatly from the technical expertise of K. BRUNNER and O. STADELMANN.

REFERENCES

[1] G. BUSCH, M. CAMPAGNA, P. COTTI and H. C. SIEGMANN: *Phys. Rev. Lett.*, **22**, 597 (1969).

[2] H. C. SIEGMANN, F. MEIER, M. ERBUDAK and M. LANDOLT: *Adv. Electron. Electron Phys.*, **62**, 2 (1984).

[3] D. T. PIERCE and F. MEIER: *Phys. Rev. B*, **13**, 5484 (1976).

[4] F. MEIER and D. PESCIA: in *Optical Orientation*, edited by F. MEIER and B. ZAKHARCHENYA (North-Holland, Amsterdam, 1984), p. 295 ff.

[5] G. L. BONA, F. MEIER, G. SCHÖNHENSE, M. AESCHLIMANN, M. STAMPANONI, G. ZAMPIERI and H. C. SIEGMANN: *Phys. Rev. B*, **34**, 7784 (1986).

[6] N. BLOEMBERGEN: in *Application of Lasers in Materials Processing*, edited by E. A. METZBOWER (American Institute of Physics, Washington, D.C., 1979), p. 1 ff.

[7] J. H. BECHTEL: *J. Appl. Phys.*, **46**, 1585 (1975).

[8] J. BOKOR, R. HAIGHT, R. H. STORZ, J. STARK, R. R. FREEMAN and P. H. BUCHSBAUM: *Phys. Rev. B*, **32**, 3669 (1985).

[9] K. GIESEN, F. HAGE, H. J. RIESS, W. STEINMANN, R. HAIGHT, R. BEIGANG, R. DREYFUS, PH. AVOURIS and F. J. HIMPSEL: *Phys. Scr.*, **35**, 578 (1987).

[10] A. M. MALVEZZI, H. KURZ and N. BLOEMBERGEN: *Appl. Phys. A*, **36**, 143 (1985).

[11] T. SUZUKI: *IEEE Trans. Magn.*, MAG-20, 675 (1984).

[12] The sample was supplied by E. MARINERO, IBM San José.

[13] A. BELL: *SPIE J.*, **382**, 1 (1983).

[14] D. PESCIA, M. STAMPANONI, G. L. BONA, A. VATERLAUS, R. F. WILLIS and F. MEIER: *J. Phys. (Paris)*, Coll. C-8, **49**, 1963 (1988).

Photoemission with Circularly Polarized Photons.

U. Heinzmann

Universität Bielefeld, Fakultät Physik - D-4800 Bielefeld 1, B.R.D.

1. – Introduction.

The concept of polarized photoelectrons ejected by circularly polarized light introduced theoretically by Fano in 1969 [1] was first experimentally confirmed for free Cs atoms in 1970 [2], for a Cs metal layer in 1971 [3], for a GaAs semiconductor in 1975 [4] and for free molecules in 1980 [5]. Spin-orbit interaction is the essential mechanism leading to a spin polarization of photoelectrons with degrees of polarization of up to 100% (optical-pumping-type process). Up to 1984 experimental analysis of the electron spin polarization in photoionization and photoemission using circularly polarized light was restricted to angle-integrated measurements without resolution of the kinetic energy of the photoelectrons ejected (for reviews see, for example, [6-8]). Parts of these have used circularly polarized synchrotron radiation emitted out of the plane of the synchrotron in Bonn [9].

With the availability of intense synchrotron radiation from electron storage rings like BESSY, fluxes of up to some 10^{11} circularly polarized VUV photons/s are available to study the photoelectron emission simultaneously resolved with respect to the photon polarization, the photon energy, the electron emission angle, the kinetic energy of the electrons and their spin polarization components. Because of the excellent UHV running conditions, the technique was open for studies not only with free atoms [10], but also with solid surfaces [11] and adsorbates [12]. Meanwhile the results of many publications demonstrate the ability of this technique and of the corresponding apparatus built at BESSY, aspects of partial fields are reviewed in detail elsewhere [13-15].

In some examples this contribution discusses some recent experimental activities of photoemission of solids [11, 16-19] and adsorbates [12] using circularly polarized synchrotron radiation [20] and also demonstrates atomic effects in condensed matter (photoemission of the dilute phase of Xe/Pd(111) [21], autoionization [13, 15] and absorption [22, 23] resonances in Xe adsorbates, influence of interference effects of partial waves of the outgoing off-normal electrons onto their spin polarization [24]) in direct comparison with the corresponding effects in free atoms [10, 25, 26].

469

2. – Experimental.

Most studies of photoionization with circularly polarized radiation are hampered by the fact that most atoms and molecules have their ionization threshold in the VUV, where conventional methods for producing circularly polarized radiation by use of a quarter-wave plate and a prism break down (photon energy > 10 eV). They can be performed, however, with synchrotron radiation, which is linearly polarized when emitted in the plane of the storage ring but is elliptically polarized with a high degree of right (left)-handed circular polarization when emitted above (below) the plane. This «source» of circularly polarized VUV radiation has been used in the energy range up to 35 eV since 1978 in Bonn[9, 27] and since 1982 at BESSY in Berlin.

The synchrotron radiation of BESSY is monochromatized by a 6.5 m N.I. UHV monochromator (fig. 1) of the Gillieson type[20] with the electron beam in

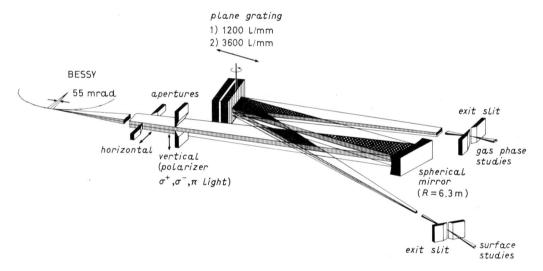

Fig. 1. – Schematic diagram of the 6.5 m N.I. monochromator for circularly polarized synchrotron radiation at BESSY.

the storage ring being the virtual entrance slit. A spherical mirror and a plane holographic grating (1200 or 3600 lines/mm) form a 1:1 image of the tangential point in one of the two exit slits. With a slit width of 2 mm a bandwidth of 0.5 nm and 0.1 nm has been achieved for the 1200 grating (first diffraction order) and for the 3600 grating (second order), respectively. Apertures movable in vertical direction are used to select radiation emitted above and below the storage ring plane, which has positive or negative helicity, respectively. The optical degrees of polarization of the synchrotron radiation have been measured[10] by means of a rotatable four-mirror analyser[9, 27]. Figure 2a) shows the results of the

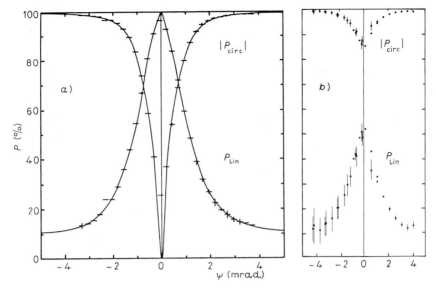

Fig. 2. – Degree of circular and linear polarization, P_{lin} and P_{circ}, measured behind the monochromator[10, 20]: *a*) as a function of the vertical angle ψ (± 0.1 mrad) for the wavelength 50 nm, *b*) the apertures accept a vertical angular range from ψ to ± 5 mrad for the wavelength 100 nm.

circular polarization P_{circ} and the linear polarization P_{lin} as functions of the vertical angle ψ (± 0.1 mrad) in comparison with theoretical predictions. Figure 2*b*) shows the optical data which are more relevant for practical purposes: the case when the apertures are opened to get a high photon flux throughput, in the vertical acceptance range from ψ above or below the storage ring plane to $+5$ mrad or -5 mrad, respectively. Note that all radiation emitted above the storage ring plane (half-cone) has an average degree of circular polarization of 85% at 100 nm. The spin-resolved photoemission experiments are usually performed with the apertures set to accept radiation above and below ± 1 mrad. Thus a photon flux of up to $5 \cdot 10^{11}$ photons s^{-1} with a degree of circular polarization of ± 93% passes the monochromator, which is about 30% of the intensity emitted in the full vertical cone.

Behind the monochromator exit slits two apparatus are in operation, which are very similar in their set-up, one for photoionization of free atoms and molecules (high-vacuum system), the other for photoemission of solids and adsorbates (UHV). As shown in fig. 3 (schematic diagram of the UHV system), the elliptically polarized VUV radiation hits the phototarget under normal incidence producing photoelectrons in a region of electric or magnetic fields. The sample is cleaned by ion bombardment, heating in oxygen and flashing in a separate preparation chamber. The crystal on top of a manipulator can be cooled by use of a liquid-He cryostat to temperatures of less than 40 K.

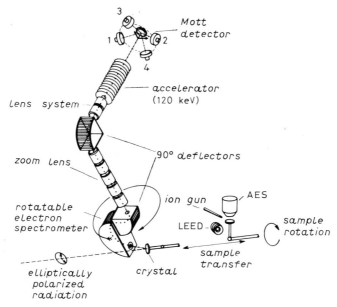

Fig. 3. – Experimental set-up of the spin-, energy- and angle-resolved photoemission experiment.

The photoelectrons emitted into a cone of $\pm 3°$ are analysed with respect to their kinetic energy by a rotatable simulated hemispherical electron spectrometer[28] and are directed by a 90° electrostatic deflection along the axis of rotation of the spectrometer. This direction, which is the normal of the reaction plane spanned by the momenta of the incoming radiation and the outgoing photoelectrons, is rotated by 45° with respect to the major (and the minor) axis of the light polarization ellipse. Thus the formulae describing the angular dependences of the spin polarization components of atomic photoelectrons become simple trigonometric relations[26, 29]. After a second deflection the electrons are accelerated to 120 keV and scattered at the gold foil of the Mott detector for the spin polarization analysis[8]. Parts of the measurements performed in the apparatus at BESSY described have used another type of electron spectrometer (CMA) and spin detector (LEED) (not shown in fig. 3)[30, 31].

3. – Symmetry-resolved band mapping by means of spin-resolved photoemission.

In electronic transitions caused by circularly polarized radiation the photoelectrons from nonferromagnetic solids are commonly spin polarized[32] due to the existence of the spin-orbit interaction. In the highly symmetrical set-up of normal light incidence and normal photoemission, the photoelectron spin polarization vector is either parallel or antiparallel to the circular light polarization and switches the sign if the helicity of the radiation is changed from σ^+ to σ^-. Whether the electron polarization is parallel (positive polarization) or

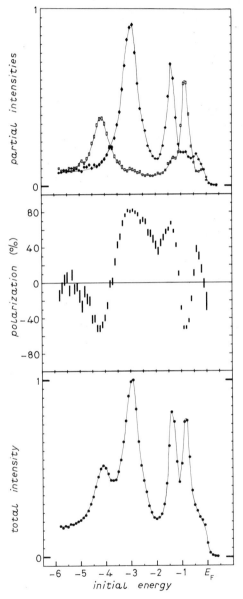

Fig. 4. – Spin-resolved photoelectron spectrum of Ir(111) in normal incidence and normal photoemission at 16 eV photon energy [17]; lower part: total (spin-independent) intensity; middle part: electron spin polarization; upper part: partial intensities for spin parallel (full symbols, I_+) and spin antiparallel (open symbols, I_-) to the light helicity.

antiparallel (negative polarization) to the light helicity depends upon the symmetry of the states involved and is described by the corresponding (relativistic) selection rules for optical dipole transitions [33-35].

The resulting photoelectron intensity spectra consist of two partial spectra correlated to a complete spin polarization $P = +1$ or $P = -1$. These partial spectra I_+ and I_- are related to the total intensity I and the spin polarization P measured as given by

$$I_{+,-} = 0.5I(1 \pm P).$$

Structures which are superimposed in the total intensity I are thus separated in the partial intensities I_+ and I_-. As an example, fig. 4 shows the spin-resolved normal photoemission spectrum of Ir(111)[17] with circularly polarized VUV radiation of 16 eV photon energy. The photoemission occurs due to direct interband transitions from four d bands below the Fermi energy with symmetries Λ_{4+5}^3, Λ_6^3, Λ_6^3, Λ_{4+5}^3. The lower part of fig. 4 shows the pronounced structure of four peaks. The sequence of the electron spin polarization sign (middle part) has been found to be negative, positive, positive, negative. Note that the high polarization of up to 80% at Ir(111) exceeds all data ever previously measured for other nonmagnetic three-dimensional solid-state systems. Combination of intensity and polarization yields the partial photoelectron spectra I_+ and I_- (upper part of fig. 4), which are used for a symmetry-resolved band mapping. It is worth noting that the positions of the peaks in the partial spectra are not identical with those in the total intensity spectrum (lower part) which shows the necessity of performing spin-resolved band mapping. This behaviour corresponds to an improvement of the photoemission resolution for the spin-resolved technique compared with the non-spin-resolved standard-type experiment. This can also be seen at the small structures close to the Fermi energy shown in the upper and middle parts of fig. 4 and connected with a weak shoulder in the lower part only. They correspond to a dipole transition into a higher final band. The improvement of the peak identification and localization is also demonstrated in fig. 5 at the photoemission spectrum at the middle-Z material Pd(111)[19], where the spin-orbit fine-structure splitting is smaller than at Ir(111). It is only seen at the different peak positions of the spin-resolved partial intensities I_+ and I_-, while the total intensity I has only one peak. The peak identification of I_- and I_+ as separate peaks is based upon the fact that the left and the right part of I have been measured to be positively and negatively polarized, respectively.

Figures 6 and 7 show the band structures of Ir(111)[17] and Pd(111)[19], respectively; the results of the symmetry-resolved band mapping procedure based on the experimental spin-resolved photoemission data are compared with the theoretical results of a self-consistent band structure calculation. The theoretically predicted sequence of bands Λ_{4+5}^3, Λ_6^3, Λ_6^1, Λ_6^3, Λ_{4+5}^3 below the Fermi

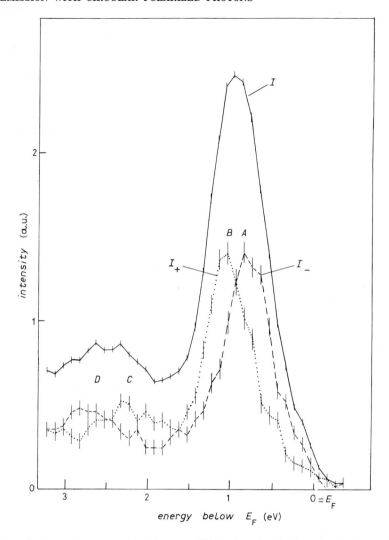

Fig. 5. – Photoelectron spectra from Pd(111) for circularly polarized synchrotron radiation at $h\nu = 16$ eV. The solid line is the photoelectron intensity I at normal photoemission, which is separated into the partial intensities I_+ and I_- by means of the spin polarization P[19].

level[11] could be experimentally verified and the bands have been mapped. Since a transition $\Lambda_6^1 \to \Lambda_6^1$ is forbidden[33], only 4 direct transitions from the initial states Λ_{4+5}^3 or Λ_6^3 did occur yielding negative or positive polarized electrons in the upper band Λ_6^1, respectively. For Ir(111) the mapping of the 4 valence bands 2, 3, 5, 6 in fig. 6 demonstrates excellent agreement with theory[36] besides an experimentally found overall shift of the two upper bands 7 and 8 by 0.8 eV (self-energy correction)[17]. The small structures in the spin-resolved

partial spectra (fig. 4 upper part) close to the Fermi energy yield a few mapping
points in the second final band No. 8 (fig. 6). It has been characterized by the
same symmetry as band No. 7 due to the spin polarization data. It is worth to
remind that without a spin polarization analysis of the photoelectrons this

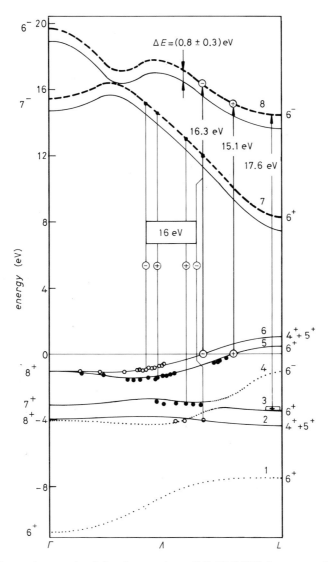

Fig. 6. – Symmetry-resolved band mapping of Ir(111)[17] in comparison with the
calculated band structure[36]. The mapping points have been obtained from the spin-
resolved partial photoelectron spectra. Bands with full symbols yield positive polarized
electrons and correspond to Λ_6 symmetry, while bands with open symbols (negative
polarization) are characterized by Λ_{4+5}. The calculated band structure has to be corrected
by a 0.8 eV broader band gap (difference between dashed and full band curves) in order to
map the uppermost band between 15.1 and 16.3 eV photon energy well.

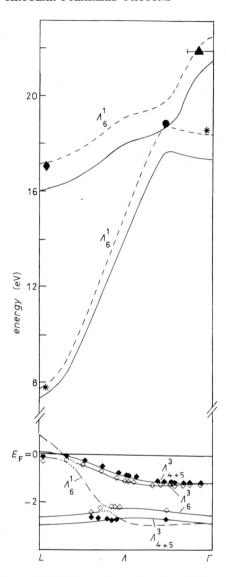

Fig. 7. – Band mapping of Pd(111) in the Λ-direction[19] in comparison with band structure calculation[37].

mapping of band No. 8 could not be performed as shown in fig. 6. Figure 7 shows similar results for Pd(111), where the spin-orbit splitting between Λ_{4+5}^3 and Λ_6^3 is narrower than in the case of Ir(111) because of the lower Z of the material. The agreement between experiment (band mapping procedure[19]) and theory[37] is also excellent; note that the self-energy correction shown in fig. 7 (difference between solid and dashed curves) is now energy dependent.

4. – Normal and off-normal spin-resolved photoemission spectra of rare-gas adsorbates.

When a rare-gas atom is ionized by 15 eV radiation the produced p hole shows a fine-structure splitting corresponding to the ionic ground-state configurations ($^2P_{3/2}$ and $^2P_{1/2}$) and resulting in two peaks in the photoelectron spectra, as shown in fig. 8 (left upper corner). The peaks are separated by the existence of the spin-

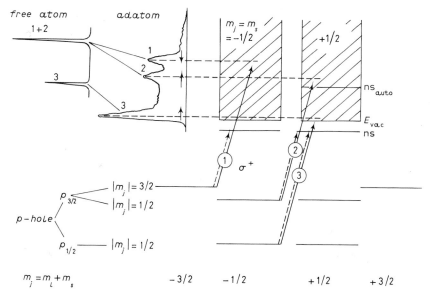

Fig. 8. – Level diagram for photoionization of rare-gas atoms and normal photoemission of rare-gas adsorbates with circularly polarized radiation (σ^+, selection rule $\Delta m_j = +1$). Arrows indicate nonvanishing transitions. The three transitions 1-3 give rise to the three peaks in the photoelectron spectrum [12], whereas peaks 1 and 2 coincide for free atoms [38].

orbit interaction. If they are resolved separately by means of an electron spectrometer, the photoionization measurements are sensitive to the spin-orbit coupling and, thus, all photoelectrons analysed are spin polarized [39]. And indeed within the experimental uncertainty the photoelectrons of peak 3 emitted into the forward or the backward direction ($\theta = 0, 180°$) have been found [10] to be completely spin polarized. This comes from the fact that at $\theta = 0$ the continuum wave functions have pure m_s character, all waves with $m_l \neq 0$ vanish there. Figure 8 demonstrates why the spin polarization is complete: taking into account the transition selection rule for right-handed circularly polarized light $\Delta m_j = +1$ the $p_{1/2}$ electron can only be photoexcited into the $m_s = +1/2$ continuum state. On the other hand, transitions from the $p_{3/2}$ $|m|_j = 3/2$ and

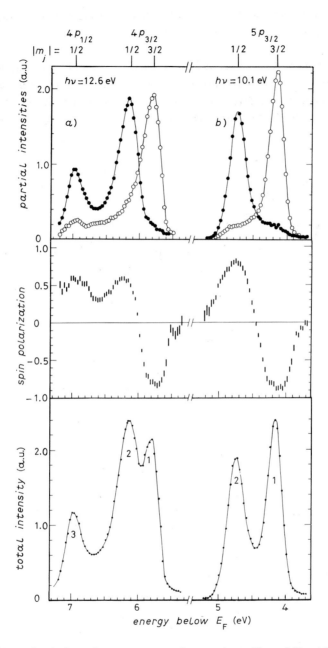

Fig. 9. – Spin-resolved photoelectron spectra of a monolayer Kr and Xe on Pt(111): a) Kr, $h\nu = 12.6$ eV; b) Xe, $h\nu = 10.1$ eV. Experimental results [12] of the total (spin-independent) intensity (lower part), electron spin polarization (middle part), partial intensities for spin up (full) and spin down (open) (upper part).

$|m|_j = 1/2$ states create photoelectrons of opposite (but complete) spin polarization. In the case of adsorbates the reduction of symmetry by the presence of the
surface induces a further splitting of the energy levels, as shown in fig. 8.
Measurements of the photoelectron spin polarization yield information about the
quantum numbers of the p-hole states, the photoelectrons come from [12].
Figure 9 (lower part) gives the photoelectron spectra of Kr and Xe on Pt(111) at
full monolayer coverage. Combination with the corresponding spin polarization
values measured (almost complete negative and positive polarizations) (middle
part) results in the partial spectra for spin up and down drawn in the upper part
of fig. 9. The peak structure is unambiguously correlated with either «spin up» or
«spin down» with the consequence that peaks 1, 2, 3 can be characterized by the
atomic hole state quantum numbers as given on top of fig. 9 and schematically set
up in fig. 8. The sequence of the m_j quantum numbers indicates that the 1-2
splitting is caused by lateral adatom interactions or crystal field splitting, but not
by other mechanisms discussed in the literature as substrate-induced relaxation
effects in the final ionic state or the formation of molecular orbitals [12, 15]. This
interpretation is supported by the further experimental result that the 1-2
splitting vanishes in the dilute adsorption system of xenon atoms on Pd(111),
where the distance between xenon adlayer neighbours is much larger [21].

Spin-resolved off-normal photoemission on solids has to solve the problem to
distinguish between spin effects of the photoelectron production process and spin
effects in the emission, diffraction and transmission process at the surface taking
into account the three-step model of photoemission of solids (excitation,
transport, transmission) [33-35]. This has been experimentally studied and
discussed in details for off-normal photoemission at Pt(111) [31, 40]. Thus the off-
normal photoemission studies of adsorbates [24] have been performed with the
centro-symmetrical crystal and middle-Z material Pd(111) as substrate in order
to disregard spin effects of photoelectron diffraction [41]. Furthermore, the
photoelectron emission was studied in the mirror plane of adsorbate and
substrate. Additionally xenon was used as adsorbate system, because it is known
from the literature [42] that even for a xenon single crystal the experimental
LEED results can be well fitted within the kinematic approximation. Finally,
the adsorbate system was the well-ordered commensurate $\sqrt{3} \times \sqrt{3} R 30°$
monolayer on Pd(111) in order to be sure that the photoelectrons studied are
produced at the outermost atoms within the interface solid/vacuum.
Photoelectron intensities and polarizations measured [24] indeed do not show
asymmetries when changing the emission angle from $+\theta$ to $-\theta$ or switching the
helicity of the incoming radiation from σ^+ to σ^- apart from the pure sign change
of the spin polarization, of course.

For the off-normal studies we have preliminarily looked at the $p_{1/2}$ peak,
because the splitting of the $p_{3/2}$ peak could be lost for angles different from $\theta = 0$
with the consequence that the polarizations of opposite sign of both $p_{3/2}$ peaks
would partly cancel one another. Figure 10 b) shows that the spin polarization of

the $p_{1/2}$ peak continuously decreases with increasing emission angle θ. At about 30° both spin directions give the same contribution, and at higher angles the antiparallel spin component dominates[24]. Taking into account that this $p_{1/2}$ peak cannot be split in further peaks of opposite spin directions, these experimental results demonstrate that the spin polarization has to be an angle-dependent mixture of quantities concerning energy degenerate final states.

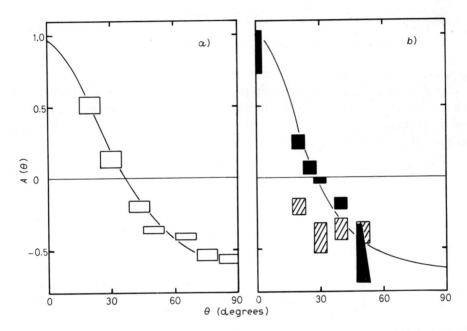

Fig. 10. – Off-normal photoelectron emission: spin polarization component $A(\theta)$ (parallel to the light helicity) as a function of the polar emission angle θ. a) For photoelectrons leaving the Xe atom in the final ionic state Xe^+ $^2P_{1/2}$, at photon energy 15.5 eV [10]. b) For Xe-adsorbate photoemission (Xe/Pd(111), $p_{1/2}$). ■ and fit curve: $h\nu = 12.5$ eV; ▨ $h\nu = 21$ eV [24]. The sizes of the symbols correspond to the experimental uncertainties.

This behaviour is well known in the spin-dependent photoionization of free atoms, e.g. in the case of Xe [10, 26], where the components of the spin polarization vector are functions of the emission angle θ for σ^+ and σ^- light, respectively:

$$A(\theta) = \pm \frac{A - \alpha \left(\frac{3}{2} \cos^2\theta - \frac{1}{2} \right)}{1 - \frac{\beta}{2} \left(\frac{3}{2} \cos^2\theta - \frac{1}{2} \right)},$$

component in the light direction;

$$P_p(\theta) = \pm \frac{-\dfrac{3}{2}\alpha\sin\theta\cos\theta}{1 - \dfrac{\beta}{2}\left(\dfrac{3}{2}\cos^2\theta - \dfrac{1}{2}\right)}\,,$$

component perpendicular to the light direction but in the reaction plane;

$$P_\perp(\theta) = \frac{2\xi\sin\theta\cos\theta}{1 - \dfrac{\beta}{2}\left(\dfrac{3}{2}\cos^2\theta - \dfrac{1}{2}\right)}\,,$$

component perpendicular to the reaction plane.

The dynamical photoionization spin parameters A, α, β, ξ are functions of the matrix elements M_s and M_d for dipole transitions from the bound p state to the s and d continuum, respectively, and the corresponding phase shift difference Δ_{sd} between the s and d continuum wave functions.

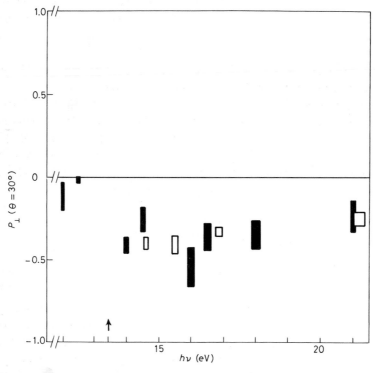

Fig. 11. – Polarization component $P_\perp$ (perpendicular to the reaction plane) for photoemission of Xe $(p_{1/2})$ as a function of photon energy at a polar emission angle of $\theta = 30°$. Filled squares show the experimental data for the adsorbate system Xe/Pd(111)[24]. Open squares show the corresponding $P_\perp$ results for the free-Xe-atom system [38, 43]. The arrow marks the photoionization threshold of the free Xe atom.

The cross-comparison of data between gas and adsorbate phase shown in fig. 10 for the spin polarization component $A(\theta)$ (a) free atom[10], b) adsorbate[24]) shows surprisingly quantitative agreement with respect to the angular dependences. The polarization being $+1$ at $\theta = 0$ for all energies drops down to the negative value -0.5. The angle, where $A(\theta)$ vanishes, increases with decreasing photon energy, in the gas phase as well as in the adsorbate case. The photoemission data of the xenon adsorbate ($A(\theta)$, fig. 10 b)) can even be fitted by use of the formula given above for the gas phase. Figure 10 also shows the results of this fit procedure as full curves. The best agreement between gas phase at photon energy 15.5 eV and adsorbate phase is at a photon energy in the photoemission experiment of about 12.5 eV. But this energy difference is not surprising because of the existing difference in the photoionization thresholds.

Even the spin polarization component perpendicular to the reaction plane exists in the photoemission of the adsorbate; note that this component, which is independent of the helicity of the radiation, is forbidden in the photoexcitation of an infinite three-dimensional solid[33-35]. Figure 11[24] shows the comparison of $P_\perp$ ($\theta = 30°$) as a function of the light energy between gas phase (open symbols) and adsorbate case (closed symbols). In the gas phase $P_\perp$ (and thus ξ according to the formula above) is a direct measure for the interference term between the s and d partial wave function; it is proportional to $M_s \cdot M_d \cdot \sin \Delta_{sd}$. The quantitative agreement between the data for free and adsorbate atoms in fig. 11 indicates that this might be true also for the adsorbate system.

Summarizing, the off-normal spin-resolved photoemission results of the Xe adsorbate show a behaviour which differs from the model for photoemission of an infinite three-dimensional crystal; the data can be well described in terms of atomic photoionization including the existence of quantum-mechanical interference between the partial continuum (final state) wave functions.

5. – Spin-resolved resonances.

In the schematic energy level diagram of fig. 8 the existence of excited atomic states below the photoionization threshold is indicated by the symbol ns which represents some Rydberg series. These excited states can be occupied by electrons which correspond to either a $p_{1/2}$ or a $p_{3/2}$ hole (dashed arrows). In fig. 8 which is of the «one-electron picture» type a discrete excitation from $p_{1/2}$ to ns energetically corresponds to a photoionization from $p_{3/2}$ to the continuum above the threshold (E_{vac}); this correspondence creates an autoionization resonance in the photoelectron emission. This is indicated by the symbol ns_{auto} in fig. 8.

In adsorbate systems resonance features have recently been observed below the adsorbate photoemission threshold[22, 23]. Figure 12 shows the results for Xe on graphite[23]. At photon energies below the first threshold (9.8 eV)

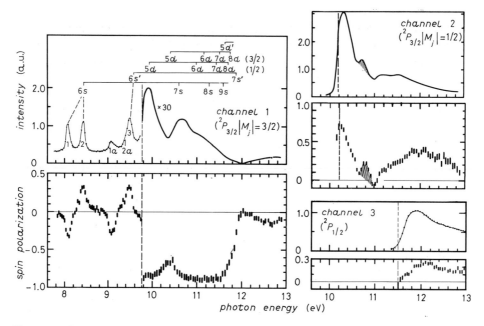

Fig 12. – Spectral dependences of electron intensity and spin polarization for normal photoemission of a monolayer xenon on graphite (0001)[23]. The vertical dashed lines indicate the photoemission thresholds. The channels 1, 2, 3 refer to the sequence of peaks in fig. 9.

resonance enhancement of the electron spectrum occurs very near to the atomic Xe $5p$-$6s$ Rydberg state transition. The resonances were detected via spin-resolved spectroscopy of electrons emitted due to a subsequent Penning-type relaxation mechanism[15]. The spin polarization results[22, 23] show again the sign sequence as in fig. 9 for the peaks 1, 2, 3 and provide thus a quantum number labelling according to fig. 9 and 8: Xe $5p^5(^2P_{3/2}|M_J| = 3/2)6s$, $5p^5(^2P_{3/2}|M_J| = 1/2)6s$, $5p^5(^2P_{1/2}|M_J| = 1/2)6s$ which is an electrically neutral complex, where no image charge screening, for example, can occur. This is evidence that an image charge mechanism must not be responsible for the M_J splitting observed (fig. 9, 12). The resonances for the Xe overlayer are closely related to the surface excitons in rare-gas crystals known in the literature[44]. This connection is supported by a systematic study of the positions of these resonances as a function of the Xe coverage on Ir(111) up to a three-dimensional crystal[23].

Above the photoemission limits fig. 12 shows the intensity and the spin polarization results for normal emission of the three photoelectron peaks (1, 2, 3 as shown in fig. 9) for a xenon monolayer on graphite[23] as a function of photon energy. Intensity and polarization demonstrate a resonance structure in all three channels, deep minima as well as small enhancements of the photoelectron intensities are correlated with pronounced variations of their spin polarizations. Positions as well as shapes of resonances in channel 2 ($^2P_{3/2}|M_J| = 1/2$) have been found to be independent of the distance between adsorbate and substrate as well as between the adatoms in the adlayer. This can be seen in fig. 13 in the comparison of this resonance in different adsorbate phases-substrate configurations. 3 monolayer regimes (commensurate $\sqrt{3} \times \sqrt{3}$ Xe layer with a 4.8 Å adatom spacing, incommensurate close-packed (h.c.p.) Xe layer with a 4.4 Å adatom spacing, both on Pt(111) as well as Xe monolayer on graphite(0001)[12, 13, 23]) are compared with 2 two-layer regimes (Xe/Xe/Pt

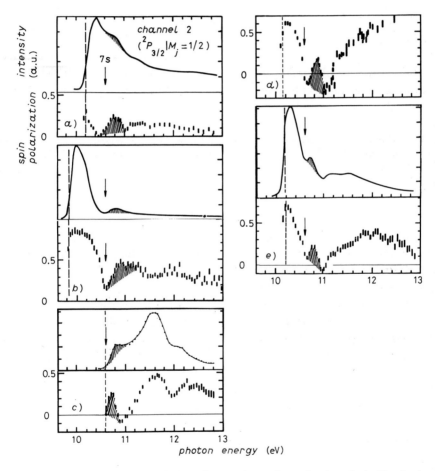

Fig. 13. – Spectral dependences of photoelectron intensity and spin polarization in channel 2 (fig. 12) for normal photoemission of xenon adsorbates in different phase structures and on different substrates close to the 7s excitation energy of a free xenon atom (arrow): a) $\sqrt{3} \times \sqrt{3}$ Xe on Pt(111), b) Xe monolayer on Pt(111), c) second-layer Xe on Pt(111), d) Xe monolayer on carbon-covered Pt(111), e) Xe monolayer on graphite (0001). Data from [12, 13, 15, 23].

and Xe/C/Pt)[15]. In all cases the positions of the resonances (hatched areas) are the same and lie close to the atomic 7s excitation indicated by an arrow in fig. 13. Because of the insensitivity of this resonance with respect to distances of the xenon atom to the next neighbours the resonance behaviour seems to be an atomic Xe-like one, maybe of the autoionization type due to the position coincidence with the 7s atomic level.

Summarizing, this section demonstrates that quantum-mechanical interferences take place in the photoemission of adsorbates not only due to the mixing of two continuum waves as discussed in the previous section but also due to the coupling of continuum states with discrete states via a resonance behaviour.

6. – Summary.

This lecture should indicate that the existence of spin-polarized photoelectrons is a common phenomenon in photoionization of atoms and photoemission of adsorbates and nonmagnetic solids rather than exceptional. It is the purpose of the angle- and spin-resolved photoelectron spectroscopy to find a set of nonredundant experimental data which characterizes the photoeffect quantum-mechanically completely. For rare-gas atoms the cross-comparison between atomic photoionization and photoemission of adsorbates could be performed and shows quantitative agreement.

* * *

The author wishes to thank N. BÖWERING, A. EYERS, CH. HECKENKAMP, B. KESSLER, H.-W. KLAUSING, M. MÜLLER, N. MÜLLER, M. SALZMANN, F. SCHÄFERS, B. SCHMIEDESKAMP, G. SCHÖNHENSE and B. VOGT for their engagement and wealth of ideas in performing the experiments at BESSY and for many intensive discussions. Thanks are due to the colleagues of the Fritz-Haber-Institute and of BESSY for friendly hospitality. Support of BMFT and MPG is gratefully acknowledged.

REFERENCES

[1] U. FANO: *Phys. Rev.*, **178**, 131 (1969).
[2] U. HEINZMANN, J. KESSLER and J. LORENZ: *Phys. Rev. Lett.*, **25**, 1325 (1970).
[3] U. HEINZMANN, J. KESSLER and B. OHNEMUS: *Phys. Rev. Lett.*, **27**, 1696 (1971).
[4] D. T. PIERCE, F. MEIER and P. ZÜRCHER: *Phys. Lett. A*, **51**, 465 (1975).
[5] U. HEINZMANN, F. SCHÄFERS and B. A. HESS: *Chem. Phys. Lett.*, **69**, 284 (1980).
[6] U. HEINZMANN: *Appl. Opt.*, **19**, 4087 (1980).

[7] F. MEYER and D. PESCIA: in *Optical Orientation*, edited by F. MEIER and B. P. ZAKHARCHENYA (North Holland, Amsterdam, 1984), p. 313 ff.

[8] J. KESSLER: *Polarized Electrons*, 2nd edition (Springer, Berlin, 1985).

[9] U. HEINZMANN: *J. Phys. B*, **13**, 4353, 4367 (1980).

[10] CH. HECKENKAMP, F. SCHÄFERS, G. SCHÖNHENSE and U. HEINZMANN: *Phys. Rev. Lett.*, **52**, 421 (1984).

[11] A. EYERS, F. SCHÄFERS, G. SCHÖNHENSE, U. HEINZMANN, H. P. OEPEN, K. HÜNLICH, J. KIRSCHNER and G. BORSTEL: *Phys. Rev. Lett.*, **52**, 1559 (1984).

[12] G. SCHÖNHENSE, A. EYERS, U. FRIESS, F. SCHÄFERS and U. HEINZMANN: *Phys. Rev. Lett.*, **54**, 547 (1985).

[13] U. HEINZMANN and G. SCHÖNHENSE: in *Polarized Electrons in Surface Physics*, edited by R. FEDER (World Scientific, Singapore, 1985), p. 467 ff.

[14] U. HEINZMANN: in *Electronic and Atomic Collisions*, edited by D. C. LORENTS, W. E. MEYERHOF and J. R. PETERSON (Elsevier Science Publishers B.V., Amsterdam, 1986), p. 37 ff.

[15] G. SCHÖNHENSE: *Appl. Phys. A*, **41**, 39 (1986).

[16] A. EYERS, G. SCHÖNHENSE, U. FRIESS, F. SCHÄFERS and U. HEINZMANN: *Surf. Sci.*, **162**, 96 (1985).

[17] N. MÜLLER, B. KESSLER, B. SCHMIEDESKAMP, G. SCHÖNHENSE and U. HEINZMANN: *Solid State Commun.*, **61**, 187 (1987).

[18] B. KESSLER, A. EYERS, K. HORN, N. MÜLLER, B. SCHMIEDESKAMP, G. SCHÖNHENSE and U. HEINZMANN: *Phys. Rev. Lett.*, **59**, 331 (1987).

[19] B. SCHMIEDESKAMP, B. KESSLER, N. MÜLLER, G. SCHÖNHENSE and U. HEINZMANN: *Solid State Commun.*, **65**, 665 (1988).

[20] F. SCHÄFERS, W. PEATMAN, A. EYERS, CH. HECKENKAMP, G. SCHÖNHENSE and U. HEINZMANN: *Rev. Sci. Instrum.*, **57**, 1032 (1986).

[21] B. VOGT, B. KESSLER, N. MÜLLER, B. SCHMIEDESKAMP, G. SCHÖNHENSE and U. HEINZMANN: *Frühjahrstagung DPG Karlsruhe 1988*, Verhandl. DPG 4/1988 0-14.1, p. 31, and to be published.

[22] G. SCHÖNHENSE, A. EYERS and U. HEINZMANN: *Phys. Rev. Lett.*, **56**, 512 (1986).

[23] G. SCHÖNHENSE, B. KESSLER, N. MÜLLER, B. SCHMIEDESKAMP and U. HEINZMANN: *Phys. Scr.*, **35**, 541 (1987).

[24] B. KESSLER, B. VOGT, B. SCHMIEDESKAMP, N. MÜLLER and U. HEINZMANN: *Frühjahrstagung DPG Karlsruhe 1988*, Verhandl. DPG 4/1988 0-14.4, p. 32, and to be published.

[25] CH. HECKENKAMP, F. SCHÄFERS, G. SCHÖNHENSE and U. HEINZMANN: *Phys. Rev. A*, **32**, 1252 (1985).

[26] CH. HECKENKAMP, F. SCHÄFERS, G. SCHÖNHENSE and U. HEINZMANN: *Z. Phys. D*, **2**, 257 (1986).

[27] U. HEINZMANN, B. OSTERHELD and F. SCHÄFERS: *Nucl. Instrum. Methods*, **195**, 395 (1982).

[28] CH. HECKENKAMP, A. EYERS. F. SCHÄFERS, G. SCHÖNHENSE and U. HEINZMANN: *Nucl. Instrum. Methods A*, **246**, 500 (1986).

[29] K. N. HUANG: *Phys. Rev. A*, **22**, 223 (1980).

[30] H. P. OEPEN, K. HÜNLICH, J. KIRSCHNER, A. EYERS, F. SCHÄFERS, G. SCHÖNHENSE and U. HEINZMANN: *Phys. Rev. B*, **31**, 6445 (1985).

[31] H. P. OEPEN, K. HÜNLICH and J. KIRSCHNER: *Phys. Rev. Lett.*, **56**, 496 (1986).

[32] U. HEINZMANN: *Phys. Scr.*, **T17**, 77 (1987).

[33] M. WÖHLECKE and G. BORSTEL: in *Optical Orientation*, edited by F. MEIER and B. P. ZAKHARCHENYA (North Holland, Amsterdam, 1984), p. 423.

[34] R. FEDER: *Polarized Electrons in Surface Physics* (World Scientific, Singapore, 1985), p. 224.

[35] J. KIRSCHNER: *Polarized Electrons at Surfaces, Springer Tracts in Modern Physics*, Vol. **106** (Springer, Heidelberg, 1985).

[36] J. NOFFKE and L. FRITSCHE: *J. Phys. F*, **12**, 921 (1982).

[37] J. NOFFKE and H. ECKARDT: private communication, the calculations are similar to those described in L. FRITSCHE, J. NOFFKE and H. ECKARDT: *J. Phys. F*, **17**, 943 (1987).

[38] U. HEINZMANN, G. SCHÖNHENSE and J. KESSLER: *Phys. Rev. Lett.*, **42**, 1603 (1979).

[39] U. HEINZMANN: in *Fundamental Processes in Atomic Collision Physics*, edited by H. KLEINPOPPEN, J. S. BRIGGS and H. O. LUTZ (Plenum Publications, New York, N.Y., 1985), p. 269 ff.

[40] H. P. OEPEN, K. HÜNLICH, J. KIRSCHNER, A. EYERS and F. SCHÄFERS: *Solid State Commun.*, **59**, 521 (1986).

[41] R. FEDER: *Solid State Commun.*, **21**, 1091 (1977).

[42] A. IGNATJEVS, J. B. PENDRY and T. N. THODIN: *Phys. Rev. Lett.*, **26**, 189 (1971).

[43] G. SCHÖNHENSE: *Phys. Rev. Lett.*, **44**, 640 (1980).

[44] V. SAILE, M. SKIBOWSKI, W. STEINMANN, P. GÜRTLER, E. E. KOCH and A. KOZEVNIKOV: *Phys. Rev. Lett.*, **37**, 305 (1976).

PROCEEDINGS OF THE INTERNATIONAL SCHOOL
OF PHYSICS «ENRICO FERMI»